Ravenna Park Publishing, Inc.
6315 22nd Ave. N.E., Seattle, WA 98115-6919
206-524-3375, e-mail: ravenna@mindspring.com

The computer programs and Electronic Text are now provided via the internet. Please send an e-mail to the above address, or look at http://www.ravennapark.com.

Graiden
February '07

Numerical Methods for Problems with Moving Fronts

Bruce A. Finlayson

Rehnberg Professor of Chemical Engineering
University of Washington

Ravenna Park Publishing, Inc.
Seattle, Washington USA

Copyright © 1992 by Ravenna Park Publishing, Inc.
All Rights Reserved. No part of this work may be reproduced or transmitted in any form or by any means, electronic or mechanical, including photocopying and recording or by any information storage or retireval system, without the written permission of the publisher, except for brief passages quoted by a reviewer.

Printed in the United States of America
10 9 8 7 6 5 4 3 2 1

Production by Frontier Publishing, Seaside, Oregon
Published by Ravenna Park Publishing, Inc.
6315 22nd Ave. N. E.
Seattle, Washington 98115-6919
206-524-3375

Library of Congress Catalog Card Number 92-93277

Finlayson, Bruce A.
    Numerical Methods for Problems with Moving Fronts

ISBN 0-9631765-0-1

A manual for instructors is available from the publisher.

CONTENTS

PREFACE   vii

1. INTRODUCTION   1

PART I. ONE-DIMENSIONAL EXAMPLES SOLVED WITH ALL METHODS

2. INTRODUCTION TO FINITE DIFFERENCE METHODS AND FINITE ELEMENT METHODS   10

2.1. Exact Solution to the Steady-State Convective Diffusion Equation   10
2.2. Finite Difference Methods for the Steady-State Convective Diffusion Equation   12
2.3. Galerkin Finite Element Methods for the Steady-State Convective Diffusion Equation   16
2.4. Applications to the Advection Equation   30
2.5. Applications to the Transient Convective Diffusion Equation   35

3. APPLICATIONS TO BURGER'S EQUATION   47

3.1. Applications to Burger's Equation Without Viscosity   47
3.2. Applications to Burger's Equation With Viscosity   50

4. ANALYSIS OF FINITE DIFFERENCE METHODS ON FIXED GRIDS   55

4.1. Accuracy, Stability, Dissipation, and Dispersion   55
4.2. Finite Difference Methods for the Advection Equation   64
4.3. Finite Difference Methods for Burger's Equation Without Viscosity   69
4.4. Finite Difference Methods for the Convective Diffusion Equation   76
4.5. Finite Difference Methods for Burger's Equation With Viscosity   88
4.6. Conclusions   89

5. ANALYSIS OF FINITE ELEMENT METHODS ON FIXED GRIDS   93

5.1. Accuracy, Stability, Dissipation and Dispersion   93
5.2. Finite Element Methods for the Advection Equation   97
5.3. Finite Element Methods for Burger's Equation Without Viscosity   102
5.4. Finite Element Methods for the Convective Diffusion Equation   110
5.5. Finite Element Methods for Burger's Equation with Viscosity   120
5.6. Conclusions   120

6. SPECIALIZED TECHNIQUES — 126

6.1. Random Choice Method — 129
6.2. Flux-corrected Transport — 137
6.3. TVD and ENO Methods — 151
6.4. Other Upstream Weighting Methods — 159
6.5. High-order Methods — 172
6.5. Orthogonal Collocation on Finite Elements — 181
6.6. Conclusion — 183

7. ADAPTIVE AND MOVING MESHES — 189

7.1. Adaptive Meshes for Steady-State Problems — 189
7.2. Moving Fronts with Analytical Placement of Nodes — 206
7.3. Moving Fronts with Placement of Nodes by Equidistribution Principles — 217
7.4. Moving Fronts with Weighted Residual Placement of Nodes — 223
7.5. Euler-Lagrange Methods — 238
7.6. Comparison of Methods — 246

PART II - APPLICATIONS AND TWO-DIMENSIONAL EXAMPLES

8. TWO-DIMENSIONAL CONSIDERATIONS — 257

8.1. Finite Difference Methods — 257
8.2. Galerkin Finite Element Method — 264
8.3. Taylor-Galerkin Methods — 268
8.4. Flux-corrected Methods — 275
8.5. Grid and Mesh Refinement — 282
8.6. Moving Grids and Finite Elements — 288
8.7. Spine Representation of Mesh — 292
8.8. Natural Boundary Conditions — 300
8.9. Group Velocity — 307

9. THE CONVECTIVE DIFFUSION EQUATION WITH ADSORPTION — 318

9.1. Linear Adsorption — 318
9.2. Chromatography with Langmuir Adsorption — 332
9.3. Two-dimensional Cross Flow — 343
9.4. A Moving Cone Problem — 346

10. CONVECTION AND DIFFUSION WITH REACTION — 355

10.1. A Model Problem With Convection — 355
10.2. A Model Problem with Convection and Diffusion — 365
10.3. Combustion in Flames — 376
10.4. Modeling a Catalytic Converter — 390
10.5. Solid-Gas Reactions — 396

11. PROBLEMS WITH PHASE CHANGE — 402

11.1. Equations — 402
11.2. One-Phase Problems in One Dimension — 405
11.3. Two-Phase Problems in One Dimension — 413
11.4. Two-dimensional Heat Conduction — 419
11.5. Two-dimensional Heat Conduction and Convection — 428

12. THE NAVIER-STOKES EQUATION — 434

12.1. Equations — 434
12.2. Finite Element Methods for Steady-State Problems — 436
12.3. Finite Difference Methods — 440
12.4. Finite Element Methods for Unsteady-State Problems — 442
12.5. The Taylor-Galerkin Method — 451

13. POLYMER FLOW — 457

13.1. Introduction — 457
13.2. Model Problem One: Stress with a Constant Gradient — 461
13.3. Model Problem Two: Stress Along a Centerline — 470
13.4. Model Problem Three: Unsteady Couette Flow — 474

14. FLOW THROUGH POROUS MEDIA — 496

14.1. Equations — 496
14.2. Miscible Displacement in One Dimension — 500
14.3. Immiscible Displacement in One Dimension — 503
14.4. Partially Saturated Flow in One Dimension — 517
14.5. Miscible Displacement in Two Dimensions — 527
14.6. Immiscible Displacement in Two Dimensions — 531

APPENDIX. DISPERSION DIAGRAMS — 542

COMPUTER PROGRAM DESCRIPTION — 579

AUTHOR INDEX — 589

SUBJECT INDEX — 595

Preface

I began this book while serving as the Gulf Professor of Chemical Engineering at Carnegie Mellon University - indeed, without that sabbatical the book could not have been written. The faculty there was very supportive of my efforts and the University provided important library and office facilities. My time there was too short to finish the book and it has been a slowly simmering project ever since.

The book was written on the Macintosh™ computer and many of the ideas developed while writing the book were possible because of the graphics programs which could be used to demonstrate the results. The equations - of which there are many - were written using the program MathWriter™. This equation writing software is so easy to use that some of the algebra was done on the computer (substituting one equation into another, etc.). There are equations in the book that have never been written down by hand. Also, since equation writing was so simple, I tended to merely repeat an equation rather than refer back to an earlier chapter for it. Thus the reader can follow the ideas more easily.

You will notice many graphs in the book; these were important in the testing of the ideas. Oftentimes an author in the literature will present a new method and present results from it, but there may be no comprehensive comparison with other methods. Thus I felt it important to try all methods on all problems, insofar as that is possible. The graphical display helps decide which methods to keep and which ones to discard. It is also possible to learn things graphically that are harder to learn otherwise. For example, the von Neumann analysis of stability can be tedious when done algebraically, so I show graphical results which present the same information. They are not merely graphical displays of the algebraic results, but are graphical presentations of the information contained in the algebraic results - without generation of the algebraic results. For example, if you want to see in what regions of the x-y plane a function $f(x,y)$ is less than one, you usually set $f=1$ and solve for $x(y)$ or $y(x)$. However, you can also plot $f(x,y)$ and see where it is 1, and you can plot contours where $f=1$. These approaches are much faster and are necessary when dealing with dozens of methods, as I was. To learn the material in the book it is necessary that you exercise the computer programs provided. I have witnessed students learning things much faster by operating the computer program, chiefly because of its graphical output. To some extent the graphical display helps you use the right side of your brain - and this is ideally suited to a generation raised on TV!

Several groups have listened to lectures on parts of the book. The graduate students and faculty in the Department of Chemical Engineering at Carnegie Mellon heard a number of lectures on the material, but the book was at an early stage then. I actually unveiled some of the material for the first time in a series of lectures at the University of Puerto Rico in March, 1989, and then presented the book in detail to my graduate class in Spring, 1989. I thank them for their patience, feedback, and comments and corrections. I was also able to use them to test out some new modes of learning to see how they reacted. Special thanks are due to my graduate students Tom Baer, Rekha Rao, Ken Westerberg, and Christophe Poulain, who worked through the entire book with me. The printed copy was made from a camera ready version produced with the program Page Maker™ on a Macintosh™ computer and printed on a Laserwriter™. My daughter Christine did an excellent job as editor and helped prepare the camera-ready copy. My son Mark also helped with some of the plotting algorithms.

Many people have contributed to this book, but my wife deserves the most thanks. Some of the book was written during the summer when she was the sole bread winner in the family. Most important, though, has been her encouragement to do something significant. The summer months I have used to write the book have been enjoyable. Seattle summers are beautiful, usually with clear blue skies and comfortable temperatures, and my office at home allows me to experience them. My faithful companion, our dog Sniffer, helped me when I was discouraged. She would join me in my office just to be there while I was working. She was old, blind and deaf at the end, but found her way to my office; since she was always there I have strong associations with her and the writing of the book.

Bruce Finlayson
Seattle, Washington

CHAPTER
ONE

INTRODUCTION

The objective of this book is to describe the modern methods that are applied to problems in which the convection term dominates or the mass (or concentration) moves along in a wave. In the early 1970's, I was faced with such problems in four different areas of application; I have been interested in methods for these problems ever since.

The first area of application is in catalytic converters. When modeling a catalytic converter for an automobile, there is a combustion zone in which the temperature rises drastically and the concentration of carbon monoxide falls dramatically. Since the car accelerates and decelerates, the flow rate through the device changes over time and this combustion zone moves back and forth. The general features of the solution are known; at any time the temperature may look like Figure 1.1. Yet the detailed features, such as the exact location of the combustion zone, are unknown. In 1972, numerical methods applied to this problem were exceedingly slow. I remember looking at tabular computer output and seeing the front move excruciatingly slowly; meanwhile, my computer budget was being used up.

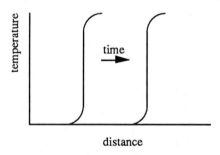

Figure 1.1. Moving Combustion Zone

Another area of application where such problems arise is in the movement of water in dry soils. In this case, the capillary pressure changes drastically in a small region of space, leading to a quite sharp front. The general features of the flow can be sketched (see Figure 1.2). The methods applied to this problem also seemed very slow, even though I was using the latest "tricks of the trade" (that was when the GEAR-type programs first became available). Again, the general features are known and the result of computation is merely to identify the details, such as where the moisture front is and how fast things happen. It seemed such a waste to have to spend large amounts of computer time to quantify what I already knew was going to happen.

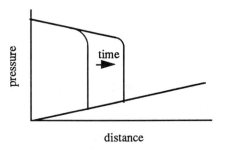

Figure 1.2. Water Pressure

The third area of application is in chemical flooding of oil fields. In this case water is injected underground and, for a period of time, a chemical is added to the water. The chemical causes more oil to be pushed out of the ground than with water flooding alone. However, the concentration of chemical is located in a relatively small region of space and moves through the porous media with time, as illustrated in Figure 1.3. The physics of the phenomena depend crucially on the concentration of chemical. Unfortunately, the numerical techniques that were used did not predict the concentration accurately, so the results were in question. Again it seems desirable to have a method that would allow us to track the concentration of this chemical very precisely, without using inordinate amounts of computer time.

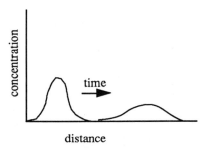

Figure 1.3. Chemical Flooding

The fourth area of application involves the underground gasification of coal, which leads to a combustion front that moves through the coal bed with time.

Again we have a moving front whose location must be predicted. Again we know the general features of the desired solution and wish to quantify our guess without enormous expenditures of computer time.

The applications of the methods given in this book are not restricted to the examples given. The same numerical difficulties arise in a variety of applications, such as bioseparations with chromatography, laser machining, electro-chemical machining, and crystal growth to make computer chips. Gupta and Arora [1988] list even more applications: metal casting, welding, food processing, freeze-coating of fibers, laser glazing, and latent heat devices for energy conservation. The equation for the spreading of a fluid on a solid is a Burger's equation under some circumstances [Teletzke, et al. 1987]; even the Fokker-Planck equation leads to equations of the type considered here [Harrison, 1988]. Clearly the methods are widely applicable.

Every problem discussed in this book can be solved given unlimited computer time. If the mesh size and time-step are reduced in the appropriate degree, the standard methods will provide an excellent solution. However, we might have to wait several years for the solution and it could cost a considerable amount of money. Computer time (or the money to pay for it) is a primary consideration. In recent years computers have become much faster, although not always cheaper on a per problem basis. Computer technology has not solved the problem for us, because during the past 15 years the problems we have wanted to solve have also gotten more complex and the faster computers are being used to solve even more difficult problems. Even though the new computers are faster, we need to be able to solve difficult problems as fast as possible so that we can examine more alternative cases (leading to a better design). Thus, the focus of this book is _efficient_ methods: to solve the problem well with a minimum of computer time. Since the computer time is proportional to the number of terms for most methods, efficiency is gained by using fewer nodes.

Historically, the methods used for these problems were finite difference methods with various forms of upstream weighting (see Chapters 2 and 4). However, these methods led to solutions that did not capture the steep front properly (see Figure 1.4). The general shape seemed reasonable, but the front was smoothed more than it should have been. In the case of chemical flooding of oil fields, that was an important limitation. Then Galerkin methods were developed for these uses, since they had been found to be good, efficient methods for solid mechanics problems. However, the solutions obtained with Galerkin methods tended to oscillate, and even a non-specialist (such as a manager) could tell us the solution didn't look like that shown in Figure 1.5. Then Petrov-Galerkin methods were developed to provide upstream weighting for Galerkin finite element methods, since this had proved useful for finite difference methods. This modification was not completely satisfactory, however (see Chapters 2 and 5). Specialized methods were also developed. After all, we know the general features of the solution; we merely have to move it along properly. The special methods

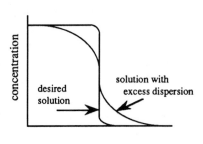
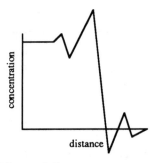

Figure 1.4. Excess Dissipation   **Figure 1.5.** Incorrect Oscillations

tried to incorporate this information to provide an improved solution: entire grids were moved; nodes were moved on a fixed grid; the moving part of the problem was solved exactly, oscillations were eliminated after the fact; and methods were developed to use one method for the convective part of the problem and another for the diffusive part of the problem. This book describes these methods and their successes in application.

The first part of the book is limited to one-dimensional problems. It gives extensive details of each of the methods discussed above and graphical displays of the results of the methods. The methods are applied to several problems. The main problem is the convective diffusion equation:

$$\frac{\partial c}{\partial t} + \text{Pe} \frac{\partial c}{\partial x} = \frac{\partial^2 c}{\partial x^2}. \tag{1.1}$$

At large Peclet numbers (Pe) this problem is dominated by convection, but diffusion is still present. This problem is linear, allowing some special methods. To see what happens when there is no diffusion (the ultimate high Peclet number problem), we drop the diffusion term to get the advection equation:

$$\frac{\partial c}{\partial t} + \text{Pe} \frac{\partial c}{\partial x} = 0. \tag{1.2}$$

This equation is still linear, however, while most applications are nonlinear. We thus also consider the nonlinear problem known as Burger's equation,

$$\frac{\partial u}{\partial t} + u \frac{\partial u}{\partial x} = \nu \frac{\partial^2 u}{\partial x^2}, \tag{1.3}$$

which we solve with and without the viscous term on the right-hand side. This problem is interesting because it is nonlinear and the solutions can also form shocks. Indeed a plot of $u(x,t)$ can steepen as time proceeds until it actually has a discontinuity in it (see Figure 1.6). The nonlinear problem thus provides a severe test case for any numerical method.

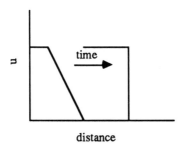

Figure 1.6. Sharpening Front

This book heavily emphasizes explicit methods. The reason is explained well in the book by Oran and Boris [1987]. If a phenomenon involves a moving concentration profile, then we generally want to follow its movement and use a small enough time-step to enable us to do so. For time-steps this small, the explicit schemes are stable and they are faster than implicit schemes for the same time-step. Implicit methods would allow a much larger time-step, but the solution would "jump over" the phenomenon we are interested in. However, in many problems there are two or more important time scales. The time scales for different phenomena are illustrated in Figure 1.7 and their ratios are given in Eq. (1.4). When diffusion and convection are important, the ratio is the Peclet

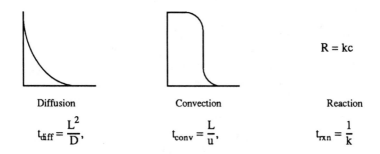

Figure 1.7. Time Scales

number, Pe. When diffusion and reaction are important, the ratio is the Thiele modulus (squared), ϕ^2. When convection and reaction are important, the ratio is the Damköhler number, Da.

$$\frac{t_{diff}}{t_{conv}} = \frac{u L}{D} = Pe, \quad \frac{t_{diff}}{t_{rxn}} = \frac{k L^2}{D} = \phi^2, \quad \frac{t_{conv}}{t_{rxn}} = \frac{L k}{u} = Da \qquad (1.4)$$

If a problem includes both convection and diffusion, the Peclet number determines if these time constants are drastically different. The problem becomes difficult when the Peclet number becomes large (e.g., 1000). The time constant for convection is then much smaller than the time constant for diffusion. A small step-size is needed to resolve the convection phenomenon and an explicit scheme may be used. Indeed, this is the prototype problem in this book.

Now let us suppose that there is also a chemical reaction, or some other phenomenon, that is important. This phenomenon has still another time constant associated with it and yet it may control the solution. If that phenomenon occurs over a very long time, then using a small time constant, dictated to resolve the convective motion, may not be practical. If we try to use a larger time-step, the explicit methods become unstable; thus implicit methods are required. We will be skipping over certain phenomena, by choice. As a rough rule of thumb, if the time scales of different phenomena are widely different, then we may wish to use an implicit method. If the time scales are about the same, we may wish to use an explicit method. In each of the chapters we will look at the time constants associated with the phenomena in order to give guidance to our numerical methods.

Part I, Chapters 2-7, treats several test problems. Each problem is solved by every method. We would like a good solution, one that maintains a steep front without oscillations, for a reasonable expenditure of computer time. By looking at the solutions in Part I, we can choose good methods, since the graphs display results for exactly the same cases. One of the advantages of the microcomputer is the ease of using graphical displays and creating graphs. You may want to xerox some of the different graphs that are printed on separate pages to have all the best methods in front of you. (Transparencies where the figures are placed one above the other are also useful.) Or you may wish to generate results yourself. The set of microcomputer programs can be used for that purpose. The program can be used to generate the solutions shown in Chapters 2-6. The program allows many methods and hundreds of possible method parameters and you can experiment with different choices, thereby obtaining immediate feedback on the impact of those choices. A brief description of the programs and their options is provided at the end of the book (beginning on page 579).

Ultimately, however, the goal is to solve problems with some application in mind. Furthermore, many of these applications are two- and three-dimensional problems. Some of the methods that worked well in one dimension may not work as well (or at all) in two dimensions. Therefore, Part II, Chapters 8-14, expands the focus to include two-dimensional problems for specific applications. Here, however, there is no need to consider all the methods since some of them did not meet our standards in Part I. Thus only the best methods are carried forth into Part II. Since so many methods are discussed, the successful ones are highlighted here: the MacCormack finite difference method (Ch. 4), the Taylor-Galerkin method (Ch. 5), the random choice method (Ch. 6), the flux-correction step (Ch. 6), the TVD and ENO methods (Ch. 6), and the moving grid/element methods (Ch.7). Most of these are used in Part II.

Part II starts with a chapter on two-dimensional problems, which answers several questions: how do they differ from one-dimensional problems? What new factors have to be considered? How do we carry out the details? In two-dimensional problems, irregular geometry is an important complication and finite element methods prove especially useful. A heat transfer problem is included to illustrate the use of natural boundary conditions, which enables us to truncate the domain in finite element methods and still obtain a good result.

The first application is in convective diffusion problems with adsorption. In the simple convective diffusion equation, we knew the velocity and some methods only worked if we knew the velocity. With adsorption, the velocity may not be known or may vary from location to location and the fluid velocity may not be equal to the front velocity. Thus, one-dimensional examples are treated with adsorption, which are prototype chromatography problems. A two-dimensional example that has become a standard test problem is the rotating cone problem. A concentration profile is begun in the shape of a cone, with a peak concentration at the center. This cone is placed in a velocity field that is a vortex about yet still another point, and the cone is rotated around a full cycle. The distortion of the cone after one revolution quickly shows the adequacy of a method. Two-dimensional examples (steady-state problems with a known velocity field) are also used in this chapter to illustrate the effect of mesh on the solution, the tendency of the solution to oscillate, the residual (how good the solution is), and the role of adaptive meshes.

The second application considers problems with a source or sink term (i.e., a chemical reaction). With such terms it is possible to get combustion zones that are very steep; in transient problems, these combustion zones move in time. Most of the applications here are for one-dimensional problems.

Heat transfer with phase change is also considered. If water is freezing (or ice melting) there is a solid-liquid front whose location changes in time. Such applications are natural problems for methods which track the solid-liquid front and/or use moving nodes. Both one- and two-dimensional problems are considered.

The Navier-Stokes equations for a Newtonian fluid are important because the form of the equations is unusual. The variables are an x- and y-velocity and pressure, and the equations are an x- and y-momentum equation and a continuity equation. It is natural to think of the x-momentum equation as an equation for the x-velocity and the y-momentum equation as an equation for the y-velocity. We might then think of the continuity equation as the equation for pressure, yet pressure does not appear in it. Consequently, special methods have been developed for such problems, for both finite difference and finite element methods.

Polymers are more complicated than Newtonian fluids since the fluid can remember, at least for awhile. A viscoelastic fluid (polymer) is a cross between a solid and a fluid, but the fluid moves much more than a solid would, and this complicates the numerical methods usually applied to solid mechanics.

For the steady-state case, some of the equations are hyperbolic and some are elliptic. Problems with strong convection are predominately hyperbolic, while problems with strong diffusion are predominately elliptic. The polymer flow equations thus provide another application of methods for problems with strong convection. Applications in this chapter include some model problems, such as stress profiles along the centerline of a tube and unsteady Couette flow. The Couette flow problem demonstrates the interplay of reaction-convection terms to give a reaction-diffusion phenomenon.

The final chapter deals with flow through porous media, both miscible displacement and immiscible displacement. Usually we consider displacing oil with water. In the case of miscible displacement, the prototype problem is merely the convective diffusion equation. In one dimension, these problems are considered throughout the book. In two dimensions, however, particularly when injecting fluid at a well, the velocity changes drastically in space and unique numerical problems occur. When the displacement is immiscible, there is a steep front (a discontinuity) that is propagated along. Before the front, the oil saturation is that of the reservoir initially, while behind the front, the oil saturation is at some other value. When the problem is solved in one dimension it is called the Buckley-Leverett problem, but we can solve it in two dimensions as well. Here again, methods with moving meshes or ones that take into account the moving front are especially useful.

There are other important areas of applications and methods that are not included in this book, such as turbulent flow, compressible flow, spectral methods, and methods for parallel computers. Turbulent flow is essentially high-speed flow, but simulation requires empirical laws, which are not discussed in this book. Compressible flow is a rich source of numerical methods. In fact, most of the methods discussed here have been applied to this field. Even though applications to compressible flow are not discussed, the methods are applied to other examples, which are increasing in importance. Spectral methods can be quite good in certain circumstances. (See the books by Gottlieb and Orszag [1977] and Canuto, *et al* [1987].) Some of the methods given here are more efficient when run on parallel computers. Unfortunately, there is not space to include these topics in addition to those discussed.

After reading this book, you should be able to pick a likely method for your moving fronts problems and see how others have treated similar problems. Hopefully, armed with this knowledge, you can solve your problems efficiently and quickly.

References

Canuto, C., Hussaini, M. Y., Quarteroni, A., Zang, T. A., *Spectral Methods in Fluid Dynamics*, Springer-Verlag [1987].

Gottlieb, D. and Orszag, S. A., *Numerical Analysis of Spectral Methods: Theory and Applications*, Soc. Ind. Appl. Math., Philadelphia [1977].

Gupta, S. C. and Arora, P. R., "Outward Spherical Solidification of a Superheated Melt with Time Dependent Boundary Flux," Appl. Sci. Res. 45 17-31 [1988].

Harrison, G. W., "Numerical Solution of the Fokker Planck Equation Using Moving Finite Elements," Num. Methods Partial Diff. Eqn. 4 219-232 [1988].

Oran, E. S. and Boris, J. P., *Numerical Simulation of Reactive Flow*, Elsevier [1987].

Teletzke, G. F., Davis, H. T., and Scriven, L. E., "How Liquids Spread on Solids," Chem. Eng. Comm. 55 41-81 [1987].

CHAPTER
TWO

INTRODUCTION TO FINITE DIFFERENCE METHODS AND FINITE ELEMENT METHODS

Before studying and developing new, specialized methods that are advantageous for problems with moving fronts, we need to see why such methods are necessary. In this chapter we will consider three simple, but very important, problems. The first one is the steady-state convective diffusion equation. It is a useful prototype problem because it has a sharp front, but is linear and permits an exact solution to both the differential equation and difference equations. For time-dependent problems we will consider the convective diffusion equation, too, since some methods are well suited to time-dependent problems. This problem can be designed to have strong convection by increasing a parameter, the Peclet number. We are interested in the changes in the solution as the Peclet number increases. The case of an infinite Peclet number is also of interest; this is called the advection equation.

In this chapter some algorithms for solving these problems with finite difference and finite element methods will be given, but we will not examine those methods in detail until Chapters 4 and 5. Here the goal is to understand the basic ideas of the methods, see how they are applied to some standard cases, and see what numerical problems arise. Literature references for these methods are cited in Chapters 4 and 5, as well as the rest of the book.

2.1. Exact Solution to the Steady-State Convective Diffusion Equation

The steady-state convective diffusion equation is

$$\text{Pe}\frac{dc}{dx} = \frac{d^2c}{dx^2}, \tag{2.1}$$

where Pe is the Peclet number (defined as Pe = u L/D), u is a velocity, L is the length of the spatial domain, and D is a diffusion or dispersion coefficient. The boundary conditions we will use are

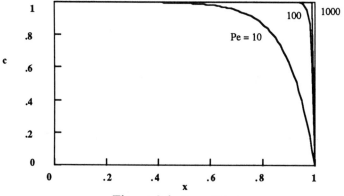

Figure 2.1. Exact Solution

$$c(0) = 1 \text{ and } c(1) = 0. \tag{2.2}$$

The boundary condition at $x = 1$ is not a realistic condition, since it is not possible to force the concentration at an outflow boundary to be a specific value. It does provide, however, a useful test problem that incorporates many of the features of the more complicated problems. The exact solution is

$$c = \frac{e^{Pe} - e^{Pe\,x}}{e^{Pe} - 1}. \tag{2.3}$$

Successive derivatives of the exact solution are

$$c^{(n)} = \frac{d^n c}{dx^n} = -\frac{1}{e^{Pe} - 1} Pe^n\, e^{Pe\,x} \tag{2.4}$$

and the norm of the n-th derivative is

$$\|c^{(n)}\| \equiv \left[\int_0^1 (c^{(n)})^2 dx\right]^{1/2} = \left[\frac{1}{2Pe}(e^{2Pe} - 1)\right]^{1/2} \frac{Pe^n}{e^{Pe} - 1}. \tag{2.5}$$

The first derivative at $x = 1$ is $-Pe$ when the Peclet number is large. Thus the larger the Peclet number, the larger the slope at $x = 1$. The solution has a boundary layer near $x = 1$ and this boundary layer becomes steeper for larger Peclet numbers. Typical solutions are shown in Figure 2.1. Each of the numerical methods must be able to approximate this solution with the boundary layer. For time-dependent problems, the boundary layer will move and the approximate solution must follow it.

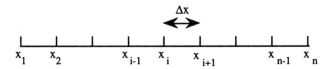

Figure 2.2. Finite Difference Grid

2.2 Finite Difference Methods for the Steady-State Convective Diffusion Equation

Finite difference methods use solutions at sets of grid points. Here we divide the domain into equidistant intervals, as shown in Figure 2.2. We let the solution at the i-th grid point be $c(x_i) = c_i$. Of course, $x_i = x_i + (i-1) \Delta x$. Let us suppose the function $c(x)$ is a continuous function of x. This may not always be true (we do consider some solutions with discontinuities), but we will develop the formulas on this basis anyway. Even the exact solution of the convective diffusion equation (see Section 2.1) has continuous derivatives, although the solution looks discontinuous at large Peclet numbers. If the function is continuous, then we can expand it in a Taylor series:

$$c(x) = c_i + c_i' (x - x_i) + c_i'' \frac{(x-x_i)^2}{2!} + c_i''' \frac{(x-x_i)^3}{3!} + c_i'''' \frac{(x-x_i)^4}{4!} + \ldots \quad (2.6)$$

We next evaluate the Taylor series at the points x_{i+1} and x_{i-1}.

$$c_{i+1} = c_i + c_i' \Delta x + c_i'' \frac{\Delta x^2}{2!} + c_i''' \frac{\Delta x^3}{3!} + c_i'''' \frac{\Delta x^4}{4!} + \ldots \quad (2.7)$$

$$c_{i-1} = c_i - c_i' \Delta x + c_i'' \frac{\Delta x^2}{2!} - c_i''' \frac{\Delta x^3}{3!} + c_i'''' \frac{\Delta x^4}{4!} + \ldots \quad (2.8)$$

These formulas can be rearranged and divided by Δx to obtain two formulas for the first derivative.

$$\frac{c_{i+1} - c_i}{\Delta x} = c_i' + \frac{\Delta x}{2} c_i'' + \ldots \quad (2.9)$$

$$\frac{c_i - c_{i-1}}{\Delta x} = c_i' - \frac{\Delta x}{2} c_i'' + \ldots \quad (2.10)$$

If we keep only the first term on the right-hand side, then the formulas are correct to $O(\Delta x)$. We say the truncation error is $O(\Delta x)$. We should note, however, that for the exact solution to the convective diffusion equation, each successive derivative is larger than the lower-ordered derivatives; thus, a very small Δx value may be necessary to make the neglected terms small. However, we can proceed by subtracting Eq. (2.7) from Eq. (2.8), dividing by Δx, and rearranging to obtain another approximation to the first derivative:

$$\frac{c_{i+1} - c_{i-1}}{2\Delta x} = c_i' + \frac{1}{3!} c_i''' \Delta x^2. \tag{2.11}$$

This formula has a truncation error of $O(\Delta x^2)$. If we add Eq. (2.7) to Eq. (2.8) and divide by Δx^2, we obtain an approximation to the second derivative.

$$\frac{c_{i+1} - 2 c_i + c_{i-1}}{\Delta x^2} = c_i'' + \frac{2}{4!} c_i'''' \Delta x^2 + \ldots \tag{2.12}$$

This is correct to $O(\Delta x^2)$.

To apply a finite difference method, we replace the differential equation (valid for all x values) by a set of difference equations, of which each member is the differential equation evaluated at a certain grid point. We will use only expressions that have a truncation error of $O(\Delta x^2)$: Eq. (2.11) and Eq. (2.12). The convective diffusion equation at the i-th grid point is then

$$\text{Pe} \frac{c_{i+1} - c_{i-1}}{2\Delta x} = \frac{c_{i+1} - 2 c_i + c_{i-1}}{\Delta x^2}. \tag{2.13}$$

This is rearranged to give

$$c_{i+1} \left(\frac{\text{Pe}\Delta x}{2} - 1 \right) + c_i (2) - c_{i-1} \left(\frac{\text{Pe}\Delta x}{2} + 1 \right) = 0, \tag{2.14}$$

where

$$\mu = \text{Pe}\, \Delta x / 2. \tag{2.15}$$

This is a second-order difference equation that can be solved exactly. We try a solution of the form

$$c_i = \phi^{i-1} \tag{2.16}$$

to obtain

$$\phi^i \left(\frac{\text{Pe}\Delta x}{2} - 1 \right) + \phi^{i-1} (2) - \phi^{i-2} \left(\frac{\text{Pe}\Delta x}{2} + 1 \right) = 0. \tag{2.17}$$

We then divide by ϕ^{i-2} to obtain

$$\phi^2 (\mu - 1) + 2 \phi - (\mu + 1) = 0 \tag{2.18}$$

14 NUMERICAL METHODS - MOVING FRONTS

and solve this equation to obtain ϕ.

$$\phi = \frac{-2 \pm 2\mu}{2(\mu-1)} = \frac{-1 \pm \mu}{\mu-1} \qquad (2.19)$$

Thus, there are two values of ϕ:

$$\phi = 1, \frac{1+\mu}{1-\mu}. \qquad (2.20)$$

The solution to the differential equation is then the combination of the two general solutions.

$$c_i = A\, \phi_1^{i-1} + B\, \phi_2^{i-1} \qquad (2.21)$$

Since ϕ_1 is 1.0, this is merely

$$c_i = A + B\, \phi_1^{i-1} = A + B\left(\frac{1+\mu}{1-\mu}\right)^{i-1}. \qquad (2.22)$$

To finish the solution, we apply the boundary conditions to determine the constants A and B.

$$\text{At } i=1, \quad c_i = 1 \text{ requires} \quad 1 = A + B. \qquad (2.23)$$

$$\text{At } i=N+1, \quad c_i = 0 \text{ requires} \quad 0 = A + B\left(\frac{1+\mu}{1-\mu}\right)^N. \qquad (2.24)$$

Solving for A and B gives

$$A = \frac{-\left(\frac{1+\mu}{1-\mu}\right)^N}{1-\left(\frac{1+\mu}{1-\mu}\right)^N}, \quad B = \frac{1}{1-\left(\frac{1+\mu}{1-\mu}\right)^N}. \qquad (2.25)$$

Eq. (2.22) with Eq. (2.25) represents the exact solution to the difference equation.

We note that A and B are constants for any particular numerical solution (Pe and μ). If $\mu < 1$, then $[(1+\mu)/(1-\mu)]$ is positive and $[(1+\mu)/(1-\mu)]^i$ is positive for all i. If $\mu > 1$, however, $[(1+\mu)/(1-\mu)]$ is negative and $[(1+\mu)/(1-\mu)]^i$ changes sign with each successive i. This means that Eq. (2.22) oscillates with successive i, since A and B are constants. The term multiplied by B will cause the solution to alternate above and below the value of A. Thus if $\mu < 1$, the solution does not

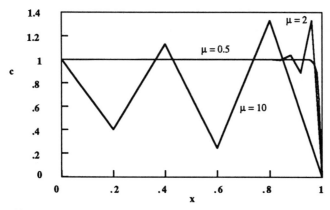

Figure 2.3. Finite Difference Solution for Pe = 100, μ = 0.5, 2, 10

oscillate. This solution is plotted in Figure 2.3 for $\mu = 0.5$ (where no oscillations are expected) and $\mu = 2$ and $\mu = 10$ (where oscillations are expected). The solution for $\mu = 10$ oscillates about the exact solution; we would like to eliminate these oscillations.

Next we solve the same problem but use an upstream difference equation for the first derivative. Putting Eq. (2.10) and Eq. (2.12) into Eq. (2.1) at the i-th grid point gives

$$\text{Pe}\frac{c_i - c_{i-1}}{\Delta x} = \frac{c_{i+1} - 2c_i + c_{i-1}}{\Delta x^2}. \tag{2.26}$$

This equation is rearranged to give

$$c_{i+1}(-1) + c_i(\text{Pe}\,\Delta x + 2) - c_{i-1}(1 + \text{Pe}\,\Delta x) = 0. \tag{2.27}$$

Assuming the same trial function, Eq. (2.16) gives the following equation for ϕ:

$$-\phi^2 + \phi(\text{Pe}\,\Delta x + 2) - (1 + \text{Pe}\,\Delta x) = 0. \tag{2.28}$$

The solutions are

$$\phi = 1 \text{ and } 1 + \text{Pe}\,\Delta x. \tag{2.29}$$

The nodal solution is then

$$c_i = A + B(1 + \text{Pe}\Delta x)^{i-1}. \tag{2.30}$$

$$A = \frac{-(1 + \text{Pe}\,\Delta x)^N}{1 - (1 + \text{Pe}\,\Delta x)^N}, \quad B = \frac{1}{1 - (1 + \text{Pe}\,\Delta x)^N}. \tag{2.31}$$

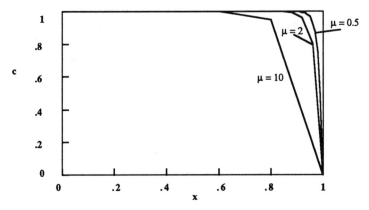

Figure 2.4. Finite Difference Solution with Upstream Derivative, Pe = 100

This solution does not oscillate from node to node for any value of Pe Δx (see Figure 2.4). The concentration profile is smooth but considerably less steep than the exact solution. We have eliminated the oscillations at the expense of spreading the solution out farther than it should be.

2.3 Galerkin Finite Element Methods for the Steady-State Convective Diffusion Equation

In a finite element method we subdivide the domain into smaller regions called elements. Then within each element we represent the unknown solution by a polynomial. Usually the simplest polynomials are used; here we will consider linear polynomials. When we develop the standard finite element equations for the Galerkin method applied to the convective diffusion equation, we see that solutions to those equations also oscillate. We then must consider ways of adjusting the method to ensure that the solution does not oscillate from node to node.

First we divide the domain $0 \leq x \leq 1$ into smaller elements, as shown in Figure 2.5a. It is not difficult to consider elements of different sizes so we do so, calling the size of the i-th element Δx_i. We note that the i-th element begins at x_i and goes to x_{i+1}. We define the trial functions as shown in Figure 2.5a: linear within an element, but zero over most of the domain. The trial function N_i takes the value 1 at x_i, 0 at x_{i+1}, and is linear between x_i and x_{i+1}; it is 0 also at x_{i-1}, and linear between x_i and x_{i+1}; it is zero outside of that domain. In two-dimensional problems it is important to develop the equations element by element; we do that here for the one-dimensional problem as well. We then define a local coordinate system in each element that goes from -1 to $+1$. Thus, in the i-th element

$$\xi = \frac{2(x - x_i)}{\Delta x_i} - 1. \tag{2.32}$$

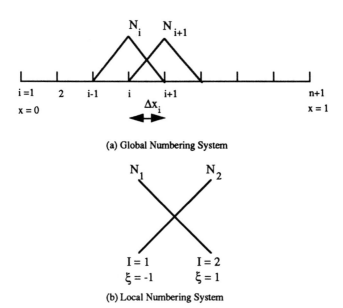

(a) Global Numbering System

(b) Local Numbering System

Figure 2.5. Galerkin Finite Element Method - Linear Functions

As ξ goes from -1 to 1, the corresponding x value goes from x_i to x_{i+1}. Within each element the trial function is linear.

$$N_I = \begin{cases} N_1(\xi) = (1 - \xi)/2 \\ \\ N_2(\xi) = (1 + \xi)/2 \end{cases} \quad (2.33)$$

This function is shown for a general element in Figure 2.5b. We identify the numbering scheme as a local one by using a capital letter I and as a global one by using a lower case letter i. We always need a conversion from i to I and this conversion depends on the element in question. For example, N_i in the global numbering scheme is $N_{I=2}$ in the (i-1)-th element but $N_{I=1}$ in the i-th element. A typical numbering scheme is shown in Figure 2.6.

With this understanding, we write the complete trial function as

$$c(x) = \sum_{i=1}^{NT} c_i N_i(x), \quad (2.34)$$

where each $N_i(x)$ is defined only on the appropriate element. Because of the way the trial functions were defined, only one N_i is non-zero at the point x_i. Thus c_i represents the value of c at the point x_i. In the Galerkin method we form the

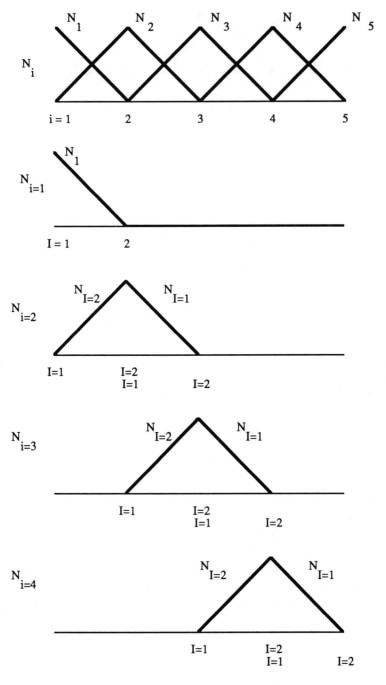

Figure 2.6. Global and Local Trial Functions

residual by substituting this trial function into the differential equation. Here we illustrate that process with the convective diffusion equation, Eq. (2.1). The residual is then

$$\text{Residual} = \sum_{i=1}^{NT} c_i \left[\text{Pe} \frac{dN_i}{dx} - \frac{d^2N_i}{dx^2} \right]. \qquad (2.35)$$

We would like to make the residual zero, because then the differential equation would be satisfied. We cannot do that in the general case, however, with only NT functions in Eq. (2.34). Thus the residual is a function of x and the parameters c_i. We could make the residual approach zero by making it orthogonal to each member of a complete set of functions. We cannot do that with a finite NT, but we can make the residual orthogonal to a set of NT functions. Here we choose the NT functions $N_j(x)$, $j = 1, \ldots$, NT. To make the residual orthogonal to one function, we multiply the residual by that function, integrate over x from 0 to 1, and set the result equal to zero.

$$\sum_{i=1}^{NT} c_i \int_0^1 N_j \left[\text{Pe} \frac{dN_i}{dx} - \frac{d^2N_i}{dx^2} \right] dx = 0 \qquad (2.36)$$
$$j = 1, \ldots, NT$$

The idea of this method is that any function that is orthogonal to each member of a complete set of functions is identically zero. Therefore, if we can make the residual orthogonal to each member of a complete set of functions, it must be zero. Then the differential equation is satisfied. Of course, there are important questions: is the residual continuous? What are the complete functions? Polynomials and sines and cosines are examples of complete sets of functions. In the numerical method we can only use a subset of the complete functions, but as we enlarge that subset, we hope the approximation improves since the residual is made orthogonal to more and more functions. This can sometimes be proven, especially for linear problems.

We now examine Eq. (2.36). It includes the second derivative of $N_i(x)$, but N_i is a linear function. In the Galerkin method we integrate such terms by parts to give

$$\sum_{i=1}^{NT} c_i \int_0^1 \left[\text{Pe} \, N_j \frac{dN_i}{dx} + \frac{dN_j}{dx} \frac{dN_i}{dx} \right] dx - \sum_{i=1}^{NT} c_i \left[N_j \frac{dN_i}{dx} \right]_0^1 = 0. \qquad (2.37)$$

Here the last term is evaluated only at $x = 0$ and $x = 1$. Yet the only non-zero $N_i(x)$ values at those points are N_1 (at $x = 0$) and N_{NT} (at $x = 1$). Thus we have

$$\sum_{i=1}^{NT} c_i \left[N_j \frac{dN_i}{dx} \right]_0^1 = -c_1 \, N_j(x{=}0) \frac{dN_1}{dx}(x{=}0) + c_{NT} \, N_j(x{=}1) \frac{dN_{NT}}{dx}(x{=}1). \quad (2.38)$$

So far we have ignored the boundary conditions. For this problem we require that

$$c_1 = 1 \text{ and } c_{NT} = 0 \quad (2.39)$$

in order to satisfy the boundary conditions. Let us use these conditions. We now have too many conditions, since Eq. (2.36) is written for NT points and we have two more conditions from the boundary conditions. Yet we have only NT variables $\{c_i\}$. So we drop two of the conditions of Eq. (2.36), and we choose the conditions with $j = 1$ and $j = NT$. In essense, we use as the weighting function in Eq. (2.36) the remaining functions in the trial function [Eq. (2.34)] that are not otherwise specified by the boundary conditions. Then the Galerkin method consists of solving

$$\sum_{i=1}^{NT} c_i \int_0^1 \left[Pe \, N_j \frac{dN_i}{dx} + \frac{dN_j}{dx} \frac{dN_i}{dx} \right] dx = 0 \quad (2.40)$$

$$j = 2, \ldots, NT{-}1$$

along with Eq. (2.39).

It is convenient to set up these equations in a way that focuses on the elements. Let us look at the last equation and focus on a particular j value. The function N_j is zero over most of the region, taking non-zero values only over the (j-1)-th element and the j-th element. Thus the integral need only be evaluated over these two elements. Let us do so in the local coordinate system. The elements in consideration are shown in Figure 2.7, with the top diagram showing the global numbering system and the bottom two diagrams showing the local numbering system. We note that in the two diagrams the ξ values are the same in both elements; only the global numbering is different. The derivatives with respect to x are given by

$$\frac{dN_j}{dx} = \frac{2}{\Delta x_e} \frac{dN_J}{d\xi}, \quad dx = \frac{\Delta x_e}{2} d\xi. \quad (2.41)$$

(node j is in the e-th element and has local node number J)

Eq. (2.40) then becomes

$$\sum_{i=1}^{NT} c_i \int_{x_{j-1}}^{x_j} \left[Pe \, N_j \frac{dN_i}{dx} + \frac{dN_j}{dx} \frac{dN_i}{dx} \right] dx + \sum_{i=1}^{NT} c_i \int_{x_j}^{x_{j+1}} \left[Pe \, N_j \frac{dN_i}{dx} + \frac{dN_j}{dx} \frac{dN_i}{dx} \right] dx = 0 \quad (2.42)$$

since the function N_j is identically zero outside of the range $x_{j-1} \le x \le x_{j+1}$. But

INTRODUCTION - FDM & FEM

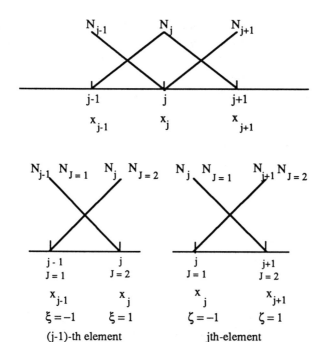

Figure 2.7. Trial Functions On an Element

the only functions N_i that are non-zero in this range are N_{j-1} and N_j in $x_{j-1} \leq x \leq x_j$ and N_j and N_{j+1} in $x_j \leq x \leq x_{j+1}$. This gives

$$\sum_{i=j-1}^{j} c_i \int_{x_{j-1}}^{x_j} \left[Pe\, N_j \frac{dN_i}{dx} + \frac{dN_j}{dx}\frac{dN_i}{dx} \right] dx + \sum_{i=j}^{j+1} c_i \int_{x_j}^{x_{j+1}} \left[Pe\, N_j \frac{dN_i}{dx} + \frac{dN_j}{dx}\frac{dN_i}{dx} \right] dx = 0. \quad (2.43)$$

Now we will use the transformation of Eq. (2.41), noting that in the first element, N_j is N_2 and in the second element, N_j is N_1.

$$c_{j-1} \int_{-1}^{1} \left[Pe\, N_2 \frac{2}{\Delta x_k} \frac{dN_1}{d\xi} + \frac{dN_2}{d\xi}\frac{dN_1}{d\xi}\frac{4}{\Delta x_k^2} \right] \frac{\Delta x_k}{2} d\xi +$$

$$+ c_j \int_{-1}^{1} \left[Pe\, N_2 \frac{2}{\Delta x_k} \frac{dN_2}{d\xi} + \frac{dN_2}{d\xi}\frac{dN_2}{d\xi}\frac{4}{\Delta x_k^2} \right] \frac{\Delta x_k}{2} d\xi +$$

$$+ c_j \int_{-1}^{1} \left[Pe\, N_1 \frac{2}{\Delta x_k} \frac{dN_1}{d\xi} + \frac{dN_1}{d\xi}\frac{dN_1}{d\xi}\frac{4}{\Delta x_k^2} \right] \frac{\Delta x_k}{2} d\xi + \quad (2.44)$$

$$+ c_{j+1} \int_{-1}^{1} \left[Pe\, N_1 \frac{2}{\Delta x_k} \frac{dN_2}{d\xi} + \frac{dN_1}{d\xi}\frac{dN_2}{d\xi}\frac{4}{\Delta x_k^2} \right] \frac{\Delta x_k}{2} d\xi = 0.$$

Of the eight integrals in Eq. (2.44), only six are unique; what is more important is that those same six integrals appear in the equations for any j value. Thus we need calculate only those six to set up the problem completely. Eq. (2.44) is the j-th equation; it is an equation involving c_{j-1}, c_j, and c_{j+1}. We have an equation like this for each j, j = 2,...,N. Combined with the boundary conditions, Eq. (2.39), we have N+1 equations for the N+1 unknowns $\{c_j\}$.

We need to have a more concise notation so that we do not have to write out Eq. (2.42) - Eq. (2.44) every time. Let us then write Eq. (2.40), but summed over each element, as occurred in Eq. (2.44):

$$\sum_{i=1}^{NT} c_i \int_0^1 \left[Pe\, N_j \frac{dN_i}{dx} + \frac{dN_j}{dx} \frac{dN_i}{dx} \right] dx =$$

$$= \sum_{e=1}^{NE} \sum_{I=1}^{2} c_I^e \int_{-1}^{1} \left[Pe\, N_J \frac{2}{\Delta x_e} \frac{dN_I}{d\xi} + \frac{dN_J}{d\xi} \frac{dN_I}{d\xi} \frac{4}{\Delta x_e^2} \right] \frac{\Delta x_e}{2} d\xi = 0. \quad (2.45)$$

The summation is over the set of I = 1 to I = 2, since there are at most two N_i values that are non-zero in the element where N_j is non-zero. We have allowed the element sizes to be different (Δx_e) and we do not require that each element have only one new node (hence the two indices, e and j). Here we could take $\Delta x_e = \Delta x_j$. We also must be able to relate c_I^e and c_i; the identification depends on e and is evident in Figure 2.6 or 2.7. The easiest way to remember this identification is to prepare a reference table that permits us to go from a local number in an element to a global number and vice versa. (This is what is done in computer programs, too.) An example is given in Table 2.1.

Table 2.1. Global-Local Number Array

Element Number e	Global Number of Local Node	
	I = 1	I = 2
1	1	2
2	2	3
3	3	4
•	•	•
99	99	100
100	100	101

Let us write Eq. (2.45) using element integrals. We define

$$A_{JI}^e = \int_{-1}^{1} \left[N_J \frac{dN_I}{d\xi} \right] d\xi, \quad B_{JI}^e = \frac{2}{\Delta x_k} \int_{-1}^{1} \frac{dN_J}{d\xi} \frac{dN_I}{d\xi} d\xi. \quad (2.46)$$

Then Eq. (2.45) can be written as

$$\sum_{e=1}^{NE}\sum_{I=1}^{2} c_I^e [\,Pe\,A_{JI}^e + B_{JI}^e\,] = 0. \tag{2.47}$$

If we use matrix notation and remember that repeated indices are summed, we can write this as

$$\sum_{e=1}^{NE} [\,Pe\,A_{JI}^e + B_{JI}^e\,]\,c_I^e = 0. \tag{2.48}$$

This represents a larger set of equations that can be written in the form

$$A_{ji} c_i = 0. \tag{2.49}$$

The equations in the form of Eq. (2.47) or Eq. (2.48) can be constructed on an element level. The summation over elements is done simply by placing the matrix elements in their proper position in the matrix represented by Eq. (2.49), using the reference table to relate local node numbers to global node numbers.

The local matrices for certain common terms are shown in Table 2.2. When the equations are more complicated, with variable coefficients or nonlinear terms, numerical quadrature is used. Normally, Gaussian quadrature is used. In this case we evaluate integrals using

$$\int_0^1 f(x)\,dx = \sum_{j=1}^{NT} W_j\,f(x_j), \tag{2.50}$$

where the quadrature points, x_j, and the weights, W_j, are listed in Table 2.3. The order of the highest polynomial that is calculated exactly is listed there as well. For the integrals in Eq. (2.46) the highest polynomial is linear in ξ, so that a single term of the quadrature is sufficient. If we had integrals involving N_J^4, where N_J was a linear function in ξ, then we would need to use three quadrature points for an exact calculation of the integral. More quadrature points are needed for nonlinear problems. When we treat two-dimensional problems, the integrals may not be exact anyway, and we must use a sufficient number of points so that the results are not effected.

The equations (2.49) are a set of N+1 equations that have a special structure. The j-th row of the matrix A_{ji} has non-zero terms in only the j-1, j, and j+1 columns. Thus the matrix is tridiagonal, as shown in Eq. (2.51) below. We solve this set of equations using an LU decomposition [Forsyth and Moler, 1967], as illustrated in Eq. (2.52).

Table 2.2. Element Trial Functions and Matrices

for linear elements

$$N_1 = \frac{1-\xi}{2} \quad N_2 = \frac{1+\xi}{2}, \quad \xi = \frac{2(x-x_i)}{\Delta x_e} - 1$$

$$\frac{dN_1}{d\xi} = -\frac{1}{2} \quad \frac{dN_2}{d\xi} = \frac{1}{2}$$

$$A_{JI}^e = \int_{-1}^{1} \left[N_J \frac{dN_I}{d\xi} \right] d\xi = \begin{bmatrix} -\frac{1}{2} & +\frac{1}{2} \\ -\frac{1}{2} & +\frac{1}{2} \end{bmatrix}$$

$$B_{JI}^e = \frac{2}{\Delta x_e} \int_{-1}^{1} \frac{dN_J}{d\xi} \frac{dN_I}{d\xi} d\xi = \frac{2}{\Delta x_e} \begin{bmatrix} \frac{1}{2} & -\frac{1}{2} \\ -\frac{1}{2} & \frac{1}{2} \end{bmatrix} = \frac{1}{\Delta x_e} \begin{bmatrix} 1 & -1 \\ -1 & 1 \end{bmatrix}$$

$$C_{JI}^e = \frac{\Delta x_e}{2} \int_{-1}^{1} N_J N_I d\xi = \frac{\Delta x_e}{2} \begin{bmatrix} \frac{2}{3} & \frac{1}{3} \\ \frac{1}{3} & \frac{2}{3} \end{bmatrix} = \Delta x_e \begin{bmatrix} \frac{1}{3} & \frac{1}{6} \\ \frac{1}{6} & \frac{1}{3} \end{bmatrix}, \quad \int_{-1}^{1} N_J d\xi = \begin{bmatrix} 1 \\ 1 \end{bmatrix}$$

Table 2.3. Quadrature Points and Weights

N	ξ	W*	N_e = degree of polynomial that is integrated exactly
1(3)	±1	0.33333 33333	3
	0.	1.33333 33333	
2	±0.57735 02692	1.0	3
3	±0.77459 66692	0.55555 55556	5
	0.	0.88888 88889	

*Sum of W's is 2.0 since element length is 2.0.

$$\begin{bmatrix} b_1 & c_1 & & & & \\ a_2 & b_2 & c_2 & & & \\ & a_3 & b_3 & c_3 & & \\ & & & \ddots & & \\ & & & a_{NT-1} & b_{NT-1} & c_{NT-1} \\ & & & & a_{NT} & b_{NT} \end{bmatrix} \begin{bmatrix} x_1 \\ x_2 \\ x_3 \\ \vdots \\ x_{NT-1} \\ x_{NT} \end{bmatrix} = \begin{bmatrix} d_1 \\ d_2 \\ d_3 \\ \vdots \\ d_{NT-1} \\ d_{NT} \end{bmatrix} \qquad (2.51)$$

We calculate in succession

$$c'_1 = \frac{c_1}{b_1}, \quad d'_1 = \frac{d_1}{b_1}$$

$$c'_{k+1} = \frac{c_{k+1}}{b_{k+1} - a_{k+1} c'_k}, \quad d'_{k+1} = \frac{d_{k+1} - a_{k+1} d'_k}{b_{k+1} - a_{k+1} c'_k} \qquad (2.52)$$

$$c_{NT} = 0, \quad x_{NT} = d'_{NT}, \quad x_k = d'_k - c'_k x_{k+1}.$$

This is the Thomas algorithm and is equivalent to a Gaussian elimination; we can rearrange it slightly to make it into an LU decomposition [Forsyth and Moler, 1967]. The number of multiplications to solve m number of such systems with N+1 equations is

$$\text{Multiplication count} = 2N + m(3N + 1). \qquad (2.53)$$

This is a significant savings over $N^3/3$, which would be required if the special structure of the matrix A was not taken into account.

Let us next build up the matrix problem, Eq. (2.49), element by element, for a specific case with Pe = 4. We divide the domain into four elements, each of width 0.25, and begin with a blank matrix A_{ji}.

$$\begin{bmatrix} 0 & 0 & 0 & 0 & 0 \\ 0 & 0 & 0 & 0 & 0 \\ 0 & 0 & 0 & 0 & 0 \\ 0 & 0 & 0 & 0 & 0 \\ 0 & 0 & 0 & 0 & 0 \end{bmatrix} \begin{bmatrix} c_1 \\ c_2 \\ c_3 \\ c_4 \\ c_5 \end{bmatrix} = \begin{bmatrix} 0 \\ 0 \\ 0 \\ 0 \\ 0 \end{bmatrix} \qquad (2.54)$$

The element matrices are

$$\text{Pe } A^e_{JI} + B^e_{JI} = \text{Pe} \begin{bmatrix} -\frac{1}{2} & +\frac{1}{2} \\ -\frac{1}{2} & +\frac{1}{2} \end{bmatrix} + \frac{2}{0.25} \begin{bmatrix} \frac{1}{2} & -\frac{1}{2} \\ -\frac{1}{2} & \frac{1}{2} \end{bmatrix} = \begin{bmatrix} +2 & -2 \\ -6 & +6 \end{bmatrix}. \qquad (2.55)$$

We assemble the first element. This means that we use the element matrices in Eq. (2.55) and place them in the appropriate place in the matrix A_{ji}. In the first element, the reference table tells us that the local nodes 1 and 2 are equivalent to the global nodes 1 and 2, so the matrices go into the 1st and 2nd positions of the matrix A_{ji}.

$$\begin{bmatrix} 2 & -2 & 0 & 0 & 0 \\ -6 & 6 & 0 & 0 & 0 \\ 0 & 0 & 0 & 0 & 0 \\ 0 & 0 & 0 & 0 & 0 \\ 0 & 0 & 0 & 0 & 0 \end{bmatrix} \begin{bmatrix} c_1 \\ c_2 \\ c_3 \\ c_4 \\ c_5 \end{bmatrix} = \begin{bmatrix} 0 \\ 0 \\ 0 \\ 0 \\ 0 \end{bmatrix} \quad (2.56)$$

Next we assemble the second element. The element matrices are the same since they do not depend on the element for this problem. The reference table now identifies the local nodes 1 and 2 with the global nodes 2 and 3, so the matrices go into the 2nd and 3rd positions of the matrix A_{ji}.

$$\begin{bmatrix} 2 & -2 & 0 & 0 & 0 \\ -6 & 8 & -2 & 0 & 0 \\ 0 & -6 & 6 & 0 & 0 \\ 0 & 0 & 0 & 0 & 0 \\ 0 & 0 & 0 & 0 & 0 \end{bmatrix} \begin{bmatrix} c_1 \\ c_2 \\ c_3 \\ c_4 \\ c_5 \end{bmatrix} = \begin{bmatrix} 0 \\ 0 \\ 0 \\ 0 \\ 0 \end{bmatrix} \quad (2.57)$$

We continue in this fashion for all the elements to obtain

$$\begin{bmatrix} 2 & -2 & 0 & 0 & 0 \\ -6 & 8 & -2 & 0 & 0 \\ 0 & -6 & 8 & -2 & 0 \\ 0 & 0 & -6 & 8 & -2 \\ 0 & 0 & 0 & -6 & 6 \end{bmatrix} \begin{bmatrix} c_1 \\ c_2 \\ c_3 \\ c_4 \\ c_5 \end{bmatrix} = \begin{bmatrix} 0 \\ 0 \\ 0 \\ 0 \\ 0 \end{bmatrix}. \quad (2.58)$$

We have not yet accounted for the boundary conditions, which require that $c_1 = 1$ and $c_5 = 0$. Because we do not need the equations corresponding to the weighting functions $j = 1$ or $j = 5$, the 1st and 5th equation can be struck from Eq. (2.58) and replaced with the boundary conditions.

$$\begin{bmatrix} 1 & 0 & 0 & 0 & 0 \\ -6 & 8 & -2 & 0 & 0 \\ 0 & -6 & 8 & -2 & 0 \\ 0 & 0 & -6 & 8 & -2 \\ 0 & 0 & 0 & 0 & 1 \end{bmatrix} \begin{bmatrix} c_1 \\ c_2 \\ c_3 \\ c_4 \\ c_5 \end{bmatrix} = \begin{bmatrix} 1 \\ 0 \\ 0 \\ 0 \\ 0 \end{bmatrix} \quad (2.59)$$

This set of equations is solved to obtain $\{c_i\}$.

It is also instructive to generate the equations for the general case. To do that we need careful accounting. We take the lower row of the matrices A^e and B^e and add them to the upper row of the matrices A^e and B^e, but displaced by one column. This corresponds to what we did for the particular case leading to Eq. (2.59). We get

$$\text{Pe}\left(-\frac{1}{2}c_{i-1}+\frac{1}{2}c_i\right)+\frac{2}{\Delta x}\left(-\frac{1}{2}c_{i-1}+\frac{1}{2}c_i\right)+ \\ \text{Pe}\left(-\frac{1}{2}c_i+\frac{1}{2}c_{i+1}\right)+\frac{2}{\Delta x}\left(\frac{1}{2}c_i-\frac{1}{2}c_{i+1}\right)=0. \quad (2.60)$$

We then multiply this by Δx and combine terms.

$$c_{i+1}\left(\frac{\text{Pe}\Delta x}{2}-1\right)+c_i(2)+c_{i-1}\left(-\frac{\text{Pe}\Delta x}{2}-1\right)=0 \quad (2.61)$$

This is the same equation [Eq. (2.14)] obtained earlier for the centered finite difference method. Thus the solution is the same as well. The solution is Eq. (2.22) and typical solutions are plotted in Figure 2.3. It does not oscillate provided that $\mu \leq 1$.

Sometimes an outflow boundary condition will be of the form

$$\left.\frac{dc}{dx}\right|_{x=1}=0. \quad (2.62)$$

This boundary condition is called a natural boundary condition because a finite element equation can force it to be satisfied as the grid is refined; it is natural to the problem. The boundary condition c = some value is an essential boundary condition; it is essential that the trial function satisfy this condition, because a finite element method will not force it to be satisfied. When we process the last element, from x_N to x_{N+1}, we obtain only the first half of Eq. (2.44).

$$c_N\int_{-1}^{1}\left[\text{Pe}\,N_2\frac{2}{\Delta x_k}\frac{dN_1}{d\xi}+\frac{dN_2}{d\xi}\frac{dN_1}{d\xi}\frac{4}{\Delta x_k^2}\right]\frac{\Delta x_k}{2}d\xi + \\ + c_{N+1}\int_{-1}^{1}\left[\text{Pe}\,N_2\frac{2}{\Delta x_k}\frac{dN_2}{d\xi}+\frac{dN_2}{d\xi}\frac{dN_2}{d\xi}\frac{4}{\Delta x_k^2}\right]\frac{\Delta x_k}{2}d\xi=0 \quad (2.63)$$

We cannot include the second half because there is no element beyond j = N+1. Then when we generate the general case, as in Eq. (2.60), we obtain only half of it, corresponding to the last two rows of the element matrices A^e and B^e.

$$c_N\left[-\frac{\text{Pe}}{2}+\frac{2}{\Delta x}\left(-\frac{1}{2}\right)\right]+c_{N+1}\left[+\frac{\text{Pe}}{2}+\frac{2}{\Delta x}\left(\frac{1}{2}\right)\right]=0 \quad (2.64)$$

When we multiply this equation by Δx and combine terms, we get

$$c_{N+1}\left(\frac{Pe\,\Delta x}{2}+1\right)-c_N\left(\frac{Pe\,\Delta x}{2}+1\right)=0 \text{ or } c_{N+1}=c_N. \tag{2.65}$$

Then we use $c_1 = 1$, Eq. (2.61) for $i = 2,..., N$, and Eq. (2.65) for $i = N+1$. For the boundary condition of Eq. (2.62) the solution is $c_i = 1$, so this result is not of interest. For the transient problem, however, the treatment of the boundary element will be needed.

We eliminated the oscillations in the centered finite difference method by changing to an upstream difference expression for the first derivative. What do we do here? We generalize the Galerkin method to the Petrov-Galerkin method. The generalization consists simply of allowing the weighting function to be a different function than the trial function. Thus, for the convective diffusion equation in steady-state, we take the Petrov-Galerkin method to be

$$\int_0^1 W_j\left[Pe\,\frac{dc}{dx}-\frac{d^2c}{dx^2}\right]dx=0, \quad j=1,...,NT. \tag{2.66}$$

We then insert the trial function, Eq.(2.34), integrate by parts, and obtain

$$\sum_{i=1}^{NT} c_i \int_0^1 \left[Pe\,W_j\frac{dN_i}{dx}+\frac{dW_j}{dx}\frac{dN_i}{dx}\right]dx=0. \tag{2.67}$$

The boundary terms are again zero, since we require that

$$\begin{aligned} W_j(x=0)&=0, \, j=2,...,NT-1 \text{ and} \\ W_j(x=1)&=0, \, j=2,...,NT-1. \end{aligned} \tag{2.68}$$

For these problems, the usual weighting function is taken as follows.

$$\begin{aligned} W_j &= N_j \pm \alpha\,F(\xi); \text{ use + for left element, – for right element;} \\ F(\xi) &= -\frac{3}{4}(\xi^2-1). \end{aligned} \tag{2.69}$$

The weighting function is illustrated in Figure 2.8. Its effect is to weight the upstream direction more heavily. The Petrov-Galerkin finite element equations

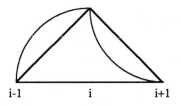

Figure 2.8. Function $W_i = N_i + \alpha\,F(\xi)$

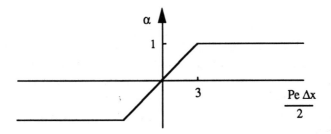

Figure 2.9. Approximation to the Optimal Weighting Function

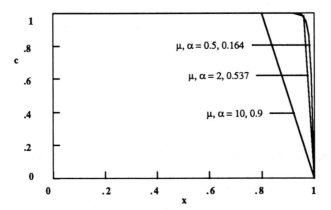

Figure 2.10. Petrov-Galerkin Solutions for Pe = 100

then become

$$\frac{Pe}{2\Delta x}[(1-\alpha)c_{i+1} + 2\alpha c_i - (1+\alpha)c_{i-1}] = \frac{1}{\Delta x^2}[c_{i+1} - 2c_i + c_{i-1}]. \qquad (2.70)$$

The diffusion term is unaffected by the new weighting function, while the convection term is affected. We choose the value of α to optimize the results. If we choose $\alpha = 1$, we get a finite difference method with upstream derivatives. The choice

$$\alpha = \coth\left(\frac{Pe\Delta x}{2}\right) - \frac{2}{Pe\Delta x} \qquad (2.71)$$

causes the solution at the nodes to be exact [Christie, *et al.*, 1976]. Sometimes this relation is approximated by the function shown in Figure 2.9 [Hughes and Brooks, 1979].

Analysis of this difference equation shows that the solution does not oscillate, provided that

$$Pe\Delta x \leq \frac{2}{1-\alpha}. \qquad (2.72)$$

For the case of Pe = 100 and various grid spacings, typical solutions with the optimal α are shown in Figure 2.10. The solution does not oscillate and the nodal values are exact; it differs from the exact solution only between nodes. The error has the smallest value that is possible on this mesh. Clearly this is a big improvement over Figure 2.3, but it depended on our knowing the exact solution so that the optimal α could be found. We, of course, want a method that works in more general cases as well.

2.4. Applications to the Advection Equation

Next let us turn to transient problems. We first consider the advection equation in order to focus on the exact feature of interest.

$$\frac{\partial c}{\partial t} + \text{Pe}\frac{\partial c}{\partial x} = 0 \tag{2.73}$$

$$c(0,t) = g(t), \quad c(x,0) = f(x) \tag{2.74}$$

To fully appreciate the methods for transient problems, we should look at methods for solving ordinary differential equations.

We take the standard problem

$$\frac{dy}{dt} = f(y), \quad y(0) = y_0. \tag{2.75}$$

In the Euler method we write a difference formula using $y^n = y(t_n) = y(n\Delta t)$.

$$\frac{y^{n+1} - y^n}{\Delta t} = f(y^n) + O(\Delta t) \tag{2.76}$$

This formula can be rearranged to solve for y^{n+1}.

$$y^{n+1} = y^n + \Delta t\, f(y^n) + O(\Delta t^2) \tag{2.77}$$

The value for y^{n+1} can be calculated directly once the value of y^n is known. Since we know the value of y_0, the calculation can begin there and proceed directly. This is called an explicit method. This particular explicit method is not very accurate, since its truncation error is proportional to Δt [see Eq. (2.76)] and a small Δt value is needed for reasonable accuracy.

An improvement on this method is the following predictor-corrector method (also called a Runge-Kutta method):

$$\frac{y^{n+1/2} - y^n}{\Delta t/2} = f(y^n) \tag{2.78}$$

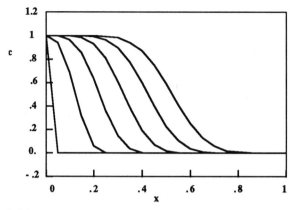

Figure 2.11. Advection Equation with Upstream Method, $\Delta x = 0.05$, $\Delta t = 0.025$
Solutions at $t = 0.1, 0.2, 0.3, 0.4, 0.5$

$$\frac{y^{n+1} - y^n}{\Delta t} = f(y^{n+1/2}). \quad (2.79)$$

The truncation error of this method can be shown to be Δt^2; it would generally be preferred over the Euler method. It requires more calculations per step, but presumably fewer steps are necessary.

An implicit method is provided by the following algorithm:

$$\frac{y^{n+1} - y^n}{\Delta t} = f(y^{n+1}) + O(\Delta t). \quad (2.80)$$

Now we must solve Eq. (2.80) to obtain y^{n+1}; this equation is generally nonlinear. Nonlinear problems are not hard to solve if only a few equations are integrated, but if thousands of equations are to be integrated, nonlinear problems can be very difficult. The advantage of implicit methods is that large time-steps will generally be more stable with Eq. (2.80) than with Eq. (2.77).

We now apply these ideas to Eq. (2.73) by using the simplest, lowest-order method: the Euler method in time and the upstream difference method in space.

$$\frac{c_i^{n+1} - c_i^n}{\Delta t} + Pe \frac{c_i^n - c_{i-1}^n}{\Delta x} = 0 \quad (2.81)$$

The solution obtained with this method when the inlet condition undergoes a step change is shown in Figure 2.11. This solution was obtained using only 21 nodes; the sharp front is smoothed over a quite large region. We recall that the exact solution is a vertical line at the position $x = t$. When more points are used (e.g., 100), the results are improved, as shown in Figure 2.12. The front is overly

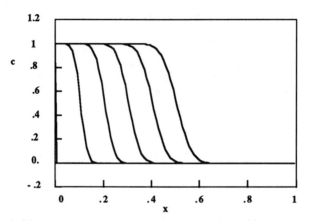

Figure 2.12. Advection Equation with Upstream Method, $\Delta x = 0.01$, $\Delta t = 0.005$
Solutions at $t = 0.1, 0.2, 0.3, 0.4, 0.5$

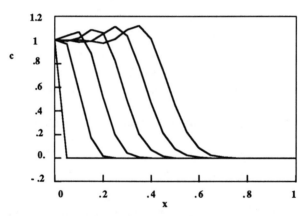

Figure 2.13. Advection Equation with Lax-Wendroff Method, $\Delta x = 0.05$, $\Delta t = 0.025$

smoothed, but no oscillations are present.

Next let us use the Lax-Wendroff method, which is one version of a predictor-corrector method. The two steps are given by

$$c_{i+1/2}^{n+1/2} = 0.5\,(c_{i+1}^n + c_i^n) - \frac{\Delta t\,Pe}{2\Delta x}(c_{i+1}^n - c_i^n) \tag{2.82}$$

$$c_i^{n+1} = c_i^n - \frac{\Delta t\,Pe}{\Delta x}(c_{i+1/2}^{n+1/2} - c_{i-1/2}^{n+1/2}). \tag{2.83}$$

The solution obtained with this method when the inlet condition undergoes a step change is shown in Figure 2.13 (for 21 nodes) and Figure 2.14 (for 100 nodes). Both solutions oscillate, but the front is steeper with 100 nodes.

The Galerkin finite element method can also be applied to this problem. The element equations are

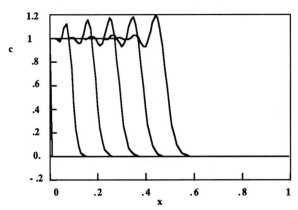

Figure 2.14. Advection Equation with Lax-Wendroff Method, $\Delta x = 0.01$, $\Delta t = 0.005$

$$C^e_{JI} \frac{dc^e_I}{dt} + Pe\, A^e_{JI} c^e_I = 0. \tag{2.84}$$

The matrix C^e_{JI} is called the mass matrix. When the element matrices are assembled from Table 2.2, the result is

$$\frac{1}{6}\frac{dc_{i+1}}{dt} + \frac{2}{3}\frac{dc_i}{dt} + \frac{1}{6}\frac{dc_{i+1}}{dt} + Pe\,\frac{c_{i+1} - c_{i-1}}{2\Delta x} = 0. \tag{2.85}$$

Now the problem has become more complicated because several time derivatives are included on the left-hand side. As a result, we cannot simply apply the Euler method and solve explicitly. The Galerkin method must be applied using methods similar to those needed for implicit methods. For the examples here, we will integrate these equations by replacing each time derivative by a first-order difference formula and solving the resulting set of equations simultaneously from step to step.

$$\frac{1}{6}\frac{c^{n+1}_{i+1} - c^n_{i+1}}{\Delta t} + \frac{2}{3}\frac{c^{n+1}_i - c^n_i}{\Delta t} + \frac{1}{6}\frac{c^{n+1}_{i-1} - c^n_{i-1}}{\Delta t} + Pe\,\frac{c^n_{i+1} - c^n_{i-1}}{2\Delta x} = 0 \tag{2.86}$$

We can then write these equations in the form

$$a_i c_{i-1} + b_i c_i + d_i c_{i+1} = f_i \quad \text{or} \quad \sum_j A_{ji} c_j = f_i. \tag{2.87}$$

This system of equations is tridiagonal, because the matrix on the left has non-zero terms on the diagonal and the two diagonal rows nearby. It can easily be solved using the Thomas algorithm or a variant of it that performs a LU decomposition [Forsyth and Moler, 1967]. Then the matrix is written in the form

$$A = LU. \tag{2.88}$$

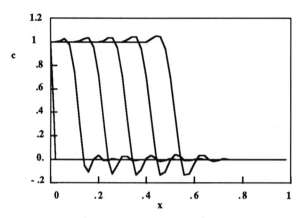

Figure 2.15. Advection Equation, Petrov-Galerkin Method, $\alpha = 1$, $\Delta x = 0.02$, $\Delta t = 0.001$

The solution for any right-hand side is

$$c = U^{-1} L^{-1} f. \tag{2.89}$$

The decomposition (giving L^{-1} and U^{-1}) need be performed only once and Eq. (2.89) can be applied for repeated changes in f_i. When there are changes in the right-hand side, the entire process does not need to be repeated. Only if the time-step is changed does the decomposition need to be performed. As a result, explicit methods are easily handled with the Galerkin method.

Another resolution to this problem is called lumping. The rows in the mass matrix are summed and the sum is placed on the diagonal while the other terms are set equal to zero. This corresponds to taking mass spread out over several nodes and placing it at the center node. The resulting equations are the same as those in the centered finite difference method (for this problem).

$$\frac{dc_i}{dt} + Pe \frac{c_{i+1} - c_{i-1}}{2\Delta x} = 0 \tag{2.90}$$

Time-dependent solutions can also oscillate and a Petrov-Galerkin method is suitable in that case, too. If the same weighting function [Eq. (2.69)] is used, the Petrov-Galerkin method gives the equation

$$\left(\frac{1}{6} - \frac{\alpha}{4}\right)\frac{dc_{i+1}}{dt} + \frac{2}{3}\frac{dc_i}{dt} + \left(\frac{1}{6} + \frac{\alpha}{4}\right)\frac{dc_{i-1}}{dt} + \frac{Pe}{2\Delta x}[(1-\alpha)c_{i+1} + 2\alpha c_i - (1+\alpha)c_{i-1}] = 0. \tag{2.91}$$

The Galerkin method is unstable for this problem (without diffusion). The lumped Galerkin method is also unstable, because of the centered difference expression for the convective term. Solutions obtained with the the Petrov-Galerkin method are shown in Figure 2.15. Clearly the methods show more

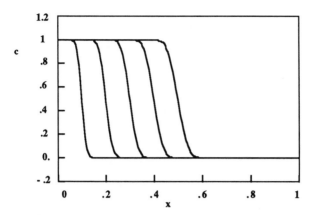

Figure 2.16. Convective Diffusion Equation, Pe = 1000, t = 0.0001 to 0.0005

oscillations than we would like to have.

We see that the methods to find solutions to the advection equation [Eq.(2.73)] all have problems: we must accept either oscillations or smearing of the front. The advection equation is the simplest of problems and we wish to avoid both oscillations and the extra dispersion for this problem, as well as for more difficult ones. In other chapters we will examine methods to do this.

2.5. Applications to the Transient Convective Diffusion Equation

When we expand the equation to add a diffusion term, we make the problem easier because the added diffusion term will tend to dampen oscillations developed by the numerical method. The transient convective diffusion equation is

$$\frac{\partial c}{\partial t} + \text{Pe} \frac{\partial c}{\partial x} = \frac{\partial^2 c}{\partial x^2} \tag{2.92}$$

along with the initial and boundary conditions

$$c(0,t) = g(t), \, c(x,0) = f(x) \tag{2.93}$$

$$c(1,t) = 0. \tag{2.94}$$

More properly, the boundary condition of [Eq. (2.94)] should be replaced by

$$\left.\frac{\partial c}{\partial x}\right|_{x=1} = 0, \tag{2.95}$$

but the difference will be seen only when the convective wave nears the point x = 1. The exact solution for a Peclet number of 1000 is shown in Figure 2.16. The finite difference method (using centered first derivatives) gives

36 NUMERICAL METHODS - MOVING FRONTS

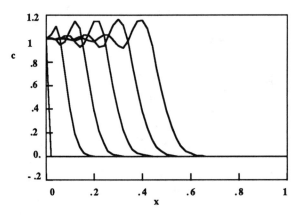

Figure 2.17. Convective Diffusion Equation, Finite Difference Method
Pe = 1000, $\Delta x = 0.02$, $\Delta t = 1.\text{E-6}$

$$\frac{dc_i}{dt} + \text{Pe} \frac{c_{i+1} - c_{i-1}}{2\Delta x} = \frac{c_{i+1} - 2c_i + c_{i-1}}{\Delta x^2}. \tag{2.96}$$

If we compare this with Eq. (2.81), we see that the first derivative is different. It is not apparent here, but if there is no diffusion term and centered differences are used in Eq. (2.81), then the method is unstable for any choice of parameters. This result is proven in Chapter 4. It does suggest possible problems, however, when solving Eq. (2.96) in a case where the Peclet number is large. When the diffusion term is present, we can use a centered derivative. The boundary conditions are

$$c(x,0) = 0, \; c(0,t) = 1. \tag{2.97}$$

An explicit method is applied to the time derivative to obtain the working equations.

$$\frac{c_i^{n+1} - c_i^n}{\Delta t} + \text{Pe} \frac{c_{i+1}^n - c_{i-1}^n}{2\Delta x} = \frac{c_{i+1}^n - 2c_i^n + c_{i-1}^n}{\Delta x^2} \tag{2.98}$$

Price, et al. [1966] prove that the solution to Eq. (2.95) does not oscillate when Pe $\Delta x \leq 2$, but that for Pe $\Delta x > 2$, damped oscillations occur with non-real eigenvalues. Thus, the same limits that we use to eliminate oscillations apply to the transient and steady-state equations for the finite difference method. For the calculations we take Pe = 1000 and various Δx values. When N = 50, then $\Delta x = 0.02$ and Pe$\Delta x/2$ is 10. Even for the steady-state solution we expect oscillations and they are apparent in Figure 2.17. To verify that the oscillations are due to the spatial discretization, the same problem is solved using N = 50 and using LSODE to integrate in time. LSODE is an integration package that guarantees the accuracy of the time-dependent simulation. The results are shown in Figure 2.18; they are nearly identical with those shown in Figure 2.17. When N = 500,

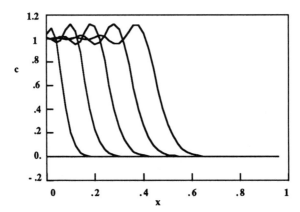

Figure 2.18. Convective Diffusion Equation, Centered Finite Difference Method, Pe = 1000, $\Delta x = 0.02$, LSODE

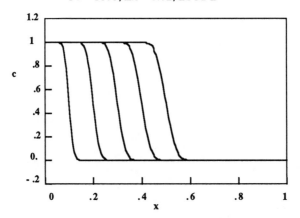

Figure 2.19. Convective Diffusion Equation, Centered Finite Difference Method, Pe = 1000, $\Delta x = 0.002$, LSODE

however, $Pe\Delta x/2 = 1$, and no oscillations are present (see Figure 2.19). Thus we can get rid of the oscillations by refining the mesh, at the cost of having more unknowns (and a more time-consuming and expensive solution).

Next we will use the upstream first derivative and replace Eq. (2.96) with

$$\frac{c_i^{n+1} - c_i^n}{\Delta t} + Pe \frac{c_i^n - c_{i-1}^n}{\Delta x} = \frac{c_{i+1}^n - 2c_i^n + c_{i-1}^n}{\Delta x^2}. \qquad (2.99)$$

The solution obtained with Eq. (2.99) is shown in Figure 2.20 for N = 50. There is excessive dispersion and the steep profile is smeared out. When the same problem is solved using N = 50 and LSODE, the results are nearly identical to those shown in Figure 2.20 (see Figure 2.21). When N = 500, the upstream dispersion is relatively unimportant and a good solution is obtained (see Figure 2.22).

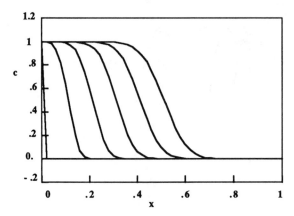

Figure 2.20. Convective Diffusion Equation, Finite Difference, Upstream Derivative
Pe = 1000, Δx = 0.02, Δt = 1.E-5, t = 0.0001 to 0.0005

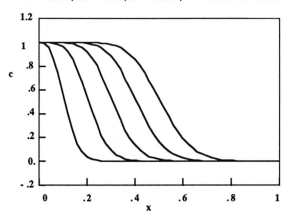

Figure 2.21. Convective Diffusion Equation, Finite Difference, Upstream Derivative
Pe = 1000, Δx = 0.02, LSODE, t = 0.0001 to 0.0005

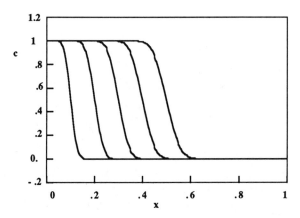

Figure 2.22. Convective Diffusion Equation, Finite Difference Method
Upstream Derivative, Pe = 1000, Δx = 0.002, LSODE, t = 0.0001 to 0.0005

INTRODUCTION - FDM & FEM

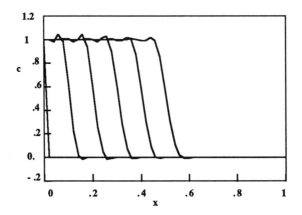

Figure 2.23. Convective Diffusion Equation, Galerkin Method
Pe = 1000, $\Delta x = 0.02$, $\Delta t = 1.E-6$

The Galerkin method involves a simple generalization of Eqs. (2.84-2.85).

$$C^e_{JI}\frac{dc^e_I}{dt} + Pe\, A^e_{JI}c^e_I = -B^e_{JI}c^e_I \qquad (2.100)$$

$$\frac{1}{6}\frac{dc_{i+1}}{dt} + \frac{2}{3}\frac{dc_i}{dt} + \frac{1}{6}\frac{dc_{i-1}}{dt} + Pe\,\frac{c_{i+1} - c_{i-1}}{2\Delta x} = \frac{c_{i+1} - 2c_i + c_{i-1}}{\Delta x^2} \qquad (2.101)$$

This leads to the general equation

$$\frac{1}{6}\frac{c^{n+1}_{i+1} - c^n_{i+1}}{\Delta t} + \frac{2}{3}\frac{c^{n+1}_i - c^n_i}{\Delta t} + \frac{1}{6}\frac{c^{n+1}_{i-1} - c^n_{i-1}}{\Delta t} + Pe\,\frac{c^n_{i+1} - c^n_{i-1}}{2\Delta x} = \frac{c^n_{i+1} - 2c^n_i + c^n_{i-1}}{\Delta x^2}. \qquad (2.102)$$

The equation for the last element is

$$\frac{1}{6}\frac{c^{n+1}_{NT-1} - c^n_{NT-1}}{\Delta t} + \frac{1}{3}\frac{c^{n+1}_{NT} - c^n_{NT}}{\Delta t} + Pe\,\frac{c^n_{NT} - c^n_{NT-1}}{2\,\Delta x} = \frac{c^n_{NT-1} - c^n_{NT}}{\Delta x^2}. \qquad (2.103)$$

Solutions obtained for Pe = 1000 and N = 50 are shown in Figures 2.23 and 2.24. Even though $Pe\Delta x/2$ is 10, the solution is quite good, with a steep front and only small oscillations. If we use N = 500 we get an even better solution (Figure 2.25), which is indistinguishable from the exact solution.

If we lump the mass matrix, the general equation replacing Eq. (2.85) is

$$\frac{dc_i}{dt} + Pe\,\frac{c_{i+1} - c_{i-1}}{2\Delta x} = \frac{c_{i+1} - 2c_i + c_{i-1}}{\Delta x^2}. \qquad (2.104)$$

Since this is the finite difference method, the results will be the same. For calculations with $Pe\Delta x/2 = 10$ we get oscillations, but for $Pe\Delta x/2 = 1$ we do not.

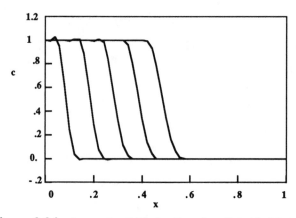

Figure 2.24. Convective Diffusion Equation, Galerkin Method
Pe = 1000, Δx = 0.02, LSODE

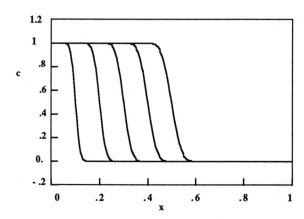

Figure 2.25. Convective Diffusion Equation, Galerkin Method
Pe = 1000, Δx = 0.002, LSODE

The Petrov-Galerkin equations are

$$\left(\frac{1}{6} - \frac{\alpha}{4}\right)\frac{dc_{i+1}}{dt} + \frac{2}{3}\frac{dc_i}{dt} + \left(\frac{1}{6} + \frac{\alpha}{4}\right)\frac{dc_{i-1}}{dt} + \frac{Pe}{2\Delta x}[(1-\alpha)c_{i+1} + 2\alpha c_i - (1+\alpha)c_{i-1}]$$
$$= \frac{c_{i+1} - 2c_i + c_{i-1}}{\Delta x^2}. \qquad (2.105)$$

The equation for the last element is

$$\left(\frac{1}{6} + \frac{\alpha}{4}\right)\frac{c_{NT-1}^{n+1} - c_{NT-1}^n}{\Delta t} + \left(\frac{1}{3} + \frac{\alpha}{4}\right)\frac{c_{NT}^{n+1} - c_{NT}^n}{\Delta t} + \frac{Pe}{2\Delta x}[(1+\alpha)c_{NT}^n - (1+\alpha)c_{NT-1}^n] =$$
$$= \frac{c_{NT-1}^n - c_{NT}^n}{\Delta x^2}. \qquad (2.106)$$

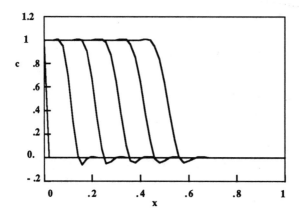

Figure 2.26. Convective Diffusion Equation, Petrov-Galerkin Method, $\alpha = 0.5164$, Pe = 1000, $\Delta x = 0.02$, $\Delta t = 1.\text{E-}6$

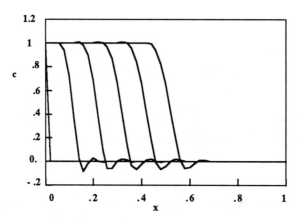

Figure 2.27. Convective Diffusion Equation, Petrov-Galerkin Method $\alpha = 0.9$, Pe = 1000, $\Delta x = 0.02$, $\Delta t = 1.\text{E-}6$

Shown are a series of figures for various α. The results obtained with $\alpha = 0$ are shown in Figure 2.23. When the same problem is solved using N = 50 and LSODE, the results are as shown in Figure 2.24; they are nearly identical. If N = 500 and LSODE is used, we get the results shown in Figure 2.25, which are excellent. For N = 50 and α = 0.5164, 0.9, and 1.0, see Figures 2.26, 2.27 and 2.28, respectively. In the progression, as α is increased, the oscillations behind the front are damped while those before the front become larger.

If we compare results from the upstream finite difference method (Figure 2.20), the Galerkin method (Figure 2.23), and the Petrov-Galerkin method (fully upstream, $\alpha = 1$, Figure 2.28), we see that the upstream finite difference results are smeared more than the other methods but that the solutions do not oscillate. Results from the Galerkin method and the Petrov-Galerkin method oscillate slightly, but in different places. The oscillations can be

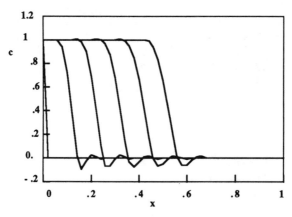

Figure 2.28. Convective Diffusion Equation, Petrov-Galerkin Method
$\alpha = 1.0$, Pe = 1000, $\Delta x = 0.02$, $\Delta t = 1.E-6$

eliminated by making Pe$\Delta x/2$ smaller than 2 (for the methods treated here). The computation time is then increased. Methods with upstream character gave results that do not oscillate (finite difference) or oscillate slightly (Petrov-Galerkin), even when Pe$\Delta x/2$ is big. The Galerkin and Petrov-Galerkin methods require a smaller time-step. We would like to find a method which does not smear the front, does not oscillate, and does not require a large number of grid points. Such methods are discussed throughout the book.

Problems

1_1. Verify that Eq. (2.3) satisfies the differential equation (2.1) and boundary conditions, Eq. (2.2).
2_1. Verify Eq. (2.5). Simplify the result for a large Peclet number. How large must Pe be for the approximation to be accurate within 1%?
3_1. Verify equations (2.17-2.25).
4_1. Verify equations (2.26-2.31).
5_2. Consider the following steady-state problem:

$$10 \frac{dc}{dx} = \frac{d^2c}{dx^2}, \quad c(0) = 1, \; c(1) = 0. \tag{2.107}$$

To see the basis for the Galerkin method without unnecessary algebra, consider the following simple approximation.

$$c = a + bx + cx^2 \tag{2.108}$$

(i) Make this function satisfy the boundary conditions.
(ii) Show that the remaining function can be written in the form

$$c = 1 - x + bx(1-x). \tag{2.109}$$

(iii) Calculate the derivatives and then evaluate the residual.

$$\text{Residual} = \frac{d^2c}{dx^2} - 10\frac{dc}{dx} \tag{2.110}$$

Note that the residual depends on the unknown parameter, b.
(iv) To simplify the algebra even further, apply the collocation method. Determine the parameter b by setting the residual equal to zero at the point x = 1/2. [The Galerkin method would make the residual orthogonal to the trial function x (1-x).]
(v) Write down the approximate solution as a polynomial in x. Plot it. The program CONVECT can also be used as a plotting program. Simply choose DATA and insert the tabular values into the program.
(vi) Use this polynomial approximation and derive the residual as a polynomial. Plot it.
(vii) Note that the residual is something that can be found after the approximate solution is derived, and you want it to be as small as possible. In this case, what is the range of values taken by the residual? If more terms were used, the solution would be better and the residual would be smaller; the residual provides a way of testing the approximate solution.

6_1. Verify the entries in Table 2.2.
7_2. Use the Gaussian quadrature formula, Eq. (2.50), together with the points and weights shown in Table 2.3 to evaluate the following integrals. Use N = 2 and N = 3.

$$\int_{-1}^{1} d\xi,\ \int_{-1}^{1} \xi\, d\xi,\ \int_{-1}^{1} \xi^2\, d\xi,\ \int_{-1}^{1} \xi^4\, d\xi,\ \int_{-1}^{1} e^{\xi}\, d\xi \tag{2.111}$$

8_1. Verify Eq. (2.54-2.59).
9_1. Derive Eq. (2.60).
10_1. Derive Eq. (2.65) from Eq. (2.63).
11_2. Derive Eq. (2.70).
12_1. Compare Eq. (2.71) with the approximation shown in Figure 2.9.
13_2. Use the three integration methods, equations (2.77-2.80), to integrate the following problem from t = 0 to t = 1.

$$\frac{dy}{dt} = -y,\ y(0) = 1 \tag{2.112}$$

Use step-sizes of Δt = 0.1, 0.05, and 0.025. Compare the accuracy of the methods by comparing the results with the exact solution. How does the error vary with Δt?
14_2. Derive Eq. (2.85) and (2.102).
15_1. Rewrite Eq. (2.86) in the form of Eq. (2.87). Give general formulas for a_i, b_i, d_i, and f_i.
16_3. Derive Eq. (2.91).
17_3. Derive Eq. (2.105).
18_1. Consider flow through the packed bed illustrated in Figure 2.29. Q represents the flow rate in m^3/s, A is the open cross sectional area, in m^2, and L is the length of the bed.

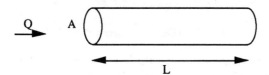

Figure 2.29. Packed bed

The convective diffusion equation is

$$\frac{\partial c}{\partial t} + u \frac{\partial c}{\partial x} = D \frac{\partial^2 c}{\partial x^2}. \tag{2.113}$$

The variable u is the velocity; $u = Q / A$.
(i) Make this equation non-dimensional by defining

$$c' = \frac{c}{c_0}, \ x' = \frac{x}{L}, \ t' = \frac{tD}{L^2}, \ Pe = \frac{uL}{D}. \tag{2.114}$$

The result is

$$\frac{\partial c'}{\partial t'} + Pe \frac{\partial c'}{\partial x'} = \frac{\partial^2 c'}{\partial x'^2}. \tag{2.115}$$

(ii) Next make the convective diffusion equation non-dimensional by defining

$$c' = \frac{c}{c_0}, \ x' = \frac{x}{L}, \ t'' = \frac{tu}{L}, \ Pe = \frac{uL}{D}. \tag{2.116}$$

The result is

$$\frac{\partial c'}{\partial t''} + \frac{\partial c'}{\partial x'} = \frac{1}{Pe} \frac{\partial^2 c'}{\partial x'^2}. \tag{2.117}$$

(iii) How are t' and t'' related?
(iv) Show the advection equation when each non-dimensionalization is used. Note that the first one uses an unknown parameter, D, to define both t' and Pe, but that by relating t' and t'', the unknown parameter can be eliminated.

19₂. (i) Consider the convective diffusion equation (2.92). Write a finite difference method using an upstream evaluation of the first derivative. Evaluate this expression to order Δx^2 by substituting Eq. (2.7-2.8) into it. The result is

$$\frac{dc_i}{dt} + Pe \, c'_i = \left(1 + \frac{Pe \, \Delta x}{2}\right) c''_i. \tag{2.118}$$

Thus the upstream finite difference equation, when evaluated to the second-order in Δx, is the centered finite difference expression with diffusion added. The non-dimensional form of the equation changes the diffusion coefficient in the amount

$$\text{from 1 to } 1 + \frac{Pe \, \Delta x}{2} \tag{2.119}$$

in the non-dimensional form of the equation and

$$\text{from D to } D + \frac{u \, \Delta x}{2} \ \text{(here x is dimensional)} \tag{2.120}$$

in the dimensional form of the equation. The same result can be derived by writing the upstream finite difference equation with the convective term and diffusion term on opposite sides of the equality sign. Then add

$$\frac{Pe \, \Delta x}{2} \frac{c_{i+1} - 2 c_i + c_{i-1}}{\Delta x^2} \tag{2.121}$$

20₂. Consider a stirred-tank model of the same packed bed shown in Figure 2.29. Divide the packed bed into a series of volumes, $V = A L / N$, where there are N stirred tanks. If each stirred tank is well mixed, giving a series of CSTRs (continuous stirred-tank reactors), the mass balance for each tank is

$$V \frac{dc_i}{dt} = Q (c_{i-1} - c_i). \qquad (2.122)$$

Relate the parameters in the differential model and the stirred-tank model using $\Delta x = L / N$. Substitute for c_{i-1} using equation (2.8), keeping terms to order Δx^2. After rearrangement we get

$$\frac{dc_i}{dt} + u\, c'_i = \frac{u\, \Delta x}{2} c''_i + O(\Delta x^2). \qquad (2.123)$$

Note: a stirred-tank model always introduces additional dispersion.

21₂. Test the computer code, CONVECT, for the advection equation. For $Pe = 2$, $\Delta x = 0.1$, and $\Delta t = 0.025$, choose the option to ignore diffusion. To obtain a finite difference method, choose the option to lump the time derivatives. Choose a step function for the initial conditions and run the code for three time-steps. Use the DATA feature to find the numerical values of the solution at that point in time. Run one more step, and get the solution at the next time-step. Verify that the predictions are in accordance with the equation for the method used. Do this for the following methods.
(i) Euler in time, upstream in space, Eq. (2.81)
(ii) Lax-Wendroff, Eqs. (2.82)-(2.83); note that only the solutions at the n-th and n+1-st level are available to you.

22₃. Repeat problem 21 but include diffusion.
(i) First do a finite difference method with $\Delta t = 0.005$; verify Eq. (2.98).
ii) Check the Galerkin method, Eq. (2.102), by substituting the solution found into the equation (do not solve the tri-diagonal matrix, but just check the solution).
(iii) Check the Petrov-Galerkin method, Eq. (2.105), for $\alpha = 0.5$ and $\alpha = 1$.
By solving problems 21 and 22, you can verify that the code is solving the problems correctly.

23₃. (Project) Solve the convective diffusion equation in the following cases by using the program CONVECT.
(i) Set $Pe = 0$ and choose the option to include diffusion. Use $\Delta x = 0.2$ for x values from 0 to 1. Integrate to $t = 0.1$ using the Euler method. First choose a step-size based on the stability criterion, $\Delta t = \Delta x^2 / 2$, which is derived in Chapter 4. Re-solve the problem using time-steps that are larger than that required for stability, in order to see what happens when the stability limit is exceeded. Compare a centered finite difference method with the Galerkin finite element method. Discuss your results.
(ii) Set $Pe = 100$ and integrate to $t = 0.005$. Pose one question to investigate. Typical questions could be:
   (1) What are the Δt requirements of the different methods?
   (2) How does the GFEM solution change as Δx is decreased?
   (3) How does the necessary Δt change for (2)?
   (4) How do the solutions differ when upstream weight parameters are different?

References

Christie, I., Griffiths, D. F., Mitchell, A. R., and Zienkiewicz, O. C., "Finite Element Methods for Second Order Differential Equations with Significant First Derivatives," Int. J. Num. Methods Engn. 10 1389-1396 [1976].

Forsyth, G. and Moler, C. B., *Computer Solution of Linear Algebraic Systems*, Prentice-Hall, Englewood Cliffs [1967].

Hughes, T. J. R. and Brooks, A., "A Multi-dimensional Upwind Scheme with No Crosswind Diffusion," *Finite Element Methods for Convection Dominated Flows*, T. J. R. Hughes (ed.), Am. Soc. Mech. Eng., New York [1979].

Price, H. S., Varga, R. S., and Warren, J. E., "Application of Oscillationa Matrices to Diffusion-Convection Equations," J. Math. & Phys. 45 301-311 [1966].

CHAPTER
THREE

APPLICATIONS TO BURGER'S EQUATION

Chapter 2 illustrated the main ideas of finite difference methods and finite element methods. The applications were for linear problems; here a nonlinear problem will be discussed. Burger's equation is a simple, prototype equation that is based on the Navier-Stokes equations without the complication of a pressure gradient. In one space dimension, Burger's equation is related to flow with a shock wave. As mentioned in Chapter 1, it is also a problem that arises for the Fokker-Planck equation and equations governing the spreading of liquids on solids. This chapter gives only a few of the methods; more methods are presented and analyzed in Chapters 4 through 7. Literature references are provided there as well.

3.1. Applications to Burger's Equation Without Viscosity

Burger's equation without viscosity is

$$\frac{\partial u}{\partial t} + \frac{\partial}{\partial x}\left(\frac{1}{2}u^2\right) = 0. \tag{3.1}$$

In this chapter we solve this equation subject to the following initial and boundary conditions:

$$\text{at } t = 0, \begin{cases} u = 0 \text{ for } x \leq 0.5 \\ u = 1 \text{ for } 0.5 < x \leq 1.5 \\ u = 0 \text{ for } 1.5 < x \end{cases} \tag{3.2}$$

$$\text{at } x = 0, \quad u = 0 \text{ for } t \geq 0.$$

The first method uses upstream differences. The equation is thus

$$\frac{du_i}{dt} + \frac{u_i^2 - u_{i-1}^2}{2\Delta x} = 0. \tag{3.3}$$

If an explicit scheme is used in time, the algorithm becomes

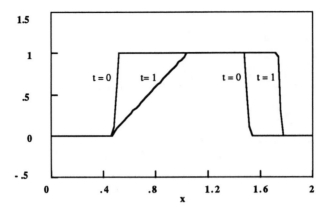

Figure 3.1. Burger's Equation Without Viscosity, Upstream Method
$\Delta x = 0.02, \Delta t = 0.01$

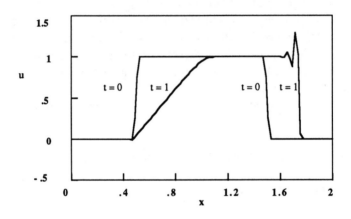

Figure 3.2. Burger's Equation Without Viscosity, Lax-Wendroff Method
$\Delta x = 0.02, \Delta t = 0.01$

$$u_i^{n+1} = u_i^n - \frac{\Delta t}{2\Delta x} [(u_i^n)^2 - (u_{i-1}^n)^2]. \qquad (3.4)$$

Since we used an upstream derivative, we expect this method to smear the fronts but not to oscillate. Results shown in Figure 3.1 confirm this: the solution is quite steep and does not oscillate.

The second method is the Lax-Wendroff method, which has two steps. When applied to Burger's equation the steps are

$$u_{i+1/2}^{n+1/2} = 0.5(u_{i+1}^n + u_i^n) - \frac{\Delta t}{4\Delta x}[(u_{i+1}^n)^2 - (u_i^n)^2] \qquad (3.5)$$

$$u_i^{n+1} = u_i^n - \frac{\Delta t}{2\Delta x}[(u_{i+1/2}^{n+1/2})^2 - (u_{i-1/2}^{n+1/2})^2]. \qquad (3.6)$$

Typical solutions are given in Figure 3.2. Oscillations occur at the leading front.

BURGER'S EQUATION

In this problem, the leading front starts out steep for the initial conditions in Figure 3.2 (it goes from 1 to 0 while x goes from 1.48 to 1.52); the effect of Burger's equation is to make the leading edge even steeper. When comparing the upstream method with the Lax-Wendroff method, the Lax-Wendroff method gives a steeper front but more oscillations. With the upstream method, the numerical dispersion smoothes the oscillations and makes the front less steep. With Burger's equation, if the leading edge goes from a high u behind the front to a low u before the front, as is the case here, then the front is self-sharpening and eventually forms a shock. When the shock is formed, the Lax-Wendroff method begins to have problems. The upstream method also has problems, but they are not as evident here because the self-sharpening feature of the solution is counteracted by the numerical dispersion of the upstream method. If a small Δt value is used in the upstream method, dispersion increases. We would like to improve on this behavior.

With the Galerkin finite element method, the details are more complicated. First we expand the unknown solution in a trial function,

$$u = \sum_{i=1}^{N} u_i N_i(x), \qquad (3.7)$$

substitute this expansion into the differential equation to form the residual, and make the weighted residual zero.

$$\int_0^1 W_j \left[\sum_{i=1}^{N} \frac{du_i}{dt} N_i(x) + \frac{1}{2} \frac{\partial}{\partial x} \left(\sum_{i=1}^{N} u_i N_i(x) \right)^2 \right] dx = 0 \qquad (3.8)$$

For the Petrov-Galerkin method we take the weighting function to be

$$W_J = \begin{cases} N_J + \alpha F(\xi) & x_{j-1} \leq x \leq x_j \\ \\ N_J - \alpha F(\xi) & x_j < x \leq x_{j+1} \end{cases}$$

$$F(\xi) = \frac{3}{4}(1 - \xi^2) \qquad \frac{dF}{d\xi} = -\frac{3}{2}\xi. \qquad (3.9)$$

The first term involving the time derivative is similar to the first term in the advection equation.

$$\Delta x \left[\left(\frac{1}{6} - \frac{\alpha}{4}\right) \frac{du_{j+1}}{dt} + \frac{2}{3} \frac{du_j}{dt} + \left(\frac{1}{6} + \frac{\alpha}{4}\right) \frac{du_{j-1}}{dt} \right] \qquad (3.10)$$

For the convective term the element integral is

$$\sum_{I,K=1}^{N} \int_{-1}^{1} W_J N_I \frac{dN_K}{d\xi} d\xi \, u_I u_K. \qquad (3.11)$$

Evaluation of these integrals gives

$$\int_{-1}^{1} N_J \, N_I \frac{dN_K}{d\xi} \, d\xi = (-1)^K \frac{1}{2} \begin{bmatrix} \frac{2}{3} & \frac{1}{3} \\ \frac{1}{3} & \frac{2}{3} \end{bmatrix} \begin{matrix} J=1 \\ J=2 \end{matrix} \qquad (3.12)$$

$$\begin{matrix} I=1 & I=2 \end{matrix}$$

$$\alpha \int_{-1}^{1} F(\xi) \, N_I \frac{dN_K}{d\xi} \, d\xi = \begin{cases} (-1)^K \frac{\alpha}{4} & x_{j-1} \leq x \leq x_j \\ \\ (-1)^{K+1} \frac{\alpha}{4} & x_j < x \leq x_{j+1}. \end{cases} \qquad (3.13)$$

How are the nonlinear terms handled? One way is to use a quadrature scheme that is sufficient to evaluate them exactly. Other options are discussed in Chapter 5. In this case the gradient is constant and the terms can be evaluated exactly. On the other hand, if quadratic terms are used for N_i, then we need to integrate a fifth-degree polynomial. A Gaussian quadrature with 3 nodes is sufficient to do this exactly. We treat the time-dependent terms using an explicit method. The algorithm is then

$$\left(\frac{1}{6} - \frac{\alpha}{4}\right)(u_{j+1}^{n+1} - u_{j+1}^{n}) + \frac{2}{3}(u_j^{n+1} - u_j^n) + \left(\frac{1}{6} + \frac{\alpha}{4}\right)(u_{j-1}^{n+1} - u_{j-1}^n) =$$
$$- \frac{\Delta t}{6\Delta x} [(u_{j+1}^n)^2 - (u_{j-1}^n)^2 + u_j^n (u_{j+1}^n - u_{j-1}^n)] + \qquad (3.14)$$
$$+ \frac{\alpha \Delta t}{4\Delta x} [(u_{j+1}^n)^2 - 2(u_j^n)^2 + (u_{j-1}^n)^2].$$

The Petrov-Galerkin weighting of the convection terms introduces a diffusion-type term on the right-hand side. We must perform one LU decomposition of the left-hand side, as indicated in Chapter 2, but other than that, the calculations are straightforward. For $\alpha = 0$, we expect the Galerkin method to be unstable, since it is unstable for the advection equation. Our results confirm this. When the Petrov-Galerkin option is exercised, the results are as shown in Figure 3.3 for $\alpha = 0.5$ and Figure 3.4 for $\alpha = 1.0$. Oscillations are evident before the front, at the leading edge, and behind the front. It should also be noted that the step-size needed for the Petrov-Galerkin method is considerably smaller than that used for finite difference methods. Based on these calculations for Burger's equation without viscosity, we would choose one of the finite difference methods since they maintain a sharp front, oscillate less, and use a larger time-step.

3.2. Applications to Burger's Equation With Viscosity

When viscosity is added to Burger's equation, it becomes

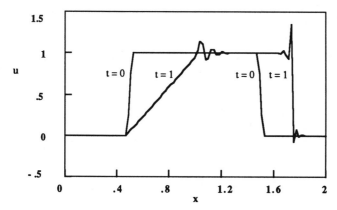

Figure 3.3. Burger's Equation Without Viscosity, Petrov-Galerkin Method, $\alpha = 0.5$, $\Delta x = 0.02$, $\Delta t = 0.004$

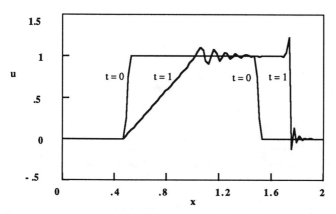

Figure 3.4. Burger's Equation Without Viscosity, Petrov-Galerkin Method, $\alpha = 1.0$, $\Delta x = 0.02$, $\Delta t = 0.004$

$$\frac{\partial u}{\partial t} + \frac{\partial}{\partial x}\left(\frac{1}{2}u^2\right) = \nu \frac{\partial^2 u}{\partial x^2}. \tag{3.15}$$

Now there is additional dissipation provided by the equation itself that does not need to be provided by the numerical method. The finite difference method that is explicit in time and upstream in space is

$$u_i^{n+1} = u_i^n - \frac{\Delta t}{2\Delta x}\left[(u_i^n)^2 - (u_{i-1}^n)^2\right] + \frac{\nu \Delta t}{\Delta x^2}(u_{i+1}^n - 2u_i^n + u_{i-1}^n). \tag{3.16}$$

The Lax-Wendroff method can be generalized to include the viscous terms in a variety of ways. Here is the method as given by Anderson *et al.* [1984].

52 NUMERICAL METHODS - MOVING FRONTS

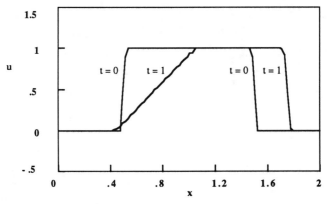

Figure 3.5. Burger's Equation, $\nu = 0.001$, Upstream Method, Δx 0.02, $\Delta t = 0.01$

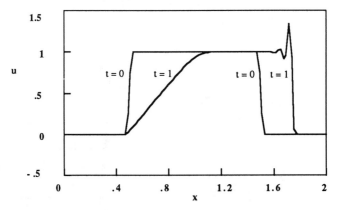

Figure 3.6. Burger's Equation, $\nu = 0.001$, Lax-Wendroff Method, Δx 0.02, $\Delta t = 0.01$

$$u_{i+1/2}^{n+1/2} = 0.5\,(u_{i+1}^n + u_i^n) - \frac{\Delta t}{4\Delta x}\,[(u_{i+1}^n)^2 - (u_i^n)^2] + $$
$$+ \frac{\nu \Delta t}{\Delta x^2}\,(u_{i+1}^n - 2\,u_i^n + u_{i-1}^n + u_{i+2}^n - 2\,u_{i+1}^n + u_i^n) \qquad (3.17)$$

$$u_i^{n+1} = u_i^n - \frac{\Delta t}{2\Delta x}\,[(u_{i+1/2}^{n+1/2})^2 - (u_{i-1/2}^{n+1/2})^2] + \frac{\nu \Delta t}{\Delta x^2}\,(u_{i+1}^n - 2\,u_i^n + u_{i-1}^n). \qquad (3.18)$$

Results obtained with these two methods are shown in Figures 3.5 and 3.6. The upstream method does not oscillate and the front is reasonably sharp. The Lax-Wendroff method exhibits oscillations at the front, but not elsewhere.

When the Galerkin finite element method is used, we have the diffusion contribution added to the right-hand side of Eq. (3.14). The diffusion contribution is from Eq. (2.102) and the Galerkin equations for Burger's equation with

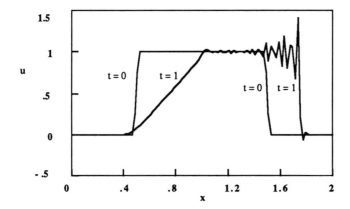

Figure 3.7. Burger's Equation, $\nu = 0.001$, Galerkin Method, $\Delta x = 0.02$, $\Delta t = 0.004$

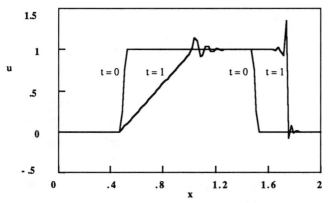

Figure 3.8. Burger's Equation, $\nu = 0.001$, Petrov-Galerkin Method, $\alpha = 0.5$, $\Delta x = 0.02$, $\Delta t = 0.004$

viscosity are

$$\left(\frac{1}{6} - \frac{\alpha}{4}\right)(u_{j+1}^{n+1} - u_{j+1}^{n}) + \frac{2}{3}(u_{j}^{n+1} - u_{j}^{n}) + \left(\frac{1}{6} + \frac{\alpha}{4}\right)(u_{j-1}^{n+1} - u_{j-1}^{n}) = \\ -\frac{\Delta t}{6\Delta x}[(u_{j+1}^{n})^2 - (u_{j-1}^{n})^2 + u_{j}^{n}(u_{j+1}^{n} - u_{j-1}^{n})] + \\ +\frac{\alpha \Delta t}{4\Delta x}[(u_{j+1}^{n})^2 - 2(u_{j}^{n})^2 + (u_{j-1}^{n})^2] + \frac{\nu \Delta t}{\Delta x^2}(u_{i+1}^{n} - 2u_{i}^{n} + u_{i-1}^{n}). \quad (3.19)$$

Results obtained with the Galerkin method are shown in Figure 3.7; the oscillations are unacceptable. Figures 3.8 and 3.9 show the Petrov-Galerkin results for $\alpha = 0.5$ and $\alpha = 1.0$, respectively. These also have too many oscillations. We therefore can conclude that neither the Galerkin nor the Petrov-Galerkin method is a good method for this problem.

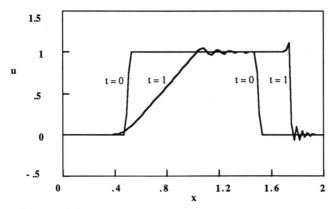

Figure 3.9. Burger's Equation, v = 0.001, Petrov-Galerkin Method
α = 1.0, Δx = 0.02, Δt = 0.004

To summarize (for nonlinear problems, too), if the mesh size is too big, oscillations develop. The various methods introduce numerical dissipation in different ways and this dissipation tends to dampen the oscillations but also to smear the front. As more mesh points are used, the time-step is also reduced, so that using more mesh points is an expensive way to solve the problem. The Galerkin method is inadequate for this problem; Petrov-Galerkin methods are only slightly better. Using an upstream finite difference method in space and Euler's method in time gives the best method for a self-sharpening front.

Problems

1_1. Redo the calculations leading to Figure 3.1, except use a smaller time-step. What is the effect of added dispersion?

2_2. Verify Eq. (3.12)-(3.13).

3_3. Verify Eq. (3.14).

References

Anderson, D. A., Tannehill, J. C., Pletcher, R. H., *Computational Fluid Mechanics and Heat Transfer*, McGraw-Hill [1984].

CHAPTER
FOUR

ANALYSIS OF FINITE DIFFERENCE METHODS ON FIXED GRIDS

The most widely used numerical methods for high-speed fluid dynamics are finite difference methods. They were developed first, and the methods have been finely tuned so that they are efficient, fast, and able to be implemented on vector computers. Presented here are methods suitable for hyperbolic equations, those with a time derivative and a convection term, but no diffusion. We will first treat the advection equation, which is linear and allows a rigorous study of stability. Then we will consider Burger's equation without viscosity, which has a nonlinear convection term (as in the Navier-Stokes equations). Next we will expand the equations by adding diffusion to obtain the convective diffusion equation and Burger's equation with viscosity. The latter case is a one-dimensional analogue of the Navier-Stokes equations. By focusing first on problems without viscosity or diffusion, we can concentrate on the central question, namely the ability of a method to model effectively the convective terms.

There are a number of important theoretical questions we must consider for any candidate method. First, how accurate is the method? The standard way of assessing the accuracy of a method is to evaluate the truncation error, which is determined by substituting a Taylor expansion into the difference equation. That is not the whole answer, however, because it tells the accuracy only for small Δt and Δx values. Sometimes in practical computations, a suitably small Δt and Δx cannot be used; in that case, the accuracy is assessed by solving standard problems, such as those displayed here. The next question is stability: under what conditions in Δt and Δx will the calculations be stable? This stability, of course, depends on the parameters of the problem. Finally, we are interested in the extent to which the method introduces additional diffusion or dissipation (and smears the front) and the extent to which it includes dispersion, making waves move with the wrong velocity. All of these questions depend on whether or not the solution oscillates in space and time.

4.1 Accuracy, Stability, Dissipation, and Dispersion

The first problem is the advection equation:

$$\frac{\partial c}{\partial t} + Pe \frac{\partial c}{\partial x} = 0$$

$$c(x, 0) = f(x) \qquad (4.1)$$

$$c(0, t) = g(t).$$

The functions f(x) and g(t) are known and determine the initial conditions and the boundary condition at x = 0. The Peclet number is retained for convenience in order to be compatible with the convective diffusion equation. It would, of course, be possible to absorb it into the time variable, using t' = t Pe. The only effect would be to change the time scale in the problem.

These concepts can be illustrated using the upstream difference expression. First, we write a forward difference in time and a backward difference in space.

$$\left. \frac{\partial c}{\partial t} \right|_i^n = \frac{c_i^{n+1} - c_i^n}{\Delta t} \qquad (4.2)$$

$$\left. \frac{\partial c}{\partial x} \right|_i^n = \frac{c_i^n - c_{i-1}^n}{\Delta x} \qquad (4.3)$$

Combining these terms gives the following algorithm:

$$c_i^{n+1} = c_i^n - \frac{\Delta t \, Pe}{\Delta x} (c_i^n - c_{i-1}^n). \qquad (4.4)$$

We determine the truncation error by substituting a Taylor series for c_i^{n+1} and c_{i-1}^n

$$c_i^{n+1} = c_i^n + \left. \frac{\partial c}{\partial t} \right|_i^n \Delta t + \left. \frac{\partial^2 c}{\partial t^2} \right|_i^n \frac{\Delta t^2}{2} + \cdots \qquad (4.5)$$

$$c_{i-1}^n = c_i^n - \left. \frac{\partial c}{\partial x} \right|_i^n \Delta x + \left. \frac{\partial^2 c}{\partial x^2} \right|_i^n \frac{\Delta x^2}{2} + \cdots \qquad (4.6)$$

and substitute into Eq. (4.4) to obtain

$$\frac{\partial c}{\partial t} + Pe \frac{\partial c}{\partial x} = \underbrace{- \frac{\partial^2 c}{\partial t^2} \frac{\Delta t}{2} + \frac{\partial^2 c}{\partial x^2} Pe \frac{\Delta x}{2}}_{\text{Remainder}}. \qquad (4.7)$$

The remainder terms are proportional to Δt and Δx, so the truncation error is $O(\Delta t)$ and $O(\Delta x)$; the method is first-order.

To study the stability of the method, we use the Fourier method. This method is applicable for linear problems defined on an infinite grid (or one that is periodic). If we are given a solution, c_j, at a series of grid points, we consider it to be the Fourier transform of a function, $\hat{c}$.

$$\hat{c}(\xi) = \sum_j c_j e^{ij\xi} \quad 0 \le \xi \le 2\pi, \ j \in \{\text{all integers}\} \tag{4.8}$$

Thus $\hat{c}$ is the function whose Fourier coefficients are c_j. Given a function, $\hat{c}$, its Fourier transform is

$$c_j = \frac{1}{2\pi} \int_0^{2\pi} \hat{c}(\xi) e^{-ij\xi} \, d\xi. \tag{4.9}$$

We will also make use of the various shift and difference operators listed in Table 4.1. Their Fourier transforms are given in Table 4.1 as well.

To interpret these terms, we suppose the solution is

$$c_{ex} = e^{i\omega t + ikx}. \tag{4.10}$$

The variable ω is the frequency, k is called the wave number, and the wavelength is $\lambda = 2\pi/k$. When $t = n\Delta t$ and $x = j\Delta x$, we have

$$c_{ex,i}^n = e^{i\omega n \Delta t + ikj\Delta x}$$
$$t = n\Delta t, \ x = j\Delta x. \tag{4.11}$$

Equation (4.10) is substituted into the advection equation (4.1) and the result is evaluated at $t = n\Delta t$ and $x = j\Delta x$.

$$i\omega \, e^{i\omega n \Delta t + ikj\Delta x} + \text{Pe} \, ik \, e^{i\omega n \Delta t + ikj\Delta x} = 0 \tag{4.12}$$

This is satisfied provided that

$$\omega = -\text{Pe} \, k. \tag{4.13}$$

For difference approximations $\omega(k)$ is called the dispersion relation, which tells how the frequency of the wave changes with the wave number.

To further explore the meaning of this analysis, let us suppose the initial condition is the oscillating function shown in Figure 4.1a. For the solution to oscillate from node to node like this, it is necessary for the wave number to be $\pi/\Delta x$. The wave length is $2\Delta x$. If we take the Fourier transform of Eq. (4.11), we get

$$\text{with } \xi = -k\Delta x \quad e^{ikj\Delta x} = e^{-ij\xi}$$
$$c_{ex,j} = e^{i\omega n \Delta t - ij\xi}$$

$$\hat{c}_{ex} = \sum_j c_{ex,j} e^{ij\xi} = e^{i\omega n \Delta t} \sum_j e^0 = \text{constant} \times e^{i\omega n \Delta t} \tag{4.14}$$

Table 4.1. Difference Operators and their Fourier Transforms

Shift Operators

$$\hat{u} = \frac{1}{\sqrt{2\pi}} \sum_j u_j e^{ij\xi}, \quad 0 \leq \xi \leq 2\pi$$

$(S_+ f)_i = f_{i+1}$ $\qquad\widehat{S_+ u} = e^{-i\xi}\hat{u}$

$(S_- f)_i = f_{i-1}$ $\qquad\widehat{S_- u} = e^{+i\xi}\hat{u}$

First Derivatives

$(D_+ f)_i = \dfrac{f_{i+1} - f_i}{\Delta x}$ $\qquad \widehat{D_+ u} = \dfrac{1}{\Delta x}(e^{-i\xi} - 1)\hat{u}$

$(D_- f)_i = \dfrac{f_i - f_{i-1}}{\Delta x}$ $\qquad \widehat{D_- u} = \dfrac{1}{\Delta x}(1 - e^{+i\xi})\hat{u}$

$(D_0 f)_i = \dfrac{f_{i+1} - f_{i-1}}{\Delta x}$ $\qquad \widehat{D_0 u} = \dfrac{1}{2\Delta x}(e^{-i\xi} - e^{+i\xi})\hat{u}$

Second Derivatives

$(D_+ D_- f)_i = \dfrac{\delta^2 f_i}{\Delta x^2} = \dfrac{f_{i+1} - 2f_i + f_{i-1}}{\Delta x^2}$ $\qquad \widehat{D_+ D_- u} = \dfrac{1}{\Delta x^2}(e^{-i\xi} - 2 + e^{+i\xi})\hat{u}$

Trigonometric Identities

$\cos\xi = 2\cos^2(\xi/2) - 1$ $\qquad \cos\xi = 1 - 2\sin^2(\xi/2)$

$1 - \cos\xi = 2\sin^2(\xi/2)$ $\qquad e^{i\xi} + e^{-i\xi} = 2\cos\xi$

$e^{i\xi} - e^{-i\xi} = 2i\sin\xi$ $\qquad e^{i\xi} + e^{-i\xi} - 2 = -4\sin^2(\xi/2)$

For the wave number $\xi = -k\Delta x$ or for Figure 4.1a, $\xi = -\pi$. Two other initial conditions are shown in Figures 4.1b and 4.1c. As can be seen, to cover all the wave numbers that are possible on this grid we must consider ξ from 0 to $-\pi$, and likewise the values from 0 to $+\pi$. Since the function $\hat{c}$ is periodic, we will generally take $0 \leq \xi \leq 2\pi$.

Suppose we have a difference scheme that takes c_j^n and computes c_j^{n+1}. We take the Fourier transform of these solutions to obtain $\hat{c}^n$ and $\hat{c}^{n+1}$ and then <u>define</u> the function ρ by the equation

$$\hat{c}^{n+1}(\xi) = \rho\,\hat{c}^n(\xi). \qquad (4.15)$$

This function is called the amplification factor. If there are multiple differential equations, we get a matrix here, which is called the amplification matrix. The amplification factor for the advection equation should retain its shape but move with frequency ω, since that is what the exact solution does.

FINITE DIFFERENCE METHODS - FIXED GRIDS

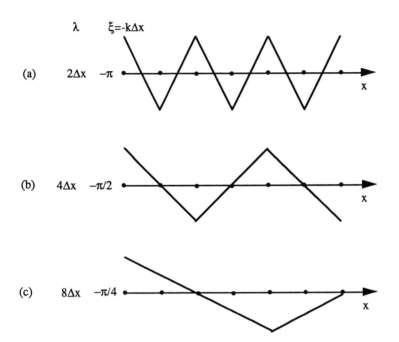

Figure 4.1. Initial Conditions with Different Wavelengths

To aid our analysis, let us review complex notation. A complex number, z, can be written in the form (where i is $\sqrt{-1}$)

$$z = x + i\, y. \tag{4.16}$$

The magnitude of the complex number is then

$$|z| = \sqrt{x^2 + y^2} \tag{4.17}$$

and the angle of the complex number with respect to the x = 0 axis is

$$\arg z = \tan^{-1} \frac{y}{x}. \tag{4.18}$$

These features are illustrated in Figure 4.2. We can also write the same complex number in the form

$$z = |z|\, e^{i\,\arg z} = r\, e^{i\theta} = r\,(\cos\theta + i\,\sin\theta), \tag{4.19}$$

where the magnitude and angle are illustrated in Figure 4.2.

The magnitude of the complex function ρ should be 1.0 in order that the solution be propagated as a wave without change. The extent to which |ρ| differs

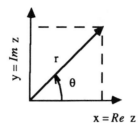

Figure 4.2. Complex Notation

from 1.0 defines the amount of dissipation introduced by the numerical method.

$$|\rho| = \frac{|\hat{c}^{n+1}|}{|\hat{c}^n|} \quad (4.20)$$

Since ξ depends on k, the wave number, the amount of dissipation will also depend on the wave number.

The argument of ρ gives the phase change when going from the time-step n to the time-step n+1. The extent to which this differs from the exact phase change defines the dispersion. The exact phase change is found by taking the exact solution [Eq. (4.11)] and looking at the difference in phase over one time-step, Δt.

$$c_{ex,i}^{n+1} = e^{i\omega(n+1)\Delta t + ikj\Delta x} \quad (4.21)$$

$$\rho_{ex} = e^{i\omega\Delta t} \quad (4.22)$$

Thus the phase change of the exact solution is $\Delta\phi = \omega\Delta t$. For this problem, $\omega = -\text{Pe k}$, so $\Delta\phi = -\text{Pe}k\Delta t$. For the function ρ we have

$$\rho_{ex} = \frac{|\hat{c}_{ex}^{n+1}|}{|\hat{c}_{ex}^n|} e^{i(\theta^{n+1} - \theta^n)} = e^{-i\,\text{Pe}\,k\,\Delta t} \quad (4.23)$$

$$\arg \frac{\hat{c}_{ex}^{n+1}}{\hat{c}_{ex}^n} = \theta^{n+1} - \theta^n = -\text{Pe}\,k\,\Delta t, \quad |\rho| = 1. \quad (4.24)$$

We can now determine the stability of Eq. (4.4) by taking the Fourier transform of that equation.

$$\hat{c}^{n+1} = [\,1 - \text{Co} + \text{Co}\,e^{i\xi}\,]\,\hat{c}^n \quad (4.25)$$

$$\text{Co} = \frac{\Delta t\,\text{Pe}}{\Delta x} \text{ is the Courant number.} \quad (4.26)$$

The function, ρ, is a circle with a radius of Co and center at 1– Co (see Figure

FINITE DIFFERENCE METHODS - FIXED GRIDS

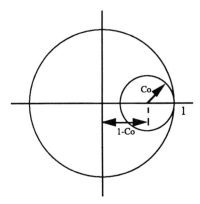

Figure 4.3. Stability Limits in Complex Plane for Explicit, Upstream Method

4.3). It intersects the real axis at $+1$ and $1-2$ Co. Thus the function $\rho(\xi)$ stays within the unit circle for all ξ, provided that Co ≤ 1; the method is stable as long as the following condition is met:

$$\text{Co} = \frac{\Delta t \, \text{Pe}}{\Delta x} \leq 1. \tag{4.27}$$

This provides an upper limit to the possible step-size in time. The stability of the method relies on the von Neumann Theorem, which states that a finite difference method is stable in the L_2 norm if and only if the von Neumann condition is satisfied. For this problem, the von Neumann condition is that $|\rho| \leq 1$ [Richtmyer and Morton, 1967].

To examine the dissipation and dispersion of this method, let us compute the magnitude of the function, ρ, and its phase angle. These are given by

$$\rho = 1 - \text{Co} + \text{Co} \, e^{i\xi}$$
$$\rho = 1 + \text{Co} \, (\cos \xi - 1) + i \, \text{Co} \sin \xi \tag{4.28}$$

$$|\rho| = \sqrt{[\,1 + \text{Co}\,(\cos \xi - 1\,)]^2 + [\,\text{Co} \sin \xi\,]^2}$$
$$|\rho| = \sqrt{1 - 2\,\text{Co} + 2\,\text{Co}^2 + 2\,\text{Co}\,(1 - \text{Co})\cos \xi} \tag{4.29}$$
$$\arg \rho = \tan^{-1} \frac{\text{Co} \sin \xi}{(1 - \text{Co}) + \text{Co} \cos \xi}.$$

These functions are plotted versus wavenumber in Figure 4.4; figure 4.4a is the absolute value of ρ, while figure 4.4b is the value of $\arg \rho / \text{Co} \, \xi$, or the relative argument. To see what these graphs mean, we take a wave with length $6\Delta x$ and a time-step such that Co= 0.5. Then the value of ξ is $\pi/3$. Eq. (4.28) and Figure 4.4a give $|\rho| = \sqrt{0.75} \approx 0.866$. If the method acts for five time-steps, then the amplitude is reduced by a factor of $(0.866)^5 \approx 0.487$. In one time-step, the exact

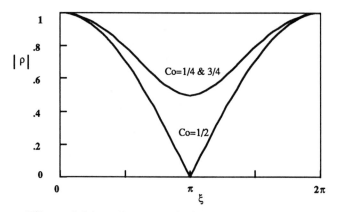

Figure 4.4a. Magnitude of ρ, Explicit, Upstream Method

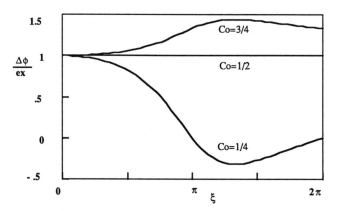

Figure 4.4b. Relative Phase Change, Explicit, Upstream Method

phase change is

$$\Delta\phi = -k\,Pe\,\Delta t = -Co\,k\,\Delta x = -Co\,\xi, \qquad (4.30)$$

which here is $-Co\,\pi/6$, or 30 degrees. Eq. (4.29) and Figure 4.4b give the relative argument as 1.0, so we get 30 degrees as well; there is no phase error. Figure 4.4b shows, however, that there is phase error for the other situations. In fact, for Co<0.5 the phase ratio is less than 1.0 (indicating that the numerical wave preceeds the true one). Starting with $c_j = \cos(\pi j/3)$, the actual numerical solution after five time-steps taken with the upstream method is plotted in Figure 4.5. The wave front retains its shape but decreases in amplitude as time proceeds, as predicted by the Fourier analysis.

The next test of the method is the result achieved by the calculations. Shown in Figure 4.6 is the solution after a square wave has been convected a distance of 1.0. The square wave has been smoothed out to be a broad change over a distance that grows with time. The solution is monotone, however, with no

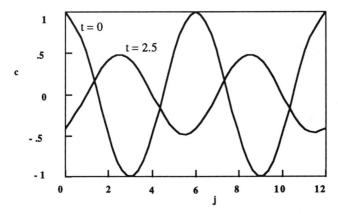

Figure 4.5. Convection of Cosine Wave with $k\Delta x = \pi/3$

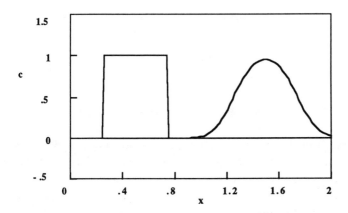

Figure 4.6. Advection Equation, Explicit, Upstream Method, $\Delta x = 0.02$, $\Delta t = 0.01$

oscillations. The square wave, of course, has all Fourier components in it, so that Figure 4.6 shows the dissipation and dispersion of the combination of all the Fourier components. The dissipation is evident in the reduction in concentration (except for the center value) and the dispersion is evident by the fact that the leading edge moves faster than it should and the trailing edge moves slower than it should.

There is another feature of the method that is of interest, named the unit CFL property by Morton and Parrott [1980]. The unit CFL property (named after Courant-Fredrichs-Lewy) says that a wave of any shape is propagated along without change if the Courant number is 1.0. If one sets Co = 1 in Eq. (4.4), the result is

$$c_i^{n+1} = c_{i-1}^n, \text{ upstream method with Co = 1.} \quad (4.31)$$

The concentration at a node moves exactly one grid spacing in one time-step, without change (see Figure 4.7). Thus the explicit, upstream method obeys the unit CFL property.

To summarize, the accuracy can be measured by determining the truncation error; the stability is determined by looking at the absolute magnitude of the amplification factor (and using the von Neumann Theorem); the dissipation is determined by looking at the magnitude of the amplification factor; and the dispersion is determined by looking at the phase angle. A scheme that obeys a unit CFL property is especially attractive.

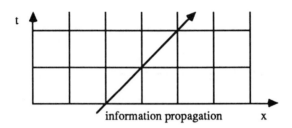

Figure 4.7. Unit CFL Property (solution propagates unchanged)

4.2. Finite Difference Methods for the Advection Equation

Several finite difference algorithms for the advection equation are listed in Table 4.2. The first one, with upstream derivatives, is discussed above. As shown previously, the truncation error was first-order in both Δt and Δx. The method is also excessively dissipative, since the amplitude of some waves is reduced by a factor of two in only five time-steps, for the particular Δt chosen. In an effort to improve upon this situation, the next three methods are all second-order in Δt and Δx. Let us consider them in turn.

Lax-Wendroff method. The Lax-Wendroff method was developed by Lax and Wendroff [1960, 1964] although the version shown in Table 4.2 comes from Richtmyer [1963]. (See Roach [1972].) To understand the motivation for the method, let us note that a second-order spatial derivative and a first-order time derivative provide an unstable method. If we use

$$\frac{c_i^{n+1} - c_i^n}{\Delta t} + \text{Pe} \frac{c_{i+1}^n - c_{i-1}^n}{2\Delta x} = 0, \qquad (4.32)$$

then we get

$$\hat{c}^{n+1} = \hat{c}^n + \frac{\text{Co}}{2}(e^{-i\xi}\hat{c}^n - e^{i\xi}\hat{c}^n)$$

$$\hat{c}^{n+1} = \hat{c}^n [1 - \text{Co} \sin \xi \, i]$$

$$|\rho| = \sqrt{1 + \text{Co}^2 \sin^2 \xi}.$$

(4.33)

FINITE DIFFERENCE METHODS - FIXED GRIDS

Table 4.2. Finite Difference Methods for the Advection Equation

Explicit, upstream convection, $O(\Delta x)$, $O(\Delta t)$

$$c_i^{n+1} = c_i^n - \frac{Pe\,\Delta t}{\Delta x}(c_i^n - c_{i-1}^n)$$

Lax-Wendroff, $O(\Delta x^2)$, $O(\Delta t^2)$

$$c_{i+1/2}^{n+1/2} = \frac{1}{2}(c_{i+1}^n + c_i^n) - \frac{Pe\,\Delta t}{2\Delta x}(c_{i+1}^n - c_i^n)$$

$$c_i^{n+1} = c_i^n - \frac{Pe\,\Delta t}{\Delta x}(c_{i+1/2}^{n+1/2} - c_{i-1/2}^{n+1/2})$$

MacCormack, $O(\Delta x^2)$, $O(\Delta t^2)$

$$c*_i^{n+1} = c_i^n - \frac{Pe\,\Delta t}{\Delta x}(c_{i+1}^n - c_i^n)$$

$$c_i^{n+1} = \frac{1}{2}(c_i^n + c*_i^{n+1}) - \frac{Pe\,\Delta t}{2\Delta x}(c*_i^{n+1} - c*_{i-1}^{n+1})$$

Leapfrog, $O(\Delta x^2)$, $O(\Delta t^2)$

$$c_i^{n+1} = \frac{\alpha}{2}(c_i^n + c_i^{n-2}) + (1-\alpha)c_i^{n-1} - \frac{Pe\,\Delta t}{\Delta x}(c_{i+1}^n - c_{i-1}^n)$$

Since the magnitude of ρ is greater than 1.0 for all values of Co, the method is unstable. We would like to have second-order accuracy in space and time, however, and so we try to expand the method in such a way to provide stability as well. We can look at the second step of the Lax-Wendroff method in Table 4.2. It is first-order in time but second-order in space, since a centered derivative is used on the right-hand side. This is, of course, the method that was unstable if the right-hand side was evaluated at the n-th time level. To obtain a stable method, we evaluate the right-hand side at the n+1/2 time level, using a predicted value of $c_j^{n+1/2}$. The predicted value is obtained by the first equation, which is also second-order in space. Although each step is first-order in time, the combination is second-order in time. To see the essence of the method, let us combine the two steps into one.

$$c_i^{n+1} = c_i^n - \frac{Co}{2}(c_{i+1}^n - c_{i-1}^n) + \frac{Co^2}{2}(c_{i+1}^n - 2c_i^n + c_{i-1}^n) \qquad (4.34)$$

The first part of the equation is a centered difference method for the first spatial derivative, while the last part is reminiscent of a diffusion term. Indeed, the truncation order can be obtained by substituting the Taylor series expressions for

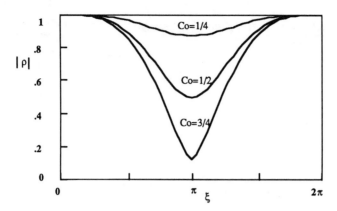

Figure 4.8a. Magnitude of ρ, Lax-Wendroff Method

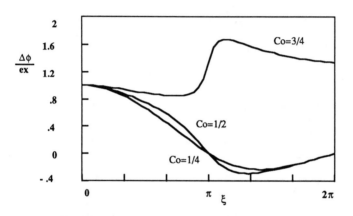

Figure 4.8b. Relative Phase Change, Lax-Wendroff Method

**Table 4.3. Dispersion and Dissipation Function
for Convective-Diffusion Equation or Advection Equation (r = 0)**

$Co = Pe \, \Delta t / \Delta x, \quad r = \Delta t / \Delta x^2$

Explicit, upstream

$$\rho(\xi) = 1 - Co - 2r + r\cos\xi + (r + Co)\cos\xi + i\, Co\sin\xi$$

Explicit, centered

$$\rho(\xi) = 1 + 2r(\cos\xi - 1) + i\, Co\sin\xi$$

Lax-Wendroff

$$\rho(\xi) = 1 - (2Co^2 + 4r)\sin^2\frac{\xi}{2} + i\, Co\sin\xi\left(-2r\sin^2\frac{\xi}{2} + 1\right)$$

MacCormack

$$\rho(\xi) = 1 - (2Co^2 + 4r)\sin^2\frac{\xi}{2} + 8r^2\sin^4\frac{\xi}{2} + i\left(Co\sin\xi - 4r\,Co\sin\xi\,\sin^2\frac{\xi}{2}\right)$$

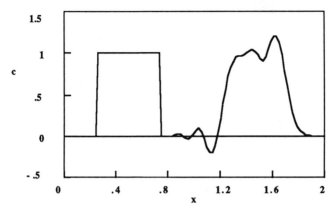

Figure 4.9a. Advection Equation, Lax-Wendroff or MacCormack Method
100 nodes, Co = 0.5

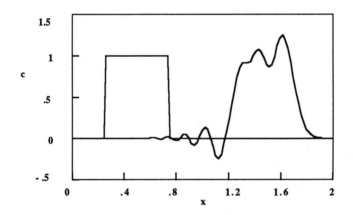

Figure 4.9b. Advection Equation, Lax-Wendroff or MacCormack Method
100 nodes, Co = 0.25

c_{i+1}^n, c_{i-1}^n, and c_i^{n+1}; it is second-order in Δt and Δx. We have added a dissipation term, which stabilizes the results. The value of $|\rho(\xi)|$ is given in Table 4.3 and is less than or equal to 1.0, provided that Co $\leq$ 1. What we are doing in the Lax-Wendroff method is solving the equation

$$\frac{\partial c}{\partial t} + Pe \frac{\partial c}{\partial x} = \frac{Pe^2 \Delta t}{2} \frac{\partial^2 c}{\partial x^2} \qquad (4.35)$$

using a second-order method in space and a first-order method in time, which is stable as long as Co $\leq$ 1. Figure 4.8a shows that the value of $|\rho(x)|$ using the Lax-Wendroff method is closer to 1.0 than the upstream, first-order method over most parameter ranges; thus, there is less dissipation with the Lax-Wendroff method. However, there is generally more phase angle error with the Lax-Wendroff method than with the upstream, first-order method, as shown in Figure 4.8b. The

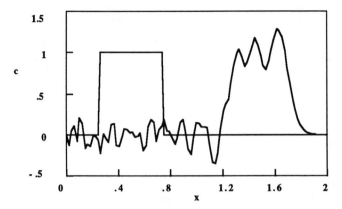

Figure 4.10a. Advection Equation, Leapfrog Method, $\alpha = 0$, 100 nodes, Co = 0.5

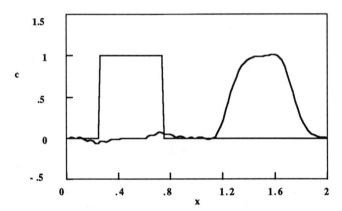

Figure 4.10b. Advection Equation, Leapfrog Method, $\alpha = 1$, 100 nodes, Co = 0.5

application shown in Figure 4.9a shows the concentration profiles to be steeper (less dissipation), but there is significant phase lag (as illustrated by the oscillations following each steep gradient). The same problem is solved using a smaller time-step (e.g., Co = 0.25) in Figure 4.9b; the results are almost identical, so there is little time-truncation error shown in Figures 4.9a and 4.9b. The Lax-Wendroff method obeys the unit CFL property. This can be seen by setting Co equal to 1.0 in Table 4.2 or Co = 1 and r = 0 in Table 4.3.

MacCormack method [MacCormack, 1969]. Table 4.2 shows that the first step is first-order in both time and space, but provides only a predicted value. This predicted value is then corrected in the second equation to give the final value. For the advection equation, this method is the same as the Lax-Wendroff method. This can be seen by combining the steps (see Problem 4.4).

Leapfrog method. The leapfrog method is shown in Table 4.2. The usual method is for $\alpha = 0$ [Roache and Dietrich, 1988]. The filtered leapfrog method, devised by Dietrich and Womeck [1985], is for $\alpha = 1$. Roache and Dietrich [1988]

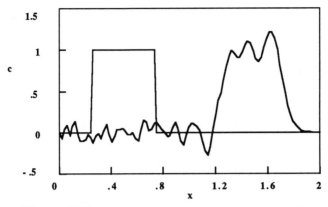

Figure 4.10c. Advection Equation, Filtered Leapfrog Method
$\alpha = 0.11$, 100 nodes, Co = 0.5

use a weighted average of these to obtain the weighted, filtered leapfrog method. When $\alpha = 0$, the truncation error is second-order in both Δt and Δx and the method is stable, provided that Co ≤ 1. When $\alpha \neq 0$, the truncation error is only first-order in Δt [Roache and Dietrich, 1988]. The stability of the leapfrog methods must be determined for multi-step methods (see [Roache and Dietrich, 1988]). Applications are shown in Figures 4.10a, 4.10b, and 4.10c. Figure 4.10a is for $\alpha = 0$, or the standard leapfrog method. The results show wild oscillations and the method is unsatisfactory. Figure 4.10b is for $\alpha = 1$, or the filtered leapfrog method. The results are very good; indeed, the best of any of the methods. Figure 4.10c is for $\alpha = 0.11$ and the results fall between those for $\alpha = 0$ and $\alpha = 1$. For this problem, the filtered leapfrog method gives the best results.

Results for a step-function initial condition can be obtained using the program CONVECT (see Problem 4.5). The upstream method gives no oscillations but the front is smoothed out; the Lax-Wendroff or MacCormack method gives oscillations; the standard leapfrog method gives wild oscillations; and the filtered leapfrog method ($\alpha = 1$) gives quite reasonable results. Based on these results, we choose the filtered leapfrog method for the advection equation.

Other methods. There are other methods of solving advection equations, and comparisons of them are given by Anderson [1974], Sod [1978], and Anderson, *et al.* [1984]. The Rusanov and Warming-Kutter-Lomax methods are suggested by Anderson, *et al.* [1984], but the results are not appreciably better than those shown here. Anderson [1974] says that MacCormack's method gives results of acceptable accuracy in most cases. Sometimes a method can be "tuned" to give good results for one class of problems. Methods that are good all-around methods are presented here.

4.3. Finite Difference Methods for Burger's Equation Without Viscosity

The same methods have also been developed for nonlinear equations.

Table 4.4. Finite Difference Methods for Burger's Equation

Explicit, upstream convection, $O(\Delta x)$, $O(\Delta t)$

$$u_i^{n+1} = u_i^n - \frac{\Delta t}{2\Delta x}[(u_i^n)^2 - (u_{i-1}^n)^2] + \frac{\upsilon \Delta t}{\Delta x^2}[u_{i+1}^n - 2u_i^n + u_{i-1}^n]$$

Lax-Wendroff, $O(\Delta x^2)$, $O(\Delta t)$

$$u_{i+1/2}^{n+1/2} = \frac{1}{2}(u_{i+1}^n + u_i^n) - \frac{\Delta t}{4\Delta x}[(u_{i+1}^n)^2 - (u_i^n)^2] + \frac{\nu \Delta t}{4 \Delta x^2}[u_{i+1}^n - 2u_i^n + u_{i-1}^n + u_{i+2}^n - 2u_{i+1}^n + u_i^n]$$

$$u_i^{n+1} = u_i^n - \frac{\Delta t}{2\Delta x}[(u_{i+1/2}^{n+1/2})^2 - (u_{i-1/2}^{n+1/2})^2] + \frac{\upsilon \Delta t}{\Delta x^2}[u_{i+1}^n - 2u_i^n + u_{i-1}^n]$$

MacCormack, $O(\Delta x^2)$, $O(\Delta t^2)$

$$u*_i^{n+1} = u_i^n - \frac{\Delta t}{2\Delta x}[(u_{i+1}^n)^2 - (u_i^n)^2] + \frac{\upsilon \Delta t}{\Delta x^2}[u_{i+1}^n - 2u_i^n + u_{i-1}^n]$$

$$u_i^{n+1} = \frac{1}{2}(u_i^n + u*_i^{n+1}) - \frac{\Delta t}{4\Delta x}[(u*_i^{n+1})^2 - (u*_{i-1}^{n+1})^2] + \frac{\upsilon \Delta t}{2 \Delta x^2}[u*_{i+1}^{n+1} - 2u*_i^{n+1} + u*_{i-1}^{n+1}]$$

DuFort-Frankel, $O(\Delta x^2)$, $O(\Delta t^2)$

$$u_i^{n+1} = \frac{\alpha}{2}(u_i^n + u_i^{n-2}) + (1-\alpha)u_i^{n-1} - \frac{\Delta t}{2\Delta x}[(u_{i+1}^n)^2 - (u_{i-1}^n)^2] + \frac{2\upsilon \Delta t}{\Delta x^2}[u_{i+1}^n - u_i^{n+1} - u_i^{n-1} + u_{i-1}^n]$$

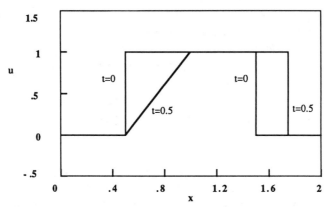

Figure 4.11. Exact Solution to Burger's Equation Without Viscosity

Burger's equation is

$$\frac{\partial u}{\partial t} + u\frac{\partial u}{\partial x} = 0, \quad u(x,0) = f(x), \quad c(0,t) = g(t). \quad (4.36)$$

The algorithms for Burger's equation are given in Table 4.4 (set $\nu = 0$). Since the equation is nonlinear, it is not possible to do a Fourier analysis, but we generally need for $|u\Delta t/\Delta x| \leq 1$ to have stability. This equation can have shocks moving in time. If the solution is a step change, the shock moves with a velocity that is the

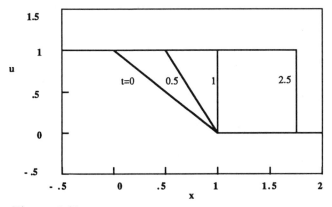

Figure 4.12. Exact Solution to Burger's Equation Without Viscosity

average value of the velocity before and after the shock (see Chapter 9 to see how to deduce this). If the step change was between 0 and 1, then the shock would move with velocity 0.5. This is illustrated in Figure 4.11. The leading edge (shock) moves with velocity = 0.5 (1 + 0). The trailing edge does not move at all, but the solution with u = 1.0 moves with velocity 1.0. Alternatively, if the solution began as a ramp function, as shown in Figure 4.12, then the leading edge of the solution does not move, while the trailing edge moves with velocity 1.0 until it reaches the leading edge. Then a shock is formed and it moves with velocity 0.5, as it did previously. Thus, this problem provides three important tests of a numerical method: the problem is nonlinear, shocks can develop from smooth solutions, and shocks can propagate in time. Solutions to the first problem are shown here; solutions to the second one can be generated using the program CONVECT (see Problem 4.14).

The first problem is solved with the following initial conditions:

$$f(x) = \begin{cases} 0.0 & x \leq 0.48 \\ \dfrac{x - 0.48}{0.52 - 0.48} & 0.48 < x \leq 0.52 \\ 1.0 & 0.52 < x < 1.48 \\ 1 - \dfrac{x - 1.48}{1.52 - 1.48} & 1.48 \leq x < 1.52 \\ 0.0 & 1.52 \leq x. \end{cases} \tag{4.37}$$

We take 100 nodes, with $\Delta x = 0.02$ and $\Delta t = 0.02$; the results are shown in Figures 4.13a-d. Again, the first-order, upstream method provides a smooth solution; the Lax-Wendroff method gives a solution with small oscillations at the leading and trailing edges; MacCormack's method gives a solution with oscillations at only the trailing edge; and the filtered leapfrog method gives a solution that oscillates excessively.

Based on these examples, we would choose the filtered leapfrog method

72 NUMERICAL METHODS - MOVING FRONTS

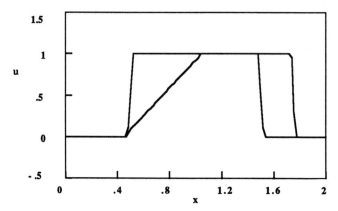

Figure 4.13a. Burger's Equation, Upstream Method, 100 nodes, $\Delta t = 0.02$

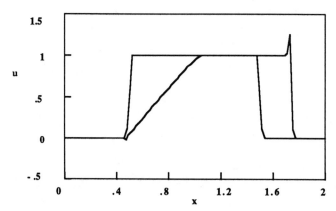

Figure 4.13b. Burger's Equation, Lax-Wendroff Method, 100 nodes, $\Delta t = 0.02$

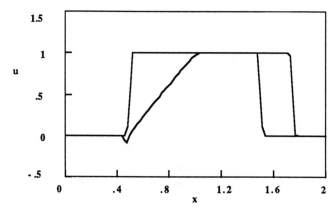

Figure 4.13c. Burger's Equation, MacCormack Method, 100 nodes, $\Delta t = 0.02$

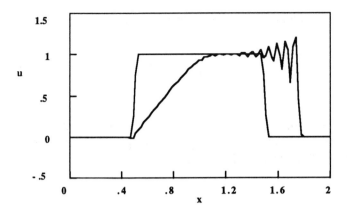

Figure 4.13d. Burger's Equation, Filtered Leapfrog Method, 100 nodes, $\Delta t = 0.02$

for linear problems and either the explicit, upstream method or the MacCormack method for nonlinear problems.

Next let us extend the methods to the general hyperbolic problem.

$$\frac{\partial u}{\partial t} = -\frac{\partial F(u)}{\partial x} \quad (4.38)$$

We expand the first derivative in a Taylor series to reduce the temporal truncation error. The result is already embedded in the Lax-Wendroff and MacCormack methods (which are already second-order in Δt), but the result is essential to the derivation of the Taylor-Galerkin methods in Chapter 5 (and throughout the book). Thus we develop the finite difference form of the Taylor series here. We differentiate Eq. (4.38) once to obtain

$$\frac{\partial^2 u}{\partial t^2} = -\frac{\partial}{\partial t}\left(\frac{\partial F(u)}{\partial x}\right) = -\frac{\partial}{\partial x}\left(\frac{\partial F(u)}{\partial t}\right) = -\frac{\partial}{\partial x}\left(\frac{dF}{du}\frac{\partial u}{\partial t}\right) = \frac{\partial}{\partial x}\left(\frac{dF}{du}\frac{\partial F(u)}{\partial x}\right). \quad (4.39)$$

We can use the Taylor series for u^{n+1} to write

$$\frac{\partial u}{\partial t} = \frac{u^{n+1} - u^n}{\Delta t} - \frac{\Delta t}{2}\frac{\partial^2 u}{\partial t^2} \quad (4.40)$$

and combine these equations to get

$$\frac{u^{n+1} - u^n}{\Delta t} = -\frac{\partial F(u)}{\partial x} + \frac{\Delta t}{2}\frac{\partial}{\partial x}\left(\frac{dF}{du}\frac{\partial F(u)}{\partial x}\right) = -\frac{\partial F(u)}{\partial x} + \frac{\Delta t}{2}\frac{\partial}{\partial x}\left[\left(\frac{dF}{du}\right)^2\frac{\partial u}{\partial x}\right]. \quad (4.41)$$

If we use only first-order terms and neglect the last term, we get the explicit, upstream method.

$$\frac{u_i^{n+1} - u_i^n}{\Delta t} = -\frac{F_i^n - F_{i-1}^n}{\Delta x} \quad (4.42)$$

If we keep the last term and apply the finite difference method to Eq. (4.41), we get

$$\frac{u_i^{n+1} - u_i^n}{\Delta t} = -\frac{F_{i+1/2}^n - F_{i-1/2}^n}{\Delta x} + \frac{\Delta t}{2 \Delta x^2}\left[\left.\frac{dF}{du}\right|_{i+1/2}^n (F_{i+1}^n - F_i^n) - \left.\frac{dF}{du}\right|_{i-1/2}^n (F_i^n - F_{i-1}^n)\right]. \quad (4.43)$$

The Lax-Wendroff and MacCormack methods are next derived from this equation.

The Lax-Wendroff method is applied to Eq. (4.38) in the two-step version [Richtmyer, 1963; Roach, 1972]. We use

$$u_{i+1/2}^{n+1/2} = \frac{1}{2}(u_{i+1}^n + u_i^n) - \frac{\Delta t}{2 \Delta x}(F_{i+1}^n - F_i^n) \quad (4.44)$$

for the predictor and

$$u_i^{n+1} = u_i^n - \frac{\Delta t}{\Delta x}(F_{i+1/2}^{n+1/2} - F_{i-1/2}^{n+1/2}) \quad (4.45)$$

for the corrector. By $F_{i+1/2}^{n+1/2}$ we mean

$$\begin{aligned} F_{i+1/2}^{n+1/2} &= F(u_{i+1/2}^{n+1/2}) \\ F_{i-1/2}^{n+1/2} &= F(u_{i-1/2}^{n+1/2}). \end{aligned} \quad (4.46)$$

We can write this as

$$\frac{u_i^{n+1} - u_i^n}{\Delta t} = -\frac{F_{i+1/2}^{n+1/2} - F_{i-1/2}^{n+1/2}}{\Delta x}. \quad (4.47)$$

Next we expand the terms appearing on the right-hand side, keeping all first-order terms in Δx or Δt.

$$F_{i \pm 1/2}^{n+1/2} = F_{i \pm 1/2}^n + \left.\frac{dF}{du}\frac{\partial u}{\partial t}\right|_{i \pm 1/2}^n \frac{\Delta t}{2} + O(\Delta t^2) \quad (4.48)$$

Then we get

$$\frac{F_{i+1/2}^{n+1} - F_{i-1/2}^{n+1}}{\Delta x} = \frac{F_{i+1/2}^n - F_{i-1/2}^n}{\Delta x} + \frac{\Delta t}{2 \Delta x}\left[\left.\frac{dF}{du}\frac{\partial u}{\partial t}\right|_{i+1/2}^n - \left.\frac{dF}{du}\frac{\partial u}{\partial t}\right|_{i-1/2}^n\right]. \quad (4.49)$$

The time derivative is evaluated using Eq. (4.38).

$$\left.\frac{\partial u}{\partial t}\right|_{i+1/2}^n = -\left.\frac{\partial F}{\partial x}\right|_{i+1/2}^n = -\frac{F_{i+1}^n - F_i^n}{\Delta x} + O(\Delta x^2) \quad (4.50)$$

$$\left.\frac{\partial u}{\partial t}\right|_{i-1/2}^n = -\left.\frac{\partial F}{\partial x}\right|_{i-1/2}^n = -\frac{F_i^n - F_{i-1}^n}{\Delta x} + O(\Delta x^2) \quad (4.51)$$

Combining Eq. (4.49) and Eq. (4.50) into Eq. (4.47) gives

$$\frac{u_i^{n+1} - u_i^n}{\Delta t} = -\frac{1}{\Delta x}[F_{i+1/2}^n - F_{i-1/2}^n] + O(\Delta x^2) +$$
$$+ \frac{\Delta t}{2\Delta x^2}\left[\frac{dF}{du}\bigg|_{i+1/2}(F_{i+1}^n - F_i^n) - \frac{dF}{du}\bigg|_{i-1/2}(F_i^n - F_{i-1}^n)\right] + O(\Delta t^2). \quad (4.52)$$

This is the same as Eq. (4.43). Thus, the Lax-Wendroff method is the same as the Taylor-finite difference method, to $O(\Delta t^2)$ and $O(\Delta x^2)$. This result is used in Chapter 5, where a finite element method is applied to Eq. (4.41) to get the analog of Eq. (4.43).

The MacCormack method applied to the general hyperbolic equation is [MacCormack, 1969; Roach, 1972]

$$u*_i^{n+1} = u_i^n - \frac{\Delta t}{\Delta x}(F_{i+1}^n - F_i^n) \quad (4.53)$$

for the predictor and

$$u_i^{n+1} = \frac{1}{2}(u_i^n + u*_i^{n+1}) - \frac{\Delta t}{2\Delta x}(F*_i^{n+1} - F*_{i-1}^{n+1}) \quad (4.54)$$

for the corrector. Problem 4.7 shows that these are equivalent to

$$u_i^{n+1} = u_i^n - \frac{\Delta t}{2\Delta x}(F_{i+1}^n - F_{i-1}^n) + \frac{\Delta t^2}{2\Delta x^2}\left[\frac{dF}{du}\bigg|_i^n(F_{i+1}^n - F_i^n) - \frac{dF}{du}\bigg|_{i-1}^n(F_i^n - F_{i-1}^n)\right]. \quad (4.55)$$

This equation is a finite difference method applied to Eq. (4.41) and (4.43) with the derivative, dF/du, evaluated at the upstream conditions. Thus, the Lax-Wendroff and MacCormack methods are both applied to Eq. (4.43) [to $O(\Delta t^2)$ and $O(\Delta x^2)$] with different ways of handling the nonlinear term, dF/du. By keeping the term multiplied by Δt^2, we add numerical diffusion to the calculation. What is the impact of handling dF/du in different ways?

To answer this question, we must notice that the added term is similar to a diffusion term with a variable diffusion coefficient. If dF/du is larger upstream, the MacCormack method gives more numerical dispersion; we would then expect the MacCormack method to have smaller oscillations than the Lax-Wendroff method, as we observed. Chapter 14 further illustrates the effect of various ways of evaluating the diffusion term and their effect on truncation error and numerical dispersion. By looking at the term

$$\frac{\partial}{\partial x}\left[k(u)\frac{\partial u}{\partial x}\right]\bigg|_i = \frac{1}{\Delta x}\left[k_{i+1/2}\left(\frac{u_{i+1} - u_i}{\Delta x}\right) - k_{i-1/2}\left(\frac{u_i - u_{i-1}}{\Delta x}\right)\right], \quad (4.56)$$

we see that there are three common ways to evaulate $k_{i+1/2}$: exact, average, and upstream. The exact meaning of $k_{i+1/2}$ is

$$k_{i+1/2} = k(u_{i+1/2}) \quad \text{(exact)}. \tag{4.57}$$

The average value is correct to $O(\Delta x^2)$:

$$k_{i+1/2} = (k_{i+1} + k_i)/2 \quad \text{(average)}. \tag{4.58}$$

The upstream value is

$$k_{i+1/2} = k_i \quad \text{when } k_i > k_{i+1} \quad \text{(upstream)}. \tag{4.59}$$

An example and demonstration is given in Chapter 14 that shows the following results. The upstream method does not oscillate, the average method oscillates only slightly, and the exact method oscillates excessively (until Δx is small). In Chapter 14 we see that the average permeability is preferred, since it is second-order and has only small oscillations. This means that the Lax-Wendroff method has some additional damping features in the evaluation of dF/du using the average values, and the MacCormack method has even more damping by using upstream values of dF/du. Since this diffusion term is added to provide damping anyway, this additional damping gives us a better solution in the sense that it oscillates less.

4.4. Finite Difference Methods for the Convective Diffusion Equation

Let us now consider the convective diffusion equation to see how the methods perform when there is both convection and diffusion. The first method is the **explicit method with upstream convection** terms and is listed in Table 4.5. The truncation error is determined by substituting in the Taylor series for c_i^{n+1} and c_{i+1}^n and c_{i-1}^n. The result, as might be expected, is $O(\Delta x)$ and $O(\Delta t)$, since the time derivative is approximated by a first-order expression and so is the convection term. Of course if the Peclet number is zero, then the truncation error is $O(\Delta x^2)$, since the second derivative is evaluated with a centered difference expression. Stabilility is evaluated by looking at the function, $\rho(\xi)$. This function is Eq. (1) in Table 4.3. When convection is absent (Co = 0), the magnitude of ρ is

$$|\rho|^2 = [1 - 2r(1 - \cos\xi)]^2. \tag{4.60}$$

Since $1 - \cos(\xi)$ goes from 0 to 2, we need $r \leq 1/2$ for stability. When there is no diffusion term (r = 0), we have

$$|\rho|^2 = 1 - 4\,\text{Co}\,(1 - \text{Co})\sin^2\frac{\xi}{2}, \tag{4.61}$$

which is less than 1.0 for any ξ provided that Co $\leq$ 1. When both terms are present, we need to keep the magnitude of ρ less than or equal to 1.0 and this gives

$$\text{Co} + 2r \leq 1. \tag{4.62}$$

The dispersion and dissipation relations are derived from Eq. (1) of Table 4.3, but they depend on two numerical parameters, Co and r, as well as ξ. To display the

Table 4.5. Finite Difference Methods for the Convective Diffusion Equation

Explicit, upstream convection, $O(\Delta x)$, $O(\Delta t)$

$$c_i^{n+1} = c_i^n - \frac{Pe\,\Delta t}{\Delta x}(c_i^n - c_{i-1}^n) + \frac{\Delta t}{\Delta x^2}(c_{i+1}^n - 2c_i^n + c_{i-1}^n)$$

Explicit, centered derivative, $O(\Delta x^2)$, $O(\Delta t)$

$$c_i^{n+1} = c_i^n - \frac{Pe\,\Delta t}{2\Delta x}(c_{i+1}^n - c_{i-1}^n) + \frac{\Delta t}{\Delta x^2}(c_{i+1}^n - 2c_i^n + c_{i-1}^n)$$

Lax-Wendroff, $O(\Delta x^2)$, $O(\Delta t)$

$$c_{i+1/2}^{n+1/2} = \frac{1}{2}(c_{i+1}^n + c_i^n) - \frac{Pe\,\Delta t}{2\Delta x}(c_{i+1}^n - c_i^n) + \frac{\Delta t}{4\Delta x^2}(c_{i+1}^n - 2c_i^n + c_{i-1}^n + c_{i+2}^n - 2c_{i+1}^n + c_i^n)$$

$$c_i^{n+1} = c_i^n - \frac{Pe\,\Delta t}{\Delta x}(c_{i+1/2}^{n+1/2} - c_{i-1/2}^{n+1/2}) + \frac{\Delta t}{\Delta x^2}(c_{i+1}^n - 2c_i^n + c_{i-1}^n)$$

MacCormack, $O(\Delta x^2)$, $O(\Delta t^2)$

$$c*_i^{n+1} = c_i^n - \frac{Pe\,\Delta t}{\Delta x}(c_{i+1}^n - c_i^n) + \frac{\Delta t}{\Delta x^2}(c_{i+1}^n - 2c_i^n + c_{i-1}^n)$$

$$c_i^{n+1} = \frac{1}{2}(c_i^n + c*_i^{n+1}) - \frac{Pe\,\Delta t}{2\Delta x}(c*_i^{n+1} - c*_{i-1}^{n+1}) + \frac{\Delta t}{2\Delta x^2}(c*_{i+1}^{n+1} - 2c*_i^{n+1} + c*_{i-1}^{n+1})$$

Filtered DuFort-Frankel, $O(\Delta x^2)$, $O(\Delta t^2)$

$$c_i^{n+1} = \frac{\alpha}{2}(c_i^n + c_i^{n-2}) + (1-\alpha)c_i^{n-1} - \frac{Pe\,\Delta t}{\Delta x}(c_{i+1}^n - c_{i-1}^n) + \frac{2\Delta t}{\Delta x^2}(c_{i+1}^n - c_i^{n+1} - c_i^{n-1} + c_{i-1}^n)$$

Implicit, upstream convection, $O(\Delta x)$, $O(\Delta t)$

$$c_i^{n+1} = c_i^n - (1-\theta)\frac{Pe\,\Delta t}{\Delta x}(c_i^n - c_{i-1}^n) - \theta\frac{Pe\,\Delta t}{\Delta x}(c_i^{n+1} - c_{i-1}^{n+1}) +$$
$$+ (1-\theta)\frac{\Delta t}{\Delta x^2}(c_{i+1}^n - 2c_i^n + c_{i-1}^n) + \theta\frac{\Delta t}{\Delta x^2}(c_{i+1}^{n+1} - 2c_i^{n+1} + c_{i-1}^{n+1})$$

Implicit, centered derivative, $O(\Delta x^2)$, $O(\Delta t)$

$$c_i^{n+1} = c_i^n - (1-\theta)\frac{Pe\,\Delta t}{2\Delta x}(c_{i+1}^n - c_{i-1}^n) - \theta\frac{Pe\,\Delta t}{2\Delta x}(c_{i+1}^{n+1} - c_{i-1}^{n+1}) +$$
$$+ (1-\theta)\frac{\Delta t}{\Delta x^2}(c_{i+1}^n - 2c_i^n + c_{i-1}^n) + \theta\frac{\Delta t}{\Delta x^2}(c_{i+1}^{n+1} - 2c_i^{n+1} + c_{i-1}^{n+1})$$

---

results, we show three-dimensional views of $|\rho|$ as a function of Co and ξ for various r values (0, 0.1, 0.25, 0.5). In addition, a contour line where

$$\max_{\xi,\, 0\le\xi\le 2\pi} |\rho(Co, r, \xi)| = 1 \tag{4.63}$$

is useful to see the region of Co-r space where a method is stable.

The plot of $|\rho|$ is shown in Figure 4.14 for the explicit method with upstream convection. The variation with ξ indicates that waves of different frequencies will be attenuated very differently. The unit CFL property corresponds to a straight line from left to right when Co = 1 (the last line in each graph). As seen in Figure 4.14, this method obeys the unit CFL property only when r = 0. When the curve is larger than 1.0, the method is unstable, and when the curve has excessive variations with ξ, it means that different wave numbers are dissipated with different magnitudes, leading to excessive smoothing in the solution. Let us consider the curve with r = 0 (Figure 4.14a) and take Co = 1, the last line in the figure. As diffusion is added (by looking at Figure 4.14b), the curve increases above 1.0, meaning the method is unstable. The time-step (Courant number) would have to be reduced. If the Courant number is kept small (near the front of each graph), then as diffusion is added there is more variation in the amplification factor with ξ. In that case the method is stable but shows excessive dissipation for some wave numbers. The region of (Co,r) space for which this method is stable is shown in Figure 4.15. This is deduced from Figure 4.14b as the largest Co value for which the amplification factor is less than or equal to 1.0 for all ξ. We can imagine the diagram as a mock-up and take a ruler along the line (Co = 0, ρ = 1) for all ξ, then move the ruler back until it touches the curve. This gives the critical Courant number for that r value (as illustrated in Figure 4.14b). The results of applying this method are shown in Figure 2.20 for a step function and Figure 4.16a for a square wave. Both of these calculations are for Co = 0.5 and r = 0.025, which is well within the stable region. The solutions show excessive dissipation, however.

Next let us consider changing the upstream derivative for a centered derivative. The **explicit method with centered derivatives** is listed in Table 4.5. The truncation error of this method is $O(\Delta t)$ and $O(\Delta x^2)$, since the time derivatives are written to first-order and all spatial derivatives are written with second-order expressions. The amplification factor is given by Eq. (2) of Table 4.3. The magnitude of $\rho(\zeta)$ is given by

$$|\rho|^2 = [1 + 2r(\cos\xi - 1)]^2 + [\text{Co} \sin\xi]^2 \qquad (4.64)$$

and this is less than 1.0 provided that

$$\text{Co}^2 \leq 2r \leq 1. \qquad (4.65)$$

Clearly, when r = 0 (no diffusion), this method is unstable. When there is no convection (Co = 0), r must be ≤ 0.5. The amplification factor is illustrated in Figure A.2 in the Appendix. The graph for r = 0 is always greater than 1.0 (except for Co = 0), so the method is unstable, as deduced above. When diffusion is added (by increasing r), the graph goes below 1.0 for small Co and eventually shows excessive dissipation for some wave numbers (Co = 0, r = 0.25 or 0.5). The results

FINITE DIFFERENCE METHODS - FIXED GRIDS

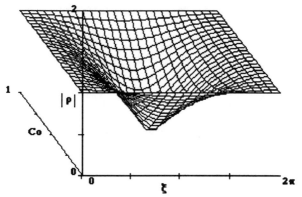

Figure 4.14a. Diagram of $|\rho|$ versus ξ and Co, r = 0.
Upstream Finite Difference Method

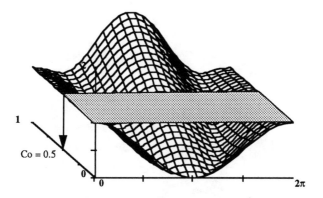

Figure 4.14b. Diagram of $|\rho|$ versus ξ and Co, r = 0.25
Upstream Finite Difference Method

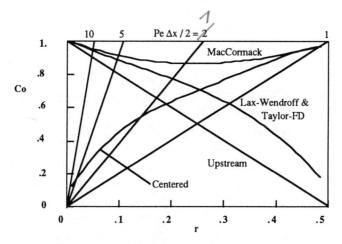

Figure 4.15. Stability Diagram for Finite Difference Methods

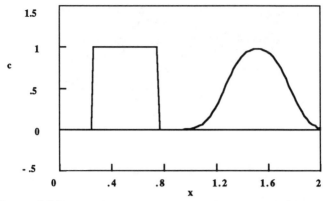

Figure 4.16a. Convective Diffusion Equation, Explicit, Upstream Method
Pe = 1000, 100 nodes, Δt = 1.E-5

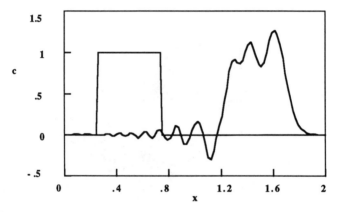

Figure 4.16b. Convective Diffusion Equation, Explicit, Centered Derivative
Pe = 1000, 100 nodes, Δt = 1.E-6

of applying this method are shown in Figure 2.17 for a step function and Figure 4.16b for a square wave. Both of these calculations are for Co = 0.05 and r = 0.0025, which is within the stable region. This solution shows excessive oscillation when Pe $\Delta x/2$ = 10, as is the case here.

The **Lax-Wendroff method** and **MacCormack's method** are listed in Table 4.5. The Lax-Wendroff version is that given by Thommen [1966], while the MacCormack version is from Anderson, *et al*. [p. 163, 1984]. MacCormack's method retains its second-order character in time, but the Lax-Wendroff method does not. The difference is in how the diffusion term is added to the methods for convection. The amplification factors for the two methods are given in Table 4.3 and are shown in Figures A.3 and A.4, respectively. For r = 0, both graphs are the same, since the methods are identical for the advection equation. The unit CFL property is illustrated by the line ρ = 1 for all ξ when Co = 1 and r = 0. For non-zero r, the amplification factor for the MacCormack method does not vary

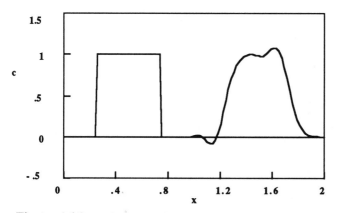

Figure 4.16c. Convective Diffusion Equation, Lax-Wendroff Method
Pe = 1000, 100 nodes, Δt = 1.E-5

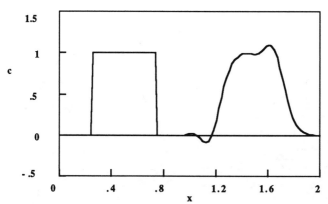

Figure 4.16d. Convective Diffusion Equation, Explicit, MacCormack Method
Pe = 1000, 100 nodes, Δt = 1.E-5

as much with ξ as it does for the Lax-Wendroff method. The amplification factor is also much bigger for the Lax-Wendroff method and this translates into a smaller region of stability, requiring a smaller Δt value. The stability regions shown in Figure 4.15 are smaller for the Lax-Wendroff method. However, both Lax-Wendroff and MacCormack methods lead to oscillations in the solutions shown in Figures 4.16c and 4.16d.

The leapfrog method is generalized to the **DuFort-Frankel method**, as given in Table 4.5. The version listed there is the one given by Anderson [p. 161, 1984], extended to include the filtering step from Dietrich and Womeck [1985]. If the diffusion term were evaluated at the n-th time level, the method would be unstable. If the middle term of the diffusion expression is changed, e. g.,

$$c_{i+1}^n - 2c_i^n + c_{i-1}^n \text{ is replaced by } c_{i+1}^n - c_i^{n+1} - c_i^{n-1} + c_{i-1}^n, \quad (4.66)$$

then the method is stable. The truncation error is $O(\Delta t^2)$ and $O(\Delta x^2)$. The stability

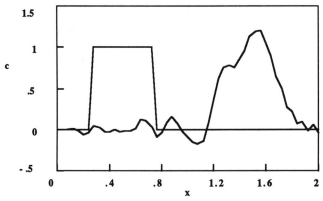

Figure 4.16e. Convective Diffusion Equation, DuFort-Frankel Method
Pe = 1000, 51 nodes, $\alpha = 0$, $\Delta t = 1.E-5$

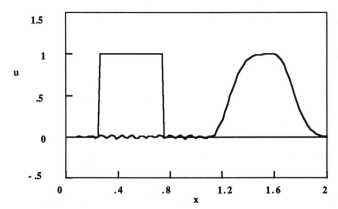

Figure 4.16f. Convective Diffusion Equation, Filtered DuFort-Frankel Method
Pe = 1000, 51 nodes, $\alpha = 1$, $\Delta t = 1.E-5$

of the method is more complicated to deduce, since the method requires several steps. For more information, you may wish to refer to Richtmyer and Morton [1967] and Sod [1985] or to Chapter 13, where a similar analysis is given for two equations. With no diffusion, the method is stable for Co < 1. With no convection, it is stable for all values of r. Typical results are given in Figures 4.16e for the standard DuFort-Frankel method; they show unacceptable oscillations. Results given in Figure 4.16f are for the filtered DuFort-Frankel method, and they are quite acceptable.

Implicit, upstream finite difference method. Next let us consider implicit methods. Table 4.5 lists an upstream method in which the convection and diffusion terms are handled implicitly. The derivation is made by writing the difference equation based on a weighted average of the convection and diffusion terms.

$$\frac{c_i^{n+1} - c_i^n}{\Delta t} = -\theta \, Pe \, \frac{\partial c}{\partial x}\bigg|^{n+1} - (1-\theta) \, Pe \, \frac{\partial c}{\partial x}\bigg|^n + \theta \, \frac{\partial^2 c}{\partial x^2}\bigg|^{n+1} + (1-\theta) \, \frac{\partial^2 c}{\partial x^2}\bigg|^n \quad (4.67)$$

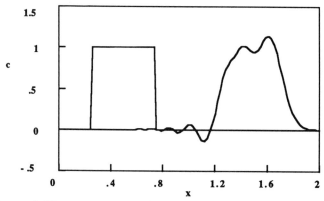

Figure 4.17a. Convective Diffusion Equation, Implicit, Centered Derivative
Pe = 1000, 100 nodes, θ = 0.5, Δt = 1.E-5

If we use upstream first derivatives, we get the expression shown in Table 4.5. Different θ give us different methods. If we use θ = 0, we have the explicit, upstream method. If we use θ = 0.5, we are using the trapezoid rule to integrate in time. If we use θ = 1.0, we are using the backward Euler method to integrate in time. Dispersion diagrams are shown in Figure A.5 for θ = 0.5 and in Figure A.6 for θ = 1.0. These can be compared with Figure A.1. When θ = 0.5 or 1.0, the methods are always stable for any $0 \leq r \leq 0.5$ and $0 \leq Co \leq 1$. The wild oscillations with changing ξ have disappeared. There is some excessive dissipation (small values of amplification factor), but there is not much dependence on the Courant number; in other words, the curves are nearly the same as we move back in each diagram. If we want the amplification factor close to 1.0, then generally as θ is increased, we can use a larger r as well. The penalty of applying an implicit method is that the calculations for each time-step are longer, since a system of equations must be solved (as described in Chapter 2). Thus, the implicit methods can only be cost-effective if they give better solutions or allow larger time-steps. Results have been given in Figure 2.20 for the explicit method with a fixed time-step and Figure 2.21 for an implicit method with a variable time-step (and controlled accuracy). These results are similar, indicating that the oscillations in the figures are due to the spatial discretization rather than the treatment of the time derivatives.

Implicit, centered finite difference method. Next we use centered difference expressions for the convective term. Dispersion diagrams are shown in Figure A.7 and A.8 for θ = 0.5 and θ = 1.0, respectively. These diagrams are similar to those with the upstream differences, except that with centered differences, there is slightly less dissipation. For r = 0 and θ = 0.5, the method obeys a unit CFL property for all Courant numbers. When r > 0, the best results are obtained with θ = 1. Results for the explicit method (θ = 0.0) are given in Figure 4.16b and show excessive oscillations. Results for θ = 0.5 and 1.0, while Δt = 1.E-5, are shown in Figures 4.17a and 4.17b, while results for the same values of θ but

84 NUMERICAL METHODS - MOVING FRONTS

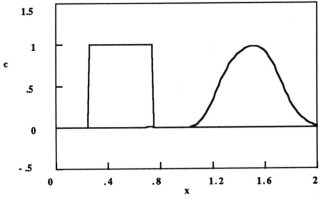

Figure 4.17b. Convective Diffusion Equation, Implicit, Centered Derivative
Pe = 1000, 100 nodes, $\theta = 1$, $\Delta t = 1.\text{E-}5$

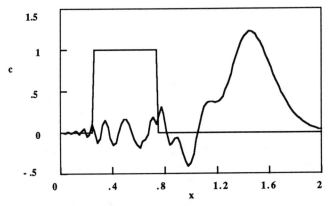

Figure 4.17c. Convective Diffusion Equation, Implicit, Centered Derivative
Pe = 1000, 100 nodes, $\theta = 0.5$, $\Delta t = 1.\text{E-}4$

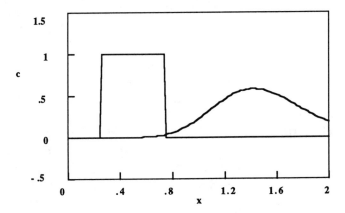

Figure 4.17d. Convective Diffusion Equation, Implicit, Centered Derivative
Pe = 1000, 100 nodes, $\theta = 1$, $\Delta t = 1.\text{E-}4$

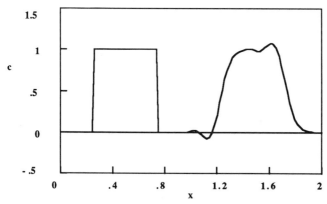

Figure 4.18a. Convective Diffusion Equation, Taylor-Finite Difference
Pe = 1000, 100 nodes, θ = 0, α = 1, Δt = 1.E-5

$\Delta t = 1.\text{E-}4$ are shown in Figures 4.17c and 4.17d. Both solutions with $\theta = 0.5$ show the solution oscillating too much (Figures 4.17a and 4.17c). When θ is increased to 1.0, giving a fully implicit method, the solution is overly smoothed by a little (for $\Delta t = 1.\text{E-}5$ in Figure 4.17b) or a lot (for $\Delta t = 1.\text{E-}4$ in Figure 4.17d). The solutions with $\Delta t = 1.\text{E-}4$ do illustrate that implicit methods are stable for large step-sizes, here for Co = 5. However, the solutions are not very good; the fully implicit method with a small step-size gave the best results (Figure 4.17b).

Implicit, Taylor-finite difference method. The Taylor term can also be added to the implicit finite difference methods.

$$\frac{c_i^{n+1} - c_i^n}{\Delta t} = -\theta \, \text{Pe} \, \frac{c_{i+1}^{n+1} - c_{i-1}^{n+1}}{2 \Delta x} - (1 - \theta) \, \text{Pe} \, \frac{c_{i+1}^n - c_{i-1}^n}{2 \Delta x} + \frac{\alpha \Delta t \, \text{Pe}^2}{2 \Delta x^2} (c_{i+1}^n - 2 c_i^n + c_{i-1}^n)$$
$$+ \frac{1}{\Delta x^2} \theta \, (c_{i+1}^{n+1} - 2 c_i^{n+1} + c_{i-1}^{n+1}) + \frac{1}{\Delta x^2}(1 - \theta) \, (c_{i+1}^n - 2 c_i^n + c_{i-1}^n)$$

(4.68)

This method is second-order in Δt only when $\alpha = 1 - 2\theta$ (see Problem 4.13). The graphs of amplification factor, when α is chosen as $1 - 2\theta$, are shown in Figures A.9-11. Solutions shown in Figure 4.18a for the explicit method ($\theta = 0$, $\alpha = 1$) are comparable to those obtained with the MacCormack method (Figure 4.16d). For $\theta = 0.5$ and $\alpha = 0$, the results are as shown in Figure 4.18b; this choice gives a second-order method. These results are also similar to those obtained with the MacCormack method. Results for $\theta = 1$ and $\alpha = -1$ (a second-order, implicit Taylor-finite differencce method) are unsatisfactory; the Δt must be too small for stability. Results for $\alpha = 1$ are shown in Figure 4.18c ($\theta = 0.5$) and Figure 4.18d ($\theta = 1.0$). Both solutions are reasonable although they are damped too much. These results suggest that it is not desirable to keep the method second-order. Instead, we should keep $\alpha = 1$ for any value of θ. The graphs of amplification factor for $\alpha = 1$ are given in Figures A.12 and A.13. There is much more variation

86 NUMERICAL METHODS - MOVING FRONTS

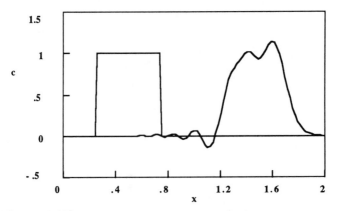

Figure 4.18b. Convective Diffusion Equation, Taylor-Finite Difference
Pe = 1000, 100 nodes, $\theta = 0.5$, $\alpha = 0$, $\Delta t = 1.\text{E-}5$

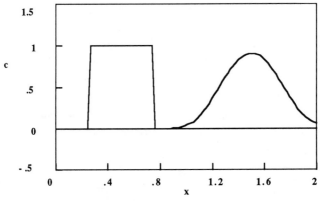

Figure 4.18c. Convective Diffusion Equation, Taylor-Finite Difference
Pe = 1000, 100 nodes, $\theta = 0.5$, $\alpha = 1$, $\Delta t = 1.\text{E-}5$

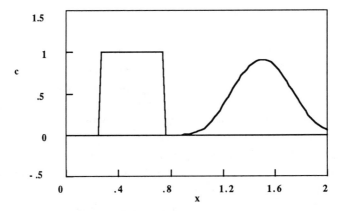

Figure 4.18d. Convective Diffusion Equation, Taylor-Finite Difference
Pe = 1000, 100 nodes, $\theta = 1$, $\alpha = 1$, $\Delta t = 1.\text{E-}5$

with ξ, but the amplification factor is less than or equal to 1.0 for all conditions shown, in contrast to Figures A.9-11. The effect of the implicit treatment of the problem is to stabilize the results for r > 0, although there is still great variation in amplification factor with ξ. It does not appear to be good strategy to use both implicit treatments and Taylor-finite difference treatments at the same time, either first- or second-order.

Figure 4.15 can be used to determine a stable step-size, since both r and Co are proportional to Δt. This most easily be done in conjunction with the value of PeΔx/2. By definition of the quantities, we have

$$\frac{Pe\Delta x}{2} = \frac{1}{2} \frac{Pe\Delta t}{\Delta x} \frac{\Delta x^2}{\Delta t} = \frac{Co}{2r} \quad \text{or} \quad Co = 2r \frac{Pe\Delta x}{2}. \quad (4.69)$$

A fixed grid size (PeΔx/2) is then a line, Co α r, with the proportionality constant related to PeΔx/2. For example, if PeΔx/2 = 10, as in most of the cases treated here, then the line is a straight line that begins at the origin and connects to the point (r = 0.05, Co = 1). If we use 100 nodes when Pe = 1000, we simply follow along this line from r = 0 until it touches the stability limit for the method we are using. The sooner it touches the stability limit, the smaller the Δt allowed. The centered finite difference method is the least stable, but the Taylor-finite difference method is a considerable improvement, since it allows a time-step about nine times larger. The MacCormack and Lax-Wendroff methods have similar results for small values of r, but the MacCormack method gives better results at larger values of r. The MacCormack method has the largest stability region of any method shown in Figure 4.15. The Taylor-finite difference method with θ = 1 and α = −1 is similar to the curve for the centered finite difference method, and is hence undesirable. The implicit finite difference methods and the implicit Taylor-finite difference methods with α = 1 are stable for all conditions shown in Figure 4.15.

The tendency to oscillate for the steady-state problem is governed by the value of Pe Δx/2. For the centered finite difference method, values of Pe Δx/2 ≤ 1 lead to non-oscillatory solutions [Price, *et al.* 1966]. Thus for the finite difference method, the region with the largest time-step that still gives non-oscillatory results is the region to the right-hand side of Figure 4.15. Yet this is also the region requiring many grid points (when Pe is large) and hence the calculation times are large. Thus the goal is to have a method that allows us to use a large Courant number and a large value of Pe Δx/2. It is clear from Figure 4.15 that the MacCormack method and Taylor finite difference method (α = 1) are both good methods.

Based on these results for the convective diffusion equation, we can conclude that most of the methods are unsatisfactory because they either have too much dissipation or severe oscillations. Implicit treatment helped some, but only if the time-step was large. The filtered DuFort-Frankel method gives reasonable results (Figure 4.16f), as does the Taylor-finite difference method with θ = 0.5 and α = 1 (Figure 4.18c).

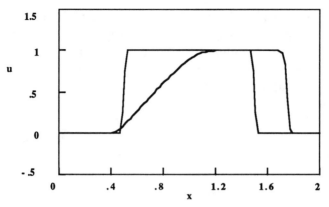

Figure 4.19a. Burger's Equation, $\nu = 0.001$, Upstream Method, 100 nodes, $\Delta t = 0.01$

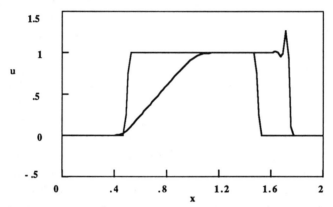

Figure 4.19b. Burger's Equation, $\nu = 0.001$, Lax-Wendroff Method 100 nodes, $\Delta t = 0.01$

4.5. Finite Difference Methods for Burger's Equation With Viscosity

The same methods can be applied to Burger's equation with viscosity, as listed in Table 4.4. Applications of the methods are shown in Figures 4.19a-d. The upstream method gives a sharp front when the phenomenon is self-sharpening; the phenomenon tends to overcome the excessive dissipation in the method. The MacCormack method is the best overall method. It minimizes the oscillations while keeping the front sharp. If oscillations cannot be present, then we must use the upstream method. It is interesting to note that Figure 4.19c for the MacCormack method shows slight oscillations when viscosity is included in the equations. The oscillations were not there in Figure 4.13c, when the MacCormack method was applied to the problem without viscosity. The reason is that Figure 4.13c was calculated with $\Delta t = \Delta x$, for a Courant number of 1.0. This value cannot be used with viscosity (it is unstable) and a smaller Δt was needed. If the problem

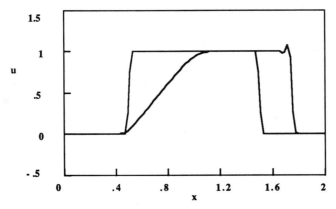

Figure 4.19c. Burger's Equation, $\nu = 0.001$, MacCormack Method, 100 nodes, $\Delta t = 0.01$

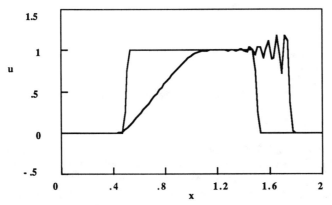

Figure 4.19d. Burger's Equation, $\nu = 0.001$, Filtered Leapfrog Method 100 nodes, $\Delta t = 0.01$

without viscosity is re-solved using the same Δt value used for Figure 4.19c with viscosity, the result shows oscillations (see Problem 4.14). The oscillations are larger when the problem is solved without viscosity than with viscosity, as long as the time-step is small enough to be stable for the problem with viscosity.

4.6. Conclusions

The advection and convective diffusion equations provide test problems that are important because they permit an analytical study of the stability of the methods. Burger's equation, with and without viscosity, provides a useful test problem because the problem is nonlinear and shocks can be formed. When treating a method without diffusion, most methods introduce some form of diffusion, called numerical diffusion, to control the growth of unwanted numerical artifacts. Thus the problems with diffusion are actually easier than those without. However, the problems without any diffusion provide a test that reveals methods without obfuscation. We found that most of the methods gave solutions that either had too much numerical diffusion or oscillation. These conclusions held whether the problem was linear or nonlinear or whether it had viscosity (or diffusion) or not. The implicit methods tended to oscillate less, allowing larger time-steps, but also included excessive dissipation. The best overall methods for linear problems are the filtered leapfrog method and the Taylor-finite difference method with $\theta = 0.5$ and $\alpha = 1$. The best method for nonlinear problems is the MacCormack method. Even these methods include oscillations in some situations. Consequently, we would like to explore other methods as well.

Problems

1_1. Derive Eq. (4.25).

2_1. Calculate the results shown in Figure 4.5.

3_1. Derive Eq. (4.34).

4_1. For the MacCormack method in Table 4.2, combine the two steps and show that they are equivalent to Eq. (4.34).

5_2. Use the program CONVECT to solve the advection equation when the initial condition is a step change in the solution. Use the same methods as those used in Figures 4.6, 4.9, and 4.10. Discuss the results.

6_2. Use the program CONVECT to solve Burger's equation when the initial condition is as illustrated in Figure 4.12. Use the same methods as those used in Figures 4.13. Use 100 nodes and a time-step such that $\Delta t = \Delta x$; integrate from time 0 to 1.0 (at which point the shock forms) and on to 2.5. The front moves with velocity 0.5 during the time interval from 1.0 to 2.5, so that it is located at position $x = 1.75$. Discuss the results.

7_2. Derive Eq. (4.55). Hint: First substitute Eq. (4.53) into Eq. (4.54). Then write $F*_i^{n+1}$ as

$$F*_i^{n+1} = F_i^n + \frac{dF}{du}\Big|_i^n (u*_i^{n+1} - u_i^n) + \ldots \qquad (4.70)$$

Use Eq. (4.53) to make this

$$F*_i^{n+1} = F_i^n - \frac{\Delta t}{\Delta x}\frac{dF}{du}\Big|_i^n (F_{i+1}^n - F_i^n) \qquad (4.71)$$

Write a similar equation for $F*_{i-1}^{n+1}$. Combine the equations to get Eq. (4.55).

8_2. Use the Taylor series to evaluate $u_{i\pm 1}$ and $k_{i\pm 1}$; substitute the expressions into Eq. (4.58) and Eq. (4.56) to find the truncation error for this formulation.

9_2. Derive the amplification factors listed in Table 4.3 for the (a) Explicit, upstream method, (b) Explicit, centered method, (c) Lax-Wendroff method, and (d) MacCormack method.

10_2. For the centered finite difference expression, derive Eq. (4.60) and Eq. (4.61) for the amplitude of the amplification factor.

11₃. Use the program CONVECT to solve the convective diffusion equation when the initial condition is a step change in the solution and the Peclet number is 1000. Use the same methods illustrated in Figures 4.16, 4.17, and 4.18. Discuss the results.

12₃. Use the program CONVECT to solve Burger's equation with viscosity when the initial condition is as illustrated in Figure 4.12. Use the same methods illustrated in Figures 4.13, but with $v = 0.001$. Discuss the results.

13₃. Derive both explicit and implicit Taylor-finite difference methods by considering the advection equation. Take $F(u) = \text{Pe } c$ and $u = c$ in Eq. (4.38). (a) The explicit, Taylor-finite difference method is then derived from Eq. (4.43) by setting

$$c_{i+1/2}^n = 0.5\,(c_{i+1}^n + c_i^n), \quad c_{i-1/2}^n = 0.5\,(c_i^n + c_{i-1}^n). \tag{4.72}$$

(b) To see how to use the Taylor term for an implicit method, Eq. (4.68), re-derive Eq. (4.43) for the advection equation. The advection equation is

$$\frac{\partial c}{\partial t} = -\text{Pe}\,\frac{\partial c}{\partial x}. \tag{4.73}$$

Differentiate this with respect to x to obtain the cross derivative. Differentiate the same equation with respect to t and substitute for the cross derivative to obtain

$$\frac{\partial^2 c}{\partial t^2} = \text{Pe}^2 \frac{\partial^2 c}{\partial x^2}. \tag{4.74}$$

Write a Taylor series for c^{n+1} and obtain the equivalent of Eq. (4.40). Evaluate the convection terms implicitly.

$$\left.\frac{\partial c}{\partial t}\right|_i^n = -\theta\,\text{Pe}\,\left.\frac{\partial c}{\partial x}\right|_i^{n+1} - (1-\theta)\,\text{Pe}\,\left.\frac{\partial c}{\partial x}\right|_i^n \tag{4.75}$$

Combine terms to get

$$\frac{c_i^{n+1} - c_i^n}{\Delta t} = -\theta\,\text{Pe}\,\left.\frac{\partial c}{\partial x}\right|_i^{n+1} - (1-\theta)\,\text{Pe}\,\left.\frac{\partial c}{\partial x}\right|_i^n + \frac{\Delta t}{2}\left.\frac{\partial^2 c}{\partial t^2}\right|_i^n. \tag{4.76}$$

In anticipation of the results given below, multiply the last term by a constant, α. Write the centered finite difference approximation for this equation and determine its truncation error. Show that the method is second-order in Δt only when

$$\alpha = 1 - 2\theta. \tag{4.77}$$

For $\theta = 0$ (explicit method), you need $\alpha = 1$; for $\theta = 0.5$, you need $\alpha = 0$; for $\theta = 1$ (implicit method), you need $\alpha = -1$. This last result is surprising, but indicates that using $\theta = 1$ introduces too much dissipation into the method; using $\alpha = -1$ removes just enough to make the method second-order in Δt. When this idea is extended to the convective diffusion equation, you get Eq. (4.68).

14₂. Use the program CONVECT to solve Burger's equation without viscosity for the following initial conditions.

$$f(x) = \begin{cases} 1 & x \leq 0 \\ 1 - x & 0 < x \leq 1 \\ 0 & 1 < x \end{cases} \tag{4.78}$$

Use MacCormack's method with $\Delta t = 0.01$; compare the results with those in Figure 4.13c, which is the same method with $\Delta t = 0.02$ and Co = 1.

References

Anderson, D. A., "A Comparison of Numerical Solutions to the Inviscid Equations of Fluid Motion," J. Comp. Phys. 15 1-20 [1974].

Anderson, D. A., Tannehill, J. C., Pletcher, R. H., *Computational Fluid Mechanics and Heat Transfer*, McGraw-Hill [1984].

Dietrich, D. E. and Wormeck, J. J., "An Optimized Implicit Scheme for Compressible Reactive Gas Flow," Num. Heat Transf. 8 335-348 [1985].

Lax, P. D. and Wendroff, B., "Systems of Conservation Laws," Comm. Pure Appl. Math. 13 217-237 [1960].

Lax, P. D. and Wendroff, B., "Difference Schemes with High Order of Accuracy for Solving Hyperbolic Equations," Comm. Pure Appl. Math. 17 381- [1964].

MacCormack, R. W., "The Effect of Viscosity in Hypervelocity Impact Cratering," AIAA Paper No. 69-354 [1969].

Morton, K. W., and Parrott, A. K., "Generalised Galerkin Methods for First-Order Hyperbolic Equations," J. Comp. Phys. 36 249-270 [1980].

Price, H. S., Varga, R. S., and Warren, J. E., "Application of Oscillationa Matrices to Diffusion-Convection Equations," J. Math. & Phys. 45 301-311 [1966].

Richtmyer, R. D., "A Survey of Difference Methods for Nonsteady Fluid Dynamics," NCAR Tech. Note, 63-2, Boulder, Colo. [1963].

Richtmyer, R. D. and Morton, K. W., *Difference Methods for Initial-Value Problems*, 2nd edition, Interscience/Wiley [1967].

Roache, P. J., *Computational Fluid Dynamics*, Hermosa Publ., Albuquerque, N. M. [1972].

Roache, P. J. and Dietrich, D. E., "Evaluation of the Filtered Leapfrog-Trapezoidal Time Integration Method," Num. Heat Transf. 14 149-164 [1988].

Sod, G. A., "A Survey of Several Finite Difference Methods for Systems of Nonlinear Hyperbolic Conservation Laws," J. Comp. Phys. 27 1-31 [1978].

Sod, G. A., *Numerical Methods in Fluid Dynamics*, Cambridge Univ. Press [1985].

Thommen, H. U., "Numerical Integration of the Navier-Stokes Equations," ZAMP 17 369-384 [1966].

CHAPTER
FIVE

ANALYSIS OF FINITE ELEMENT METHODS ON FIXED GRIDS

The Galerkin finite element method was developed after the finite difference method and it was initially applied to elliptic problems (such as steady-state heat conduction) rather than flow problems. Upon application to flow problems, it was soon realized that the standard Galerkin finite element method did not work well when the problem had strong convective terms; as a result, various formulations of the Petrov-Galerkin method were developed. Here we will study these formulations in detail, including the truncation error and stability of each formulation. We will perform a Fourier analysis for the linear problems in order to determine the combination of parameters that make the formulations stable.

5.1 Accuracy, Stability, Dissipation and Dispersion

To study the stability and accuracy of finite element methods, let us consider the convective diffusion equation:

$$\frac{\partial c}{\partial t} + \text{Pe}\frac{\partial c}{\partial x} = \frac{\partial^2 c}{\partial x^2}. \qquad (5.1)$$

If we apply the Petrov-Galerkin method, we obtain the equation (2.105):

$$\left(\frac{1}{6} - \frac{\alpha}{4}\right)\frac{dc_{i+1}}{dt} + \frac{2}{3}\frac{dc_i}{dt} + \left(\frac{1}{6} + \frac{\alpha}{4}\right)\frac{dc_{i-1}}{dt} = -\frac{\text{Pe}}{2\Delta x}[(1-\alpha)c_{i+1} + 2\alpha c_i \\ - (1+\alpha)c_{i-1}] + \frac{1}{\Delta x^2}[c_{i+1} - 2c_i + c_{i-1}]. \qquad (5.2)$$

We then consider an explicit method for solving this equations, which leads to

$$\left(\frac{1}{6} - \frac{\alpha}{4}\right)(c_{i+1}^{n+1} - c_{i+1}^n) + \frac{2}{3}(c_i^{n+1} - c_i^n) + \left(\frac{1}{6} + \frac{\alpha}{4}\right)(c_{i-1}^{n+1} - c_{i-1}^n) = \\ -\frac{\text{Pe}\,\Delta t}{2\Delta x}[(1-\alpha)c_{i+1}^n + 2\alpha c_i^n - (1+\alpha)c_{i-1}^n] + \frac{\Delta t}{\Delta x^2}[c_{i+1}^n - 2c_i^n + c_{i-1}^n]. \qquad (5.3)$$

We obtain the Galerkin method by taking $\alpha = 0$ and the advection equation by setting $r = 0$. Let us consider the accuracy and stability of these equations.

The truncation error in Δx is clear, since the same equations resulted from finite difference methods. We recognize that the truncation error on the right-hand side is $O(\Delta x^2)$ if we use the Galerkin method ($\alpha = 0$), but it is only $O(\Delta x)$ if we use the Petrov-Galerkin method. The truncation error of the time-dependent term is obtained by expanding terms such as c_{i+1}^{n+1} in a Taylor series.

$$c_{i+1}^{n+1} = c_{i+1}^n + c'{}_{i+1}^n \Delta t + c''{}_{i+1}^n \frac{\Delta t^2}{2!} \tag{5.4}$$

$$c_{i-1}^{n+1} = c_{i-1}^n + c'{}_{i-1}^n \Delta t + c''{}_{i-1}^n \frac{\Delta t^2}{2!} \tag{5.5}$$

$$c_i^{n+1} = c_i^n + c'{}_i^n \Delta t + c''{}_i^n \frac{\Delta t^2}{2!} \tag{5.6}$$

Combining various terms gives us

$$\left(\frac{1}{6} - \frac{\alpha}{4}\right)\frac{c_{i+1}^{n+1} - c_{i+1}^n}{\Delta t} + \frac{2}{3}\frac{c_i^{n+1} - c_i^n}{\Delta t} + \left(\frac{1}{6} + \frac{\alpha}{4}\right)\frac{c_{i-1}^{n+1} - c_{i-1}^n}{\Delta t} = \left.\frac{\partial c}{\partial t}\right|_i +$$

$$\frac{1}{6}\left(\left.\frac{\partial c}{\partial t}\right|_{i+1} - 2\left.\frac{\partial c}{\partial t}\right|_i + \left.\frac{\partial c}{\partial t}\right|_{i-1}\right) + \frac{\Delta t}{12}\left(\left.\frac{\partial^2 c}{\partial t^2}\right|_{i+1} - 2\left.\frac{\partial^2 c}{\partial t^2}\right|_i + \left.\frac{\partial^2 c}{\partial t^2}\right|_{i-1}\right) + \tag{5.7}$$

$$\left.\frac{\partial^2 c}{\partial t^2}\right|_i \frac{\Delta t}{2} - \frac{\alpha}{2}\left[\frac{\partial^2 c}{\partial t \partial x}\Delta x + \frac{\partial^3 c}{\partial x \partial t^2}\frac{\Delta t \Delta x}{2}\right]\bigg|_i.$$

The second term on the right-hand side can be further expanded to give

$$c'{}_{i+1} - 2 c'{}_i + c'{}_{i-1} = c'{}_{xx} \Delta x^2 + O(\Delta x^4). \tag{5.8}$$

$$c''{}_{i+1} - 2 c''{}_i + c''{}_{i-1} = c''{}_{xx} \Delta x^2 + O(\Delta x^4). \tag{5.9}$$

The second term has a truncation error of $O(\Delta x^2)$, the third term has a truncation error of $O(\Delta t \Delta x^2)$, and the fourth term has a truncation error of $O(\Delta t)$. We neglect the truncation error of the third term because it is small. Thus, the Galerkin method ($\alpha = 0$) has a truncation error of $O(\Delta x^2)$ and $O(\Delta t)$. When the method is Petrov-Galerkin ($\alpha \neq 0$), the truncation error is reduced to $O(\Delta x)$ and $O(\Delta t)$.

The stability of the Petrov-Galerkin method is obtained by taking the discrete Fourier transform of Eq. (5.3). Using Table 4.1 we obtain

$$\left(1 - \frac{6\alpha}{4}\right) e^{-i\xi} \hat{c}^{n+1} + 4 \hat{c}^{n+1} + \left(1 + \frac{6\alpha}{4}\right) e^{i\xi} \hat{c}^{n+1} = \left(1 - \frac{6\alpha}{4}\right) e^{-i\xi} \hat{c}^n + 4 \hat{c}^n +$$
$$\left(1 + \frac{6\alpha}{4}\right) e^{i\xi} \hat{c}^n - 3 \operatorname{Co} (e^{-i\xi} - e^{i\xi}) \hat{c}^n + (6 r + 3 \operatorname{Co} \alpha) (e^{-i\xi} - 2 + e^{i\xi}) \hat{c}^n. \tag{5.10}$$

After writing the exponential terms in terms of sines and cosines and collection of terms we can write Eq. (5.10) as

$$(4 + 2 \cos \xi + 3 \alpha i \sin \xi) \hat{c}^{n+1} =$$
$$\left[4 + 2 \cos \xi + 3 \alpha i \sin \xi + 6 \text{Co} \, i \sin \xi - 12 \alpha \text{Co} \sin^2 \frac{\xi}{2} + 6 r \left(-4 \sin^2 \frac{\xi}{2} \right) \right] \hat{c}^n. \quad (5.11)$$

For the Galerkin method ($\alpha = 0$), the amplification factor is

$$\rho_r = \frac{4 + 2 \cos \xi - 24 r \sin^2 (\xi/2)}{4 + 2 \cos \xi}, \quad (5.12)$$

$$\rho_i = \frac{6 \text{Co} \sin \xi}{4 + 2 \cos \xi}. \quad (5.13)$$

This amplification factor is a more complicated function than that for finite difference methods, because of the multiple terms appearing on the left-hand side of Eq. (5.2). Some special cases are worth mentioning here. If the convection term is absent ($\text{Co} = 0$), the amplification factor is a real number. The critical condition is when $\rho = -1$. This gives

$$\rho_r = \frac{4 + 2 \cos \xi - 24 r \sin^2 (\xi/2)}{4 + 2 \cos \xi} = -1 \quad (5.14)$$

$$r \leq \frac{\frac{1}{3} + \frac{1}{6} \cos \xi}{\sin^2 \frac{\xi}{2}}. \quad (5.15)$$

The right-hand side is a minimum when $\xi = -\pi$. Thus, for the Galerkin finite element method applied to the diffusion equation, the time-step must be smaller ($r \leq 0.1667$) than for the centered finite difference method ($r \leq 0.5$). If the diffusion term is absent ($r = 0$), the real part of the amplification factor is 1.0 and the imaginary part is non-zero. Consequently, the amplification factor has a magnitude greater than 1.0 and the method is unstable. This is the same result that we obtained for the finite difference method with centered difference expressions for the second derivative. Figure 5.1 shows the magnitude of ρ as a function of the wave-number. For $r = 0.1$ and $\text{Co} = 0.5$ or 0.75, the magnitude is greater than 1.0 for some wave numbers and the method is unstable. For $r = 0.1$ and $\text{Co} = 0.25$, the magnitude is less than 1.0 for all wave numbers and the method is stable. For $r = 0.25$ and any Courant number, the magnitude is greater than 1.0 for some wave numbers and the method is unstable.

For the Petrov-Galerkin method, the amplification factor depends on the value of α. Plots of the amplification factor are shown in Figure 5.2 for $\alpha = 0.5$ and in Figure 5.3 for $\alpha = 1.0$. In both cases, when $r = 0$ the method is unstable. This can be seen by evaluating the amplification factor at small ξ values. However, the values of the amplification factor are very close to 1.0 (e.g., 1.001), so growth is not rapid. For $r > 0$, the method is stable for some Courant numbers,

96 NUMERICAL METHODS - MOVING FRONTS

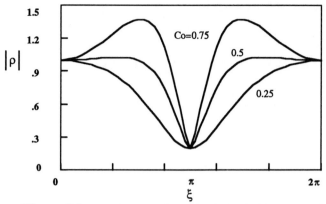

Figure 5.1. Amplification Factor for Galerkin Method, r = 0.1

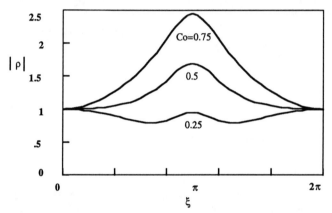

Figure 5.2. Amplification Factor for Petrov-Galerkin Method, $\alpha = 0.5$, r = 0.1

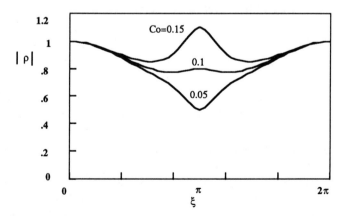

Figure 5.3. Amplification Factor for Petrov-Galerkin Method, $\alpha = 1$, r = 0.1

as shown in Figures 5.2 and 5.3. Figure 5.2 shows r = 0.1 and α = 0.5; the method is stable for Co = 0.26, but unstable for Co = 0.27. Figure 5.3 shows r = 0.1 and α = 1; the method is stable for Co = 0.13, but unstable for Co = 0.14. Thus the Petrov-Galerkin method does not stabilize the results for the advection equation, and only stabilizes the results for the convective diffusion equation when when very small Courant numbers are used. This is because of the treatment of the time-dependent term, since the finite difference method with upstream weighting (α = 1) is stable for larger values of Courant numbers (see Figure 4.15).

These results suggest that using the Petrov-Galerkin method in time-dependent terms may not be desirable with explicit schemes, since the method is unstable for r = 0, albeit just barely unstable. If we lump the time derivative (and use α = 1), we get the upstream finite difference method, which is stable. Thus the difficulty occurs because of the time-dependent terms in the Petrov-Galerkin formulation. This feature is clarified in the stability diagrams presented in the Appendix.

5.2. Finite Element Methods for the Advection Equation

Shown in Table 5.1 are several methods for solving the advection equation using finite element methods. The first method is the explicit method described above. The Galerkin method applied to the advection equation (r = 0) is unstable for any Courant number. The Petrov-Galerkin method applied to the advection equation is also unstable, but the amplification factor is only slightly greater than 1.0 for small Courant numbers. We next derive the Taylor-Galerkin method, which uses a Taylor series to evaluate the time derivative [Donea, 1984; Donea, et al., 1984].

We first take the advection equation

$$\frac{\partial c}{\partial t} + \text{Pe} \frac{\partial c}{\partial x} = 0 \tag{5.16}$$

and write a standard explicit scheme in time.

$$\left.\frac{\partial c}{\partial t}\right|^n = -\text{Pe} \left.\frac{\partial c}{\partial x}\right|^n \tag{5.17}$$

Next we use the Taylor series to evaluate c^{n+1}:

$$c^{n+1} = c^n + \left.\frac{\partial c}{\partial t}\right|^n \Delta t + \left.\frac{\partial^2 c}{\partial t^2}\right|^n \frac{\Delta t^2}{2} + O(\Delta t^3). \tag{5.18}$$

Donea [1984] discusses higher-order terms, but these are not as useful for general equations, so they are not included here. We rearrange Eq. (5.18) into the form

$$\left.\frac{\partial c}{\partial t}\right|^n = \frac{c^{n+1} - c^n}{\Delta t} - \left.\frac{\partial^2 c}{\partial t^2}\right|^n \frac{\Delta t}{2} + O(\Delta t^2). \tag{5.19}$$

Table 5.1. Galerkin Methods for the Advection Equation

Explicit, Petrov-Galerkin

$$\left(\frac{1}{6} - \frac{\alpha}{4}\right)(c_{i+1}^{n+1} - c_{i+1}^{n}) + \left(\frac{2}{3}\right)(c_{i}^{n+1} - c_{i}^{n}) + \left(\frac{1}{6} + \frac{\alpha}{4}\right)(c_{i-1}^{n+1} - c_{i-1}^{n}) =$$
$$-\frac{Pe\Delta t}{2\Delta x}[(1-\alpha)c_{i+1}^{n} + 2\alpha c_{i}^{n} - (1+\alpha)c_{i-1}^{n}]$$

Taylor-Galerkin

$$\frac{1}{6}(c_{i+1}^{n+1} - c_{i+1}^{n}) + \frac{2}{3}(c_{i}^{n+1} - c_{i}^{n}) + \frac{1}{6}(c_{i-1}^{n+1} - c_{i-1}^{n}) = -\frac{Pe\Delta t}{2\Delta x}(c_{i+1}^{n} - c_{i-1}^{n}) + \frac{Pe^2\Delta t^2}{2\Delta x^2}(c_{i+1}^{n} - 2c_{i}^{n} + c_{i-1}^{n})$$

Leapfrog-Galerkin

$$\frac{1}{6}(c_{i+1}^{n+1} - c_{i+1}^{n-1}) + \frac{2}{3}(c_{i}^{n+1} - c_{i}^{n-1}) + \frac{1}{6}(c_{i-1}^{n+1} - c_{i-1}^{n-1}) = -\frac{Pe\Delta t}{\Delta x}(c_{i+1}^{n} - c_{i-1}^{n})$$

Morton-Parrott-Galerkin

$$\frac{1}{6}\left(1 - \frac{Pe\Delta t}{\Delta x}\right)(c_{i+1}^{n+1} - c_{i+1}^{n}) + \left(\frac{2}{3} + \frac{Pe\Delta t}{3\Delta x}\right)(c_{i}^{n+1} - c_{i}^{n}) + \frac{1}{6}\left(1 - \frac{Pe\Delta t}{\Delta x}\right)(c_{i-1}^{n+1} - c_{i-1}^{n}) =$$
$$-\frac{Pe\Delta t}{2\Delta x}(c_{i+1}^{n} - c_{i-1}^{n}) + \frac{Pe^2\Delta t^2}{2\Delta x^2}(c_{i+1}^{n} - 2c_{i}^{n} + c_{i-1}^{n})$$

The successive differentiation of Eq. (5.16) gives us

$$\frac{\partial^2 c}{\partial t^2} = -Pe \frac{\partial^2 c}{\partial t \partial x} = -Pe \frac{\partial}{\partial x}\left(\frac{\partial c}{\partial t}\right) = -Pe \frac{\partial}{\partial x}\left(-Pe \frac{\partial c}{\partial x}\right) = Pe^2 \frac{\partial^2 c}{\partial x^2}. \quad (5.20)$$

We put this equation into Eq. (5.17) to obtain

$$\frac{c^{n+1} - c^n}{\Delta t} = -Pe \frac{\partial c^n}{\partial x} + \frac{Pe^2 \Delta t}{2} \frac{\partial^2 c^n}{\partial x^2}. \quad (5.21)$$

When we apply the Galerkin finite element method, we obtain the usual three terms on the left-hand side. We thus propose the following explicit method for the advection equation:

$$\frac{1}{6}(c_{i+1}^{n+1} - c_{i+1}^{n}) + \frac{2}{3}(c_{i}^{n+1} - c_{i}^{n}) + \frac{1}{6}(c_{i-1}^{n+1} - c_{i-1}^{n}) = -\frac{Pe\Delta t}{2\Delta x}(c_{i+1}^{n} - c_{i-1}^{n}) +$$
$$+ \frac{Pe^2\Delta t^2}{2\Delta x^2}(c_{i+1}^{n} - 2c_{i}^{n} + c_{i-1}^{n}). \quad (5.22)$$

The equation for the last element is

$$\frac{1}{3}(c_{NT}^{n+1} - c_{NT}^{n}) + \frac{1}{6}(c_{NT-1}^{n+1} - c_{NT-1}^{n}) = -\frac{Pe\Delta t}{2\Delta x}(c_{NT}^{n} - c_{NT-1}^{n}) + \frac{Pe^2\Delta t^2}{2\Delta x^2}(-c_{NT}^{n} + c_{NT-1}^{n}). \quad (5.23)$$

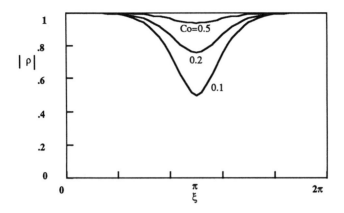

Figure 5.4. Amplification Factor for Taylor-Galerkin Method, r = 0.0

Comparison with Eq. (4.34) shows that this is an extension of the Lax-Wendroff method to finite element methods. The only difference is in the multiple time derivatives on the left-hand side that come from the finite element method. Since the MacCormack method and Lax-Wendroff method are the same for the advection equation, Eq. (5.22) is also an extension of the MacCormack method to the finite element method. The Taylor-Galerkin method is listed in Table 5.1 and its truncation error is $O(\Delta x^2)$ and $O(\Delta t^2)$. The stability diagram is shown in Figure 5.4. Among other conditions, we must satisfy Co < 0.578 (for r = 0). Since the method was derived using a Taylor series, Donea [1984] called it a Taylor-Galerkin method. This method is an improvement on the Petrov-Galerkin method, which is slightly unstable when r = 0 and α = 1.

We can also extend the leapfrog method as a finite element method. We write the equation in the form

$$\frac{c^{n+1} - c^{n-1}}{2 \, \Delta t} = - \text{Pe} \, \frac{\partial c^n}{\partial x} \tag{5.24}$$

and apply the Galerkin method to it. The result is the leapfrog-Galerkin method listed in Table 5.1. Its truncation error is also $O(\Delta x^2)$ and $O(\Delta t^2)$.

The success of these methods in practice is illustrated in Figures 5.5a-g. All solutions use 100 nodes, with Δx = 0.02. The Petrov-Galerkin method shows an instability for Co = 0.1 (Figure 5.5a) and Co = 0.05 (Figure 5.5b). These figures are further evidence that the explicit Petrov-Galerkin method is unstable without diffusion. The Taylor-Galerkin method gives the results shown in Figures 5.5c and 5.5d. With a Courant number of 0.5, there are significant oscillations, but when the Courant number is reduced to 0.25, the oscillations are much smaller, giving an acceptable solution. We can compare Figure 5.5c (for the Taylor-Galerkin method) with Figure 4.9a (for the MacCormack-finite difference method). These two figures both use a Courant number of 0.5 and the solutions

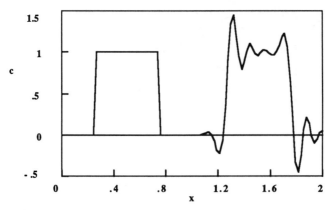

Figure 5.5a. Advection Equation, Petrov-Galerkin Method, 100 nodes, Co = 0.1

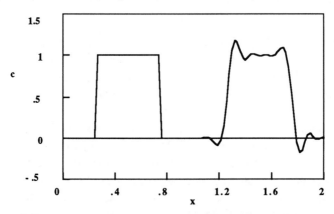

Figure 5.5b. Advection Equation, Petrov-Galerkin Method, 100 nodes, Co = 0.05

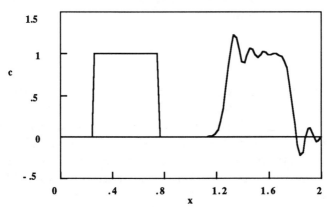

Figure 5.5c. Advection Equation, Taylor-Galerkin Method, 100 nodes, Co = 0.5

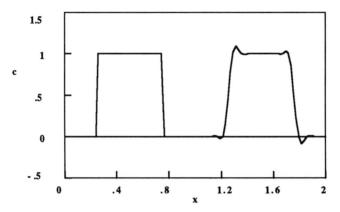

Figure 5.5d. Advection Equation, Taylor-Galerkin Method, 100 nodes, Co = 0.25

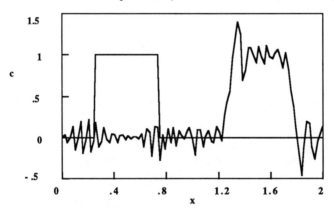

Figure 5.5e. Advection Equation, Leapfrog-Galerkin Method, 100 nodes, Co = 0.5

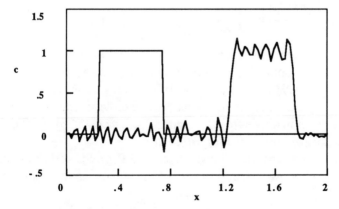

Figure 5.5f. Advection Equation, Leapfrog-Galerkin Method, 100 nodes, Co = 0.25

are both unacceptable. If the time-step is reduced (e.g., Co = 0.25), however, the Taylor-Galerkin method gives the results shown in Figure 5.5d, which are much better than those from the MacCormack method at the same Courant number (Figure 4.9b). Figures 5.5e and 5.5f show that the leapfrog-Galerkin method is not a good method.

5.3. Finite Element Methods for Burger's Equation Without Viscosity

Next we consider a nonlinear problem, Burger's equation without viscosity. We will present the methods for a general equation, but do calculations for the specific case of Burger's equation. We take the general equation to be

$$\frac{\partial u}{\partial t} = -\frac{\partial F(u)}{\partial x}, \qquad (5.25)$$

where F is a function of u. For Burger's equation, $F = u^2/2$. We apply the Petrov-Galerkin method and the Taylor-Galerkin method to this equation. For the Taylor-Galerkin method, we differentiate the equation once to obtain

$$\frac{\partial^2 u}{\partial t^2} = -\frac{\partial}{\partial t}\left(\frac{\partial F(u)}{\partial x}\right) = -\frac{\partial}{\partial x}\left(\frac{\partial F(u)}{\partial t}\right) = -\frac{\partial}{\partial x}\left(\frac{dF}{du}\frac{\partial u}{\partial t}\right) = \frac{\partial}{\partial x}\left(\frac{dF}{du}\frac{\partial F(u)}{\partial x}\right). \qquad (5.26)$$

We use the Taylor series expression

$$\frac{\partial u}{\partial t} = \frac{u^{n+1} - u^n}{\Delta t} - \frac{\Delta t}{2}\frac{\partial^2 u}{\partial t^2} \qquad (5.27)$$

and combine all terms to get

$$\frac{u^{n+1} - u^n}{\Delta t} = -\frac{\partial F(u)}{\partial x} + \frac{\Delta t}{2}\frac{\partial}{\partial x}\left(\frac{dF}{du}\frac{\partial F(u)}{\partial x}\right). \qquad (5.28)$$

This is the governing equation for the Taylor-Galerkin method, which has been applied to nonlinear convection problems by a variety of authors [Löhner, *et al.*, 1985; Oden, *et al.*, 1986; Baker and Kim, 1987; Chung, *et al.*, 1987]. Baker and Kim [1987] analyze of a great many variations and show several one-dimensional comparisons. Four methods are presented below for Eq. (5.25) and four Taylor-Galerkin methods are presented for Eq. (5.28).

The first method to be considered is a Petrov-Galerkin method that ignores the last term involving Δt. We expand the solution in a trial function,

$$u = \sum u_j N_j(x), \qquad (5.29)$$

substitute into the differential equation, and make the weighted residual zero.

$$\int_0^1 W_i(x) \sum \frac{du_j}{dt} N_j(x)\, dx = -\int_0^1 W_i \frac{\partial F(\Sigma u_j N_j)}{\partial x}\, dx \qquad (5.30)$$

For the Petrov-Galerkin method we take the weighting function to be

$$W_I = \begin{cases} N_I + \alpha F(\xi) & x_{i-1} \leq x \leq x_i \\ N_I - \alpha F(\xi) & x_i \leq x \leq x_{i+1} \end{cases}$$

$$F(\xi) = \frac{3}{4}(1-\xi^2), \quad \frac{dF}{d\xi} = -\frac{3}{2}\xi. \tag{5.31}$$

The first term, involving the time derivative, is

$$\left(\frac{1}{6} - \frac{\alpha}{4}\right)\frac{dc_{i+1}}{dt} + \frac{2}{3}\frac{dc_i}{dt} + \left(\frac{1}{6} + \frac{\alpha}{4}\right)\frac{dc_{i-1}}{dt}. \tag{5.32}$$

For the convective term, we integrate by parts, giving

$$-\int_0^1 N_i(x) \frac{\partial F}{\partial x} dx = -[N_i F]_0^1 + \int_0^1 F \frac{\partial N_i}{\partial x} dx. \tag{5.33}$$

We can recall that the trial function, N_i, is non-zero at only the i-th node; it is zero at all other nodes. The boundary term for $x = 0$ then involves only N_1; however, at $x = 0$ we specify the value of u, so that we do not need to write the equation for $i = 1$. For $i = 2, \ldots, N$, the function N_i is zero at $x = 0$ and $x = 1$. When $i = N+1$, we add the contribution

$$-\delta_{i,N+1} F|_{x=1}. \tag{5.34}$$

The complete convection term is then

$$-\delta_{i,N+1} F|_{x=1} + \sum_e \int_{-1}^1 F[u(\xi)] \frac{dW_I}{d\xi} d\xi. \tag{5.35}$$

How do we calculate these integrals? In specific cases, the function F might permit an analytical integration, but it sometimes does not. Thus we need to use numerical quadrature to calculate the integrals. We thus rewrite Eq. (5.35) as

$$-\delta_{i,N+1} F|_{x=1} + \sum_e \sum_{IG=1}^{NG} WG_{IG} F[u(\xi_{IG})] \frac{dW_I}{d\xi}\bigg|_{\xi_{IG}}. \tag{5.36}$$

The complete equation is then

Method 1 - Petrov-Galerkin

$$\left(\frac{1}{6} - \frac{\alpha}{4}\right)\frac{u_{i+1}^{n+1} - u_{i+1}^n}{\Delta t} + \frac{2}{3}\frac{u_i^{n+1} - u_i^n}{\Delta t} + \left(\frac{1}{6} + \frac{\alpha}{4}\right)\frac{u_{i-1}^{n+1} - u_{i-1}^n}{\Delta t} =$$

$$= \frac{1}{\Delta x} \sum_e \sum_{IG=1}^{NG} WG_{IG} F[u^n(\xi_{IG})] \frac{dW_I}{d\xi}\bigg|_{\xi_{IG}}. \tag{5.37}$$

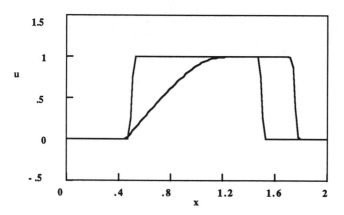

Figure 5.6. Burger's Equation, Petrov-Galerkin Method, lumped
$\alpha = 1$, $\Delta t = 0.004$, 100 nodes

The equation for the last node is

$$\left(\frac{1}{3} + \frac{\alpha}{4}\right)\frac{u_{NT}^{n+1} - u_{NT}^n}{\Delta t} + \left(\frac{1}{6} + \frac{\alpha}{4}\right)\frac{u_{NT-1}^{n+1} - u_{NT-1}^n}{\Delta t} =$$
$$= -\frac{F(u_{NT}^n)}{\Delta x} + \frac{1}{\Delta x}\sum_{IG=1}^{NG} WG_{IG} \, F[u^n(\xi_{IG})]\frac{dW_I}{d\xi}\bigg|_{\xi_{IG}}. \quad (5.38)$$

For the square wave, the Petrov-Galerkin method gives the results shown in Figure 3.4 and the lumped Petrov-Galerkin method gives the results shown in Figure 5.6. These can be compared with Figure 4.13a for the upstream finite difference method; the finite difference method is preferred. We thus have three methods: the Petrov-Galerkin method ($\alpha = 1$), the Petrov-Galerkin method with lumping of the time-derivative terms ($\alpha = 1$), and the finite difference method with upstream derivatives. Lumping is always preferred in the Petrov-Galerkin method. The upstream finite difference method is preferred over the Petrov-Galerkin method for applications such as Burger's equation without viscosity.

With a finite element method we have several other options as well. We can carry out the integration using other quadrature schemes, such as the trapezoid rule. Another option is to take the nonlinear terms and interpolate them onto the same finite element basis set, and then derive the equation for the interpolated functions. We can also evaluate the convective term in an upstream manner, still using the Galerkin method. We will try each of those options below.

First, we use the trapezoid rule for integration of the terms in Eq. (5.30) and Eq. (5.35). The result of Problem 5.6 gives **Method 2 - Trapezoid Rule:**

$$\frac{u_i^{n+1} - u_i^n}{\Delta t} = \frac{1}{\Delta x}\left[\left(-\frac{1}{2} + \frac{3\alpha}{2}\right)F_{i+1}^n - 3\alpha F_i^n + \left(\frac{1}{2} + \frac{3\alpha}{2}\right)F_{i-1}^n\right]. \quad (5.39)$$

Because of the quadrature error, the 3 should actually be a 1. Thus, we limit α

≤ 1/3. The equation for the last point is

$$\frac{u_{NT}^{n+1} - u_{NT}^n}{\Delta t} = \frac{1}{\Delta x}\left[F(u_{NT})\left(\frac{1}{2} - \frac{3\alpha}{2}\right) + F(u_{NT-1})\left(\frac{1}{2} + \frac{3\alpha}{2}\right)\right] - \frac{F(u_{NT})}{\Delta x}. \quad (5.40)$$

If we use $\alpha = 1/3$, this is the same as the upstream finite difference method; the results are shown in Figure 4.13a.

The next method is obtained by taking the nonlinear term, evaluating it at the nodes, and then representing it by a finite element representation. Thus within the element we take

$$F[u(\xi)] = \sum_k F_K N_K. \quad (5.41)$$

The integral we must evaluate becomes

$$\int_{-1}^{1} \sum_k F_K N_K \frac{dW_I}{d\xi} d\xi = \sum_k F_K \int_{-1}^{1} N_K \frac{dW_I}{d\xi} d\xi. \quad (5.42)$$

The integrals are evaluated in Problem 5.7, giving

Method 3 - Finite Element Representation.

$$\left(\frac{1}{6} - \frac{\alpha}{4}\right)\frac{u_{i+1}^{n+1} - u_{i+1}^n}{\Delta t} + \frac{2}{3}\frac{u_i^{n+1} - u_i^n}{\Delta t} + \left(\frac{1}{6} + \frac{\alpha}{4}\right)\frac{u_{i-1}^{n+1} - u_{i-1}^n}{\Delta t} =$$

$$= \frac{1}{\Delta x}\left[\left(-\frac{1}{2} + \frac{\alpha}{2}\right)F_{i+1}^n - \alpha F_i^n + \left(\frac{1}{2} + \frac{\alpha}{2}\right)F_{i-1}^n\right]. \quad (5.43)$$

The last equation is

$$\left(\frac{1}{3} + \frac{\alpha}{4}\right)\frac{u_{NT}^{n+1} - u_{NT}^n}{\Delta t} + \left(\frac{1}{6} + \frac{\alpha}{4}\right)\frac{u_{NT-1}^{n+1} - u_{NT-1}^n}{\Delta t} =$$

$$= \frac{1}{\Delta x}\left[F(u_{NT})\left(\frac{1}{2} - \frac{\alpha}{2}\right) + F(u_{NT-1})\left(\frac{1}{2} + \frac{\alpha}{2}\right)\right] - \frac{F(u_{NT})}{\Delta x}. \quad (5.44)$$

With the Galerkin time derivative and $\alpha = 1$, this method is not satisfactory. With $\alpha = 1$ and mass lumping on the time derivative terms, it is the same as Method 2 with $\alpha = 1/3$: the upstream finite difference method illustrated in Figure 4.13a.

We can also use an upstream evaluation of the convective terms (i.e., use $\alpha = 1$) and use a standard Galerkin method.

Method 4 - Upstream

$$\frac{1}{6}\frac{u_{i+1}^{n+1} - u_{i+1}^n}{\Delta t} + \frac{2}{3}\frac{u_i^{n+1} - u_i^n}{\Delta t} + \frac{1}{6}\frac{u_{i-1}^{n+1} - u_{i-1}^n}{\Delta t} = \frac{1}{\Delta x}(F_{i-1}^n - F_i^n) \quad (5.45)$$

Method 4 gives an upstream Galerkin method. Calculations are not satisfactory, however; the results are worse than those shown in Figure 4.13a, which were obtained by lumping the equations.

To summarize the results for the first four methods, the good ones all have $\alpha = 1$ and the time derivatives lumped. Thus, Method 1 (using Gaussian quadrature) is acceptable (Figure 5.6). Methods 2 and 3 are finite difference methods when $\alpha = 1$ and the time derivatives are lumped; these are the best methods. In all cases, the methods with lumped time derivatives are better than those with a Galerkin time derivative.

The next four methods are Taylor-Galerkin methods, which include the added term multiplied by Δt on the right-hand side of Eq. (5.28). They differ only in how this term is evaluated. Since the term is added to provide numerical dispersion, we do not include the Petrov-Galerkin method in these cases. The added term is

$$-\frac{\Delta t}{2}\int_0^1 \frac{dN_i}{dx}\left(\frac{dF}{du}\right)^2 \frac{\partial u}{\partial x} dx = -\frac{\Delta t}{2}\sum_e \frac{1}{\Delta x_e}\int_{-1}^1 \frac{dN_I}{d\xi}\left(\frac{dF}{du}\right)^2 \frac{\partial u}{\partial \xi} d\xi =$$

$$= -\frac{\Delta t}{2}\sum_e \frac{1}{\Delta x_e}\sum_J u_J \int_{-1}^1 \frac{dN_I}{d\xi}\left(\frac{dF}{du}\right)^2 \frac{dN_J}{d\xi} d\xi . \quad (5.46)$$

One option is to use Gaussian quadrature to evaluate the nonlinear integral.

$$-\frac{\Delta t}{2}\int_0^1 \frac{dN_i}{dx}\left(\frac{dF}{du}\right)^2 \frac{\partial u}{\partial x} dx = -\frac{\Delta t}{2}\sum_e \frac{1}{\Delta x_e}\sum_J u_J \sum_{iG=1}^{NG} WG_{IG}\left[\frac{dN_I}{d\xi}\left(\frac{dF}{du}\right)^2 \frac{dN_J}{d\xi}\right]_{\xi_{IG}} \quad (5.47)$$

This leads to

Method 5 - Taylor-Galerkin I.

$$\frac{1}{6}\frac{u_{i+1}^{n+1} - u_{i+1}^n}{\Delta t} + \frac{2}{3}\frac{u_i^{n+1} - u_i^n}{\Delta t} + \frac{1}{6}\frac{u_{i-1}^{n+1} - u_{i-1}^n}{\Delta t} =$$

$$= \frac{1}{\Delta x}\sum_e \sum_{IG=1}^{NG} WG_{IG}\ F[u^n(\xi_{IG})]\frac{dW_I}{d\xi}\bigg|_{\xi_{IG}}$$

$$-\frac{\Delta t}{2}\frac{1}{\Delta x^2}\sum_e \sum_J u_J^e \sum_{IG=1}^{NG} WG_{IG}\left[\frac{dN_I}{d\xi}\left(\frac{dF}{du}\right)^2 \frac{dN_J}{d\xi}\right]_{\xi_{IG}} \quad (5.48)$$

The success of Method 5 in practice is shown in Figure 5.7a, when the time derivative is not lumped, and in Figure 5.7b, when the time derivative is lumped. Clearly, lumping is desirable. Comparison with Figure 4.13b for the Lax-Wendroff finite difference method shows that there is not much difference. The inclusion of the extra terms in Eq. (5.48) is an improvement over the Galerkin method without those terms, since such a method is unstable. It is still not as good as the Petrov-Galerkin method, which gives the results shown in Figure 5.6.

Another option is to use trapezoid rule for the quadrature, which gives terms such as

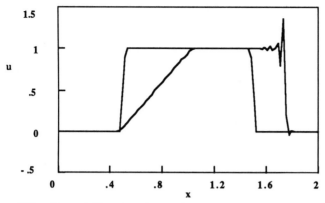

Figure 5.7a. Burger's Equation, Taylor-Galerkin Method-I, $\Delta t = 0.005$, 100 nodes

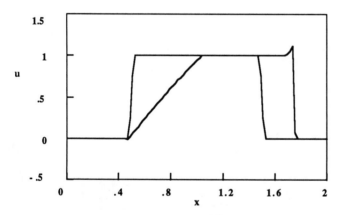

Figure 5.7b. Burger's Equation, Taylor-Galerkin Method-I, lumped $\Delta t = 0.02$, 100 nodes

$$\int_{-1}^{1} \frac{dN_I}{d\xi}\left(\frac{dF}{du}\right)^2 \frac{dN_J}{d\xi} d\xi = \left[\frac{dN_I}{d\xi}\left(\frac{dF}{du}\right)^2 \frac{dN_J}{d\xi}\bigg|_{\xi=1} + \frac{dN_I}{d\xi}\left(\frac{dF}{du}\right)^2 \frac{dN_J}{d\xi}\bigg|_{\xi=-1}\right]. \quad (5.49)$$

The full equation is then (see Problem 5.8)

Method 6 - Taylor-Galerkin II

$$\frac{u_i^{n+1} - u_i^n}{\Delta t} = \frac{\Delta t}{2\Delta x}[-F_{i+1}^n + F_{i-1}^n] +$$

$$+ \frac{\Delta t}{4\Delta x^2}\left\{\left[\left(\frac{dF}{du}\right)_{i+1}^2 + \left(\frac{dF}{du}\right)_i^2\right][u_{i+1}^n - u_i^n] - \left[\left(\frac{dF}{du}\right)_i^2 + \left(\frac{dF}{du}\right)_{i-1}^2\right][u_i^n - u_{i-1}^n]\right\}. \quad (5.50)$$

Results obtained with Method 6 are shown in Figure 5.7c. This is the best Galerkin result so far: the solutions are better and the time-step is much larger. Furthermore, this is better than the finite difference version shown in Figures 4.13b and 4.13c.. Thus the use of the Taylor terms in the Galerkin method is an

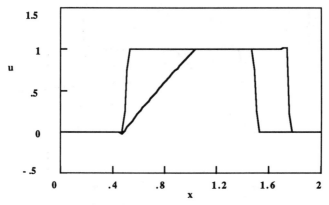

Figure 5.7c. Burger's Equation, Taylor-Galerkin Method-II, 100 nodes, $\Delta t = 0.02$

improvement.

If we interpolate the nonlinear function,

$$\left(\frac{dF}{du}\right)^2 \approx \sum_K F2_K N_K, \tag{5.51}$$

we have terms such as

$$-\frac{\Delta t}{2}\int_0^1 \frac{dN_i}{dx}\left(\frac{dF}{du}\right)^2 \frac{\partial u}{\partial x} dx = -\frac{\Delta t}{2}\sum_e \frac{1}{\Delta x_e}\int_{-1}^1 \frac{dN_I}{d\xi}\left(\frac{dF}{du}\right)^2 \frac{\partial u}{\partial \xi} d\xi =$$

$$= -\frac{\Delta t}{2}\sum_e \frac{2}{\Delta x_e}\sum_J\sum_K u_J F2_K \int_{-1}^1 N_K \frac{dN_I}{d\xi}\frac{dN_J}{d\xi} d\xi . \tag{5.52}$$

The method using these terms gives the following equation (see Problem 5.9):

Method 7 - Taylor-Galerkin III

$$\frac{1}{6}\frac{u_{i+1}^{n+1} - u_{i+1}^n}{\Delta t} + \frac{2}{3}\frac{u_i^{n+1} - u_i^n}{\Delta t} + \frac{1}{6}\frac{u_{i-1}^{n+1} - u_{i-1}^n}{\Delta t} = \frac{1}{2\Delta x}[-F_{i+1}^n + F_{i-1}^n] +$$

$$+ \frac{\Delta t}{4\Delta x^2}\left\{\left[\left(\frac{dF}{du}\right)_{i+1}^2 + \left(\frac{dF}{du}\right)_i^2\right][u_{i+1}^n - u_i^n] - \left[\left(\frac{dF}{du}\right)_i^2 + \left(\frac{dF}{du}\right)_{i-1}^2\right][u_i^n - u_{i-1}^n]\right\}. \tag{5.53}$$

The last equation is

$$\frac{1}{3}\frac{u_{NT}^{n+1} - u_{NT}^n}{\Delta t} + \frac{1}{6}\frac{u_{NT-1}^{n+1} - u_{NT-1}^n}{\Delta t} = \frac{1}{2\Delta x}[F(u_{NT}) + F(u_{NT-1})] - \frac{F(u_{NT})}{\Delta x} +$$

$$+ \frac{\Delta t}{4\Delta x^2}\left[\left(\frac{dF}{du}\right)_{NT}^2 + \left(\frac{dF}{du}\right)_{NT-1}^2\right][-u_{NT}^n + u_{NT-1}^n]. \tag{5.54}$$

The Taylor-Galerkin III, or Method 7, can be regarded as the finite element

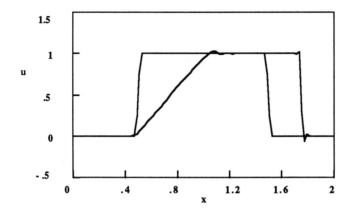

Figure 5.7d. Burger's Equation, Taylor-Galerkin-III, $\Delta t = 0.01$, 100 nodes

extension of the finite difference expression, Eq. (4.52), which is the analog of the Lax-Wendroff method. It might also be called a one-step finite element Lax-Wendroff method. Results obtained with this method are shown in Figure 5.7d. The Taylor-Galerkin III method also shows an improvement over the finite difference counterpart. This method is a straight Galerkin method (all terms are handled in the direct Galerkin fashion) and provides a quite good solution.

For the final method, we simply evaluate the nonlinear term by using its value at the upstream edge of the element. Upstream is defined as where dF/du is largest. If dF/du increases when x decreases we get

$$-\frac{\Delta t}{2}\int_0^1 \frac{dN_i}{dx}\left(\frac{dF}{du}\right)^2 \frac{\partial u}{\partial x}\,dx = -\frac{\Delta t}{2}\sum_e \frac{1}{\Delta x_e}\int_{-1}^1 \frac{dN_I}{d\xi}\left(\frac{dF}{du}\right)^2_{\xi=-1} \frac{\partial u}{\partial \xi}\,d\xi k. \quad (5.55)$$

The result of Problem 5.10 is:

Method 8 - Taylor-Galerkin IV

$$\frac{1}{6}\frac{u_{i+1}^{n+1}-u_{i+1}^n}{\Delta t} + \frac{2}{3}\frac{u_i^{n+1}-u_i^n}{\Delta t} + \frac{1}{6}\frac{u_{i-1}^{n+1}-u_{i-1}^n}{\Delta t} =$$

$$= \frac{1}{\Delta x}\sum_e \sum_{IG=1}^{NG} WG_{IG}\, F[u^n(\xi_{IG})]\left.\frac{dW_I}{d\xi}\right|_{\xi_{IG}} +$$

$$+ \frac{\Delta t}{4\Delta x}\left[\left(\frac{dF}{du}\right)^2_i (u_{i+1}^n - u_i^n) - \left(\frac{dF}{du}\right)^2_{i-1}(u_i^n - u_{i-1}^n)\right]$$

(5.56)

Method 8, Taylor-Galerkin IV, is the Galerkin analog to the MacCormack method, since the nonlinear term is handled in exactly the same way as in MacCormack's finite difference method. The only difference in the final result is that the time derivatives are combined on the left-hand side for the Taylor-

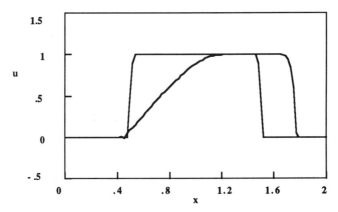

Figure 5.7e. Burger's Equation, Taylor-Galerkin-IV, $\Delta t = 0.005$, 100 nodes

Galerkin method, whereas they are mass lumped for the finite difference MacCormack method. Results obtained with this method are shown in Figure 5.7e. The results do not show oscillations. They are, however, smoothed more than desired. [The results shown in Figure 5.7e use Eq. (5.56) as written, despite the fact that the "upstream" direction behind the front is in the other direction. This presumably accounts for some of the excessive smoothing.] Compared with the MacCormack finite difference results shown in Figure 4.13c, the Galerkin results are too smooth.

Of the Petrov-Galerkin methods (i.e., Methods 1-4 without the extra terms), only a few were good, and these had the time derivative lumped. Even the best of these was inferior to finite difference methods. When the extra terms were included, all the Galerkin methods (Methods 5-8) were good, although the one with finite element interpolation was best. When Galerkin methods are compared with finite difference methods, though, finite difference methods (either MacCormack or explicit, upstream) are preferred. However, the use of these extra terms does provide a viable Galerkin method, with the best one being the Taylor-Galerkin-III, or Method 7. This is the method that is used throughout the rest of the book as the Taylor-Galerkin method.

5.4 Finite Element Methods for the Convective Diffusion Equation

Methods for the convective diffusion equation are a simple extension to those for the advection equation. The equations have been derived before [see Eq. (2.105) and Table 5.2]. In contrast to the advection equation, however, the problem is stable without using the Petrov-Galerkin method. The stability diagram for the Galerkin method is shown in Figure A.14. We can note the much lower r value (1/6) than in finite difference methods (1/2). Generally the same features are displayed as in Figure A.2 for the centered finite difference method, except that both r and Co must be smaller in the Galerkin method. For the Galerkin method at low r values, the wavelengths with $\xi = \pi$ ($\lambda = 2\Delta x$) are

Table 5.2. Galerkin Methods for the Convection-Diffusion Equation

Explicit, Petrov-Galerkin

$$\left(\frac{1}{6} - \frac{\alpha}{4}\right)(c_{i+1}^{n+1} - c_{i+1}^{n}) + \frac{2}{3}(c_{i}^{n+1} - c_{i}^{n}) + \left(\frac{1}{6} + \frac{\alpha}{4}\right)(c_{i-1}^{n+1} - c_{i-1}^{n}) =$$

$$-\frac{Pe\,\Delta t}{2\Delta x}[(1-\alpha)c_{i+1}^{n} + 2\alpha c_{i}^{n} - (1+\alpha)c_{i-1}^{n}] + \frac{\Delta t}{\Delta x^2}[c_{i+1}^{n} - 2c_{i}^{n} + c_{i-1}^{n}]$$

Taylor-Galerkin (including Taylor-Galerkin-I, III, and IV)

$$\frac{1}{6}(c_{i+1}^{n+1} - c_{i+1}^{n}) + \frac{2}{3}(c_{i}^{n+1} - c_{i}^{n}) + \frac{1}{6}(c_{i-1}^{n+1} - c_{i-1}^{n}) =$$

$$-\frac{Pe\,\Delta t}{2\Delta x}(c_{i+1}^{n} - c_{i-1}^{n}) + \left(\frac{\Delta t}{\Delta x^2} + \frac{Pe^2\Delta t^2}{2\Delta x^2}\right)(c_{i+1}^{n} - 2c_{i}^{n} + c_{i-1}^{n})$$

Implicit, Galerkin

$$\frac{1}{6}(c_{i+1}^{n+1} - c_{i+1}^{n}) + \frac{2}{3}(c_{i}^{n+1} - c_{i}^{n}) + \frac{1}{6}(c_{i-1}^{n+1} - c_{i-1}^{n}) =$$

$$-\frac{Pe\,\Delta t}{2\Delta x}[1-\theta][c_{i+1}^{n} - c_{i-1}^{n}] - \frac{Pe\,\Delta t}{2\Delta x}\theta[c_{i+1}^{n+1} - c_{i-1}^{n+1}]$$

$$+ \frac{\Delta t}{\Delta x^2}[1-\theta][c_{i+1}^{n} - 2c_{i}^{n} + c_{i-1}^{n}] + \frac{\Delta t}{\Delta x^2}\theta[c_{i+1}^{n+1} - 2c_{i}^{n+1} + c_{i-1}^{n+1}]$$

Implicit, Petrov-Galerkin

$$\left(\frac{1}{6} - \frac{\alpha}{4}\right)(c_{i+1}^{n+1} - c_{i+1}^{n}) + \left(\frac{2}{3}\right)(c_{i}^{n+1} - c_{i}^{n}) + \left(\frac{1}{6} + \frac{\alpha}{4}\right)(c_{i-1}^{n+1} - c_{i-1}^{n}) =$$

$$-\frac{Pe\,\Delta t}{2\Delta x}[1-\theta][(1-\alpha)c_{i+1}^{n} + 2\alpha c_{i}^{n} - (1+\alpha)c_{i-1}^{n}] - \frac{Pe\,\Delta t}{2\Delta x}\theta[(1-\alpha)c_{i+1}^{n+1} + 2\alpha c_{i}^{n+1} - (1+\alpha)c_{i-1}^{n+1}]$$

$$+ \frac{\Delta t}{\Delta x^2}[1-\theta][c_{i+1}^{n} - 2c_{i}^{n} + c_{i-1}^{n}] + \frac{\Delta t}{\Delta x^2}\theta[c_{i+1}^{n+1} - 2c_{i}^{n+1} + c_{i-1}^{n+1}]$$

Implicit, Taylor-Galerkin

$$\frac{1}{6}(c_{i+1}^{n+1} - c_{i+1}^{n}) + \frac{2}{3}(c_{i}^{n+1} - c_{i}^{n}) + \frac{1}{6}(c_{i-1}^{n+1} - c_{i-1}^{n}) = -\frac{Pe\,\Delta t}{2\Delta x}[1-\theta][c_{i+1}^{n} - c_{i-1}^{n}] - \frac{Pe\,\Delta t}{2\Delta x}\theta[c_{i+1}^{n+1} - c_{i-1}^{n+1}]$$

$$+ \frac{\Delta t}{\Delta x^2}[(1-\theta)(c_{i+1}^{n} - 2c_{i}^{n} + c_{i-1}^{n}) + \theta(c_{i+1}^{n+1} - 2c_{i}^{n+1} + c_{i-1}^{n+1})] + \alpha\frac{Pe^2\Delta t^2}{2\Delta x^2}(c_{i+1}^{n} - 2c_{i}^{n} + c_{i-1}^{n})$$

---

damped more than the other wavelengths, but as r increases, these become the crucial wavelengths (for Co = 0). Looking at r = 1/6, when Co = √(1/3) ≈ 0.5773, we get a straight line; thus, there is no dispersion. This is a "magic" Courant number for the Galerkin method and is the analog to the unit CFL property of finite difference methods. A concentration front is translated without change, even though there is diffusion (and we expect some smoothing). However, if our

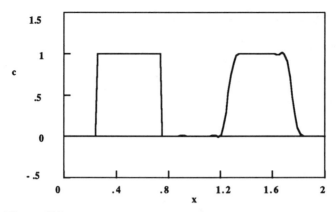

Figure 5.8a. Convective Diffusion Equation, Galerkin Method, Pe = 1000
Co = 0.05, 100 nodes

objective is to retain the shape of a concentration profile, then Co = $\sqrt{(1/3)}$ would be a good choice. We note that

$$\frac{Co}{2r} = \frac{Pe \, \Delta x}{2}. \tag{5.57}$$

Thus r = 1/6 and Co = $\sqrt{(1/3)}$ correspond to Pe Δx /2 = 1.73 and should give excellent results. For the sample calculation with Pe = 1000, if we use 100 nodes, then Pe Δx/2 = 5 and Co = 10r. When r = 0.1, Co = 1 and the method is unstable; when r = 0.05, Co = 0.5 and the method is also unstable. For this Δx and Pe combination we need even smaller r values (hence Δt). Convection of the square wave is illustrated in Figure 5.8a, for Co = 0.05 and r = 0.0025 (i.e., very small values). However, the solution is excellent. Use of a Δt that is twice as large gives unacceptable results. Figure 5.8a for the Galerkin method can be compared with Figure 4.16b for the centered finite difference method at the same Co, r parameters. The finite difference result is unacceptable. The solution for a step change in the concentration is shown in Figure 2.23 for the Galerkin method. For this case Co = 0.05, and the solution is reasonably good.

The results for the Petrov-Galerkin method ($\alpha = 0.5$) are shown in Figure A.15. The stability region (Co, r) is even smaller than for the Galerkin method. Results obtained with the square wave are shown in Figure 5.8b and for the step change in Figure 2.26. The oscillations are slightly larger when α is increased and they become worse if Δt is increased. Similar results are achieved with $\alpha = 1$ (see Figures A.16, 5.8c, and 2.28). In fact, the oscillations are larger with larger α values; this is also demonstrated in the magnitude of ρ values, which increase with α when Co is large. The Petrov-Galerkin method does not improve on the Galerkin method, at least for these parameters (Pe Δx/2 = 10).

The stability results for the Taylor-Galerkin method are shown in Figure A.17. When the function F(u) is linear in u (as it is for the convective diffusion equation), the Taylor-Galerkin-III and Taylor-Galerkin-IV methods are the same.

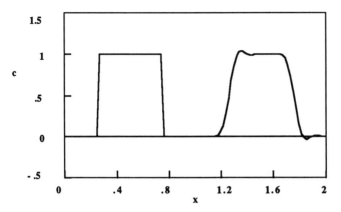

Figure 5.8b. Convective Diffusion Equation, Petrov-Galerkin Method
Pe = 1000, α = 0.5, Co = 0.05, 100 nodes

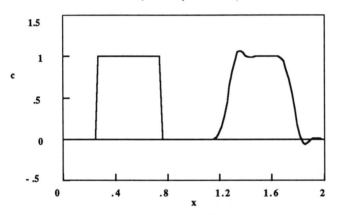

Figure 5.8c. Convective Diffusion Equation, Petrov-Galerkin Method
Pe = 1000, α = 1, Co = 0.05, 100 nodes

The Taylor-Galerkin amplification factors are also large (for large Co), as in the Petrov-Galerkin method, but they have a large range of Co values for which the method is stable. For r = 0.1 and Co = 0.35, the magnitude of ρ is relatively flat (it falls to a minimum of about 0.6 for ξ = 2); this corresponds to Pe Δx/2 = 1.75. Sample calculations are shown in Figure 5.8d for the square wave when Co = 0.5 (near the limit of stability). When the Courant number is reduced to 0.25, we get the results shown in Figure 5.8e, which are excellent. We can compare these with Figures 4.16c and 4.16d for two finite difference methods, the Taylor-Galerkin method gives much better results. To see if this is because of the smaller Δt value, we repeat the calculations with MacCormack method and the same Δt; the results shown in Figure 5.8f are not much better than those in Figures 4.16c and 4.16d. Thus, we can conclude that the Taylor-Galerkin method is a much better method than the MacCormack finite difference method. By comparing Figure 5.8e with Figure 5.8a, we see that the Taylor-Galerkin method gets results as good as the

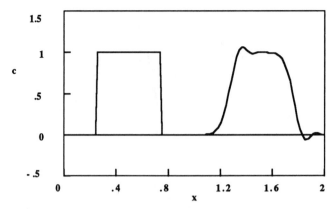

Figure 5.8d. Convective Diffusion Equation, Taylor-Galerkin Method
Pe = 1000, Co = 0.5, 100 nodes

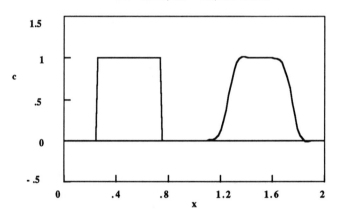

Figure 5.8e. Convective Diffusion Equation, Taylor-Galerkin Method
Pe = 1000, Co = 0.25, 100 nodes

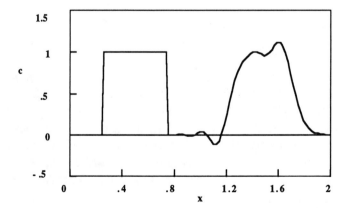

Figure 5.8f. Convective Diffusion Equation, MacCormack Method
Pe = 1000, Co = 0.25, 100 nodes

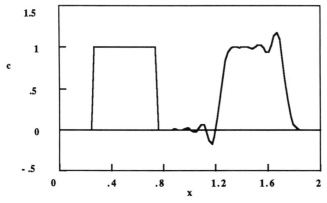

Figure 5.9a. Advection Equation, Morton-Parrott Method, Co = 0.5, 100 nodes

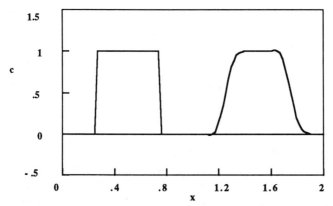

Figure 5.9b. Convective Diffusion Equation, Morton-Parrott Method
Pe = 1000, Co = 0.5, 100 nodes

Galerkin method with a time-step five times larger.

There is apparently no analog to the DuFort-Frankel method using the Galerkin method. Thus, the leapfrog-Galerkin option is not available for the convective diffusion equation.

The Morton-Parrott method is developed in Section 6.3 as a Galerkin method (or Petrov-Galerkin method) using a special weighting function. It is constructed to obey the unit CFL property. The method gives the stability results shown in Figure A.18. It is stable for large Courant numbers, as long as r < 0.16. Calculations with the Morton-Parrott method are shown in Figure 5.9b for the convective diffusion equation. The results are similar to those from the Taylor-Galerkin method (Figure 5.8e) and better than those from the Galerkin method (Figure 5.8a), even with a time-step ten times larger. The Morton-Parrott results are also better than the Petrov-Galerkin results, Figures 5.8b and 5.8c. Next we consider the advection equation with 100 nodes. The results shown in Figure 5.9a show some oscillations for the Morton-Parrott method, but the results are better

116 NUMERICAL METHODS - MOVING FRONTS

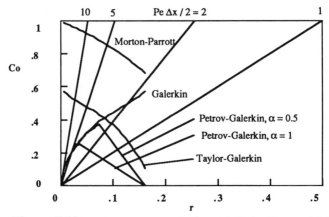

Figure 5.10. Stability Diagram for Explicit Finite Element Methods

than those with the Galerkin method (unstable) or the Petrov-Galerkin method (Figure 5.5b). The Morton-Parrott method also allows a time-step ten times larger than the Petrov-Galerkin method. Results can also be compared with results from the Taylor-Galerkin method (Figure 5.5d). The Morton-Parrott method is better than the Petrov-Galerkin method and comparable to the Taylor-Galerkin method.

The stability region in Co-r space is shown in Figure 5.10 for all the methods. The lines of constant Pe $\Delta x/2$ are also shown. For a fixed grid (i.e., Pe $\Delta x/2$) as Δt increases, we follow the line from the origin. It is clear that if Pe $\Delta x/2$ is large (e.g. 10) then the largest Co (Δt) value can be used with the Morton-Parrott method and the next largest with the Taylor-Galerkin method; the Petrov-Galerkin method requires the smallest time-step. The Galerkin method has very tight stability restrictions when there is little diffusion and Pe$\Delta x/2$ is large. For smaller Pe $\Delta x/2$ values (e.g. 2) the Galerkin method is relatively better and allows the same Δt (or larger) as the Taylor-Galerkin method or the Petrov-Galerkin method. For Pe $\Delta x/2 = 2$, the Galerkin method is stable at Co = 0.48 (and unstable at 0.52) while the Taylor-Galerkin method is stable at Co = 0.36 (and unstable at 0.40). The Petrov-Galerkin method is especially limiting, since for $\alpha = 1$ it is stable at Co = 0.2 and unstable at Co = 0.24. If the criterion determining a method is the largest step-size (Δt), then it is clear that the Galerkin method would not be the method of choice if there is little or no diffusion; either the Morton-Parrott method or Taylor-Galerkin method would be preferred. If there is diffusion, however, the Morton-Parrott method is still preferred but the Galerkin method is better than the Taylor-Galerkin method. In all cases, the Petrov-Galerkin method is a distant last, except when compared with the Galerkin method at very small r values.

Let us consider the implicit methods listed in Table 5.2, but limit consideration to $\theta = 0.5$ (the trapezoid rule) and $\theta = 1.0$ (the backward Euler method). The **implicit, Galerkin method** has stability diagrams shown in Figure

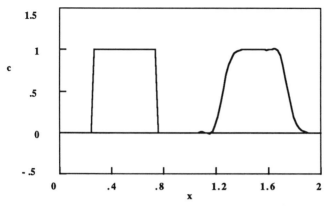

Figure 5.11a. Convective Diffusion Equation, Galerkin Method, Implicit
Pe = 1000, θ = 0.5, Co = 0.5, 100 nodes

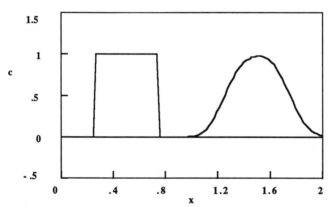

Figure 5.11b. Convective Diffusion Equation, Galerkin Method, Implicit
Pe = 1000, θ = 1, Co = 0.5, 100 nodes

A.19 for the trapezoid rule and A.20 for the backward Euler method in time; results for the explicit Euler method are shown in Figure A.14. The effect of the implicit treatment is strongly stabilizing, since all the implicit methods are stable for Co ≤ 1 and r ≤ 0.5. The trapezoid rule gives the results shown in Figure 5.11a, which compare favorably with the explicit results shown in Figure 5.8a. The results are very similar, but the trapezoid rule can use a time-step ten times larger (thus reducing the computation time by an order of magnitude). The backward Euler method is strongly dissipative and gives results that are smoothed too much (see Figure 5.11b). Thus, the trapezoid rule, along with the Galerkin method, gives a viable method. If we compare the implicit, Galerkin method (θ = 0.5, Figure 5.11a) with the explicit, Taylor-Galerkin method (Figure 5.8e) we see that they give comparable results, with small oscillations in different places.

Next we consider the **implicit, Petrov-Galerkin method**. The stability diagram for the explicit Petrov-Galerkin method is shown in Figure A.16, and

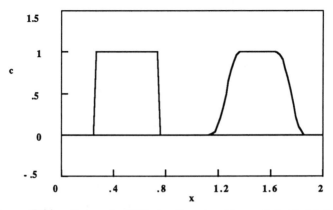

Figure 5.11c. Convective Diffusion Equation, Petrov-Galerkin Method, Implicit
Pe = 1000, $\alpha = 1$, $\theta = 0.5$, Co = 0.5, 100 nodes

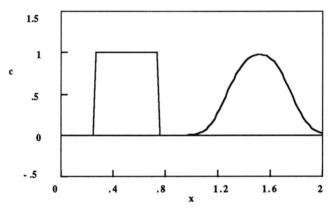

Figure 5.11d. Convective Diffusion Equation, Petrov-Galerkin Method, Implicit
Pe = 1000, $\alpha = 1$, $\theta = 1$, Co = 0.5, 100 nodes

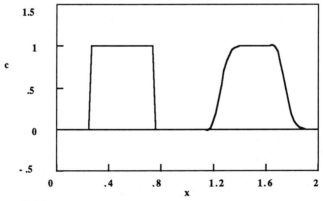

Figure 5.11e. Convective Diffusion Equation, Taylor-Galerkin Method, Implicit
Pe = 1000, $\alpha = 0$, $\theta = 0.5$, Co = 0.5, 100 nodes

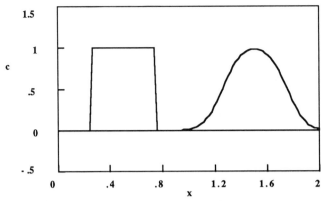

Figure 5.11f. Convective Diffusion Equation, Taylor-Galerkin Method, Implicit Pe = 1000, $\alpha = 1$, $\theta = 0.5$, Co = 0.5, 100 nodes

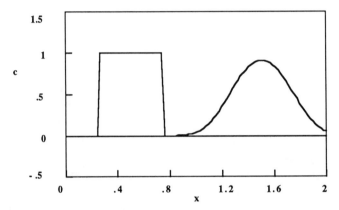

Figure 5.11g. Convective Diffusion Equation, Taylor-Galerkin Method, Implicit Pe = 1000, $\alpha = 1$, $\theta = 1$, Co = 0.5, 100 nodes

those for the trapezoid rule and backward Euler method are shown in Figures A.20 and A.21, respectively. Results using explicit integration are shown in Figure 5.8b, while results for the trapezoid rule and the backward Euler method are shown in Figures 5.11c and 5.11d, respectively. The trapezoid rule gives results which are slightly better than those using explicit integration, with a time-step five times larger. The backward Euler method has too much dispersion and the fronts are smoothed excessively.

Finally, we consider the **implicit, Taylor-Galerkin method**. First we show calculations when the method is made to be second-order ($\theta = 0.5$, $\alpha = 0$). Figure 5.11e shows results that are as good as the explicit Taylor-Galerkin results (Figure 5.8e). Results for the case $\theta = 1$ and $\alpha = -1$ are not satisfactory. Thus we do not automatically choose to make the methods second-order. The stability diagram for the explicit Euler method, trapezoid rule, and backward Euler method when combined with the Taylor-Galerkin method ($\alpha = 1$) are shown in Figures A.23, A.24, and A.25, respectively. (Figure A.17 for the explicit Taylor-Galerkin method is for smaller values of r.) The implicit nature is strongly dissipative provided Co < 0.5, and the stability diagrams show significant variation with ξ. This is reflected in the results shown in Figure 5.11f (for the trapezoid rule) and Figure 5.11g (for the backward Euler method). Based on these results, the implicit Taylor-Galerkin method with $\theta = 0.5$ and $\alpha = 0$ is as good a method as the explicit Taylor-Galerkin method. We should note that when $\alpha = 0$, the method is a Galerkin method, not a Taylor-Galerkin method. The explicit Galerkin method gives results that are almost as good, but the time-step is five to ten times smaller. Thus, the best of these methods are the explicit Taylor-Galerkin method and the implicit ($\theta = 0.5$) Galerkin method.

5.5. Finite Element Methods for Burger's Equation with Viscosity

These finite element methods for problems with viscosity involve a simple extension to the equations given in Section 5.3. The complete equations are listed in Table 5.3 for Burger's equation; here we present solutions with $v = 0.001$. Figure 5.12a is a solution for the Galerkin method; it shows serious oscillations. Figure 5.12b shows the results from the Petrov-Galerkin method; these oscillate, too, but less so. If the time derivatives are lumped in the Petrov-Galerkin method, the solution improves (See Figure 5.12c).

The Galerkin methods with extra terms added are demonstrated in Figures 5.12d-g. Figure 5.12d is for the Taylor-Galerkin-I method and the solution has oscillations; it is almost identical to the solution when using the MacCormack finite difference method (Figure 4.19c). Figure 5.12e shows the Taylor-Galerkin-II method using trapezoid quadrature. This method gives results that are a slight improvement over those obtained with Taylor-Galerkin-I. Figure 5.12f is for the Taylor-Galerkin-III method using a finite element representation of the nonlinear terms. This method also gives good results. The Taylor-Galerkin-IV method, illustrated in Figure 5.12g, shows that the method suffers from too much added dispersion. The Galerkin feature does, however, eliminate the oscillations apparent in the MacCormack finite difference method (Figure 4.19c). Implicit Galerkin methods and the Morton-Parrott method have not been applied to Burger's equation.

The conclusions about the best method for Burger's equation with viscosity are similar to those for Burger's equation without viscosity. The best method is the Taylor-Galerkin-III method (with finite element interpolation of the nonlinear terms). This method gives results very similar to the MacCormack finite difference method, except that the oscillations are before the front instead of after the front.. The other Taylor-Galerkin methods are almost as good.

5.6 Conclusions

For linear problems, such as the advection equation or the convective

Table 5.3. Galerkin Methods for Burger's Equation

Explicit, Petrov-Galerkin

$$\left(\frac{1}{6}-\frac{\alpha}{4}\right)\frac{u_{i+1}^{n+1}-u_{i+1}^{n}}{\Delta t}+\frac{2}{3}\frac{u_{i}^{n+1}-u_{i}^{n}}{\Delta t}+\left(\frac{1}{6}+\frac{\alpha}{4}\right)\frac{u_{i-1}^{n+1}-u_{i-1}^{n}}{\Delta t}=$$

$$=\frac{1}{\Delta x}\sum_{e}\sum_{IG=1}^{NG} WG_{IG}\, F[u^n(\xi_{IG})]\left.\frac{dW_I}{d\xi}\right|_{\xi_{IG}}+\frac{\nu}{\Delta x^2}(u_{i+1}^n-2u_i^n+u_{i-1}^n)$$

For the last element

$$\left(\frac{1}{3}+\frac{\alpha}{4}\right)\frac{u_{NT}^{n+1}-u_{NT}^{n}}{\Delta t}+\left(\frac{1}{6}+\frac{\alpha}{4}\right)\frac{u_{NT-1}^{n+1}-u_{NT-1}^{n}}{\Delta t}=$$

$$=-\frac{F(u_{NT}^n)}{\Delta x}+\frac{1}{\Delta x}\sum_{IG=1}^{NG} WG_{IG}\, F[u^n(\xi_{IG})]\left.\frac{dW_I}{d\xi}\right|_{\xi_{IG}}+\frac{\nu}{\Delta x^2}(u_{NT-1}^n-u_{NT}^n)$$

Taylor-Galerkin (also called Taylor-Galerkin-III in Chapter 5)

$$\frac{1}{6}\frac{u_{i+1}^{n+1}-u_{i+1}^{n}}{\Delta t}+\frac{2}{3}\frac{u_{i}^{n+1}-u_{i}^{n}}{\Delta t}+\frac{1}{6}\frac{u_{i-1}^{n+1}-u_{i-1}^{n}}{\Delta t}=\frac{1}{2\Delta x}[-F_{i+1}^n+F_{i-1}^n]+$$

$$+\frac{\Delta t}{4\Delta x^2}\left\{\left[\left(\frac{dF}{du}\right)_{i+1}^2+\left(\frac{dF}{du}\right)_{i}^2\right][u_{i+1}^n-u_i^n]-\left[\left(\frac{dF}{du}\right)_{i}^2+\left(\frac{dF}{du}\right)_{i-1}^2\right][u_i^n-u_{i-1}^n]\right\}$$

$$+\frac{\nu}{\Delta x^2}(u_{i+1}^n-2u_i^n+u_{i-1}^n)$$

For the last element

$$\frac{1}{3}\frac{u_{NT}^{n+1}-u_{NT}^{n}}{\Delta t}+\frac{1}{6}\frac{u_{NT-1}^{n+1}-u_{NT-1}^{n}}{\Delta t}=\frac{1}{2\Delta x}[F(u_{NT})+F(u_{NT-1})]-\frac{F(u_{NT})}{\Delta x}+$$

$$+\frac{\Delta t}{4\Delta x^2}\left[\left(\frac{dF}{du}\right)_{NT}^2+\left(\frac{dF}{du}\right)_{NT-1}^2\right][-u_{NT}^n+u_{NT-1}^n]+\frac{\nu}{\Delta x^2}(u_{NT-1}^n-u_{NT}^n)$$

---

diffusion equation, we found the Taylor-Galerkin method to be excellent. It usually allowed a larger step-size than was allowed by the standard Galerkin method. The Morton-Parrott method and the implicit Galerkin method ($\theta = 0.5$) gave results that were nearly equivalent: the oscillations were the same magnitude and the time-steps were larger than for the other good methods. For nonlinear problems, there are several Taylor-Galerkin methods, depending on how the quadratures were done; these are better than the standard Petrov-Galerkin method or the standard Taylor-Galerkin method I. The best method is with a finite element representation of the nonlinear terms and exact quadrature of the resulting

122 NUMERICAL METHODS - MOVING FRONTS

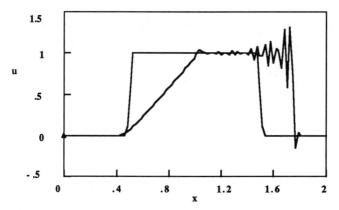

Figure 5.12a. Burger's Equation, $\nu = 0.001$, Galerkin Method, $\Delta t = 0.004$, 100 nodes

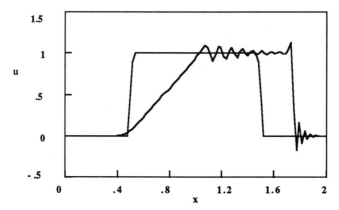

Figure 5.12b. Burger's Equation, $\nu = 0.001$, Petrov-Galerkin Method
$\alpha = 1$, $\Delta t = 0.005$, 100 nodes

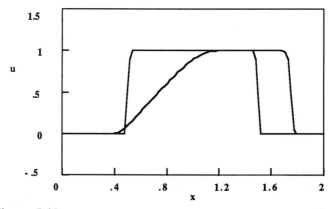

Figure 5.12c. Burger's Equation, $\nu = 0.001$, Petrov-Galerkin Method, Lumped
$\Delta t = 0.01$, 100 nodes

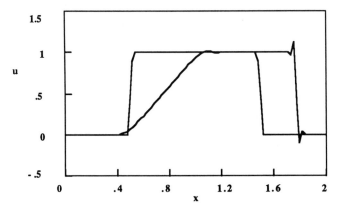

Figure 5.12d. Burger's Equation, $\nu = 0.001$, Taylor-Galerkin Method-I
$\Delta t = 0.01$, 100 nodes

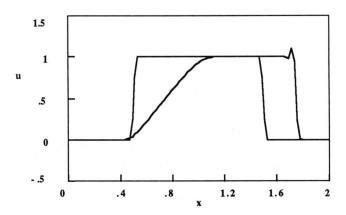

Figure 5.12e. Burger's Equation, $\nu = 0.001$, Taylor-Galerkin Method-II
$\Delta t = 0.01$, 100 nodes

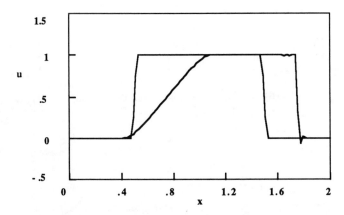

Figure 5.12f. Burger's Equation, $\nu = 0.001$, Taylor-Galerkin Method-III
$\Delta t = 0.005$, 100 nodes

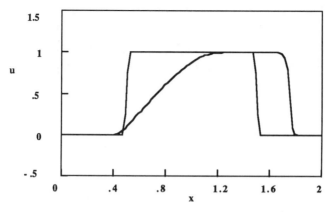

Figure 5.12g. Burger's Equation, $v = 0.001$, Taylor-Galerkin Method-IV
$\Delta t = 0.005$, 100 nodes

equations.

For linear problems, the Taylor-Galerkin method is very good and dramatically better than the finite difference methods. For nonlinear problems, the Taylor-Galerkin-III is a good method, but it is not clearly better than the MacCormack finite difference method. The differences between these two methods are small.

For a method that works for both the linear and nonlinear cases, the Taylor-Galerkin-III, Method 7, is preferred. Throughout the rest of the book, this will be referred to as simply the Taylor-Galerkin method.

Problems

1_1. Derive Eq. (5.7).

2_2. Derive Eq. (5.11).

3_1. Derive Eq. (5.12-5.13).

4_3. Evaluate Eq. (5.11) for the Petrov-Galerkin method when there is no diffusion ($r = 0$) and ξ is small. When the Courant number is positive, verify that the amplification factor has a magnitude greater than 1.0. [Hint: expand for small Co and small ξ].

5_2. Use the program CONVECT to solve the advection equation when the initial condition is a step change in the solution. Use the same methods used in Figures 5.5a-g. Discuss the results.

6_2. Derive Eq. (5.39) from Eq. (5.30) and Eq. (5.35); apply the trapezoid rule for integration. [Hint: the Petrov-Galerkin addition to the weighting function is zero at the end points of the interval.]

7_3. Derive Eq. (5.43) [Hint: for problems 5.6-5.10 you need to carefully evaluate the terms, put them into matrix notation, and then combine the matrix terms appropriately.].

8_3. Derive Eq. (5.50).

9_3. Derive Eq. (5.53).

10_3. Derive Eq. (5.56).

11_3. Use the program CONVECT to solve Burger's equation without viscosity when the initial condition is the ramp function defined in Problem 4.14. Use the same methods illustrated in Figure 5.12. Discuss the results.

12_3. Use the program CONVECT to solve the convective diffusion equation when the Peclet number is 1000 and the step-size is chosen to give the "magic" Courant number. Use the initial condition shown in Figure 5.8 and a step change in the solution.

13_3. Use the program CONVECT to solve the convective diffusion equation when the initial condition is a step change in the solution and the Peclet number is 1000. Use the same methods illustrated in Figure 5.8. Discuss the results.

14_3. Use the program CONVECT to solve Burger's equation with viscosity ($v = 0.001$) when the initial condition is the ramp function defined in Problem 4.14. Use the same methods illustrated in Figure 5.12. Discuss the results.

15_2. Use the program CONVECT to solve Burger's equation with viscosity ($v = 0.001$). Investigate one of the following questions:
    (1) Is lumping the time derivative a good idea?
    (2) Is the trapezoid rule for integration better than Gaussian quadrature?
    (3) Is adding Taylor terms to a finite difference method a good idea?
    (4) Is adding the Taylor terms to the Galerkin method a good idea?
    (5) Is the Petrov-Galerkin method an improvement over the Galerkin method?

16_2. For the advection equation, which is the best finite difference method and the best finite element method? Answer the same question for the convective diffusion equation, Burger's equation without viscosity, and Burger's equation with viscosity. Cite figures from the book or show other ones that you have obtained.

References

Baker, A. J. and Kim, J. W. "A Taylor Weak-statement Algorithm for Hyperbolic Conservation Laws," Int. J. Num. Methods Engn. 7 489-520 [1987].

Chung, T. J., Kim, Y. M., Sohn, J. L., "Finite Element Analysis in Combustion Phenomena," Int. J. Num. Methods Fluids 7 989-1012 [1987].

Donea, J., "A Taylor-Galerkin Method for Convective Transport Problems," Int. J. Num. Methods Eng. 20 101-119 [1984].

Donea, J., Giuliani, S., Laval, H., Quartapelle, L., "Time-Accurate Solution of the Advection-Diffusion Problems by Finite Elements," Comp. Meth. Appl. Mech. Eng. 45 123-145 [1984].

Löhner, R., Morgan, K., and Zienkiewicz, O. C., "An Adaptive Finite Element Procedure for Compressible High Speed Flows," Comp. Methods. Appl. Mech. Eng. 51 441-465 [1985].

Oden, J. T., Strouboulis, T., Devloo, P., "Adaptive Finite Element Methods for the Analysis of Inviscid Compressible Flow: Part 1. Fast Refinement/Unrefinement and Moving Mesh Methods for Unstructured Meshes." Comp. Meth. Appl. Mech. Eng. 59 327-362 [1986].

CHAPTER
SIX

SPECIALIZED TECHNIQUES

This chapter begins with a summary of the best methods presented in Chapters 2-5. We have discussed linear and nonlinear problems and problems with and without viscosity or diffusivity for a variety of initial conditions. For linear problems, the concentration distribution is convected along; for nonlinear problems, some solutions are self-sharpening while others are spreading.

Figures that have been presented previously will be shown first in reduced size: Figure 6.1 is for linear problems and Figure 6.2 is for nonlinear problems. In each figure, finite difference methods are on the left, finite element methods are on the right, problems without diffusion or viscosity are at the top, and problems with diffusion and viscosity are at the bottom. Shown here are only the best methods of those discussed in Chapters 2-5. In cases where the results are similar, only one graph is shown but the names of several methods are applied to it.

For the linear problems shown in Figure 6.1, the explicit, upstream finite difference method smooths the solution too much. The filtered leapfrog method gives an acceptable solution with only small oscillations. When there is no viscosity or diffusivity, the best method is the Taylor-Galerkin method. With viscosity or diffusivity, the best method is either the Taylor-Galerkin method or the Galerkin method. The implicit Petrov-Galerkin method, Morton-Parrott method, and implicit Taylor-Galerkin method give comparable results. In summary, the leapfrog method and all these finite element methods provide good solutions with steep fronts but without excessive oscillations.

For nonlinear problems, the solutions to Burger's equation are shown in Figure 6.2. If the function u decreases with increasing x, shocks are formed; if the function u increases with increasing x, the solution is spread out; if the function u equals zero at any point, it does not move at all. The finite difference methods give quite good results. The explicit, upstream method is good, especially when the front is self-sharpening, such as in a shock. MacCormack's method gives solutions with only small oscillations. The Taylor-Galerkin III method uses a finite element representation of the added dispersion terms. It gives acceptable results for problems both with and without viscosity. When there is

SPECIALIZED TECHNIQUES 127

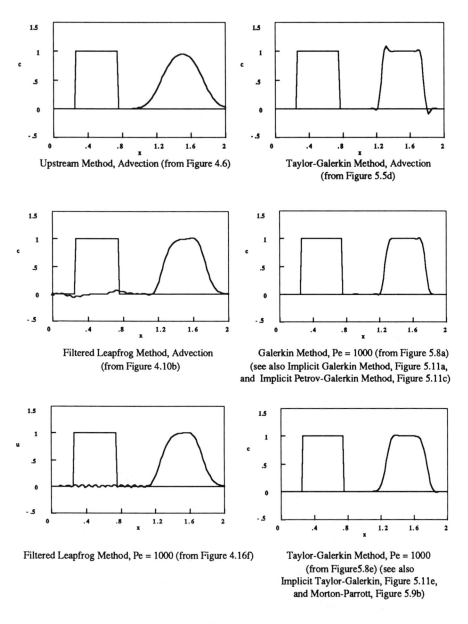

Figure 6.1. Best Methods for the Advection Equation and the Convective Diffusion Equation, 100 nodes

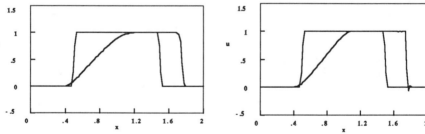

Figure 6.2. Best Methods for Burger's Equation, 100 nodes

viscosity, the lumped Petrov-Galerkin finite element method is also acceptable. These finite element methods give solutions that are just as good (i.e., with only small oscillations) as those provided by the MacCormack finite difference method. Thus, for Burger's equation we have several methods that we can use: the explicit upstream finite difference method, MacCormack's finite difference method, the Taylor-Galerkin finite element method, and (if there is viscosity) the Galerkin method.

In summary, we have methods that do not smear the solution excessively and are economical to run, but which cause small oscillations in the solutions. In an effort to eliminate even the small oscillations, special methods have been designed. Several of these special methods will be presented in this chapter. The random choice method is a method that solves for the location of the front on a fixed grid so that grid distortion is not a problem. The front remains as sharp as the resolution allows. The flux-corrected transport method introduces some dispersion to dampen the oscillations and then removes enough dispersion to keep the front sharp without oscillations. It can be applied as a post-processing step to any method and is applied here to both finite difference methods and finite element methods. A rational explanation of the Petrov-Galerkin method as applied to steady-state problems will be given. Several special methods are based on higher-order expressions for the convection term: upstream convection terms that are second-order, the QUICK method with a third-order upstream convection term, and special methods devised by both Laumbach and Rosenberg. Finally, the orthogonal collocation method is applied on finite elements, giving a high-order representation to the convection terms. In this chapter the methods listed above and their application to one-dimensional problems will be described.

6.1. Random Choice Method

The random choice method developed from an initial paper by Gudunov [1959], as reported by Holt [1984]. In this paper, Gudunov wanted to develop a method that was simple, like the method of characteristics, yet avoided the grid distortion that can be caused by convected meshes. Since the objective is to solve problems dominated by convection, it must be possible to include discontinuities in the solution. The version presented here is modified, using the ideas attributed by Holt [1984] to Glimm [1955], Chorin [1976, 1977], Concus and Proskurowski [1979], Sod [1978], and Collella [1982]. The article by Sod has a complete description of the method as applied to gas dynamics equations. The method is sometimes referred to as Glimm's method because of his development of the method.

Let us consider the purely hyperbolic problem with no dissipation:

$$\frac{\partial u}{\partial t} + \frac{\partial F(u)}{\partial x} = 0. \tag{6.1}$$

By using $F = u^2/2$, we get Burger's equation without viscosity while by using $F = u$, we get the advection equation. If we solved this problem using the method of characteristics, we would put it in the form

$$\frac{\partial u}{\partial t} + \frac{dF}{du}\frac{\partial u}{\partial x} = 0 \qquad (6.2)$$

and solve the following equations:

$$\frac{dx}{ds} = \frac{dF}{du}, \quad \frac{dt}{ds} = 1, \text{ and } \frac{du}{ds} = 0. \qquad (6.3)$$

Thus we can take $t = s$ and solve

$$\frac{dx}{dt} = \frac{dF(u(x,t))}{du}, \quad \frac{du}{dt} = 0 \qquad (6.4)$$

along with the initial conditions

$$u = u_1, x = x_1 \text{ at } t = 0. \qquad (6.5)$$

The solution tells us the location (x,t) of a solution whose value was u_1 at the point $(x = x_1, t = 0)$. The difficulty with this form of solution is that the positions keep moving and grids (in two dimensions) quickly become distorted. The random choice method was developed as a way to find the location on a fixed grid.

We define a Riemann problem as the differential equation, Eq. (6.1), combined with the initial conditions

$$u = \begin{cases} u_0 & x < x_1 + \dfrac{\Delta x}{2} \quad \text{at } t = t_1 \\ \\ u_1 & x \geq x_1 + \dfrac{\Delta x}{2} \quad \text{at } t = t_1. \end{cases} \qquad (6.6)$$

Thus, the initial conditions have a discontinuity, which is to be propagated in time. We assume that this problem is soluble and we call its solution $v(x,t)$. In order to keep the notation clean, we will not include the values of x_1, t_1, u_0, or u_1, but we must remember that any $v(x,t)$ has to be defined in terms of a discontinuity at some location at a specific time.

In the course of solving the problem, let us suppose the solution at time t^n is given at a set of points, x_i, as shown in Figure 6.3. Define the following Riemann problem, Eq. (6.1) with the initial conditions

$$u = \begin{cases} u_i^n & x < x_i + \dfrac{\Delta x}{2} \quad \text{at } t = t_1 \\ \\ u_{i+1}^n & x \geq x_i + \dfrac{\Delta x}{2} \quad \text{at } t = t_1. \end{cases} \qquad (6.7)$$

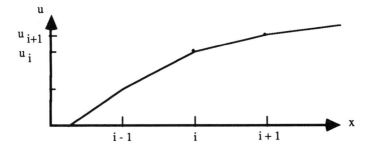

Figure 6.3. Solution at Time t^n

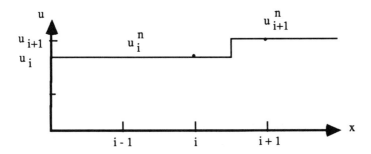

Figure 6.4. Initial Conditions for the Riemann Problem

The initial conditions then look as shown in Figure 6.4. We solve Eq. (6.2) from time t^n to time $t^{n+1/2}$ (subject to these initial conditions), calling the solution $v(x,t)$. We take θ_n as a random variable with a uniform probability distribution from -0.5 to $+0.5$. Then we look at the point P ($x = x_{i+1/2} + \theta_n \Delta x$, $t = t^n + \Delta t/2$). We next assign $u_{i+1/2}^{n+1/2}$ as the Riemann solution at the point P:

$$u_{i+1/2}^{n+1/2} = v\left(x_{i+1/2} + \theta_n \Delta x, \, t^n + \frac{\Delta t}{2}\right). \tag{6.8}$$

Thus u (at the i+1/2 point and n+1/2 time) is set equal to the solution of the Riemann problem at point P, where the Riemann problem is defined with initial conditions based on a discontinuity at the point i+1/2 and time n. The point $x = x_{i+1/2} + \theta_n \Delta x$, $t = t^{n+1/2}$ is a randomly selected point somewhere in the interval shown in Figure 6.5. We take the point P and project the solution back to the time level n and use the Riemann solution to find the value of the solution.

$$u_{i+1/2}^{n+1/2} = \begin{cases} u_i^n & \theta_n \Delta x < \text{speed } \Delta t/2 \\ \\ u_{i+1}^n & \theta_n \Delta x \geq \text{speed } \Delta t/2 \end{cases} \tag{6.9}$$

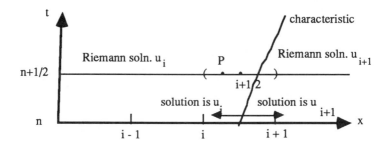

Figure 6.5. Movement of Discontinuity in the Random Choice Method

We do this for every point i; then we repeat the process to go from the staggered grid (i+1/2) at time (n+1/2) to the regular grid (i) and the new time (n+1). For simple advection this means that in one half-time-step the front moves from x to a point $x_c \pm \theta_n \Delta x$, where x_c is the new position of x as predicted by the method of characteristics. The point is in the correct place on average over many time-steps. The front moves along in jerky steps, due to the random choice of θ_n. However, the front is sharp and remains sharp. To see this, we follow the Riemann solution discontinuity from point $x_i + \Delta x/2$ at time t^n to time $t^{n+1/2}$ by following the characteristic through the point $x_{i+1/2}$. If the random point P is to the left of this point, the solution is taken as u_i, whereas if it is to the right of that point, it is taken as u_{i+1} (see Figure 6.5).

For the next half-step, the same procedure is followed, as illustrated in Figure 6.6. Now, however, the point Q surrounds a point that we must take as x_i. The effect of this is to offset the grid $\Delta x/2$ at the first half-time-step and $-\Delta x/2$ at the second half-time-step.

$$u_i^{n+1} = v(x_{i+1} + \theta_n \Delta x, t^n + \Delta t) \qquad (6.10)$$

For Godunov's method we would evaluate the Riemann solution at the discontinuity. Then the discontinuity might move with a speed that on average is incorrect. The improvement provided by Glimm was to use a random point on either side of the discontinuity so that the average speed of the front is correct.

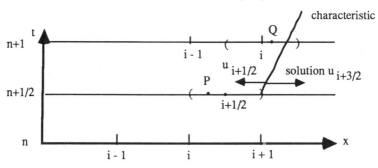

Figure 6.6. Movement Over One Complete Time-step

The choice of θ_n is also important. Chorin [1976] suggests choosing only one θ_n per time level (i.e., we only choose θ_n and use the same θ_n value for all values of i when going from the n to the n+1/2 time level). Sod recommends using the same θ_n for both half-steps. To make the front move with the right average velocity, the choices of θ_n must be a randomly distributed variable between –0.5 to 0.5 with a uniform distribution [Holt, 1984]. Collella [1982] provides a more elaborate choice using what is called the van der Corput sequence. We tried both the uniform distribution and the van der Corput sequence and found that the van der Corput sequence worked much better. In the van der Corput sequence, if we compute for 2N half-steps before displaying the solution again, we choose the θ_i values as follows for each of the 2N half-steps:

$$\theta_i = a^{(i)} - \frac{1}{2}, \quad i = 1,\ldots,2N. \tag{6.11}$$

The functions $a^{(i)}$ are obtained from the series

$$a^{(i)} = \sum_{k=0}^{m} \frac{i_k}{2^{k+1}}. \tag{6.12}$$

The i_k are the numbers 0 or 1, obtained from the binary representation of the number i:

$$i = \sum_{k=0}^{m} i_k 2^k, \quad i_k = 0 \text{ or } 1. \tag{6.13}$$

The value of m is as large as is necessary. Thus for i = 1, the binary representation is 1, so that m = 0 and i_0 = 1. For i = 2, the binary representation is 10, so m = 1, i_0 = 0, and i_1 = 1. For i = 3, the binary representation is 11, so m = 1, i_0 = 1, and i_1 = 1. This gives in Eq. (6.12) $a^{(1)}$ = 0.5, $a^{(2)}$ = 0.25, $a^{(3)}$ = 0.75, $a^{(4)}$ = 0.125, $a^{(5)}$ = 0.625, $a^{(6)}$ = 0.375, $a^{(7)}$ = 0.875, and $a^{(8)}$ = 0.0625. This gives in turn θ_i = 0, –0.25, 0.25, –0.375, 0.125, –0.125, 0.375, and –0.4375.

In order to use the random choice method, we must be able to solve the Riemann problem exactly. For whatever problem we solve, we must be able to solve for the convection of a step change in the initial data. Furthermore, we must be able to do this a number of times for each time-step, since it is necessary for each grid point at each time-step. Thus the solution must be efficient, which usually means analytical. When the problem is the advection equation, this is easy.

$$v(x/t, v_L, v_R) = \begin{cases} v_L & x/t < Pe \\ v_R & x/t \geq Pe \end{cases} \tag{6.14}$$

The step change in initial data moves with velocity Pe. If $c(x,t^n) = f(x)$, then the solution to the Riemann problem is $v(x,t) = f(x - Pe\ t)$. When the problem is

Burger's equation, we use a solution of the linearized Burger's equation; this solution is described by Sod [1985] and is attributed to Roe [1981a] {see also Roe [1981b]}. We can solve this exactly for a shock:

$$v(x/t, v_L, v_R) = \begin{cases} v_L & x/t < (v_L + v_R)/2 \\ v_R & x/t \geq (v_L + v_R)/2 \end{cases} \quad \text{for } v_L \geq v_R \quad (6.15)$$

and approximately for a rarefaction wave:

$$v(x/t, v_L, v_R) = \begin{cases} v_L & x/t \leq v_L \\ \frac{v_L + v_R}{2} & v_L < x/t \leq v_R \\ v_R & x/t > v_R \end{cases} \quad \text{for } v_L < v_R. \quad (6.16)$$

The elements v_L and v_R are the initial values for the Riemann problem to the left and right of the discontinuity. These approximate solutions are acceptable in practice because they are only used to go from one time level to another. At the new time level, Burger's equation is made linear again [although the solution looks the same in Eq. (6.15)-(6.16)].

The use of the equality sign in Eqs. (6.14) and (6.15) deserves discussion since the literature is contradictory about it. For the advection equation and convective diffusion equation the best solutions were obtained with the equality as shown, compared with using the equality for v_L [Poulain, 1991]. However, for nonlinear problems comparable results were achieved with the equality sign used with either v_L or v_R.

When a problem has several variables, such as a gas dynamics equation problem, the random choice method can be applied to each variable in turn. When the equations are not in conservative form, modifications must be made. For example, Holt gives the following algorithm, based on initial work done by Sod [1977]. For the problem

$$\frac{\partial u}{\partial t} + \frac{\partial}{\partial r}[F(u)] = -W(u), \quad (6.17)$$

we first solve the problem

$$\frac{\partial u}{\partial t} + \frac{\partial}{\partial r}[F(u)] = 0 \quad (6.18)$$

for two half-steps, using the random choice method to obtain $u*_i^{n+1}$. Then we obtain u_i^{n+1} from

$$\frac{u_i^{n+1} - u*_i^{n+1}}{\Delta t} = -W(u*_i^{n+1}). \quad (6.19)$$

We repeat the steps using Eq. (6.18)-(6.19) but use the initial data u_i^{n+1}. Other variations are also suggested by Holt [1984]. Li and Holt [1981] solve Eq. (6.18)

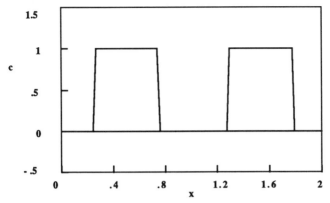

Figure 6.7. Advection Equation, Random Choice Method, $\Delta x = 0.02$, Co = 0.5

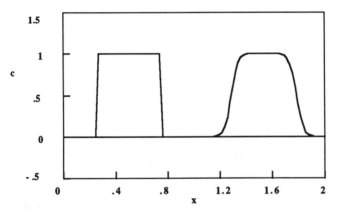

Figure 6.8. Convective Diffusion Equation, Random Choice Method
Pe = 1000, $\Delta x = 0.02$, $\Delta t = 1.\text{E-}5$

for one half-step, then Eq. (6.19) for a second half-step, and then repeat the process for another half-step. Such approaches can handle diffusion terms and other nonconvective terms. In this method we are essentially using operator splitting, in that we solve two subproblems — one subproblem with convection and one subproblem without convection — and then combine the results to get the final solution. Each of these subproblems can be solved using methods ideally suited to it.

For the advection problem, we solve for the convection term using Eq. (6.14) and call the solution $c*_i^{n+1}$. Then for the convective diffusion equation we add the step

$$\frac{c_i^{n+1} - c*_i^{n+1}}{\Delta t} = \frac{c*_{i+1}^{n+1} - 2\,c*_i^{n+1} + c*_{i-1}^{n+1}}{\Delta x^2}. \tag{6.20}$$

For Burger's equation we also solve for the convection term using Eq. (6.15)-

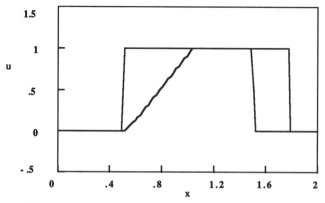

Figure 6.9. Burger's Equation, Random Choice Method, $v = 0$, $\Delta x = 0.02$, $\Delta t = 0.02$

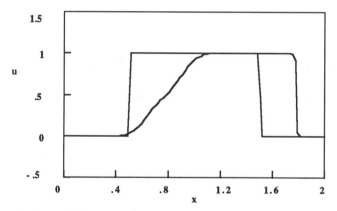

Figure 6.10. Burger's Equation, Random Choice Method, $v = 0.001$ $\Delta x = 0.02$, $\Delta t = 0.02$

(6.16). Then if viscosity is non-zero, we add the step

$$\frac{u_i^{n+1} - u*_i^{n+1}}{\Delta t} = v \frac{u*_{i+1}^{n+1} - 2u*_i^{n+1} + u*_{i-1}^{n+1}}{\Delta x^2}. \qquad (6.21)$$

The results of applying the random choice method to the advection equation are shown in Figure 6.7. The fronts essentially are as sharp as the mesh refinement allows them to be, although the location of the front is slightly off. When diffusion is added, the results are as shown in Figure 6.8; here the fronts are still sharp but have rounded edges. These are clearly the best solutions of the advection equation and convective diffusion equation. Some results of applying the random choice method to Burger's equation without viscosity are given by Glimm, *et al.* [1980]. The method described here gives results as shown in Figure 6.9. In areas where the front is not steep, there are slight wiggles in it. When viscosity is added, we get results as shown in Figure 6.10. Here some of the wiggles are mitigated. These results are the best solutions of Burger's equation that have been presented so far.

When applying the random choice method, it is still necessary to maintain the Courant number (Co) as less than or equal to 1.0. Furthermore, it should be clear that the best results will be obtained if several time-steps occur before examining the solution, so the random choice of θ_i can be averaged out; this keeps the front in the correct location.

6.2. Flux-corrected Transport

The flux-corrected transport method is a way of keeping the front sharp on a fixed grid; a flux-correction step can be added to any method. The method employs an anti-diffusion step to overcome the effects of excessive smearing. The method was devised by Boris and Book [1973] while modern versions are given by Book [1981] and Sod [1978]. The first treatment given here follows work by Sod [1978] and Book, *et al* [1975]. The form of the method devised by Zalesak [1979] is then given; it is easier to extend to two-dimensional problems and finite element methods. A version of flux-correction that applies to finite element methods is also given; still another finite element version is given in Chapter 8. The idea of flux-correction is closely related to the idea of total variation non-increasing methods, and TVI methods are discussed in the next section.

Let us consider the problem

$$\frac{\partial u}{\partial t} + \frac{\partial}{\partial x}\left(\frac{1}{2}u^2\right) = 0, \tag{6.22}$$

which is solved using the MacCormack method.

$$u_i^{*n+1} = u_i^n - \frac{\Delta t}{2\Delta x}\left[(u_{i+1}^n)^2 - (u_i^n)^2\right] \tag{6.23}$$

$$\tilde{u}_i^{n+1} = \frac{1}{2}(u_i^n + u_i^{*n+1}) - \frac{\Delta t}{4\Delta x}\left[(u_i^{*n+1})^2 - (u_{i-1}^{*n+1})^2\right] \tag{6.24}$$

We apply the basic method, as shown, but obtain an intermediate solution as denoted by the tilde overbar. We define the intermediate quantity

$$\hat{\Delta}_{i+1/2} = \eta\,(\tilde{u}_{i+1}^{n+1} - \tilde{u}_i^{n+1}), \tag{6.25}$$

which will indicate the direction of the flux according to the sign of Eq. (6.25). The parameter η is adjustable, but is often taken to be 1/8. We then compute a further intermediate solution, which adds extraneous diffusion

$$\hat{u}_i^{n+1} = \tilde{u}_i^{n+1} + \eta\,(u_{i+1}^n - 2\,u_i^n + u_{i-1}^n). \tag{6.26}$$

The intermediate quantity

$$\Delta_{i+1/2} = \hat{u}_{i+1}^{n+1} - \hat{u}_i^{n+1} \tag{6.27}$$

is based on the new variable. Finally, the flux is corrected according to the prescription

$$f^c_{i+1/2} = \text{sign}(\hat{\Delta}_{i+1/2}) \max \left\{ \begin{array}{l} 0 \\ \min \left\{ \begin{array}{l} \text{sign}(\hat{\Delta}_{i+1/2}) \Delta_{i-1/2} \\ |\hat{\Delta}_{i+1/2}| \\ \text{sign}(\Delta_{i+1/2}) \Delta_{i+3/2} \end{array} \right. \end{array} \right. \qquad (6.28)$$

and the final solution is

$$u_i^{n+1} = \hat{u}_i^{n+1} - (f^c_{i+1/2} - f^c_{i-1/2}). \qquad (6.29)$$

The interpretation of these steps is that Eq. (6.26) adds diffusion, which helps dampen any oscillations that might develop, while Eq. (6.29) removes some of the diffusion, but only enough to sharpen the front; the fluxes are limited so they will not aggravate any oscillations that may be part of the true solution. As presented here, the method does have a parameter, η, that can be adjusted to improve the results in special cases. Here we use only $\eta = 1/8$, which is the number usually recommended [Boris and Book, 1973]. In this form, the flux-correction step can be applied to any method, not only the MacCormack method.

Sod [1978] explains the method as follows. The basic method is written in the form of Eq. (4.41):

$$\frac{\partial u}{\partial t} + \frac{\partial F(u)}{\partial x} = \Delta t \frac{\partial}{\partial x} \left[g\left(u, \frac{\Delta t}{\Delta x}\right) \frac{\partial u}{\partial x} \right]. \qquad (6.30)$$

The modification given above is represented by

$$\frac{\partial u}{\partial t} + \frac{\partial F(u)}{\partial x} = \Delta t \frac{\partial}{\partial x} \left\{ \left[g\left(u, \frac{\Delta t}{\Delta x}\right) - r\left(u, \frac{\Delta t}{\Delta x}\right) \right] \frac{\partial u}{\partial x} \right\}, \qquad (6.31)$$

where r is a positive function that gives the "anti-diffusion" term. When

$$g\left(u, \frac{\Delta t}{\Delta x}\right) - r\left(u, \frac{\Delta t}{\Delta x}\right) \geq 0, \qquad (6.32)$$

the combined scheme is stable, and Eq. (6.32) presents a limitation on $\Delta t/\Delta x$ that is more stringent than the limitation imposed by the basic method. Thus, the acceptable step-size must sometimes be found by trial and error.

Results obtained using the MacCormack method with flux-correction, as applied to the advection equation [F(u) = u], are shown in Figure 6.11. The program CONVECT was used with the Mac-FCT option; the Courant number was 0.5. The results demonstrate great improvement over the straight MacCormack method (see Figures 4.9a and 4.9b). The profiles are not quite as steep as in the random choice method but are excellent nonetheless. When diffusivity is added, the results are as shown in Figure 6.12. These results are much improved over the

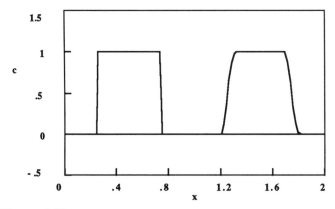

Figure 6.11. Advection Equation, MacCormack Method with Flux-correction
$\Delta x = 0.02$, $\Delta t = 0.01$

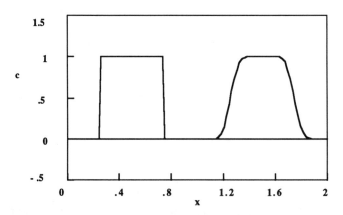

Figure 6.12. Convective Diffusion Equation, MacCormack Method with Flux-correction, Pe = 1000, $\Delta x = 0.02$, $\Delta t = 0.01$

straight MacCormack method (see Figure 4.16d) because they have no oscillations. Thus for linear problems, the flux-corrected method is extremely good.

For a nonlinear problem, Burger's equation, the MacCormack method with flux-correction gives results as shown in Figure 6.13 when $\Delta t = \Delta x/2$. These results are excellent. When viscosity is included, the results are just as good (see Figure 6.14). The flux-correction step gives results that are a slight improvement over the MacCormack method alone (see Figures 4.13c and 4.19c). The MacCormack method with flux-correction is an excellent method, giving steep profiles that have few or no oscillations.

Next let us consider an alternative formulation of the flux-correction step. This is the form used for problems in two dimensions by Zalesak [1979] and for finite elements [Löhner, et al., 1987]. We consider the equation

$$\frac{\partial u}{\partial t} + \frac{\partial F(u)}{\partial x} = \nu \frac{\partial^2 u}{\partial x^2}. \qquad (6.33)$$

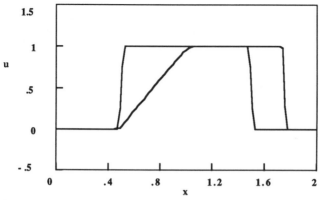

Figure 6.13. Burger's Equation, MacCormack Method with Flux-correction
$v = 0$, $\Delta x = 0.02$, $\Delta t = 0.01$

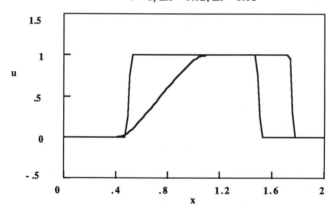

Figure 6.14. Burger's Equation, MacCormack Method with Flux-correction
$v = 0.001$, $\Delta x = 0.02$, $\Delta t = 0.01$

An upstream method would give us

$$\frac{u_i^{n+1} - u_i^n}{\Delta t} = -\frac{F_i^n - F_{i-1}^n}{\Delta x} + \frac{v}{\Delta x^2}(u_{i+1}^n - 2u_i^n + u_{i-1}^n). \tag{6.34}$$

This can be written as

$$u_i^{n+1} = u_i^n - \frac{\Delta t}{\Delta x}(F_i^n - F_{i-1}^n) + \frac{v \Delta t}{\Delta x^2}(u_{i+1}^n - 2u_i^n + u_{i-1}^n). \tag{6.35}$$

We use the upstream method as the low-order method since it does not oscillate:

$$u_i^L = u_i^n - [F_{i+1/2}^L - F_{i-1/2}^L], \quad \text{where } F_{i+1/2}^L = \frac{\Delta t}{\Delta x} F_i^n - \frac{v \Delta t}{\Delta x^2}(u_{i+1}^n - u_i^n). \tag{6.36}$$

For the last point we use

$$F_{NT+1/2}^L = \frac{\Delta t}{\Delta x} F_{NT}^n + \frac{v \Delta \tau}{\Delta x^2}(u_{NT}^n). \tag{6.37}$$

We present two different methods for the high-order method. A centered finite difference method gives

$$\frac{u_i^{n+1} - u_i^n}{\Delta t} = -\frac{F_{i+1}^n - F_{i-1}^n}{2\Delta x} + \frac{v}{\Delta x^2}(u_{i+1}^n - 2u_i^n + u_{i-1}^n). \tag{6.38}$$

This can be written as

$$u_i^{n+1} = u_i^n - \frac{\Delta t}{2\Delta x}(F_{i+1}^n - F_{i-1}^n) + \frac{v\Delta t}{\Delta x^2}(u_{i+1}^n - 2u_i^n + u_{i-1}^n). \tag{6.39}$$

We write the centered method in the form

$$u_i^H = u_i^n - [F_{i+1/2}^H - F_{i-1/2}^H], \tag{6.40}$$

where

$$F_{i+1/2}^H = \frac{\Delta t}{2\Delta x}[F_{i+1}^n + F_i^n] - \frac{v\Delta t}{\Delta x^2}(u_{i+1}^n - u_i^n). \tag{6.41}$$

For the last point we use a false solution outside the domain.

$$F_{NT+1/2}^H = \frac{\Delta t}{2\Delta x}(F_{NT+1}^n + F_{NT}^n) - \frac{v\Delta t}{\Delta x^2}(u_{NT+1}^n - u_{NT}^n) \tag{6.42}$$

Alternatively, we can use the MacCormack method as the high-order method. The MacCormack method is a two-step method, which needs to be put into the form of Eq. (6.40). To achieve this form, we combine the steps. First we have

$$u{*}_i^{n+1} = u_i^n - \frac{\Delta t}{\Delta x}(F_{i+1}^n - F_i^n) + \frac{v\Delta t}{\Delta x^2}(u_{i+1}^n - 2u_i^n + u_{i-1}^n). \tag{6.43}$$

The second step is

$$u_i^{n+1} = \frac{1}{2}(u_i^n + u{*}_i^{n+1}) - \frac{\Delta t}{2\Delta x}(F{*}_i^{n+1} - F{*}_{i-1}^{n+1}) + \frac{v\Delta t}{2\Delta x^2}(u{*}_{i+1}^{n+1} - 2u{*}_i^{n+1} + u{*}_{i-1}^{n+1}). \tag{6.44}$$

Combining these gives us

$$u_i^{n+1} = \frac{1}{2}\left[u_i^n + u_i^n - \frac{\Delta t}{\Delta x}(F_{i+1}^n - F_i^n) + \frac{v\Delta t}{\Delta x^2}(u_{i+1}^n - 2u_i^n + u_{i-1}^n)\right] \\ - \frac{\Delta t}{2\Delta x}(F{*}_i^{n+1} - F{*}_{i-1}^{n+1}) + \frac{v\Delta t}{2\Delta x^2}(u{*}_{i+1}^{n+1} - 2u{*}_i^{n+1} + u{*}_{i-1}^{n+1}). \tag{6.45}$$

Thus we use the following in Eq. (6.40):

$$F_{i+1/2}^H = \frac{\Delta t}{2\Delta x}(F_{i+1}^n + F{*}_i^{n+1}) - \frac{v\Delta t}{2\Delta x^2}(u_{i+1}^n - u_i^n + u{*}_{i+1}^{n+1} - u{*}_i^{n+1}). \tag{6.46}$$

142 NUMERICAL METHODS - MOVING FRONTS

The flux-correction method then involves the following six steps:

Step 1. We evaluate the low-order flows, F^L, in Eq. (6.36).

Step 2. We evaluate the high-order flows, F^H, in Eq. (6.46).

Step 3. We define the following "anti-diffusion" terms:

$$A_{i+1/2} = F^H_{i+1/2} - F^L_{i+1/2}. \qquad (6.47)$$

These terms represents the difference between flows evaluated with the high-order method and flows evaluated with the low-order method.

Step 4. We calculate the low-order solution using Eq. (6.36).

Step 5. We perform the flux-limiting step, using the following algorithm to find the anti-diffusion correction factors, $A^c_{i+1/2}$:

$$\begin{aligned} u^a_i &= \max(u^n_i, u^L_i) \\ u^{max}_i &= \max(u^a_{i-1}, u^a_{i+1}, u^a_i) \\ u^b_i &= \min(u^n_i, u^L_i) \\ u^{min}_i &= \min(u^b_{i-1}, u^b_{i+1}, u^b_i). \end{aligned} \qquad (6.48)$$

If the point does not exist (e.g., u_{NT+1}), we leave it out of these formulas. The algorithm is constructed to limit the u^{n+1}_i to be below u^{max}_i and above u^{min}_i.

$$\begin{aligned} P^+_i &= \max(0, A_{i-1/2}) - \min(0, A_{i+1/2}) \\ Q^+_i &= u^{max}_i - u^L_i \\ R^+_i &= \begin{cases} \min(1, Q^+_i / P^+_i) & \text{if } P^+_i > 0 \\ 0 & \text{if } P^+_i = 0 \end{cases} \end{aligned} \qquad (6.49)$$

$A_{i+1/2}$ is defined as the <u>difference</u> between flows with the high-order scheme and flows with the low-order scheme. To avoid oscillations, we want to limit that difference. The value of P^+_i is then taken as the maximum difference between flows that can be removed from node i without inducing oscillations. If $A_{i-1/2}$ is positive, it represents flow into node i and adds to the amount that can be removed. If $A_{i-1/2}$ is negative, it represents flow away from node i; in this case, it is replaced by 0 in the definition of P^+_i so that it will not cause the value of u to be above its desired maximum. The same considerations apply for $A_{i+1/2}$, except that a positive $A_{i+1/2}$ value represents flow away from node i. Q^+_i gives the maximum change of u that will keep the value of u below the desired maximum value. Similarly, we define

$$P_i^- = \max(0, A_{i+1/2}) - \min(0, A_{i-1/2})$$
$$Q_i^- = u_i^L - u_i^{min}$$
$$R_i^- = \begin{cases} \min(1, Q_i^-/P_i^-) & \text{if } P_i^- > 0 \\ 0 & \text{if } P_i^- = 0, \end{cases} \quad (6.50)$$

$$C_{i+1/2} = \begin{cases} \min(R_{i+1}^-, R_i^+) & A_{i+1/2} < 0 \\ \min(R_i^-, R_{i+1}^+) & A_{i+1/2} \geq 0, \end{cases} \quad (6.51)$$

$$A_{i+1/2}^c = C_{i+1/2} A_{i+1/2}, \quad 0 \leq C_{i+1/2} \leq 1. \quad (6.52)$$

If the correction factor involves points outside the domain, we leave them out of the formulas.

Step 6. We obtain the final solution as a correction or limitation to the low-order solution.

$$u_i^{n+1} = u_i^L - (A_{i+1/2}^c - A_{i-1/2}^c) \quad (6.53)$$

In the algorithm [Eq. (6.36)-(6.52)] we calculate the low-order solution, u^L, and add flux-limited flows in Eq. (6.53). If $C = 1$ for all values of i, then Eq. (6.53) is exactly the high-order solution. If $C = 0$ for all i, then Eq. (6.53) is exactly the low-order solution (see Problem 6.2). By limiting the value of C, we add just enough of the difference between the high-order solution and the low-order solution to keep the front sharp without inducing oscillations.

In the algorithm, Eq. (6.23)-(6.29), we use a high-order method to calculate $\tilde{u}$. We then add an arbitrary amount of diffusion so that the function $\hat{u}$ is monotone. Adding diffusion smooths the sharp profile. Finally, we use Eq. (6.29) to add some limited flows and sharpen the front without inducing oscillations.

We next consider simplifications for the convective diffusion equation. With Taylor terms added to the centered finite difference method, Eq. (6.41) becomes

$$F_{i+1/2}^H = \frac{Pe \, \Delta t}{2 \, \Delta x}(c_{i+1}^n + c_i^n) - \left(\frac{\Delta t}{\Delta x^2} + \frac{Pe^2 \Delta t^2}{2 \, \Delta x^2}\right)(c_{i+1}^n - c_i^n). \quad (6.54)$$

This is the form of the equations used in CONVECT when choosing the following options: (a) flux-correction, centered derivative and (b) lumped. For the MacCormack method we use the following equation in place of Eq. (6.43)-(6.46):

$$c_i^{*n+1} = c_i^n - \frac{Pe \, \Delta t}{\Delta x}(c_{i+1}^n - c_i^n) + \frac{\Delta t}{\Delta x^2}(c_{i+1}^n - 2c_i^n + c_{i-1}^n) \quad (6.55)$$

144 NUMERICAL METHODS - MOVING FRONTS

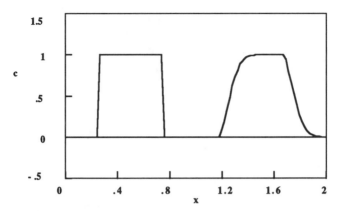

Figure 6.15a. Advection Equation, MacCormack Method with Second Flux-correction Version, $\Delta x = 0.02$, $\Delta t = 0.01$

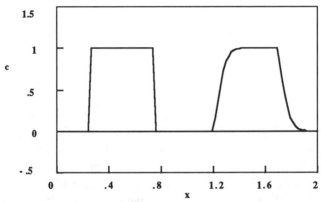

Figure 6.15b. Convective Diffusion Equation, MacCormack Method with Second Flux-correction Version, Pe = 1000, $\Delta x = 0.02$, $\Delta t = 0.01$

$$F^H_{i+1/2} = \frac{Pe\,\Delta t}{2\,\Delta x}(c^n_{i+1} + c^{*n+1}_i) - \frac{\Delta t}{2\,\Delta x^2}(c^n_{i+1} - c^n_i + c^{*n+1}_{i+1} - c^{*n+1}_i). \tag{6.56}$$

This is the method that is used in CONVECT under the flux-correction, MacCormack option. For the advection equation, these methods are the same (see Problem 6.4).

Results obtained using the flux-correction, MacCormack option are shown in Figures 6.15a and 6.15b. For the advection equation, the profile has no oscillations but does have extra smoothing at the corners. This smoothing is enhanced when diffusion is added, as shown in Figure 6.15b. This version of the flux-correction method, Eq. (6.36)-(6.53), is not as good the first one, Eq. (6.25)-(6.29). Results when the centered finite difference method is used as the high-order method are shown in Figure 6.16. Results for the advection equation are

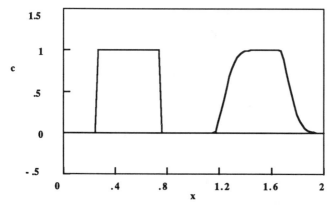

Figure 6.16. Convective Diffusion Equation, Centered Finite Difference Method with Flux-correction, Pe = 1000, $\Delta x = 0.02$, Co = 0.5

not shown since they are the same as displayed in Figure 6.15a. The Taylor finite difference flux-corrected method is also a reasonable method.

Next let us consider the flux-correction step when applied to a finite element method. Finite element methods use the form of the algorithm given in Eq. (6.36)-(6.53); the treatment here is based on work by Löhner, et al. [1987]. Because the flux-correction step is based on corrections to the element flows, rather than the nodal flows, it is necessary to carefully re-derive the following equations so as to identify the contributions to each element. The flux-correction step will be illustrated using the convective diffusion equation. The low-order method is a lumped, Petrov-Galerkin method (or upstream finite difference method), while the high-order method is a Taylor-Galerkin method.

The Petrov-Galerkin method is derived as described below. The contribution from the element to the left of node i is

$$\frac{\Delta x}{6}\frac{dc_{i-1}}{dt} + \frac{\Delta x}{3}\frac{dc_i}{dt} + \frac{\alpha \Delta x}{4}\left[\frac{dc_{i-1}}{dt} + \frac{dc_i}{dt}\right] = -\text{Pe}\left[\left(-\frac{1}{2}-\frac{\alpha}{2}\right)c_{i-1} + \left(\frac{1}{2}+\frac{\alpha}{2}\right)c_i\right]$$
$$+ \frac{1}{\Delta x}[-c_i + c_{i-1}], \quad (6.57)$$

while the contribution from the element to the right of node i is

$$\frac{\Delta x}{3}\frac{dc_i}{dt} + \frac{\Delta x}{6}\frac{dc_{i+1}}{dt} - \frac{\alpha \Delta x}{4}\left[\frac{dc_i}{dt} + \frac{dc_{i+1}}{dt}\right] = -\text{Pe}\left[\left(-\frac{1}{2}+\frac{\alpha}{2}\right)c_i + \left(\frac{1}{2}-\frac{\alpha}{2}\right)c_{i+1}\right]$$
$$+ \frac{1}{\Delta x}[c_{i+1} - c_i]. \quad (6.58)$$

Combined, these give

$$\left(\frac{1}{6}-\frac{\alpha}{4}\right)\frac{dc_{i+1}}{dt} + \frac{2}{3}\frac{dc_i}{dt} + \left(\frac{1}{6}+\frac{\alpha}{4}\right)\frac{dc_{i-1}}{dt} = -\frac{\text{Pe}}{2\Delta x}[(1-\alpha)c_{i+1} + 2\alpha c_i - (1+\alpha)c_{i-1}]$$
$$+ \frac{1}{\Delta x^2}[c_{i+1} - 2c_i + c_{i-1}]. \quad (6.59)$$

The low-order method is equivalent to Eq. (6.59) with $\alpha = 1$ and lumped time derivative terms. It is given here as

$$\frac{dc_i}{dt} = -\frac{Pe}{\Delta x}[c_i - c_{i-1}] + \frac{1}{\Delta x^2}[c_{i+1} - 2c_i + c_{i-1}]. \quad (6.60)$$

When a first-order method is used in time, we obtain

$$c_i^L = c_i^n - \frac{Pe\,\Delta t}{\Delta x}[c_i - c_{i-1}] + \frac{\Delta t}{\Delta x^2}[c_{i+1}^n - 2c_i^n + c_{i-1}^n]. \quad (6.61)$$

We define the low-order element contributions (LECs) as

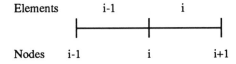

Figure 6.17. Element and Nodal Notation

$$\begin{aligned}
LEC_i^{i-1} &= \frac{Pe\,\Delta t}{\Delta x} c_{i-1}^n + \frac{\Delta t}{\Delta x^2}[-c_i^n + c_{i-1}^n] \\
LEC_i^i &= -\frac{Pe\,\Delta t}{\Delta x} c_i^n + \frac{\Delta t}{\Delta x^2}[c_{i+1}^n - c_i^n] \\
LEC_{NT}^{NT-1} &= \frac{Pe\,\Delta t}{\Delta x}(c_{NT-1}^n - c_{NT}^n).
\end{aligned} \quad (6.62)$$

The superscript is the element number and the subscript is the node number (see Figure 6.17). Eq. (6.61) can be rewritten as

$$c_i^L = c_i^n + \sum_e LEC_i^e = c_i^n + LEC_i^{i-1} + LEC_i^i. \quad (6.63)$$

Next let us consider the high-order method, a Taylor-Galerkin method. The finite element formulation is

$$C_{JI}^e \frac{dc_I^e}{dt} + Pe\, A_{JI}^e c_I^e = \left(1 + \frac{Pe^2 \Delta t}{2}\right) B_{JI}^e c_I^e. \quad (6.64)$$

The element matrices for the Galerkin method are given in Table 2.2. We consider the contributions to this equation from the element to the left of node i:

$$\frac{\Delta x}{6}\frac{dc_{i-1}}{dt} + \frac{\Delta x}{3}\frac{dc_i}{dt} = Pe\left(\frac{1}{2}c_{i-1} - \frac{1}{2}c_i\right) + \left(\frac{1}{\Delta x} + \frac{Pe^2 \Delta t}{2\Delta x}\right)(-c_i + c_{i-1}). \quad (6.65)$$

The contribution from the element to the right of node i is

$$\frac{\Delta x}{3}\frac{dc_i}{dt} + \frac{\Delta x}{6}\frac{dc_{i+1}}{dt} = \text{Pe}\left(\frac{1}{2}c_i - \frac{1}{2}c_{i+1}\right) + \left(\frac{1}{\Delta x} + \frac{\text{Pe}^2 \Delta t}{2\Delta x}\right)(c_{i+1} - c_i). \quad (6.66)$$

Combined, these give

$$\frac{1}{6}\frac{dc_{i+1}}{dt} + \frac{2}{3}\frac{dc_i}{dt} + \frac{1}{6}\frac{dc_{i-1}}{dt} = -\frac{\text{Pe}}{2\Delta x}(c_{i+1} - c_{i-1}) + \left(\frac{1}{\Delta x^2} + \frac{\text{Pe}^2 \Delta t}{2\Delta x^2}\right)(c_{i+1} - 2c_i + c_{i-1}). \quad (6.67)$$

When a first-order method is used for the time derivative, we get

$$\frac{1}{6}(c_{i+1}^H - c_{i+1}^n) + \frac{2}{3}(c_i^H - c_i^n) + \frac{1}{6}(c_{i-1}^H - c_{i-1}^n) = -\frac{\text{Pe}\,\Delta t}{2\Delta x}(c_{i+1}^n - c_{i-1}^n) +$$

$$+ \left(\frac{\Delta t}{\Delta x^2} + \frac{\text{Pe}^2 \Delta t^2}{2\Delta x^2}\right)(c_{i+1}^n - 2c_i^n + c_{i-1}^n). \quad (6.68)$$

We define the contributions to the i-th node from the (i-1)-th and i-th elements as

$$\text{RHS}_i^{i-1} = \frac{\text{Pe}\,\Delta t}{2\Delta x}(c_{i-1}^n + c_i^n) + \frac{\Delta t}{\Delta x^2}\left(1 + \frac{\text{Pe}^2 \Delta t}{2}\right)(-c_i^n + c_{i-1}^n)$$

$$\text{RHS}_i^i = -\frac{\text{Pe}\,\Delta t}{2\Delta x}(c_i^n + c_{i+1}^n) + \frac{\Delta t}{\Delta x^2}\left(1 + \frac{\text{Pe}^2 \Delta t}{2}\right)(c_{i+1}^n - c_i^n) \quad (6.69)$$

$$\text{RHS}_{NT}^{NT-1} = \frac{\text{Pe}\,\Delta t}{2\Delta x}(c_{NT-1}^n - c_{NT}^n) + \frac{\Delta t}{\Delta x^2}\left(1 + \frac{\text{Pe}^2 \Delta t}{2}\right)(-c_{NT}^n + c_{NT-1}^n).$$

The complete equation form of Eq. (6.67) can then be rewritten as

$$\frac{1}{6}(c_{i+1}^H - c_{i+1}^n) + \frac{2}{3}(c_i^H - c_i^n) + \frac{1}{6}(c_{i-1}^H - c_{i-1}^n) = \sum_e \text{RHS}_i^e = \text{RHS}_i^{i-1} + \text{RHS}_i^i. \quad (6.70)$$

This equation is also rewritten as

$$C_{ij}(c_j^H - c_j^n) = \text{RHS}_i^{i-1} + \text{RHS}_i^i, \quad (6.71)$$

where the matrix C_{ij} is given by

$$C_{ij} = \begin{cases} 1/6 & j = i \pm 1 \\ 2/3 & j = i \\ 0 & \text{otherwise.} \end{cases} \quad (6.72)$$

The lumped version of the matrix C_{ij} is

$$C_{ij}^L = \begin{cases} 1 & j = i, i \neq NT \\ 0.5 & j = i, i = NT \\ 0 & \text{otherwise.} \end{cases} \quad (6.73)$$

We can rewrite Eq. (6.71) as

$$C_{ij}^L (c_j^H - c_j^n) = (C_{ij}^L - C_{ij})(c_j^H - c_j^n) + RHS_i^{i-1} + RHS_i^i. \qquad (6.74)$$

When we compare flows, it is necessary to use the element contribution to the right-hand side of Eq. (6.74); i.e., we use the lumped version (since that is how the LECs are calculated). Thus we define

$$HEC_i^{i-1} = \frac{1}{6}(c_i^H - c_i^n) - \frac{1}{6}(c_{i-1}^H - c_{i-1}^n) + \frac{\Delta t\, Pe}{2\, \Delta x}(c_{i-1}^n + c_i^n) +$$

$$+ \frac{\Delta t}{\Delta x^2}\left(1 + \frac{Pe^2\, \Delta t}{2}\right)(-c_i^n + c_{i-1}^n)$$

$$\qquad (6.75)$$

$$HEC_i^i = \frac{1}{6}(c_i^H - c_i^n) - \frac{1}{6}(c_{i+1}^H - c_{i+1}^n) - \frac{\Delta t\, Pe}{2\, \Delta x}(c_i^n + c_{i+1}^n) +$$

$$+ \frac{\Delta t}{\Delta x^2}\left(1 + \frac{Pe^2\, \Delta t}{2}\right)(c_{i+1}^n - c_i^n).$$

For the last node, Eq. (6.74) becomes

$$\frac{1}{2}(c_{NT}^H - c_{NT}^n) = -\frac{1}{6}(c_{NT-1}^H - c_{NT-1}^n) + \frac{1}{6}(c_{NT}^H - c_{NT}^n) + RHS_{NT}^{NT-1} \qquad (6.76)$$

and we use

$$HEC_{NT}^{NT-1} = -\frac{1}{3}(c_{NT-1}^H - c_{NT-1}^n) + \frac{1}{3}(c_{NT}^H - c_{NT}^n) + 2\, RHS_{NT}^{NT-1}. \qquad (6.77)$$

The flux-corrected transport finite element method is then:

Step 1. We evaluate the low-order element contributions, LECs, in Eq. (6.62).

Step 2. We solve Eq. (6.70) for the high-order solution and evaluate the high-order element contributions, HECs, in Eq.(6.75).

Step 3. We define the folowing "anti-diffusion" terms for each node in each element:

$$AEC_i^e = HEC_i^e - LEC_i^e. \qquad (6.78)$$

Each such term represents the difference between contributions evaluated with the high-order method and contributions evaluated with the low-order method.

Step 4. We calculate the low-order solution using Eq. (6.63).

Step 5. We perform the flux-limiting step using the following algorithm to find the anti-diffusion correction factors, AEC_i^e.

$$c_i^a = \max(c_i^n, c_i^L)$$

$$c_e^* = \max_{\text{all nodes in element } e}(c_i^a)$$

$$c_i^{\max} = \max_{\text{all elements touching node } i}(c_e^*) \quad (6.79)$$

$$c_i^b = \min(c_i^n, c_i^L)$$

$$c_e^\dagger = \min_{\text{all nodes in element } e}(c_i^b)$$

$$c_i^{\min} = \min_{\text{all elements touching node } i}(c_e^\dagger) \quad (6.80)$$

If the point does not exist (e.g., c_{NT+1}), we leave it out of these formulas.

$$P_i^+ = \sum_e \max(0, AEC_i^e) = \max(0, AEC_i^{i-1}) + \max(0, AEC_i^i)$$

$$Q_i^+ = c_i^{\max} - c_i^L$$

$$R_i^+ = \begin{cases} \min(1, Q_i^+/P_i^+) & \text{if } P_i^+ > 0 \\ 0 & \text{if } P_i^+ = 0 \end{cases} \quad (6.81)$$

$$P_i^- = \sum_e \min(0, AEC_i^e) = \min(0, AEC_i^{i-1}) + \min(0, AEC_i^i)$$

$$Q_i^- = c_i^{\min} - c_i^L$$

$$R_i^- = \begin{cases} \min(1, Q_i^-/P_i^-) & \text{if } P_i^- < 0 \\ 0 & \text{if } P_i^- = 0 \end{cases} \quad (6.82)$$

$$a_1 = \begin{cases} R_i^+ & \text{if } AEC_i^i > 0 \\ R_i^- & \text{if } AEC_i^i < 0 \end{cases} \qquad a_2 = \begin{cases} R_{i+1}^+ & \text{if } AEC_{i+1}^i > 0 \\ R_{i+1}^- & \text{if } AEC_{i+1}^i < 0 \end{cases}$$

$$(6.83)$$

$$C^i = \min(a_1, a_2), \quad 0 \leq C^i \leq 1, \text{ for the i-th element}$$

$$(AEC^c)_i^i = C^i \, AEC_i^i, \quad (AEC^c)_i^{i-1} = C^{i-1} \, AEC_i^{i-1} \quad (6.84)$$

Step 6. We obtain the final solution as a correction or limitation to the low-order solution.

$$c_i^{n+1} = c_i^L + \sum_e (AEC^c)_i^e = c_i^L + (AEC^c)_i^{i-1} + (AEC^c)_i^i \quad (6.85)$$

Results obtained with this version of the flux-correction step for the Galerkin method are quite good, as shown in Figures 6.17a and 6.17b. These

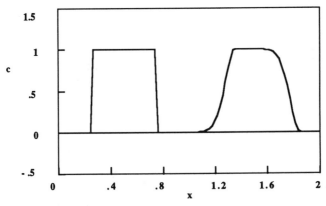

Figure 6.17a. Advection Equation, Galerkin Method with Flux-correction
$\Delta x = 0.02$, $\Delta t = 0.01$

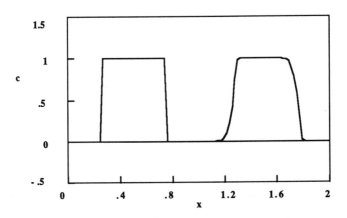

Figure 6.17b. Convective Diffusion Equation, Galerkin Method with Flux-correction
Pe = 1000, $\Delta x = 0.02$, $\Delta t = 1.\text{E-}5$

results were obtained by using CONVECT with the option "flux-correction, centered, not lumped." For the advection equation, the flux-correction step takes an unstable method and eliminates the oscillations. The original version, Eq. (6.25)-(6.29), gives the best results, but it is not suitable for finite element methods.

For the low-order scheme, Löhner et al. [1987] use a lumped-mass, Taylor-Galerkin method with enough added diffusion to make it monotone. Then the second-order fully-consistent Taylor-Galerkin method is used as the second-order method. How much added diffusion is necessary to ensure that the low-order method is monotone? Figure 4.18a and Problem 5.15-Part 3 indicate that the lumped-mass Taylor-Galerkin method will oscillate. Thus we use the following lumped-mass, Taylor-Galerkin method:

$$\frac{dc_i}{dt} + \frac{\text{Pe}}{2\Delta x}(c_{i+1}^n - c_{i-1}^n) = \left(\frac{\text{Pe}^2 \Delta t}{2} + c_d\right)\frac{1}{\Delta x^2}(c_{i+1}^n - 2c_i^n + c_{i-1}^n). \qquad (6.86)$$

To ensure that no oscillations are present, we use

$$\frac{Pe^2 \Delta t}{2} + c_d = \frac{Pe \Delta x}{2}. \qquad (6.87)$$

This turns the equation into

$$\frac{dc_i}{dt} + \frac{Pe}{\Delta x}(c_i^n - c_{i-1}^n) = 0, \qquad (6.88)$$

which does not oscillate. Eq. (6.88) is a finite element scheme (and finite difference scheme) that can be used as the low-order scheme that does not oscillate. The Taylor-Galerkin method can then be used as the high-order scheme, which can oscillate. The flux-correction step provides a blend of these two methods that maintains the sharp front but does not oscillate.

6.3. TVD and ENO Methods

The flux-correction methods discussed in the previous section are closely related to total variation non-increasing (TVNI) methods, or TVI methods. These methods also employ flux-correction to limit the introduction of oscillations. The key feature is the definition of the total variation:

$$TV(u) = \sum_i |u_{i+1} - u_i|. \qquad (6.89)$$

A finite difference method is said to be total variation non-increasing if it obeys

$$TV(u^{n+1}) \le TV(u^n). \qquad (6.90)$$

The choice of flux-limiter is then the key to proving that a method obeys the TVNI property. The original articles by Harten [1983] and Yee, *et al.* [1985] as well as current literature on compressible flow calculations discuss these concepts in more detail. Harten [1983], Yee, *et al.* [1985] and Sweby [1984] provide the basis of the methods. More recent work by Shu and Osher [1988, 1989] and Chen, *et al.*, [1991] is described here.

TVD Method. The first method is a second-order TVD method. The proof that the method is TVD lies in earlier papers, but the most clear exposition of the method is in Chen, *et al.*, [1991] and is summarized here. The method is described for the hyperbolic equation

$$\frac{\partial u}{\partial t} + \frac{\partial F(u)}{\partial x} = 0 \qquad (6.91)$$

Define the intermediate quantity

$$\Delta_i = \text{sign}\,(u_{i+k1}^n - u_{i-1+k1}^n)\max \begin{cases} 0 \\ \min \begin{cases} |u_{i+k1}^n - u_{i-1+k1}^n| \\ (u_{i+1+k1}^n - u_{i+k1}^n)\,\text{sign}\,(u_{i+k1}^n - u_{i-1+k1}^n), \end{cases} \end{cases} \quad (6.92)$$

$$k1 = \begin{cases} 0 & \text{if } \partial F/\partial u \geq 0 \\ 1 & \text{if } \partial F/\partial u < 0. \end{cases}$$

The values at $i = 1$ and NT are needed and are zero.

$$\Delta_1 = 0, \quad \Delta_{NT} = 0 \qquad (6.93)$$

The first equation can be derived by taking $u_0 = u_1$ and evaluating Eq. (6.92). The second equation is derived in the same way by taking $u_{NT} = u_{NT+1}$. The value at the mid-point is given by

$$u_{i+1/2}^n = u_i^n + \frac{1}{2}\Delta_i. \qquad (6.94)$$

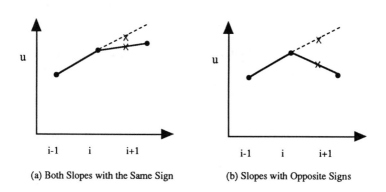

(a) Both Slopes with the Same Sign   (b) Slopes with Opposite Signs

Figure 6.18. Rationale for Flux-correction Step

The explanation of what is being done is clear from Figure 6.18, as described by Chen, et al. [1991]. In Figure 6.18 the value of the mid-point can be obtained by either interpolating between the points i and i+1 or between the points i-1 and i and extrapolation to i+1/2. In Figure 6.18a the slopes are of the same sign. Equation (6.91) then selects the slope having the smallest absolute value for use in Eq. (6.94). The value chosen is then the lowest one, which does not lead to an extraneous maximum (the lower x in Figure 6.18a). If the slopes are of opposite sign, as in Figure 6.18b, then there is a local extremum and we get $\Delta = 0$. In this case the mid-point value will be taken as u_i; the extremum is not made any worse, although the front can become sharper. When $u_i = u_{i-1}$ then the smallest slope in absolute value is zero. Thus we use

$$\text{sign}(0) = 1. \tag{6.95}$$

Next we evaluate the flux.

$$f^n_{i+1/2} = F(u^n_{i+1/2}) \tag{6.96}$$

and step forward to obtain a projected value.

$$u_i^* = u_i^n - \frac{\Delta t}{\Delta x}(f^n_{i+1/2} - f^n_{i-1/2}) \tag{6.97}$$

We then repeat the operation using u^*.

$$\Delta_i^* = \text{sign}(u^*_i - u^*_{i-1}) \max \begin{cases} 0 \\ \min \begin{cases} |u^*_i - u^*_{i-1}| \\ (u^*_{i+1} - u^*_i)\,\text{sign}(u^*_i - u^*_{i-1}) \end{cases} \end{cases} \tag{6.98}$$

$$u^*_{i+1/2} = u^*_i + \frac{1}{2}\Delta_i^* \tag{6.99}$$

$$f^*_{i+1/2} = F(u^*_{i+1/2}) \tag{6.100}$$

The final calculation obtains the new value.

$$u_i^{n+1} = u_i^n - \frac{\Delta t}{2\Delta x}(f^n_{i+1/2} - f^n_{i-1/2}) - \frac{\Delta t}{2\Delta x}(f^*_{i+1/2} - f^*_{i-1/2}) \tag{6.101}$$

Next consider the case when diffusion is important. The equation is now

$$\frac{\partial u}{\partial t} + \frac{\partial F(u)}{\partial x} = \nu \frac{\partial^2 u}{\partial x^2}. \tag{6.102}$$

In TVD we replace Eq. (6.96) with

$$f^n_{i+1/2} = F(u^n_{i+1/2}) - \nu \frac{u^n_{i+1} - u^n_i}{\Delta x} \tag{6.103}$$

and make a similar change in Eq. (6.100); otherwise use the same algorithm.

To see how to add source terms, it is convenient to write the scheme in a form which clearly shows that it is a second-order Runge-Kutta method for ordinary differential equaitons. Take the equation

$$\frac{\partial u}{\partial t} + \frac{\partial F(u)}{\partial x} = \nu \frac{\partial^2 u}{\partial x^2} + G(u) \tag{6.104}$$

and write

$$\Delta \hat{f}(u_j) = f^n_{j+1/2}(u) - f^n_{j-1/2}(u) \tag{6.105}$$

where the right-hand side is evaluated using Eq. (6.103). Then write the method

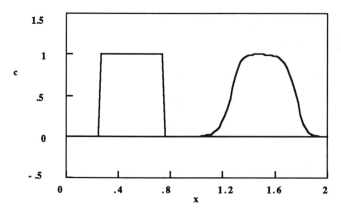

Figure 6.19. Advection Equation, TVD Method, $\Delta x = 0.02$, $\Delta t = 0.01$

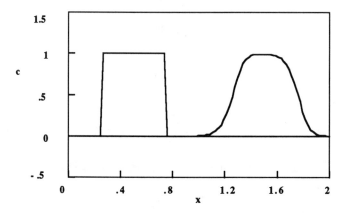

Figure 6.20. Convective Diffusion Equation, TVD Method, Pe = 1000
$\Delta x = 0.02$, $\Delta t = 0.01$ 10^{-5}

as

$$k_1 = -\Delta t \frac{\Delta \hat{f}(u^n)}{\Delta x} + \Delta t\, G(u^n),$$

$$k_2 = -\Delta t \frac{\Delta \hat{f}(u^n + k_1)}{\Delta x} + \Delta t\, G(u^n + k_1), \qquad (6.106)$$

$$u_j^{n+1} = u_j^n + \frac{1}{2}(k_1 + k_2).$$

Problem 6.24 proves that this is the same as Eq. (6.101) when there are no source terms. Problem 6.25 proves that it is correct to second order in Δt.

Application to the advection equation is shown in Figure 6.19. The solution, while good, is not as good as that shown for the MacCormack method with flux-correction in Figure 6.11. When diffusion is added (see Figure 6.20) the solution is almost the same as for the advection equation; these results are as

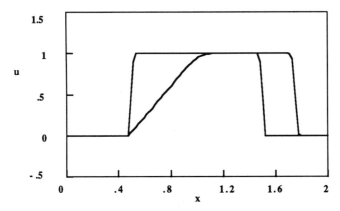

Figure 6.21. Burger's Equation, TVD Method, $v = 0$, $\Delta x = 0.02$, $\Delta t = 0.01$

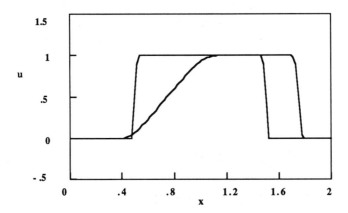

Figure 6.22. Burger's Equation, TVD Method, $v = 0.001$, $\Delta x = 0.02$, $\Delta t = 0.01$

good as those with the MacCormack method with flux-correction (see Figure 6.12). The results for Burger's equation are excellent; Figure 6.21 shows results when viscosity is zero and Figure 6.22 shows results with viscosity. Both figures are comparable to Figure 6.13-6.14 for the MacCormack method with flux-correction. For non-linear problems the TVD method is excellent.

ENO Method. The essentially-non-oscillatory method (ENO) of Shu and Osher [1988, 1989] is a multi-order scheme. What is described here is the third-order version of that scheme. There are actually several versions, ENO-Roe, ENO-RF, and ENO-LLF. We give here the simpliest, ENO-Roe. The other schemes have slightly more numerical viscosity for damping oscillations [Shu and Osher, 1989]. The stability of the ENO schemes has not been proved, but they work well in practice.

We first compute a divided difference table [Carnahan, Luther, and Wilkes, 1969]. Starting with values of f(x) at different x we compute in turn the differences.

$$fl_{i+1/2} = \frac{f_{i+1} - f_i}{x_{i+1} - x_i}, \quad i = 1, \ldots, NT-1 \qquad (6.107)$$

$$f2_i = \frac{fl_{i+1/2} - fl_{i-1/2}}{x_{i+1} - x_{i-1}}, \quad i = 2, \ldots, NT-1 \qquad (6.108)$$

Equations (6.107)-(6.108) can be combined to give

$$f2_i = \frac{\dfrac{f_{i+1} - f_i}{x_{i+1} - x_i} - \dfrac{f_i - f_{i-1}}{x_i - x_{i-1}}}{x_{i+1} - x_{i-1}}, \qquad (6.109)$$

which shows that the first divided difference is a representation of the first derivative and the second divided difference is a representation of the second derivative divided by 2! [see Eq. (7.15)]. To apply ENO to Eq. (6.91) we first calculate

$$f_i = F(u_i). \qquad (6.110)$$

Then we test

$$DF = \frac{f(u_{j+1}) - f(u_j)}{u_{j+1} - u_j}. \qquad (6.111)$$

If $DF \geq 0$ then use $k1 = 0$;

if $DF < 0$ then use $k1 = 1$. $\qquad (6.112)$

Next we define Q1.

$$Q1_j(x) = f_{j+k1}(x - x_{j+k1-1/2}) \qquad (6.113)$$

The derivative is needed below.

$$\left. \frac{dQ1_j}{dx} \right|_{x = x_{j+1/2}} = f_{j+k1} \qquad (6.114)$$

To get the second-order terms, we define

$$a_j^{(2)} = \frac{1}{2} fl_{j+k1+1/2}, \quad b_j^{(2)} = \frac{1}{2} fl_{j+k1-1/2} \qquad (6.115)$$

and apply the following test.

If $|a_j^{(2)}| \geq |b_j^{(2)}|$ then use $c_j^{(2)} = b_j^{(2)}$ and $k2 = k1 - 1$;

else use $c_j^{(2)} = a_j^{(2)}$ and $k2 = k1$. (6.116)

Then the function Q2 is defined as

$$Q2_j(x) = Q1_j(x) + c_j^{(2)} (x - x_{j+k1-1/2})(x - x_{j+k1+1/2}),$$ (6.117)

and its derivative is

$$\left.\frac{dQ2_j}{dx}\right|_{x=x_{j+1/2}} = \left.\frac{dQ1_j}{dx}\right|_{x=x_{j+1/2}} + c_j^{(2)}\left[(x_{j+1/2} - x_{j+k1+1/2}) + (x_{j+1/2} - x_{j+k1-1/2})\right].$$ (6.118)

To get the third-order terms, we define

$$a_j^{(3)} = \frac{1}{3} f2_{j+k2+1}, \quad b_j^{(3)} = \frac{1}{3} f2_{j+k2}$$ (6.119)

and apply the following test.

If $|a_j^{(3)}| \geq |b_j^{(3)}|$ then use $c_j^{(3)} = b_j^{(3)}$;

else use $c_j^{(3)} = a_j^{(3)}$. (6.120)

The function Q3 is defined as

$$Q3_j(x) = Q2_j(x) + c_j^{(3)}(x - x_{j+k2-1/2})(x - x_{j+k2+1/2})(x - x_{j+k2+3/2})$$ (6.121)

and its derivative is

$$\left.\frac{dQ3_j}{dx}\right|_{x=x_{j+1/2}} = \left.\frac{dQ2_j}{dx}\right|_{x=x_{j+1/2}} + c_j^{(3)}\Big[(x_{j+1/2} - x_{j+k2+1/2})(x_{j+1/2} - x_{j+k2+3/2}) +$$
$$+ (x_{j+1/2} - x_{j+k2-1/2})(x_{j+1/2} - x_{j+k2+3/2}) + (x_{j+1/2} - x_{j+k2-1/2})(x_{j+1/2} - x_{j+k2+1/2})\Big].$$ (6.122)

Finally we define the flux as

$$\hat{f}_{j+1/2} = \left.\frac{dQ3_j}{dx}\right|_{x=x_{j+1/2}}.$$ (6.123)

For the third-order ENO method we use the algorithm (6.110)-(6.123) frequently. We calculate in turn

$$u_i^1 = u_i^n - \frac{\Delta t}{\Delta x}(\hat{f}_{i+1/2} - \hat{f}_{i-1/2})$$

$$u_i^2 = \frac{3}{4} u_i^n + \frac{1}{4} u_i^1 - \frac{\Delta t}{4\Delta x}(\hat{f}_{i+1/2}^1 - \hat{f}_{i-1/2}^1)$$ (6.124)

$$u_i^{n+1} = \frac{1}{3} u_i^n + \frac{2}{3} u_i^2 - \frac{2\Delta t}{3\Delta x}(\hat{f}_{i+1/2}^2 - \hat{f}_{i-1/2}^2).$$

The definition of $\hat{f}^1$ and $\hat{f}^2$ are the fluxes resulting from the algorithm (6.110)-

(6.123) when applied to the functions u^1 and u^2, respectively.

The algorithm given above is suitable when the boundary conditions are periodic, which is true for all the cases treated by Shu and Osher [1988, 1989]. When there is inflow and outflow special equations must be used for the outflow point (i = NT) and the points just inside the boundary (i = 2 or NT-1). This complication arises because the flux equations need information that is outside the domain. The approach taken here is to use a lower order system at these special points. Thus for the points i = 2 or NT-1 we use flux terms only through Q2 and replace Eq. (6.123) with

$$\hat{f}_{j+1/2} = \left.\frac{dQ2_j}{dx}\right|_{x=x_{j+1/2}}. \quad (6.125)$$

For the last point we use only a first order system and replace Eq. (6.123) with

$$\hat{f}_{j+1/2} = \left.\frac{dQ2_j}{dx}\right|_{x=x_{j+1/2}}. \quad (6.126)$$

For the ENO version applied to Eq. (6.104) we replace Eq. (6.123) with

$$\hat{f}_{j+1/2} = \left.\frac{dQ3_j}{dx}\right|_{x=x_{j+1/2}} - v\frac{u_{j+1}^n - u_j^n}{\Delta x}. \quad (6.127)$$

Then the method is written in the form

$$k_1 = -\Delta t\, \frac{\Delta \hat{f}(u^n)}{\Delta x} + \Delta t\, G(u^n),$$

$$k_2 = -\Delta t\, \frac{\Delta \hat{f}(u^n + k_1)}{\Delta x} + \Delta t\, G(u^n + k_1),$$

$$k_2 = -\Delta t\, \frac{\Delta \hat{f}\left(u^n + \frac{1}{4}k_1 + \frac{1}{4}k_2\right)}{\Delta x} + \Delta t\, G\left(u^n + \frac{1}{4}k_1 + \frac{1}{4}k_2\right), \quad (6.128)$$

$$u_j^{n+1} = u_j^n + \frac{1}{6}(k_1 + k_2) + \frac{2}{3}k_3.$$

Problem 6.26 proves that this is the same as Eq. (6.124) when there are no source terms. Problem 6.27 proves that it is correct to third order in Δt.

Applications to the advection and convective diffusion equations are shown in Figures 6.23 and 6.24, respectively. The quality of the solution is similar to that of the TVD method. The results for Burger's equation are excellent; Figure 6.25 shows results when viscosity is zero and Figure 6.26 shows results with viscosity. Both figures are comparable to Figure 6.13-6.14 for the MacCormack method with flux-correction. For non-linear problems the ENO method is excellent. The TVD and ENO methods also have the advantage that they are second-order and third-order, respectively, away from the front. Of course the MacCormack method with the flux-correction scheme given in Section 6.2 is also

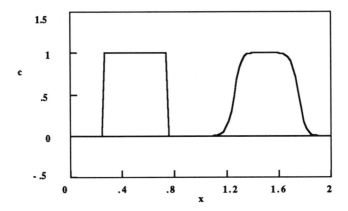

Figure 6.23. Advection Equation, ENO Method, $\Delta x = 0.02$, $\Delta t = 0.01$

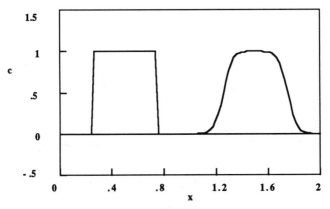

Figure 6.24. Convective Diffusion Equation, ENO Method, Pe = 1000
$\Delta x = 0.02$, $\Delta t = 0.01$

second-order away from sharp fronts.

6.4. Other Upstream Weighting Methods

In the application of Galerkin finite element methods to the convective diffusion equation, many different ways have been devised to eliminate oscillations while keeping the front steep. In this section we will consider some of these ways and compare the results achieved with them to the results already discussed. In the Petrov-Galerkin method, we make the residual orthogonal to a set of weighting functions, which are chosen to dampen the oscillations in some way. In this section we will first consider upstream quadrature and streamline upstream weighting, as well as a different weighting function in the Petrov-Galerkin method. We then develop an alternative justification of the Petrov-Galerkin method, using a variational principle for the steady-state convective diffusion

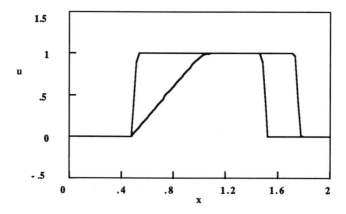

Figure 6.25. Burger's Equation, ENO Method, $\nu = 0$, $\Delta x = 0.02$, $\Delta t = 0.01$

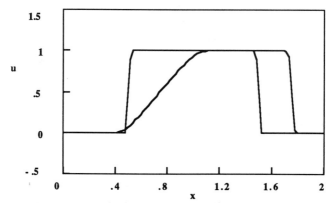

Figure 6.26. Burger's Equation, ENO Method, $\nu = 0.001$, $\Delta x = 0.02$, $\Delta t = 0.01$

equation. This provides a much firmer foundation for the Petrov-Galerkin method.

Let us consider the transient convective diffusion equation:

$$\frac{\partial c}{\partial t} + \text{Pe} \frac{\partial c}{\partial x} = \frac{\partial^2 c}{\partial x^2}. \tag{6.129}$$

The trial function is taken as

$$c = \sum_{i=1}^{NT} c_i \, N_i(x). \tag{6.130}$$

The Petrov-Galerkin method multiplies the residual by a weighting function and integrates over the domain, giving

$$\int_0^1 W_j \frac{\partial c}{\partial t} dx + Pe \int_0^1 W_j \frac{\partial c}{\partial x} dx = \int_0^1 W_j \frac{\partial^2 c}{\partial x^2} dx. \qquad (6.131)$$

The last term is integrated by parts and the entire equation is expressed in terms of the trial function

$$\sum_{i=1}^{NT} \int_0^1 W_j N_i\, dx \frac{dc_i}{dt} + Pe \sum_{i=1}^{NT} \int_0^1 W_j \frac{dN_i}{dx} dx\, c_i = -\sum_{i=1}^{NT} \int_0^1 \frac{dW_j}{dx} \frac{dN_i}{dx} dx\, c_i. \qquad (6.132)$$

On an element basis this is

$$\sum_e \sum_I \frac{\Delta x^e}{2} \int_{-1}^1 W_J N_I\, d\xi \frac{dc_I^e}{dt} + Pe \sum_e \sum_I \int_{-1}^1 W_J \frac{dN_I}{d\xi} d\xi\, c_I^e =$$
$$= -\sum_e \sum_I \frac{2}{\Delta x^e} \int_{-1}^1 \frac{dW_J}{d\xi} \frac{dN_I}{d\xi} d\xi\, c_I^e. \qquad (6.133)$$

The standard Galerkin method uses $W_J = N_J$ to obtain

$$\sum_e \sum_I \left\{ C_{JI}^e \frac{dc_I^e}{dt} + Pe\, A_{JI}^e c_I^e = B_{JI}^e c_I^e \right\}. \qquad (6.134)$$

This equation is the same as Eq. (2.100); the resulting general equation is

$$\frac{1}{6} \frac{dc_{i-1}}{dt} + \frac{2}{3} \frac{dc_i}{dt} + \frac{1}{6} \frac{dc_{i+1}}{dt} + Pe \frac{c_{i+1} - c_{i-1}}{2\Delta x} = \frac{c_{i+1} - 2c_i + c_{i-1}}{\Delta x^2}. \qquad (6.135)$$

There are several possible choices of weighting functions for the Petrov-Galerkin method, and these choices are analyzed below.

Reduced Quadrature. The reduced quadrature method was developed by Hughes [1978] and involves a special quadrature for the convective terms. The convection integrals are evaluated with a one-term quadrature that uses a location that is chosen adroitly.

$$A_{JI}^e = \int_{-1}^1 N_J \frac{dN_I}{d\xi} d\xi \approx N_J(\bar\xi) \frac{dN_I}{d\xi}(\bar\xi)\, 2 \qquad (6.136)$$

The factor 2 is used in Eq. (6.136) because the standard element has length 2.0. When the terms are evaluated using Table 2.2, we obtain

$$A_{JI}^e = \begin{bmatrix} -\frac{1}{2} & \frac{1}{2} \\ -\frac{1}{2} & \frac{1}{2} \end{bmatrix} + \bar\xi \begin{bmatrix} \frac{1}{2} & -\frac{1}{2} \\ -\frac{1}{2} & \frac{1}{2} \end{bmatrix}. \qquad (6.137)$$

The second term leads to an additional term in Eq. (6.135):

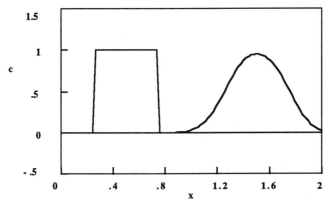

Figure 6.27. Convective Diffusion Equation, Quadrature Weighting
Pe = 1000, $\bar{\xi}$ = 0.9, Δx = 0.02, Co = 0.25

$$\frac{Pe\,\bar{\xi}}{\Delta x}\left(-\tfrac{1}{2}c_{i-1} + c_i - \tfrac{1}{2}c_{i+1}\right) = -\frac{Pe\,\bar{\xi}}{2\Delta x}(c_{i-1} - 2c_i + c_{i+1}). \tag{6.138}$$

The final result is then

$$\frac{1}{6}\frac{dc_{i+1}}{dt} + \frac{2}{3}\frac{dc_i}{dt} + \frac{1}{6}\frac{dc_{i+1}}{dt} + \frac{Pe}{2\Delta x}(c_{i+1} - c_{i-1}) =$$
$$= \left(1 + \frac{Pe\,\Delta x}{2}\bar{\xi}\right)\frac{c_{i+1} - 2c_i + c_{i-1}}{\Delta x^2}. \tag{6.139}$$

In this method we are thus adding a diffusion term multiplied by $Pe\Delta x\bar{\xi}/2$. The equation is the same as Eq. (2.105) but with $\alpha = \bar{\xi}$ in the convective term and $\alpha = 0$ in the time derivative terms. Hughes [1978] suggests taking

$$\bar{\xi} = \alpha = \coth(\mu) - \frac{1}{\mu}, \qquad \mu = \frac{Pe\,\Delta x}{2} \tag{6.140}$$

so that the steady-state solution is exact at the nodes. This, of course, is the same choice made in Chapter 2 for the Petrov-Galerkin method presented there.

Sample results are shown in Figure 6.27. There is excessive dissipation for these parameters. The problem is easy to find: the term multiplying the diffusion term in Eq. (6.139) is changed from 1.0 to 10.0 by the Petrov-Galerkin addition (with $\bar{\xi} = 0.9$). Thus there is much more diffusion than is reasonable. This solution should be compared with Figure 4.16a for a finite difference method (the lumped Petrov-Galerkin method with $\alpha = 1$) and with Figure 5.8c for the Petrov-Galerkin method with $\alpha = 1$. The methods with lumped time derivatives show too much dissipation. If we reduce $\bar{\xi}$ to 0.25 the value of the coefficient multiplying the diffusion term in Eq. (6.139) is exactly the same as in the Taylor-Galerkin method. Thus the results shown in Figure 5.8e are also for the quadrature-weighted Petrov-Galerkin method and are quite good. Thus the proper amount of added diffusion ($\tilde{\xi}$) depends on how the time derivatives are handled.

Streamwise Upwinding. The streamwise upwinding method was developed by Hughes and Brooks [1979]. It is most useful in two-dimensional cases, but we present it here for the one-dimensional case to show its relationship with the other methods. In this method we solve an augmented equation with an additional diffusion term:

$$\int_0^1 W_j \frac{\partial c}{\partial t} dx + Pe \int_0^1 W_j \frac{\partial c}{\partial x} dx = -[1 + \tilde{D}] \int_0^1 \frac{dW_j}{dx} \frac{\partial c}{\partial x} dx . \qquad (6.141)$$

The additional diffusion term is taken as

$$\tilde{D} = \frac{Pe \, \Delta x}{2} \alpha = \mu \alpha, \qquad \alpha = \coth(\mu) - \frac{1}{\mu}, \qquad \mu = \frac{Pe \, \Delta x}{2}. \qquad (6.142)$$

The equations can be derived by taking Eq. (2.102) and multiplying the diffusion term by $1 + \tilde{D}$. The result is Eq. (6.139). The parameter α is then derived from Eq. (6.140). Thus streamwise upwinding in the one-dimensional case gives the same results as upstream quadrature.

Augmented Weighting Function. In this method we use the weighting function

$$W_j = N_j + \frac{\Delta x}{2} \frac{dN_j}{dx}. \qquad (6.143)$$

Then Eq. (6.131) becomes

$$\int_0^1 N_j \frac{\partial c}{\partial t} dx + Pe \int_0^1 N_j \frac{\partial c}{\partial x} dx - \int_0^1 N_j \frac{\partial^2 c}{\partial x^2} dx = $$
$$= -\frac{\Delta x}{2} \int_0^1 \frac{dN_j}{dx} \frac{\partial c}{\partial t} dx - \frac{Pe \, \Delta x}{2} \int_0^1 \frac{dN_j}{dx} \frac{\partial c}{\partial x} dx + \frac{\Delta x}{2} \int_0^1 \frac{dN_j}{dx} \frac{\partial^2 c}{\partial x^2} dx. \qquad (6.144)$$

The terms in the first term of the weighting function are taken to the left-hand side of the equation. They are the standard Galerkin terms. The terms on the right-hand side come from the second term in the weighting function. The third term on the left-hand side is integrated by parts. The resulting boundary term at $x = 0$ vanishes for $j = 2$ to $j = NT$; when $j = 1$, the equation is not used. The boundary term at $x = 1$ vanishes because $\partial c/\partial x = 0$ at $x = 1$. We also note that for linear trial functions, the last term on the right-hand side is zero. The next to last term on the right-hand side is the same as a diffusion term. The first term on the right-hand side is

$$-\sum_{i=1}^{NT} \frac{\Delta x}{2} \int_0^1 \frac{dN_j}{dx} N_i \, dx \, \frac{dc_i}{dt} = -\sum_e \sum_I \frac{\Delta x_e}{2} \int_{-1}^1 \frac{dN_J}{d\xi} N_I \, d\xi \, \frac{dc_I^e}{dt} . \qquad (6.145)$$

This can be written as

$$-\sum_e \sum_I \frac{\Delta x_e}{2} A^e_{IJ} \frac{dc^e_I}{dt}, \qquad (6.146)$$

where the matrix A^e_{IJ} is the transpose of the matrix A^e_{JI} in Table 2.2.

$$A^e_{IJ} = \begin{bmatrix} -\frac{1}{2} & -\frac{1}{2} \\ \frac{1}{2} & \frac{1}{2} \end{bmatrix} \qquad (6.147)$$

Putting Eq. (6.147) into Eq. (6.146) gives

$$-\frac{\Delta x_e}{2}\left[\frac{1}{2}\frac{dc_{i-1}}{dt} - \frac{1}{2}\frac{dc_{i+1}}{dt}\right]. \qquad (6.148)$$

When all these results are combined (in the case when $\Delta x_e = \Delta x$), we get

$$\left(\frac{1}{6}+\frac{1}{4}\right)\frac{dc_{i+1}}{dt} + \frac{2}{3}\frac{dc_i}{dt} + \frac{1}{6} + \left(\frac{1}{6}-\frac{1}{4}\right)\frac{dc_{i+1}}{dt} +$$
$$+ \frac{\text{Pe}}{2\Delta x}(c_{i+1} - c_{i-1}) = \left(1 + \frac{\text{Pe}\,\Delta x}{2}\right)\frac{c_{i+1} - 2c_i + c_{i-1}}{\Delta x^2}. \qquad (6.149)$$

This is the equation for the Petrov-Galerkin method with $\alpha = 1$ [see Eq. (2.105)]. The results obtained with the augmented weighting function, Eq.(6.143), are thus the same as those in Fig. 2.28 for the Petrov-Galerkin method with $\alpha = 1$. Small oscillations are evident, although the steepness of the front is preserved. Yu and Heinrich [1986] agreed that the augmented weighting function method is inadequate for the unsteady-state equation; they added additional terms to the weighting function to try to improve the results.

Rational Basis. A rational basis for the Petrov-Galerkin method (i.e., the choice of the weighting functions) is presented here and illustrated for the convective diffusion equation. The choice of weighting functions is closely related to whether a given differential equation is derivable from a variational principle; with the correct choice of weighting function, a symmetric, bilinear norm is produced.

Let us first consider the convective diffusion equation in one dimension. The operator, L, is

$$L(c) \equiv \frac{d^2c}{dx^2} - \text{Pe}\frac{dc}{dx} = 0. \qquad (6.150)$$

Whether or not a variational principle exists for this equation is a question that can be answered by using Fréchet differentials. The theory outlined by Finlayson [1972, Section 9.1] indicates that a variational principle exists if the operator has a symmetric Fréchet differential. This operator is not symmetric; indeed, the adjoint operator, L*, is [Finlayson, 1972, p. 308]:

$$L^*(c) \equiv \frac{d^2c}{dx^2} + \text{Pe}\frac{dc}{dx} = 0. \qquad (6.151)$$

A second-order ordinary differential equation can always be transformed to a form for which a variational principle exists [see Finlayson, 1972, p. 309]. In this case we consider the expanded operator

$$L_2 = g(x) L(x) \qquad (6.152)$$

and we find that $g(x) = e^{\text{Pe}\,x}$. The variational integral is then

$$I[c] = \int_0^1 e^{-\text{Pe}\,x} \left(\frac{dc}{dx}\right)^2 dx. \qquad (6.153)$$

Eq. (6.153) gives a symmetric, bilinear form that forms the basis of a norm:

$$\left\| e^{-\text{Pe}\,x/2} \frac{dc}{dx} \right\|. \qquad (6.154)$$

The value of having a variational principle is that the variational integral provides a norm that can be minimized (if the principle is a minimum principle). Barrett and Morton [1980] search for a norm in a different way. Their initial steps lead to the norm given above and then a method is provided that leads to a norm in more general cases.

First we factor the equation

$$L(c) = \frac{d}{dx}\left(\frac{dc}{dx} - \text{Pe}\,c\right). \qquad (6.155)$$

We then wish to find an operator, N, that makes the following bilinear form B symmetric. We thus look at the two possibilities

$$\text{i: } B[u, Nv] = \int_0^1 \frac{du}{dx}\left(\frac{d}{dx} + \text{Pe}\right) Nv\, dx$$

$$\text{ii: } B[u, Nv] = \int_0^1 \left(\frac{d}{dx} - \text{Pe}\right) u\, \frac{d}{dx}(Nv)\, dx. \qquad (6.156)$$

Since we wish the bilinear form to be symmetric, we need the following conditions:

$$\text{i: } B[u, Nv] = B_E[u,v] = \int_0^1 \frac{du}{dx} \rho \frac{dv}{dx} dx$$

$$\text{ii: } B[u, Nv] = B_E[u,v] = \int_0^1 \left[\left(\frac{d}{dx} - \text{Pe}\right)u\right] \rho \left[\left(\frac{d}{dx} - \text{Pe}\right)v\right] dx. \qquad (6.157)$$

The positive weighting function is introduced for convenience. If these equalities are to hold, then the operator N must satisfy

$$\begin{aligned} &\text{i: } \left(\frac{d}{dx} + \text{Pe}\right) Nv = \rho \frac{dv}{dx} \\ &\text{ii: } \frac{d}{dx} Nv = \rho \left(\frac{d}{dx} - \text{Pe}\right) v. \end{aligned} \tag{6.158}$$

Since the original problem involved two boundary conditions, the weighting function must satisfy two boundary conditions as well.

$$Nv(0) = Nv(1) = 0 \tag{6.159}$$

Let us consider the second equation and rewrite it using differentiation by parts as

$$\frac{d}{dx} Nv = \frac{d}{dx}[\rho\, v] - v \frac{d\rho}{dx} - \rho\, \text{Pe}\, v. \tag{6.160}$$

We can easily solve for Nv, provided that we can make the last two terms cancel. This happens if

$$Nv = v\rho, \quad \frac{d\rho}{dx} = -\rho\, \text{Pe}, \quad \rho = e^{-\text{Pe}\, x}. \tag{6.161}$$

Thus we choose $Nv = e^{(-\text{Pe}\, x)} v$. The corresponding bilinear form is

$$B_E[u, Nv] = \int_0^1 \frac{du}{dx} e^{-\text{Pe}\, x} \frac{dv}{dx} dx \tag{6.162}$$

and the associated norm is

$$\left\| e^{-\text{Pe}\, x/2} \frac{dv}{dx} \right\|. \tag{6.163}$$

This is the same norm that comes from the variational principle. We note that the same operator (Nv) satisfies the first equation, too, so there is only one result. Once we have a norm, we know that the resulting Petrov-Galerkin method will give the best approximation in that norm.

Least Squares Method. This approach, derived either through a variational principle or by factoring the equation and making the weak form of the equation symmetric, leads to a norm in this case. However, all equations cannot be make symmetric so easily (or possibly at all). Furthermore, for large Peclet numbers the calculations become ill-conditioned because of the exponential term. Thus, we need to look at other ways of making the bilinear form symmetric. For the least squares method we choose the norm

$$I[c] = \int_0^1 \left[\frac{d^2 c}{dx^2} - \text{Pe} \frac{dc}{dx}\right]^2 dx. \tag{6.164}$$

This method works for any equation, but has higher-order derivatives in Eq.

(6.164) than usually occur in the Galerkin method. This is disadvantageous because it increases the complexity of the finite elements, especially for multi-dimensional problems.

An alternative method has been provided by Barrett and Morton [1980]. They offer a way to obtain an approximate weighting function that approximates the above treatment, gives well-conditioned equations, and has the weighting function defined over only a few elements. Unfortunately, it has not been possible to do the same thing for multi-dimensional problems [Demkowicz and Oden, 1986]. It is significant, however, that the weighting functions derived by Barrett and Morton have similar functional forms to those used in upstream weighting.

Comparison and Use of Methods. To use this theory, we expand the trial function, as in Eq. (6.130). This is substituted into the variational integral, Eq. (6.153), which is then minimized with respect to the parameters. The result is

$$\frac{1}{2}\frac{\partial I}{\partial c_j} = \int_0^1 e^{-Pe\,x}\frac{dc}{dx}\frac{dN_j}{dx}\,dx = 0. \tag{6.165}$$

This integral is written on an element basis as

$$\sum_e \sum_I \frac{2}{\Delta x_e}\int_{-1}^1 e^{-Pe\,x}\frac{dN_I}{d\xi}\frac{dN_J}{d\xi}\,d\xi\, c_I = 0. \tag{6.166}$$

Eq. (6.166) is evaluated in Problem 6.6. The result is

$$c_{i-1} - c_i(1 + e^{-2\mu}) + c_{i+1}e^{-2\mu} = 0. \tag{6.167}$$

We note that in the diffusion limit, $\mu = 0$, we get

$$c_{i-1} - 2c_i + c_{i+1} = 0, \tag{6.168}$$

which is the Galerkin difference expression for the diffusion equation. In the convection limit, $\mu \to \infty$, we get

$$c_{i-1} - c_i = 0. \tag{6.169}$$

This is the proper equation for convection only. Thus, this equation has proper limits at the two extremes.

The advantage of the variational principle is that we minimize a functional and we know that the results obtained will be optimal; in other words, they will be the best possible for that trial function, since they minimize the variational integral. [This assumes that "goodness" is measured using the same norm, Eq. (6.163)]. Furthermore, the results from the variational method will not oscillate. This can be seen by considering an oscillating solution; the derivatives are larger than for a smooth "mean" curve drawn through the points and thus the variational integral is larger. Since the variational method finds the function that minimizes the variational integral, it will choose the smooth curve instead of the

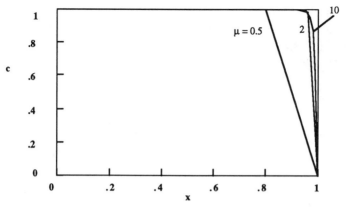

Figure 6.28. Variational Solution for Pe = 100

oscillating curve. Thus, the variational method provides the best method and gives a goal to be achieved by any method.

This method can be applied to the steady convective diffusion equation with the boundary conditions

$$c(0) = 1 \quad c(1) = 0. \tag{6.170}$$

The result is shown in Figure 6.28 for Pe = 100 and several different mesh sizes, such that $\mu = 0.5$, 2, and 10. We note that none of the solutions oscillate, even when the value of μ is greater than 1.0. The front is also as steep as the mesh allows. (We should note that the first curve only uses six element nodes.) We can contrast this with the finite difference solution shown in Figure 2.3 (in this case the standard Galerkin method, too). For these methods, the solutions oscillate when $\mu > 1$. Thus the variational approach is clearly superior. It is even superior to a finite difference method with upstream weighting, whose results are shown in Figure 2.4; the variational solution gives results that are as steep as the mesh allows. Finally, we can compare the results to those in Figure 2.10 for the Petrov-Galerkin method with optimal weighting. The results look very similar.

To examine the similarity between the Petrov-Galerkin method and the variational method (the curves are not identical), we look at an analysis by Hughes and Atkinson [1980]. The variational method is contained in Eq. (6.165), which is further expanded here to include a trial function:

$$\sum_{i=1}^{NT} \int_0^1 e^{-Pe\,x} \frac{dN_j}{dx} \frac{dN_i}{dx} dx\, c_i = 0. \tag{6.171}$$

The Petrov-Galerkin method is contained in Eq. (6.131), which also is further expanded to include a trial function:

$$\sum_{i=1}^{NT} \int_0^1 \left(Pe\, W_j \frac{dN_i}{dx} + \frac{dW_j}{dx} \frac{dN_i}{dx} \right) dx\, c_i = 0. \tag{6.172}$$

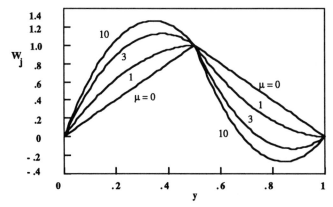

Figure 6.29. Weighting Function for Petrov-Galerkin Method Elements from 0 to 0.5, 0.5 to 1.0

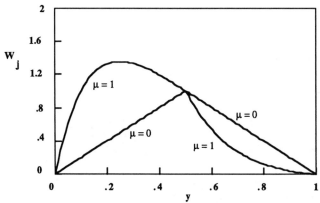

Figure 6.30. Weighting Function for Variational Method Elements from 0 to 0.5, 0.5 to 1.0

If we take

$$W_j = e^{-Pe\,x} N_j, \qquad (6.173)$$

we can show that the two equations are the same (see Problem 6.7). Thus the Petrov-Galerkin method using the weighting function Eq. (6.173) is the proper weighting function and leads to optimal results.

How well does the Petrov-Galerkin method presented here meet this goal? The Petrov-Galerkin method is defined by the weighting function [Eq. (2.69)] shown in Figure 2.8. In Figure 6.29 the weighting functions for specific values of $\mu = 0, 1, 3$, and 10 are shown. For the node at $y = 0.5$, the elements on either side extend from 0 to 0.5 and 0.5 to 1.0. We notice that as μ increases, the weighting function increases on the upstream side and decreases on the downstream side (even taking negative values). The weighting function is the same for all elements. The comparable figures for the optimal weighting function, Eq.

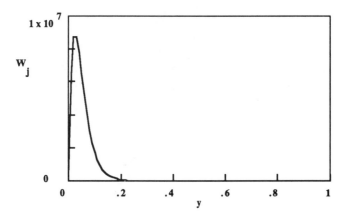

Figure 6.31. Weighting Function for Variational Method, $\mu = 10$

(6.173), are Figure 6.30 for $\mu = 0$ and 1 and Figure 6.31 for $\mu = 10$. To the right of the center node (y = 0.5) we have

$$e^{-Pe\,x} N_j, \quad \xi = \frac{2(x-x_j)}{\Delta x} - 1, \quad x = x_j + \frac{(\xi+1)}{2}\Delta x, \quad N_j = N_1 = \frac{1-\xi}{2} \quad (6.174)$$

and the weighting function is

$$\exp\left[-Pe\left(x_j + \frac{\Delta x\,(\xi+1)}{2}\right)\right] N_1 = e^{-Pe\,x_j} \exp\left[-\frac{Pe\,\Delta x\,(\xi+1)}{2}\right]\frac{1-\xi}{2}. \quad (6.175)$$

Thus we plot

$$\exp\left[-\frac{Pe\,\Delta x\,(\xi+1)}{2}\right]\frac{1-\xi}{2}. \quad (6.176)$$

To the left of the center node at y = 0.5 we have

$$e^{-Pe\,x} N_j, \quad \xi = \frac{2(x-x_{j-1})}{\Delta x} - 1, \quad x = x_{j-1} + \frac{(\xi+1)}{2}\Delta x, \quad N_j = N_2 = \frac{1-\xi}{2} \quad (6.177)$$

and the weighting function is

$$\exp\left[-Pe\left(x_{j-1} + \frac{\Delta x\,(\xi+1)}{2}\right)\right] N_2 = e^{-Pe\,x_j} \exp\left[-\frac{Pe\,\Delta x\,(\xi-1)}{2}\right]\frac{1+\xi}{2}. \quad (6.178)$$

Thus we plot

$$\exp\left[-\frac{Pe\,\Delta x\,(\xi-1)}{2}\right]\frac{1+\xi}{2}. \quad (6.179)$$

For the variational weighting function, the shift to upstream weighting is much more dramatic; for $\mu = 10$ it was so dramatic that the scale had to be expanded. However, these figures do show that the weighting functions used in the Petrov-Galerkin method, namely Eq. (2.69) and the equivalent upstream quadrature, streamwise upwinding, or augmented weighting function, all approximate the optimal results derived from the variational method. Certainly they approximate the optimal results better than the standard Galerkin method by itself.

Unfortunately, these improvements cannot be applied to transient problems. There is no variational principle for the transient problem [Finlayson, 1972, p. 313] so no symmetric bilinear norm is possible [in analogy to Eq. (6.154)]. As we have seen, the Petrov-Galerkin method does not do a very good job for the transient convective diffusion equation. However, Morton and Parrott [1980] have devised a special weighting function for the Petrov-Galerkin method that results in the scheme shown in Table 5.1. They used the weighting function

$$W_j = (1 - v) N_j + v N_j^+, \qquad (6.180)$$

where the function N_j^+ is given by

$$N_j^+ = 4 - 6\xi, \quad x_j \le x \le x_{j+1}, \quad \xi = \frac{2(x - x_j)}{\Delta x_j} - 1. \qquad (6.181)$$

The parameter v is chosen to have a "unit CFL property". For the advection equation it is taken as

$$v = \frac{Pe \, \Delta t}{\Delta x}. \qquad (6.182)$$

The unit CFL property is one that ensures for a Courant number of 1.0 that a wave is advected without change, giving the solution

$$c_i^{n+1} = c_i^n. \qquad (6.183)$$

Most of the finite difference methods have this property, but this is the only Galerkin method that does. If we put Pe $\Delta t = \Delta x$ in Table 5.1 for the Morton-Parrott-Galerkin method, the result is Eq. (6.183). We note also that the Morton-Parrott-Galerkin method includes extra diffusion terms on the right-hand side, here derived through the weighting function rather than through a Taylor series, as in the case of the other Galerkin methods listed in Table 5.1. The Morton-Parrott method applied to Burger's equation without viscosity gave sharp fronts but the wrong shock speed [Morton and Parrott, 1980].

For the steady-state convective diffusion equation in more than one dimension, there is a variational principle only when the velocity is derivable from a potential [Finlayson, 1972, p. 313]. A velocity is derivable from a potential if it is the gradient of a function. Even in that case a variational method suffers the same difficulty in two dimensions that it does in one dimension: the variational integrals are ill-conditioned for large Peclet numbers. Thus, the variational methods using the exact variational integral are not useful for the steady-state convective diffusion equation at large Peclet numbers. An approximation to the variational method has been developed in a finite element form [Barrett and Morton, 1984], but not too successfully in two dimensions [Demkowicz and Oden, 1986].

6.5. High-order Methods

Specialty methods have been developed for specific equations. The use of such methods for the convective diffusion equation will be discussed in this section. Often the methods can only be applied to that particular equation, but they do offer insight into solving the convective diffusion equation at large Peclet numbers. These specialty methods include the three-point upstream first derivative method, the QUICK method, and methods developed by Lambauch and Rosenberg.

Three-point upstream first derivative. The three-point upstream method incorporates an upstream first derivative for the convection term. We have done that before [see Eq. (2.99)], but the earlier expression had a truncation error of only Δx. Here we wish to derive an upstream first derivative that has a higher truncation order.

If we use the Taylor expansion

$$c_{i-1} = c_i - c_i' \Delta x + c_i'' \frac{\Delta x^2}{2} - c_i''' \frac{\Delta x^3}{6}, \tag{6.184}$$

we get the first-order derivative

$$\frac{c_i - c_{i-1}}{\Delta x} = c_i' - c_i'' \frac{\Delta x}{2}. \tag{6.185}$$

In the convective term this gives

$$\text{Pe} \frac{c_i - c_{i-1}}{\Delta x} = \text{Pe}\, c_i' - \frac{1}{2} c_i'' \, \text{Pe}\, \Delta x. \tag{6.186}$$

We notice that the truncation error is multiplied by the Peclet number, suggesting that the truncation error will be large when the Peclet number is large. An alternative is to derive a more accurate first derivative, but use only points upstream. If we use

$$c_{i-2} = c_i - c_i'\, 2\Delta x + c_i'' \frac{(2\Delta x)^2}{2} - c_i''' \frac{(2\Delta x)^3}{6}, \tag{6.187}$$

we get the following expression for the convection term:

$$\text{Pe} \frac{c_{i-2} - 4 c_{i-1} + 3 c_i}{2\Delta x} = \text{Pe}\, c_i' - \frac{1}{3} c_i''' \, \text{Pe}\, \Delta x^2. \tag{6.188}$$

The truncation error is still multiplied by the Peclet number, but it is also multiplied by one more Δx than in Eq. (6.186). This is the basis for the method developed by Price, *et al.* [1968]. For the interior points they use

$$\frac{dc_i}{dt} = -\text{Pe} \frac{c_{i-2} - 4 c_{i-1} + 3 c_i}{2\Delta x} + \frac{c_{i-1} - 2 c_i + c_{i+1}}{\Delta x^2}. \tag{6.189}$$

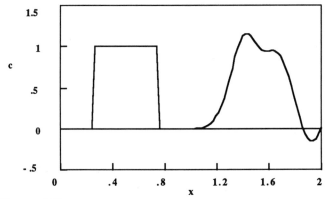

Figure 6.32. Convective Diffusion Equation, 3-point Upstream Method
Pe = 1000, $\Delta x = 0.02$, $\Delta t = 1.\text{E-}6$

When this equation is applied at $i = 2$, the value of c_0 is needed; however, this does not exist (the boundary point is $i = 1$). Thus, for this point they use the first-order expression to obtain

$$\frac{dc_2}{dt} = -\text{Pe}\,\frac{c_2 - c_1}{\Delta x} + \frac{c_1 - 2c_2 + c_3}{\Delta x^2}. \tag{6.190}$$

At the right-hand boundary the first derivative is zero; as a result, the convective term is absent. The diffusion term is handled using a false boundary. The boundary condition gives

$$\left.\frac{\partial c}{\partial x}\right|_{x=1} = 0, \quad c_{NT+1} = c_{NT-1} \tag{6.191}$$

and the diffusion term is then

$$\frac{\partial^2 c}{\partial x^2} = \frac{2(c_{NT-1} - c_{NT})}{\Delta x^2}. \tag{6.192}$$

Thus the equation for $i = NT$ is

$$\frac{dc_{NT}}{dt} = \frac{2(c_{NT-1} - c_{NT})}{\Delta x^2}. \tag{6.193}$$

If this set of equations is integrated using an explicit first-order method in time, the results are as shown in Figure 6.32. These should be compared with results from the centered finite difference method (Figure 4.16b) or with first-order upstream derivatives (Figure 4.16a). The results obtained with three-point upstream derivative have less artificial diffusion than those obtained with the first-order upstream derivative. They also have fewer oscillations than the centered finite difference method. Thus, the use of a three-point upstream derivative is better than a first-order upstream derivative or a centered derivative. However, the results using the three-point upstream derivative, shown in Figure 6.32, are inadequate when compared with the best results displayed in this chapter. (See also the comparison by Zurigat and Ghajar [1990].)

QUICK. The QUICK method uses an even higher truncation error. As developed by Leonard [1979a], it is

$$\frac{dc_i}{dt} = -\text{Pe}\left[\frac{c_{i+1} - c_{i-1}}{2\,\Delta x} - \frac{c_{i+1} - 3c_i + 3c_{i-1} - c_{i-2}}{6\,\Delta x}\right] + \frac{c_{i-1} - 2c_i + c_{i+1}}{\Delta x^2}. \quad (6.194)$$

We cannot use this equation for $i = 2$, so we use Eq. (6.190) instead. For the last point, $i = \text{NT}$, we use Eq. (6.193). The truncation error of the convective term can be checked using a Taylor series expansion (see Problem 6.8). The result is

$$\frac{c_{i+1} - c_{i-1}}{2\,\Delta x} - \frac{c_{i+1} - 3c_i + 3c_{i-1} - c_{i-2}}{6\,\Delta x} = c_i' + \frac{1}{12} c_i'''' \,\Delta x^3. \quad (6.195)$$

Thus the truncation error is proportional to $\text{Pe}\Delta x^3$, which appears to be better than the second-order result, Eq. (6.188). Evaluation of the actual truncation term, using the exact solution for steady flow [Eq. (2.3)], gives us

$$\text{second-order first derivative:} \quad -\frac{e^{\text{Pe}\,x}}{e^{\text{Pe}} - 1} \frac{1}{3} \text{Pe}^4 \,\Delta x^2 \quad (6.196)$$

$$\text{third-order first derivative:} \quad +\frac{e^{\text{Pe}\,x}}{e^{\text{Pe}} - 1} \frac{1}{12} \text{Pe}^5 \,\Delta x^3. \quad (6.197)$$

Because each higher derivative is multiplied by one more Peclet number, the ratio of these truncation errors is $\text{Pe}\Delta x/4$. Thus $\text{Pe}\Delta x/2$ must be ≤ 2.0 to have a smaller truncation error with a higher-order term. However, results obtained with large $\text{Pe}\Delta x$ values do not always follow the theory derived for small $\text{Pe}\Delta x$ values. We look instead at the results achieved with the QUICK method in Figure 6.33; they are better than those from the standard finite difference methods (see Figure 4.16). The steepness of the front is equivalent to that obtained with the centered finite difference method, yet the oscillations are much smaller. The QUICK method was unstable, however, for the time-step used by the other methods ($\Delta t = 10^{-5}$) and had to be reduced to 5×10^{-7}. The use of a method with a higher-order truncation error for the convective term seems promising, however.

Other results for the advection equation using the QUICK method are shown in Figure 6.34. The results have oscillations that grow slightly with time, but they are comparable to the oscillations observed with the Lax-Wendroff finite difference method (see Figure 4.9). The QUICK method might work better with an implicit treatment of the time derivatives, since the time-step has to be so small. Overall, though, the QUICK method is as good as or better than the finite difference methods and comparable to the better finite element methods. Below we will consider the method when applied to nonlinear problems.

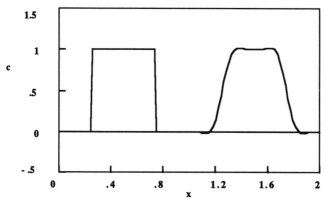

Figure 6.33. Convective Diffusion Equation, QUICK Method, Pe = 1000
$\Delta x = 0.02$, $\Delta t = 5E-7$

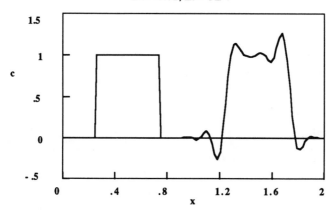

Figure 6.34. Advection Equation, QUICK Method, $\Delta x = 0.02$, $\Delta t = 0.001$

The dispersion diagrams for the QUICK method are shown in Figure A.26. The stability line for the QUICK method is shown on the Co-r diagram in Figure 6.35. The stability region is generally smaller than the Lax-Wendroff, Taylor-finite difference, or centered finite difference methods, except for a region near r = 0.25. However, the solutions using QUICK are much better than those methods.

When applied to the nonlinear equation

$$\frac{\partial u}{\partial t} + \frac{\partial}{\partial x}[F(u)] = v \frac{\partial^2 u}{\partial x^2}, \qquad (6.198)$$

the QUICK method is

$$\frac{du_i}{dt} + \frac{F_{i+1} - F_{i-1}}{2\Delta x} - \frac{F_{i+1} - 3F_i + 3F_{i-1} - F_{i-2}}{6\Delta x} = v \frac{u_{i+1} - 2u_i + u_{i-1}}{\Delta x^2}. \qquad (6.199)$$

Results when the method is applied to Burger's equation are shown in Figures

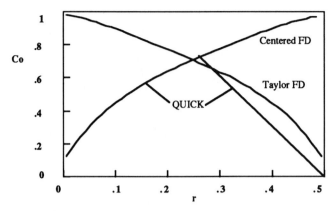

Figure 6.35. Stability Diagram for the QUICK Method

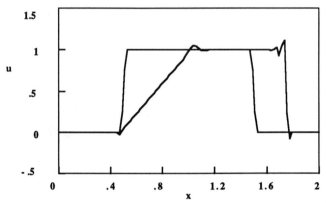

Figure 6.36a. Burger's Equation, QUICK Method, $v = 0$, $\Delta x = 0.02$, $\Delta t = 0.005$

6.36a and 6.36b. These figures indicate that the QUICK method oscillates more than the MacCormack method (compared with Figure 4.13c) and that the QUICK method requires a smaller time-step. Figure 6.36b shows that the results are not sensitive to the time-step in the range used. When viscosity is included, the method performs better, as illustrated in Figure 6.37. The oscillations now are comparable to those obtained with the MacCormack method, although a smaller time-step is still needed than that used for MacCormack's method. The QUICK method gives results which are inferior to the best finite element methods for Burger's equation (see Figure 6.2).

There are other versions of the QUICK method. Leonard [1979b] uses

$$\frac{c_i^{n+1} - c_i^n}{\Delta t} + \frac{Pe}{8 \Delta x}(c_{i-2}^n - 7 c_{i-1}^n + 3 c_i^n + 3 c_{i+1}^n) = \frac{1}{\Delta x^2}(c_{i+1}^n - 2 c_i^n + c_{i-1}^n). \quad (6.200)$$

This method is only second-order in Δx. A finite element formulation of QUICK

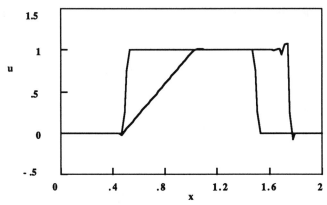

Figure 6.36b. Burger's Equation, QUICK Method, $\nu = 0$, $\Delta x = 0.02$, $\Delta t = 0.0025$

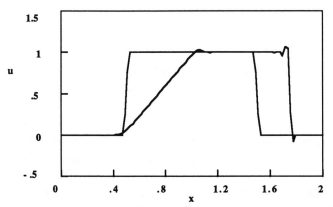

Figure 6.37. Burger's Equation, QUICK Method, $\nu = 0.001$, $\Delta x = 0.02$, $\Delta t = 0.005$

is constructed by Steffler [1989]; the algorithm is not merely a finite element variation of the finite difference formulation, but a complete reconstruction using the ideas motivating QUICK. Steffler uses a quadratic interpolation over two adjacent elements, extending into the upstream element. The steady convective diffusion equation is then represented as

$$\frac{Pe}{12 \Delta x} (c_{i-2}^n - 9 c_{i-1}^n + 3 c_i^n + 5 c_{i+1}^n) = \frac{1}{\Delta x^2} (c_{i+1}^n - 2 c_i^n + c_{i-1}^n). \quad (6.201)$$

The results appear to be no better than those described here, although Steffler's method should be easier to extend to two dimensions. Barga [1990] discusses a variety of upstream methods that use similar ideas. Lin and Chieng [1991] compare QUICK, flux-corrected, and TVD methods.

Laumbach's method. An implicit method that combines aspects of the finite element formulation and the finite difference formulation was derived by Laumbach [1975]. The derivation of this method was designed to make the truncation error smaller. The result of Lambauch's method is the following equation (see Problem 6.9):

$$(1-w)\frac{c_i^{n+1} - c_i^n}{\Delta t} + \frac{w}{2}\frac{c_{i+1}^{n+1} - c_{i+1}^n + c_{i-1}^{n+1} - c_{i-1}^n}{\Delta t} =$$
$$- \frac{Pe}{4\Delta x}[c_{i+1}^{n+1} - c_{i-1}^{n+1} + c_{i+1}^n - c_{i-1}^n] + \qquad (6.202)$$
$$+ \frac{1}{2\Delta x^2}[c_{i+1}^{n+1} - 2c_i^{n+1} + c_{i-1}^{n+1} + c_{i+1}^n - 2c_i^n + c_{i-1}^n].$$

The truncation error is (see Problem 6.9)

$$\frac{\partial c}{\partial t} + Pe\frac{\partial c}{\partial x} - \frac{\partial^2 c}{\partial x^2} = \frac{\partial^3 c}{\partial x^3}Pe\,\Delta x^2\left(-\frac{Co^2}{12} + \frac{w}{2} - \frac{1}{6}\right) + \frac{\partial^4 c}{\partial x^4}\Delta x^2\left(\frac{1}{12} + \frac{Co^2}{8}\right). \qquad (6.203)$$

If we choose

$$w = \frac{1}{3} + \frac{Co^2}{6}, \quad Co = \frac{Pe\,\Delta t}{\Delta x}, \qquad (6.204)$$

the first term of the truncation error in Eq. (6.203) vanishes. The remaining term is not multiplied by a Peclet number, so the truncation error does not become unbounded when the Peclet number becomes large.

Laumbach's method is implicit, and the solution method is given by Eq. (2.87)-(2.89). We notice that as μ approaches zero, the parameter w approaches 1/3, and the time derivatives are handled like the Galerkin method. The results (Figure 6.38) are better than achieved with either finite difference methods (Figure 4.16) or the Taylor-Galerkin method (Figure 5.8e). They are also better than the results obtained with the QUICK method, in that the oscillations are slightly smaller, the front is steeper, and the time-step allowed is larger. For this time-step, w is approximately equal to 1/3. Thus, the improvement over the Galerkin methods must result from averaging the diffusion and convection terms. When applied to the advection equation, the results are not nearly as good (see Figure 6.39). In this case, the oscillations are larger than for finite difference methods. Since Laumbach's method is restricted to problems for which truncation analysis can be done (usually linear ones) and is an improvement in only some of the cases, we will consider it no further.

Rosenberg's method [1986] is a linear combination of the equations from a centered finite difference method and an upstream finite difference method, both of which are treated implicitly (see Problem 6.10). This method is similar to the flux-correction schemes in that parts of a high-order solution are added in a special way to a low-order solution in order to eliminate oscillations. The method is

$$\frac{1}{\Delta t}\left\{c_i^{n+1} - c_{i-1}^n + \frac{R}{4}[3c_{i-1}^n + c_{i+1}^{n+1} + c_{i+1}^n - c_{i-1}^{n+1} - 4c_i^n]\right\} =$$
$$+ \frac{1}{2\Delta x^2}[c_{i+1}^{n+1} - 2c_i^{n+1} + c_{i-1}^{n+1} + c_{i+1}^n - 2c_i^n + c_{i-1}^n]. \qquad (6.205)$$

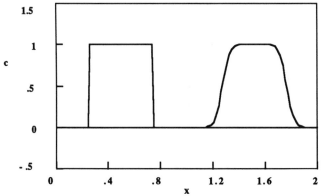

Figure 6.38. Convective Diffusion Equation, Laumbach Method, Pe = 1000 $\Delta x = 0.02$, $\Delta t = 1.E-6$

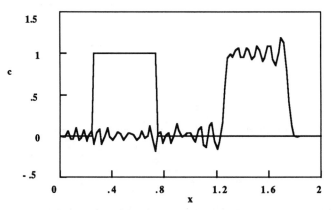

Figure 6.39. Advection Equation, Laumbach Method, $\Delta x = 0.02$, $\Delta t = 0.002$

We limit the time-step to

$$\frac{Pe \, \Delta t}{\Delta x} = 1 \text{ or } \Delta t = \frac{\Delta x}{Pe}; \quad \text{let } R = \frac{2}{Pe \, \Delta x}; \quad \text{then } \frac{\Delta t}{\Delta x^2} = \frac{R}{2}. \quad (6.206)$$

Rearrangement gives the final representation of the method as

$$(2+R) \, c_i^{n+1} = (2-R) \, c_{i-1}^n + R \, c_{i-1}^{n+1} + R \, c_i^n. \quad (6.207)$$

We note that as Pe approaches infinity, R approaches zero, and the equation reduces to

$$2 \, c_i^{n+1} = 2 \, c_{i-1}^n, \quad (6.208)$$

as it should for an infinite Peclet number.

The difference expression, Eq. (6.205), is a special one; we note that it cannot be used for steady-state cases. However, it is a very accurate method for

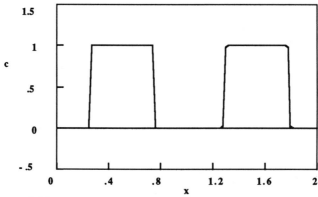

Figure 6.40. Convective Diffusion Equation, Rosenberg Method, Pe = 1E5
$\Delta x = 0.02, \Delta t = 2E-7$

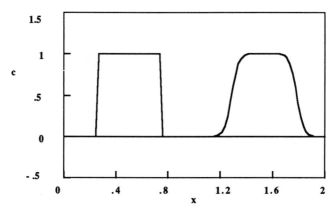

Figure 6.41. Convective Diffusion Equation, Rosenberg Method, Pe = 1000
$\Delta x = 0.02, \Delta t = 2E-5$

the time-dependent problem (see Figures 6.40 and 6.41). This method has the steepest front, has no oscillations and allows the largest time-step. The reason the method is so accurate in time can be seen by calculating the dispersion relation. We take the finite Fourier transform of Eq. (6.207):

$$(2+R)\hat{c}^{n+1} = (2-R)e^{i\xi}\hat{c}^n + R e^{i\xi}\hat{c}^{n+1} + R \hat{c}^n. \tag{6.209}$$

The dispersion relation is

$$[2+R(1-e^{i\xi})]\hat{c}^{n+1} = [2+R(1-e^{i\xi})]\hat{c}^n \quad \text{or } \rho = 1. \tag{6.210}$$

Thus, any waveform is convected without error in the wavespeed or the phase lag. This makes it an excellent method. Despite the positive aspects of this method, it has yet to be extended to nonlinear problems, although Rosenberg has tried to do so. El-Ageli and Chierici [1988] have extended it to multi-dimensional problems for the convective diffusion equation. Because of the method's limited applicability, however, we will consider it no further.

Of all the special methods treated in this section, the QUICK method is the only general one (applicable to all the problems treated) that is a good method. Without diffusion or viscosity, the method is not as good as the corresponding finite difference methods, but with diffusion or viscosity, the method is better. It is, however, inferior to the best finite element methods for Burger's equation. It is also inferior to the random choice method and the MacCormack method with flux-correction. If we are determined to use finite difference methods, it is probably better for us to use the flux-correction method, but the QUICK method is also a possible choice, provided there is some viscosity or diffusion.

6.5 Orthogonal Collocation on Finite Elements

The method of orthogonal collocation was developed by Villadsen and Stewart [1967] {see also Villadsen and Michelsen [1978]}. When the Peclet number is large, the orthogonal collocation method will not provide good results, since the degree of polynomial must be very large to approximate a solution with large gradients. In that case we are better off using the method of orthogonal collocation on finite elements, as developed by Carey and Finlayson [1975]. In this method, the domain is divided into elements and the method of orthogonal collocation is applied within each element. Within an element, the solution is expressed as a high-order polynomial. We must specify the number of elements and the number of collocation points within an element. Then the equations are derived by applying the collocation method at a set of points within each element. Since the method is thoroughly explained in another book by Finlayson [1980], only the results will be given here. The method is also described in "Orthogonal Collocation on Finite Elements," which is provided as a MacWrite™ document on the computer diskette.

Results are given in this chapter for the orthogonal collocation method, but using the implicit method (which was readily available). When two interior collocation points are used, the solution is expressed as a cubic polynomial within each element. For the convective diffusion equation, the collocation results for 25 elements and 2 interior points are as shown in Figure 6.42a. There are small oscillations. Jensen and Finlayson [1980] have shown that oscillations do not occur in the steady-state problem provided that

$$\frac{Pe\Delta x}{3.464} \leq 1. \qquad (6.211)$$

This parameter is comparable with $Pe\Delta x/2$ in the Galerkin method. For the calculation shown in Figure 6.42a, the quantity in Eq. (6.211) takes the value 11.5. The Galerkin method with $Pe\Delta x/2 = 10$ gives the results shown in Figure 5.8a; the Taylor-Galerkin method with $Pe\Delta x/2 = 10$ gives the results shown in Figure 5.8e. There are small oscilllations in those solutions, too. When more elements are

182 NUMERICAL METHODS - MOVING FRONTS

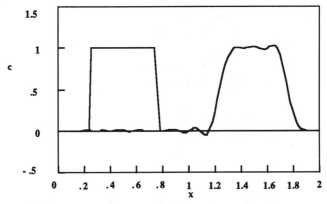

Figure 6.42a. Convective Diffusion Equation, Orthogonal Collocation on Finite Elements, Pe = 1000, Ne = 25, NCOL = 2, Δt = 2E-6

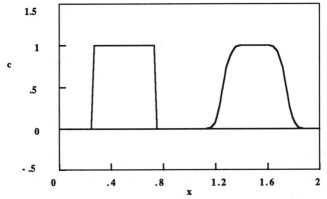

Figure 6.42b. Convective Diffusion Equation, Orthogonal Collocation on Finite Elements, Pe = 1000, Ne = 50, NCOL = 2, Δt = 2E-6

used, such as 50 elements and 2 interior collocation points, the method of orthogonal collocation on finite elements gives the results shown in Figure 6.42b. In this case there are no oscillations. The value of PeΔx/2 is 5. The Galerkin method and the Taylor-Galerkin method for PeΔx/2 = 5 gives results that are also good (see Problem 6.11). We thus conclude that the method of orthogonal collocation can be used for this problem, but we already have methods that are just as good.

The method of orthogonal collocation on finite elements has been applied to the convective diffusion equation using a mesh moving with the velocity of the flow [Chawla, *et al.* 1984]. Excellent results are achieved even when the Peclet number is as large as 160,000 and only 40 elements are used.

6.6 Conclusion

Presented in this chapter are some excellent methods, such as the random choice method, MacCormack's method with flux-correction, and the TVD and ENO methods. The Petrov-Galerkin methods were not nearly as good for transient problems, despite their firm grounding for steady-state problems. Several high-order methods, including the QUICK, Laumbach, and orthogonal collocation on finite elements methods, gave excellent solutions for linear problems. Thus, there are excellent methods available on fixed meshes. Chapter 7 considers the improvements that are necessary when the mesh is allowed to move.

Problems

1_1. Use the program CONVECT for the explicit Taylor-finite difference method to show that this method oscillates. Use a step change for initial conditions.

2_1. Prove that the flux-correction step in Eq. (6.53) and preceding equations gives the low-order solution if all the correction factors are 0 and the high-order solution if all the correction factors are 1.

3_1. Prove that the flux-correction step in Eq. (6.85) and preceding equations gives the low-order solution if all the correction factors are 0 and the high-order solution if all the correction factors are 1.

4_1. For the advection equation, combine the steps in Eq. (6.55)-(6.56) and show that the result is the same as in Eq. (6.54). [Note: In CONVECT the first point (boundary condition) is handled differently in the MacCormack method and the centered finite difference method; slight discrepancies may be noted for this point. In the MacCormack method, $c^{*\,n+1}_1$ can be calculated if there is no diffusion, but not if there is diffusion. In the code CONVECT this value is always taken as $c^{*\,n+1}_1 = c_1^{n+1}$, the inlet condition, for the MacCormack method.]

5_3. Consider the finite element flux-correction method in Eq. (6.85); apply lumping to the time-derivative terms [including those in Eq. (6.71)] and ignore the Taylor terms. Prove that this method is the same as the centered finite difference method described by Eq. (6.53).

6_2. Evaluate Eq. (6.166). (Note: the derivatives of the trial functions are constant within an element.)

7_1. Show that the variational method in Eq. (6.171) is the same as the Petrov-Galerkin method in Eq. (6.134) when the weighting function is taken as Eq. (6.173).

8_1. Evaluate the truncation error of the QUICK equation, Eq. (6.194).

9_3. Derive the truncation error of the Laumbach method, Eq. (6.202). [Hint: evaluate it at the point (i,n+1/2)].

10_2. Derive the Rosenberg method, Eq. (6.205). Write one equation using a centered difference expression and the trapezoid rule in time. Write a second equation using an

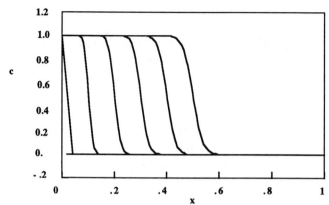

Figure 6.43. Convective Diffusion Equation, Orthogonal Collocation on Finite Elements, Pe = 1000, Ne = 50, NCOL = 2, Δt = 2E-6

explicit, upstream first derivative and the centered second derivative with the trapezoid rule. Multiply the first equation by (1-R) and the second equation by R and add the equations to get Eq. (6.205).

11$_2$. Use the program CONVECT to solve the convective diffusion equation when using the Galerkin and Taylor-Galerkin methods. Study the case of a step change for initial conditions when Pe = 1000 and PeΔx/2 = 5. Compare with Figure 6.43, obtained with the method of orthogonal collocation on finite elements under similar conditions.

12$_2$. Use the program CONVECT to solve the advection equation when using the random choice method and the various flux-correction methods. Study the case of a step change for initial conditions. Compare with the results derived in Problems 4.5 and 5.5.

13$_2$. Use the program CONVECT to solve the convective diffusion equation when using the random choice method and the various flux-correction methods. Use a step change for initial conditions and Pe = 1000. Compare with the results derived in Problems 4.11 and 5.13.

14$_2$. Use the program CONVECT to solve the advection equation when using the Morton-Parrott method. Use a step change for initial conditions. Compare with the results derived in Problem 5.12.

15$_2$. Use the program CONVECT to solve the convective diffusion equation when using the Morton-Parrott method. Use a step change for initial conditions and Pe = 1000. Compare with the results derived in Problems 5.11, 5.13, or 5.14.

16$_2$. Use the program CONVECT to solve the advection equation using explicit and implicit Petrov-Galerkin methods. Study a step change for the initial conditions and write an essay comparing the explicit and implicit methods. Also compare results obtained by the implicit Galerkin method with those obtained by Petrov-Galerkin methods.

17$_2$. Use the program CONVECT to solve the convective diffusion equation using explicit and implicit Petrov-Galerkin methods. Study a step change for the initial conditions and Pe = 1000. Write an essay comparing the explicit and implicit methods. Also compare results obtained by the implicit Galerkin method with those obtained by Petrov-Galerkin methods.

18$_2$. Do problem 6.16, except with the Taylor-Galerkin method instead of the Petrov-Galerkin method.

19$_2$. Do problem 6.17, except with the Taylor-Galerkin method instead of the Petrov-Galerkin method.

20$_2$. Use the program CONVECT to solve the convective diffusion equation using any method. Study a step change for the initial conditions when Pe = 1000 and 100 nodes. Calculate the solution at x = 0.5, t = 0.5/Pe and compare your results with the following table:

Table 6.1. Slope of Solution at x = 0.5, t = 0.5/Pe; Pe = 1000

Method	Slope of Solution
1st-order upstream, FD	-4.8
3-point upstream, FD	-7.3
Centered FD	-8.3
QUICK	-8.7
Laumbach	-11.5
Rosenberg	-12.0
Galerkin FE	-13.4
Exact	-14

21$_2$. Use the program CONVECT to solve Burger's equation without viscosity when using the random choice method and the flux-correction method. Take the initial condition as the ramp function, as defined in Problem 4.14. Compare with the results derived in Problems 4.6 and 5.11.

22$_2$. Use the program CONVECT to solve Burger's equation with viscosity ($\nu = 0.001$) when using the random choice method and the flux-correction method. Take the initial condition as the ramp function, as defined in Problem 4.14. Compare with the results derived in Problems 4.12 and 5.14.

23$_1$. Verify Eq. (6.93).

24$_2$. Prove that Eq. (6.106) gives the same results as Eq. (6.101) when there are no source terms.

25$_3$. Prove that Eq. (6.106) is second order in Δt and second order in Δx when there are no steep fronts.

26$_2$. Prove that Eq. (6.128) gives the same results as Eq. (6.124) when there are no source terms.

27$_3$. Prove that Eq. (6.128) is third order in Δt and second order in Δx when there are no steep fronts.

References

Barga, W., "On the Use of Some Weighted Upwind Schemes for Strongly Convective Flows," Num. Heat Trans., Part B. 18 43-60 [1990].

Barrett, J. W. and Morton, K. W., "Optimal Finite Element Solutions to Diffusion-Convection

Problems in One Dimension," Int. J. Num. Methods Eng. 15 1457-1474 [1980].

Barrett, J. W. and Morton, K. W., "Approximate Symmetrization and Petrov-Galerkin Methods for Diffusion-Convection Problems," Comp. Meth. Appl. Mech. Eng. 45, 97-122 [1984].

Book, D. L., *Finite-Difference Techniques for Vectorized Fluid Dynamics Calculations*, Springer-Verlag [1981].

Book, D. L., Boris, J. P., and Hain, K. H., "Flux-Corrected Transport-II: Generalizations of the Method," J. Comp. Phys. 18 248-283 [1975].

Boris, J. P. and D. L. Book, "Flux-Corrected Transport. I. Shasta, A Fluid Transport Algorithm That Works," J. Comp. Phys. 11, 38-69 [1973].

Carey, G. F. and Finlayson, B. A., "Orthogonal Collocation on Finite Elements," Chem. Eng. Sci. 30 587-596 [1975].

Carnahan, B., Luther, H. A., and Wilkes, J. O., *Applied Numerical Methods*, Wiley [1969].

Chawla, T. C., Leaf, G., and Minkowycz, W. J., "A Collocation Method for Convection Dominated Flows," Int. J. Num. Methods Fluids 4 271-281 [1984].

Chen, W. H., Durlofsky, L. J., Engquist, B., and Osher, S., "Minimization of Grid Orientation Effects Through Use of Higher Order Finite Difference Methods," Soc. Pet. Eng. paper 22887 [1991].

Chorin, A. J., "Random Choice Solution of Hyperbolic Systems," J. Comp. Phys. 22 517-533 [1976].

Chorin, A. J., "Random Choice Methods with Applications to Reacting Gas Flows," J. Comp. Phys. 25 253-272 [1977].

Collella, P., " Glimm's Method for Gas Dynamics," SIAM J. Sci. Comput. 3 76-110 [1982].

Concus, P. and Proskurowski, W., "Numerical Solution of a Nonlinear Hyperbolic Equation by the Random Choice Method," J. Comp. Phys. 30 153-166 [1979].

Demkowicz, L. and Oden, J. T., "An Adaptive Characteristic Petrov-Galerkin Finite Element Method for Convection-dominated Linear and Nonlinear Parabolic Problems in Two Space Variables," Comp Meth Appl Mech Eng. 55 63-87 [1986].

El-Ageli, M. A. and Chierici, G., "An Extension of the One-dimensional von Rosenberg Finite Difference Scheme to Multidimensional Problems of the Convective Dispersion Equation," Num. Methods Part. Diff. Eqn. 4 1-13 [1988].

Finlayson, B. A., *The Method of Weighted Residuals and Variational Principles*, Academic Press [1972].

Finlayson, B. A., *Nonlinear Analysis in Chemical Engineering*, McGraw-Hill [1980].

Glimm, J., "Solutions in the Large for Nonlinear Hyperbolic Systems of Equations," Comm. Pure Appl. Math. 18 697-715 [1955].

Glimm, J., Marchesin, D., McBryan, O., "Subgrid Resolution of Fluid Discontinuities, II", J. Comp. Phys. 37 336-354 [1980].

Gudunov, S. K., "Finite Difference Methods for Numerical Computation of Discontinuous Solutions of the Equations of Fluid Dynamics," Mat. Sbornik 47 271 [1959].

Harten, A., "High Resolution Schemes for Hyperbolic Conservation Laws," J. Comp. Phys. 49 357-393 [1983].

Holt, M., *Numerical Methods in Fluid Dynamics*, Springer-Verlag, 2nd edition [1984].

Hughes, T. J. R., "A Simple Scheme for Developing 'Upwind' Finite Elements," Int. J. Num. Methods Eng. 12 1359-1365 [1978].

Hughes, T. J. R. and Brooks, A., "A Multi-dimensional Upwind Scheme with no Crosswind Diffusion," pp. 19-35 in *Finite Element Methods for Convection Dominated Flows*, AMD, Vol. 34, ASME [1979].

Hughes, T. J. R. and Atkinson, J. D., "A Variational Basis for "Upwind" Finite Elements", pp.387-391 in *Variational Methods in the Mechanics of Solids*, ed. Nemat-Nassar, Pergamon Press [1980].

Jensen, O. K. and Finlayson, B. A., "Oscillation Limits for Weighted Residual Methods," Int. J. Num. Methods Eng. 15 1681-1689 [1980].

Laumbach, D. D. "A High-Accuracy Finite-Difference Technique for Treating the Convection-Diffusion Equation," Soc. Pet. Eng. J. 15 517-531 [1975].

Leonard,B.P., "A Survey of Finite Differences of Opinion on Numerical Muddling of the incomprehensible Defective Confusion Equation," pp. 1-17 in *Finite Element Methods for Convection Dominated Flows*," ed. T. J. R. Hughes, AMD, Vol. 34, ASME [1979a].

Leonard,B.P., "A Stable and Accurate Convective Modelling Procedure Based on Quadratic Upstream Interpolation," Comp. Method. Appl. Mech. Eng. 19 59-98 [1979b].

Li, K. M. and Holt, M., "Numerical Solutions to Water Waves Generated by Shallow Underwater Explosions," Phys. Fluids 24 816-824 [1981].

Lin, H. and Chieng, C. C., "Characteristic-based Flux Limiters of an Essentially Third-Order Flux-splitting Method for hyperbolic Conservation Laws," Int. J. Num. Methods Fluids 13 287-307 [1991].

Löhner, R., Morgan, K., Peraire, J. and Vahdati, M., "Finite Element Flux-corrected Transport (FEM-FCT) for the Euler and Navier-Stokes Equations," Int. J. Num. Methods Fluids 7 1093-1109 [1987].

Morton, K. W. and Parrott, A. K., "Generalized Galerkin Methods for First-Order Hyperbolic Equations," J. Comp. Phys. 36 249-270 [1980].

Poulain, C., private communication [1991].

Price, H. S., Cavendish, J. C. and Varga, R. S., "Numerical Methods of Higher-Order Accuracy for Diffusion-Convection Equations," Soc. Pet. Eng. J. 8 293-303 [1968].

Roe, P. L., "The Use of the Riemann Problem in Finite Difference Schemes," Proc. Seventh Int. Conf. Num. Methods Fluid Dynamics, Lecture Note in Physics, Vol. 141, W. C. Reynolds, and R. W. MacCormack (ed.), Springer-Verlag [1981a].

Roe, P. L., "Approximate Riemann Solvers, Parameter Vectors, and Difference Schemes," J. Comp. Phys. 43 357-372 [1981b].

Rosenberg, D. U. von, "An Explicit Finite Difference Solution to the Convection-Dispersion Equations," Num. Meth. Partial Diff. Eqn. 2 229-237 1986].

Shu, C. W. and Osher, S., "Efficient Implementation of Essentially Non-oscillatory Shock-Capturing Schemes" J. Comp. Phys. 77 439-471 [1988].

Shu, C. W. and Osher, S., "Efficient Implementation of Essentially Non-oscillatory Shock-Capturing Schemes, II," J. Comp. Phys. 83 32-78 [1989].

Sod, G. A., "A Numerical Study of a Converging Cylindrical Shock," J. Fluid Mech. 83 785-794 [1977].

Sod, G. A., "A Survey of Several Finite Difference Methods for Systems of Nonlinear Hyperbolic Conservation Laws," J. Comp. Phys. 27 1-31 [1978].

Sod, G. A., *Numerical Methods in Fluid Dynamics,* Cambridge Univ. Press [1985].

Steffler, P. M., "Upwind Basis Finite Elements for Convection-dominated Problems," Int. J. Num. Methods Fluids 9, 385-403 [1989].

Sweby, P. K., "High Resolution Scheme Using Flux Limiters for Hyperbolic Conservation Laws," SIAM J. Num. Anal. 21 995-111 [1984].

Villadsen, J. and Michelsen, M. L., *Solution of Differential Equation Models by Polynomial Approximation,* Prentice-Hall [1978].

Villadsen, J.V. and Stewart, W. E., "Solutions of Boundary-Value Problems by Orthogonal Collocation," Chem. Eng. Sci. 22 1483-1501 [1967].

Yee, H. C., Warming, R. F., Harten, A., "Implicit Total Variation Diminishing (TVD) Schemes for Steady-State Calculations," J. Comp. Phys 57 327-360 [1980].

Yu, C.-C. and Heinrich, J. C., "Petrov-Galerkin Methods for the Time-dependent Convective Transport Equation," Int. J. Num. Methods Engn. 23 883-901 [1986].

Zalesak, S. T., "Fully Multidimensional Flux-Corrected Transport Algorithms for Fluids," J. Comp. Phys. 31 335-362 [1979].

Zuragat, Y. H. and Ghajar, A. J., "Comparative Study of Weighted Upwind and Second Order Upwind Difference Schemes," Num. Heat Trans., Part B: Fund 18 61-79 [1990].

CHAPTER
SEVEN

ADAPTIVE AND MOVING MESHES

In preceding chapters we have seen that one way to eliminate the oscillations in a solution is to use a small mesh. This may be prohibitively expensive, however, since using a small mesh will increase the amount of computer time necessary to solve a problem. It is apparent that the small mesh is only needed in certain regions - if we only knew where! In this chapter we will consider several ways to adapt the mesh to the problem, using small elements or small meshes where needed and allowing larger elements elsewhere. Thus, we can obtain both the improved accuracy of a small mesh and the improved economy of a small number of nodal points. We will first consider steady-state problems and learn to adjust the mesh to improve results. Then we will consider unsteady-state problems, those problems with a moving front, and learn how to move the mesh so that we have a small mesh only where we need one.

7.1. Adaptive Meshes for Steady-State Problems

For a steady-state problem we need a method that will take an approximate solution and determine where smaller mesh sizes are needed. We then refine the grid there and re-solve the problem. The refinement of the grid is simple for one-dimensional problems since we merely add nodes. In this section we will concentrate on the criteria used to decide if a smaller mesh is needed in some region. Four different criteria are presented here, three of which work well. Some of these are *ad hoc*, but others are firmly grounded in mathematical theory. Before presenting the different criteria, however, let us look at an example that demonstrates the advantages of adaptive mesh refinement.

We take the steady convective diffusion equation:

$$\text{Pe}\frac{dc}{dx} = \frac{d^2c}{dx^2}. \qquad (7.1)$$

This problem is solved in Chapter 2 [see Eq. (2.3)] and is displayed in Figure 2.1. We concentrate here on the case where the Peclet number = 100, which is severe enough to demonstrate the important problems but still allows reasonable

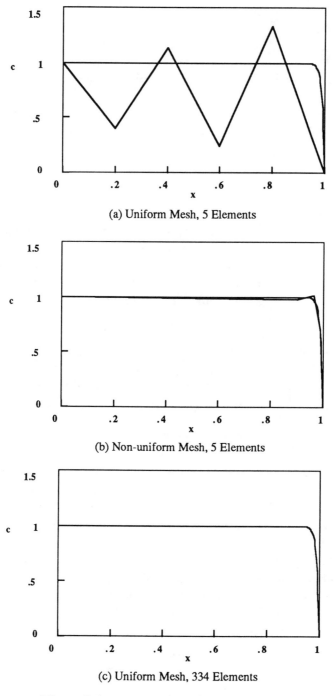

Figure 7.1. Convective Diffusion Equation, Pe = 100

graphical presentation of the results. The centered finite difference method was applied, with the results for different meshes as shown in Figure 2.3. We take here the case with six nodes, $\Delta x = 0.2$, and Pe $\Delta x/2 = 10$. The solution oscillates wildly and is completely unsatisfactory, as shown in Figure 7.1a. Next let us solve the problem using the same number of grid points (six) but using a variable mesh. To do so we need to develop equations that apply to problems in which the elements (or the distances between grid points) are not all the same size. We do that here for the transient convective diffusion equation.

The Galerkin method applied to the transient problem gives

$$\sum_{i=1}^{NT} \int_0^1 N_j N_i \, dx \, \frac{dc_i}{dt} + \text{Pe} \sum_{i=1}^{NT} \int_0^1 N_j \frac{dN_i}{dx} \, dx \, c_i = -\sum_{i=1}^{NT} \int_0^1 \frac{dN_j}{dx} \frac{dN_i}{dx} \, dx \, c_i. \qquad (7.2)$$

If we use

$$\frac{dN_j}{d\xi} = \frac{2}{\Delta x_e} \frac{dN_J}{d\xi}, \qquad dx = \frac{\Delta x_e}{2} d\xi \qquad (7.3)$$

and revert to an element notation, we get

$$\sum_e \sum_I \frac{\Delta x_e}{2} \int_{-1}^1 N_J N_I \, d\xi \, \frac{dc_I^e}{dt} + \text{Pe} \sum_e \sum_I \int_{-1}^1 N_J \frac{dN_I}{d\xi} \, d\xi \, c_I^e =$$

$$= -\sum_e \sum_I \frac{2}{\Delta x_e} \int_{-1}^1 \frac{dN_J}{d\xi} \frac{dN_I}{d\xi} \, d\xi \, c_I^e. \qquad (7.4)$$

Rearrangement gives

$$\sum_e \frac{\Delta x_e}{2} \sum_I \int_{-1}^1 N_J N_I \, d\xi \, \frac{dc_I^e}{dt} + \text{Pe} \sum_e \sum_I \int_{-1}^1 N_J \frac{dN_I}{d\xi} \, d\xi \, c_I^e +$$

$$+ \sum_e \frac{2}{\Delta x_e} \sum_I \int_{-1}^1 \frac{dN_J}{d\xi} \frac{dN_I}{d\xi} \, d\xi \, c_I^e = 0. \qquad (7.5)$$

Using the matrices in Table 2.2 makes this

$$\sum_e \frac{\Delta x_e}{2} \begin{bmatrix} \frac{2}{3} & \frac{1}{3} \\ \frac{1}{3} & \frac{2}{3} \end{bmatrix} \frac{dc_I^e}{dt} + \text{Pe} \sum_e \begin{bmatrix} -\frac{1}{2} & \frac{1}{2} \\ -\frac{1}{2} & \frac{1}{2} \end{bmatrix} c_I^e + \sum_e \frac{2}{\Delta x_e} \begin{bmatrix} \frac{1}{2} & -\frac{1}{2} \\ -\frac{1}{2} & \frac{1}{2} \end{bmatrix} c_I^e = 0. \qquad (7.6)$$

Next let us consider the assembly for the j-th equation. A typical grid is shown in Figure 7.2. When we process the left-hand element, j in the global nodal numbering system equals 2 in the local numbering system; thus, we use the lower row of the matrix equations. When we process the right-hand element, j in the global nodal numbering system equals 1 in the local numbering system; thus, we use the upper row of the matrix equations. The combined result is

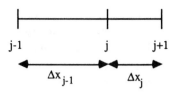

Figure 7.2. Variable Mesh

$$\frac{\Delta x_{j-1}}{6}\frac{dc_{j-1}}{dt} + \frac{\Delta x_{j-1}}{3}\frac{dc_j}{dt} + \frac{Pe}{2}(-c_{j-1}+c_j) + \frac{1}{\Delta x_{j-1}}(-c_{j-1}+c_j) +$$
$$+ \frac{\Delta x_j}{3}\frac{dc_j}{dt} + \frac{\Delta x_j}{6}\frac{dc_{j+1}}{dt} + \frac{Pe}{2}(-c_j+c_{j+1}) + \frac{1}{\Delta x_j}(c_j-c_{j+1}) = 0. \quad (7.7)$$

If we have a natural boundary condition at $x = 1$, then we have only the first half of this equation. Collecting terms for the general case gives

$$\frac{\Delta x_{j-1}}{6}\frac{dc_{j-1}}{dt} + \frac{1}{3}(\Delta x_{j-1}+\Delta x_j)\frac{dc_j}{dt} + \frac{\Delta x_j}{6}\frac{dc_{j+1}}{dt} + \frac{Pe}{2}(c_{j+1}-c_{j-1}) +$$
$$+ \frac{1}{\Delta x_{j-1}}(-c_{j-1}+c_j) + \frac{1}{\Delta x_j}(c_j-c_{j+1}) = 0. \quad (7.8)$$

If all the element sizes are the same, we obtain Eq. (2.101), as expected. For the steady-state problem we obtain

$$c_{j-1}\left(\frac{1}{\Delta x_{j-1}} + \frac{Pe}{2}\right) + c_j\left(-\frac{1}{\Delta x_{j-1}} - \frac{1}{\Delta x_j}\right) + c_{j+1}\left(\frac{1}{\Delta x_j} - \frac{Pe}{2}\right) = 0. \quad (7.9)$$

The boundary conditions are

$$c_1 = 1, \, c_{NT} = 0. \quad (7.10)$$

The same equation can be derived using the centered finite difference method (see Finlayson [1980, p.69]).

Eqs. (7.9)-(7.10) are solved using $Pe = 100$, $\Delta x = 0.2$, and $NT = 6$. Then $Pe\,\Delta x/2 = 10$, which is above the limit of 1.0, so oscillations cannot be ruled out. Figure 7.1a shows that the solution oscillates. Next we solve the problem with the nodes placed at 0, 0.9, 0.964, 0.988, 0.997, and 1.0 (the reason for these numbers is given below). This solution is quite good, as is shown in Figure 7.1b. This solution takes slightly more computer time to solve than the case with constant grid spacing, but the results are dramatically improved. We should note that although $Pe\,\Delta x/2 = 45$ in the first element, the results are good. This illustrates that a smaller mesh is needed in regions where the solution changes sharply or where the front is steep, but not everywhere. To demonstrate the accuracy of the approximation with carefully chosen nodes, we also solve the problem using a fine mesh ($\Delta x = 0.003$) everywhere. That solution is plotted in Figure 7.1c, too; we see that the solution is very much like that using only six nodes. Thus we have solved the problem with the same accuracy as a 334 node solution, but using the computer time associated with a six node solution, for a time savings of 98%. Our task, then, is to find automatically the region where small elements need to be located.

Let us consider the different ways in which we can locate the region where small meshes are needed. The first method of locating the nodes is to put more nodes where we know the solution is steep. Although this method assumes some knowledge of the solution, it can often be used. In this problem we know from the exact solution that the steep front is located at $x = 1$. This information could also have been obtained by solving a singular perturbation problem to the first approximation. The exact nodes used in the example were chosen as follows. Since the exact solution remains very close to 1.0 for most of the region, we put the second node at $x = 0.9$ (the exact solution was used to deduce the value of 0.9). Then we put the three nodes between $x = 0.9$ and $x = 1.0$ in such a way that the intervals are equal exponential values. In other words, we make the last element proportional to e, the previous one to e^2, the previous one to e^3 and the fourth element from the end proportional to e^4. The sum of the element lengths must be 0.1; we determine k from

$$(e^1 + e^2 + e^3 + e^4)k = 0.1. \tag{7.11}$$

This then gives the values of the elements and the corresponding nodal values as shown in Table 7.1.

Table 7.1. Examples of Element and Nodal Locations

i	Δx_i	x_i
5	0.003206	0.99679
4	0.008714	0.98808
3	0.023688	0.96439
2	0.064391	0.90000

Such a method of solving the problem may seem unimportant since the exact solution was known; the nodes, therefore, will be well-placed. It does however show that if the nodes are properly placed, good results can be achieved. We next turn to other methods that predict the region where small elements are needed.

After a solution is obtained, it is necessary to have some independent way to decide where to refine the mesh. In one of the early studies, Pearson [1968] used a centered finite difference method to find the solution and then required that the difference in solution values at successive nodes be made uniform over space. This is closely related to making the gradient of the solution uniform over space. Other early applications to finite difference methods are by Pereyra and Sewell [1975] (who distributed the local truncation error equally) and Russell [1977]. Several different strategies are reviewed by Russell and Christiansen [1978]; these include the global error, the truncation error, and derivatives of the solution. White [1979, 1982] distributed equally the curvature of the solution. The object is to make whichever function that is chosen somewhat uniform over space. Babuska and Miller [1984] advocate using a criterion that is related to some property of the solution we are interested in. In the method of weighted residuals, the objective is to make the residual small. Sometimes the residual can be related to the error in the solution [Ferguson and Finlayson, 1972, Finlayson, 1972, p. 388], and in these cases the residual is a natural choice. If it can somehow be reduced by optimal placement of the nodes, the error is thereby reduced. Even in cases where the relation between residual and error has not been shown, the principle has proved useful [Carey and Finlayson, 1975; Carey, 1979; and Carey and Humphrey, 1981]. Another criterion is the truncation error; this has been used by Christiansen and Russell [1978] as the basis for the program COLSYS, which uses the collocation method with an adaptive mesh to solve boundary value problems. Blottner [1978] and Carey and Plover [1983] have applied adaptive mesh strategies to the convective diffusion equation; we do this below in a slightly different way. We will consider in turn using the residual, the estimated error (which turns out to be proportional to the second derivative), and the gradient (or the first derivative). We then will look at the use of upstream methods in conjunction with adaptive mesh strategies.

Let us first consider using the residual to locate the nodes. The residual was defined in Chapter 2 as the differential equation with the approximate solution substituted in. We would like to make it zero everywhere: this is the basis of the method of weighted residuals. We never achieve that goal, however, except in special cases. Instead, we make the residual smaller by adding more and more terms. We can also look at the residual as a function of position and add mesh points where the residual is larger than in other regions. To do that we first need to be able to calculate the residual.

The residual for the steady-state convective diffusion equation is

$$R = \text{Pe}\frac{dc}{dx} - \frac{d^2c}{dx^2}. \tag{7.12}$$

The problem is that the approximation is linear and has no second derivatives. We need to develop the second derivative from the information at hand, which is the value of c at each node. Let us consider the grid identification shown previously in Figure 7.2.

If we use linear interpolation between grid points, the values of c at the two x-locations are (as shown in Figure 7.3):

$$\text{left x} \quad c_{Lx} = \frac{c_j + c_{j-1}}{2} \quad x_{Lx} = \frac{x_j + x_{j-1}}{2}$$

$$\text{right x} \quad c_{Rx} = \frac{c_j + c_{j+1}}{2} \quad x_{Rx} = \frac{x_j + x_{j+1}}{2}. \tag{7.13}$$

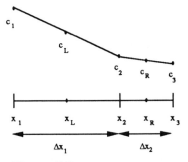

Figure 7.3. Local Interpolation

The first derivative is

$$\left.\frac{dc}{dx}\right|_j = \frac{c_{j-1} - c_{j+1}}{x_{j-1} - x_{j+1}}, \quad x_{j+1} - x_{j-1} = \Delta x_{j-1} + \Delta x_j. \tag{7.14}$$

The second derivative is (see Problem 7.1)

$$\left.\frac{d^2c}{dx^2}\right|_j = \frac{\frac{c_{j+1} - c_j}{x_{j+1} - x_j} - \frac{c_j - c_{j-1}}{x_j - x_{j-1}}}{\frac{1}{2}(x_{j+1} - x_{j-1})}. \tag{7.15}$$

The residual is

$$R = \text{Pe} \left.\frac{dc}{dx}\right|_j - \left.\frac{d^2c}{dx^2}\right|_j. \tag{7.16}$$

Collection of terms gives

$$\frac{(x_{j+1} - x_{j-1})}{2} R = \frac{\text{Pe}}{2}(c_{j+1} - c_{j-1}) - \left(\frac{c_{j+1} - c_j}{\Delta x_j} - \frac{c_j - c_{j-1}}{\Delta x_{j-1}}\right). \tag{7.17}$$

Eq. (7.17) is proportional to the equation used to solve for the solution in the first place, Eq. (7.9). Since Eq. (7.9) equals zero, Eq. (7.17) equals zero as well. Thus, the residual is zero; since we cannot use a zero residual, we need another way to calculate the residual.

A residual can be obtained by interpolating the solution in these two elements using a quadratic expression that goes through the three points. Then the solution has both first and second derivatives at the node i. The interpolation is obtained by taking the quadratic expression

$$c = a + bx + cx^2 \tag{7.18}$$

and evaluating it at the points $x = 0$ (without loss of generality), $x = \Delta x_1$, and $x = \Delta x_1 + \Delta x_2$ (see Figure 7.3).

$$c_1 = a$$
$$c_2 = a + b\,\Delta x_1 + c\,\Delta x_1^2 \qquad (7.19)$$
$$c_3 = a + b\,(\Delta x_1 + \Delta x_2) + c\,(\Delta x_1 + \Delta x_2)^2$$

Solving for a, b, and c gives

$$c_1 = a$$
$$b = \frac{c_2 - c_1 - c\,\Delta x_1^2}{\Delta x_1} = \frac{c_2 - c_1}{\Delta x_1} - c\,\Delta x_1 \qquad (7.20)$$
$$c_3 - c_2 - (c_2 - c_1)\frac{\Delta x_2}{\Delta x_1} = c\,(\Delta x_2^2 + \Delta x_1 \Delta x_2).$$

The first and second derivatives at the point $x = \Delta x_1$ are

$$\left.\frac{dc}{dx}\right|_2 = \frac{c_2 - c_1}{\Delta x_1} + \frac{(c_3 - c_2)\,\Delta x_1 - (c_2 - c_1)\,\Delta x_2}{\Delta x_2^2 + \Delta x_1 \Delta x_2}$$

$$\left.\frac{d^2c}{dx^2}\right|_2 = 2\,\frac{(c_3 - c_2) - (c_2 - c_1)\dfrac{\Delta x_2}{\Delta x_1}}{\Delta x_2^2 + \Delta x_1 \Delta x_2}. \qquad (7.21)$$

The residual at node 2 is then

$$R = \mathrm{Pe}\left(\frac{c_2 - c_1}{\Delta x_1} + \frac{(c_3 - c_2)\,\Delta x_1 - (c_2 - c_1)\,\Delta x_2}{\Delta x_2^2 + \Delta x_1 \Delta x_2}\right) - 2\,\frac{(c_3 - c_2) - (c_2 - c_1)\dfrac{\Delta x_2}{\Delta x_1}}{\Delta x_2^2 + \Delta x_1 \Delta x_2}. \qquad (7.22)$$

Basically we wish to have a greater number of nodes in the regions that have large residuals. How many nodes are added for residuals of a certain size is a matter of preference. Here let us calculate the residual at all the nodes and average them. Then we will use the following criteria:

$$\begin{aligned}
&\text{res}_j < 0.1 \text{ avg. res} &&\text{remove node } j\\
&0.1 \text{ avg.res} \le \text{res}_j \le 10 \text{ avg. res.} &&\text{keep node } j\\
&10 \text{ avg. res.} < \text{res}_j \le 100 \text{ avg. res.} &&\text{add 1 node}\\
&100 \text{ avg. res.} < \text{res}_j \le 1000 \text{ avg. res.} &&\text{add 2 nodes}\\
&1000 \text{ avg. res.} < \text{res}_j &&\text{add 3 nodes.}
\end{aligned} \qquad (7.23)$$

Finally, we smooth the nodal placement using

$$x_k = \frac{1}{2}(x_k + x_{k+1}), \quad k = (\text{NT} - 1) \text{ to } 2. \qquad (7.24)$$

This process represents the adaptive mesh method using the residual as the criterion. Six methods will be presented using this strategy with the residual and other criteria.

To illustrate the method using the residual, we start with six nodes, distributed equidistantly. On the first solution, the process adds one node and moves all seven of them towards $x = 1$, as shown in Figure 7.4, Step 2. In this figure and in all succeeding figures in this section, the exact solution is the smooth curve in the diagram. On successive solutions, the number of nodes does not change drastically, but the nodes are moved towards $x = 1$, as shown for Step 5 in Table 7.2. The last solution is quite good: the elements are exactly where they are needed. Despite this, the average residual increases as the mesh is moved towards $x = 1$. This is because the residual is only defined at the nodes and they are moved to a location where the terms are large and well-represented.

Table 7.2. Node Placement for Step 5

Criterion	Figure	x-positions of nodes
Residual placement	7.4	0.000 0.940 0.965 0.980 0.988 0.995 1.000
Estimated error placement	7.5	0.000 0.920 0.960 0.970 0.980 0.988 0.993 0.997 1.000
Placement to meet est. error	7.6	0.000 0.710 0.850 0.930 0.960 0.970 0.978 0.984 0.9885 0.9926 0.9956 0.9983 1.000
Gradient error placement	7.7	0.000 0.910 0.960 0.977 0.984 0.990 0.994 0.998 1.000
Variational placement	7.8	0.000 0.560 0.800 0.975 1.000
Estimated error, upstream	7.9	0.000 0.900 0.940 0.960 0.970 0.980 0.990 0.996 1.000
Estimated error, more nodes	7.10	0.000 0.840 0.915 0.937 0.956 0.971 0.977 0.981 0.985 0.988 0.990 0.992 0.994 0.996 0.997 0.9984 0.99936 1.000
Est. error, upstream, more nodes	7.11	0.000 0.863 0.934 0.950 0.961 0.969 0.974 0.978 0.981 0.984 0.986 0.988 0.990 0.992 0.993 0.995 0.996 0.997 0.998 0.999 0.99955 1.000

A second method of locating the nodes tries to minimize the estimated error in the solution. For linear trial functions, the error is proportional to the second derivative of the solution [Prenter 1975].

$$\text{error}_j = C \, \Delta x_j^2 \left\| \frac{d^2 c}{dx^2} \right\|_j \tag{7.25}$$

We thus try to make this error value small and uniform throughout. We use the same expression for the second derivative, Eq. (7.15), and try to make uniform the value of

$$\Delta x_j^2 \left| \frac{d^2 c}{dx^2} \right|_j, \tag{7.26}$$

where the second derivative is given in Eq. (7.15). The strategy given in Eq. (7.23) is used, except that Eq. (7.26) replaces the residual. The results from this

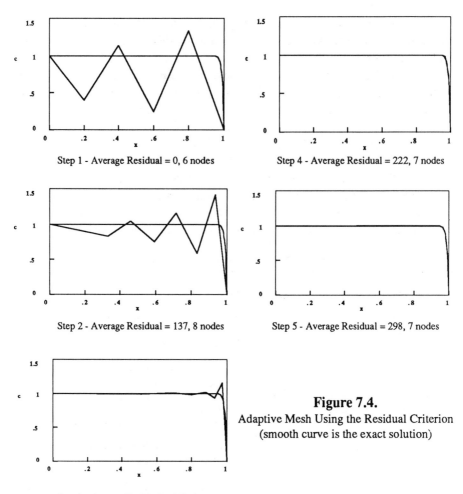

Step 1 - Average Residual = 0, 6 nodes

Step 2 - Average Residual = 137, 8 nodes

Step 3 - Average Residual = 178, 8 nodes

Step 4 - Average Residual = 222, 7 nodes

Step 5 - Average Residual = 298, 7 nodes

Figure 7.4.
Adaptive Mesh Using the Residual Criterion (smooth curve is the exact solution)

method are quite good after a few steps, as shown in Figure 7.5 and Table 7.2.

The next method uses the same estimate of the error as Eq. (7.26) but adds enough elements each step to satisfy Eq. (7.25). Thus, Eq. (7.23) is not used. This method results in more elements, but the solutions are excellent, as shown in Figure 7.6.

The fourth method tries to make the elements small where the slope is large. We calculate the mean square derivative over an element. The criterion is now taken as

$$\text{Res}_j = \left[\int_{x_{j-1}}^{x_j} \left(\frac{dc}{dx} \right)^2 dx \right]^{1/2} = \left[\Delta x_{j-1} \left(\frac{c_j - c_{j-1}}{\Delta x_{j-1}} \right)^2 \right]^{1/2} = \frac{|c_j - c_{j-1}|}{\sqrt{\Delta x_{j-1}}}. \quad (7.27)$$

The same strategy is used as in Eq. (7.23), except that Eq. (7.27) replaces the residual. The results from this method are also quite good after the first few steps,

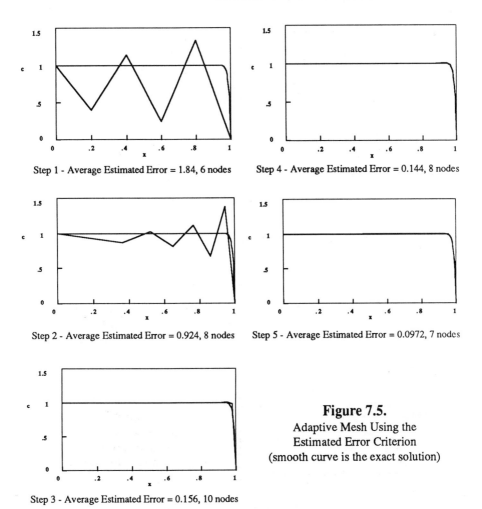

Figure 7.5.
Adaptive Mesh Using the Estimated Error Criterion
(smooth curve is the exact solution)

as shown in Figure 7.7 and Table 7.2. In fact, the placement of nodes is very similar to that in Figure 7.5, which was obtained using the second-derivative criterion.

The fifth method is based on the proper norm for this problem (see Section. 6.4):

$$I = \int_0^1 e^{-Pe\,x} \left(\frac{dc}{dx}\right)^2 dx. \qquad (7.28)$$

The error measured in this norm is

$$\text{Error} = \int_0^1 e^{-Pe\,x} \left(\frac{dc}{dx} - \frac{dc_{ex}}{dx}\right)^2 dx. \qquad (7.29)$$

This is evaluated in Problem 7.2 to give

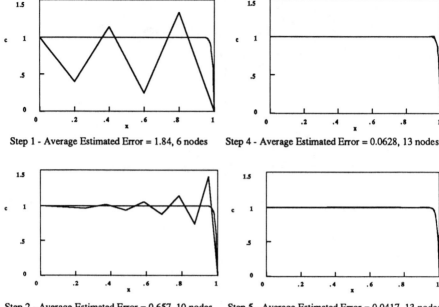

Step 1 - Average Estimated Error = 1.84, 6 nodes

Step 4 - Average Estimated Error = 0.0628, 13 nodes

Step 2 - Average Estimated Error = 0.657, 10 nodes

Step 5 - Average Estimated Error = 0.0417, 13 nodes

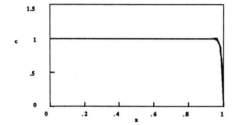

Step 3 - Average Estimated Error = 0.0714, 18 nodes

Figure 7.6.
Adaptive Mesh Using a
Required Error Criterion
(smooth curve is the exact solution)

$$\text{Error} = \int_0^1 e^{-\text{Pe}\,x} \left(\frac{dc}{dx}\right)^2 dx + 2\,\frac{dc_{ex}}{dx}(0) - \frac{dc_{ex}}{dx}(0). \qquad (7.30)$$

Minimizing the function, Eq. (7.28), also minimizes the error since the boundary terms are constants (although unknown). Here we use this same norm in our criterion:

$$\text{Res}_j = \left[\int_{x_{j-1}}^{x_j} e^{-\text{Pe}\,x} \left(\frac{dc}{dx}\right)^2 dx\right]^{1/2} = \left[\frac{(c_j - c_{j-1})^2}{\Delta x_{j-1}^2} \int_{x_{j-1}}^{x_j} e^{-\text{Pe}\,x}\, dx\right]^{1/2}. \qquad (7.31)$$

Evaluation of the terms gives

ADAPTIVE AND MOVING MESHES

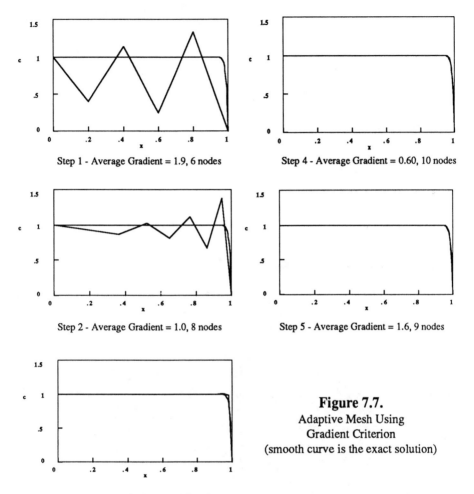

Figure 7.7.
Adaptive Mesh Using
Gradient Criterion
(smooth curve is the exact solution)

$$\text{Res}_j = \frac{|c_j - c_{j-1}|}{\Delta x_{j-1}} \left[\frac{e^{-\text{Pe} x_{j-1}} - e^{-\text{Pe} x_j}}{\text{Pe}} \right]^{1/2} =$$

$$= \frac{|c_j - c_{j-1}|}{\sqrt{\Delta x_{j-1}}} \left[\frac{e^{-\text{Pe} x_{j-1}}}{\text{Pe} \Delta x_j} (1 - e^{-\text{Pe} \Delta x_j}) \right]^{1/2}. \quad (7.32)$$

The same strategy is used as in Eq. (7.23), except that Eq. (7.32) replaces the residual. When PeΔx is small, Eq. (7.32) becomes

$$\text{Res}_j = \frac{|c_j - c_{j-1}|}{\sqrt{\Delta x_{j-1}}} [e^{-\text{Pe} x_{j-1}}]^{1/2}, \quad (7.33)$$

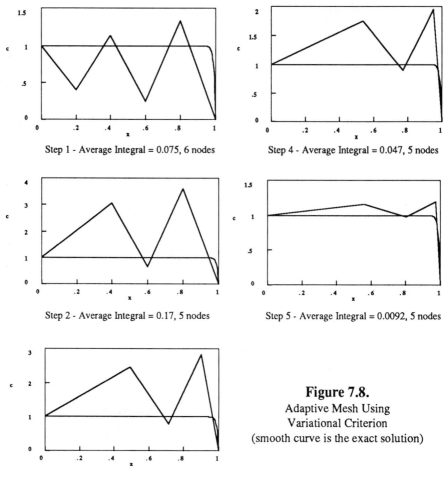

Figure 7.8.
Adaptive Mesh Using Variational Criterion
(smooth curve is the exact solution)

which is proportional to the norm used in Eq. (7.27). The results from this method are unsatisfactory, as shown in Figure 7.8. The reason these results are so poor is that the factor exp(-Pe x) is so small near x = 1 that the gradients in that region have minimal importance. The factor exp(-Pe x) tends to emphasize the region x = 0, yet nodes in that region contribute little to a good solution.

To summarize these results, the residual can be used to locate the nodes. The use of the estimated error (the second derivative) or the first derivative as the criterion works best, however. The use of the variational integral does not work well at all. Thus, the most mathematically rigorous method does not work in practice. We should notice that the final solution has only one or two more grid points than the initial guess. The final solutions shown in Figures 7.4-7.7 all represent quite good solutions that were easy and quick to obtain.

An alternative approach to solving Eq. (7.1) is to use an upstream method.

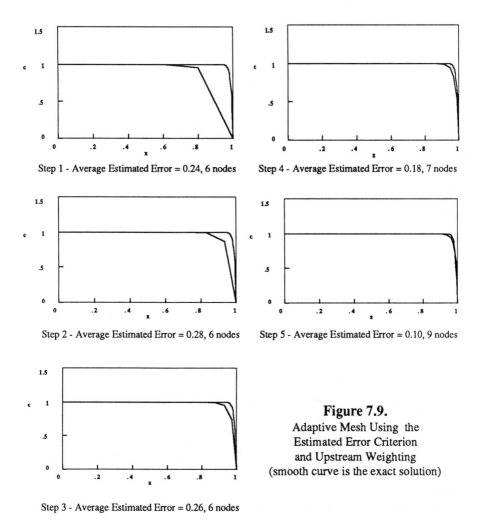

Figure 7.9.
Adaptive Mesh Using the Estimated Error Criterion and Upstream Weighting (smooth curve is the exact solution)

In this case, the solution does not oscillate. As the mesh is refined, the truncation error decreases, so the extra dispersion in the upstream method disappears. Here we use the equation

$$\frac{\Delta x_{j-1}}{2}\left(\frac{1}{3}+\frac{\alpha}{2}\right)\frac{dc_{j-1}}{dt}+\left[\frac{1}{3}(\Delta x_{j-1}+\Delta x_j)+\frac{\alpha}{4}(-\Delta x_{j-1}+\Delta x_j)\right]\frac{dc_j}{dt}+\frac{\Delta x_j}{2}\left(\frac{1}{3}-\frac{\alpha}{2}\right)\frac{dc_{j+1}}{dt}+$$
$$+\frac{Pe}{2}[(1-\alpha)c_{j+1}+2\alpha c_j-(1+\alpha)c_{j-1}]+\frac{1}{\Delta x_{j-1}}(-c_{j-1}+c_j)+\frac{1}{\Delta x_j}(c_j-c_{j+1})=0, \quad (7.34)$$

which is the extension to the variable element sizes of Eq. (2.105). The steady-state version is

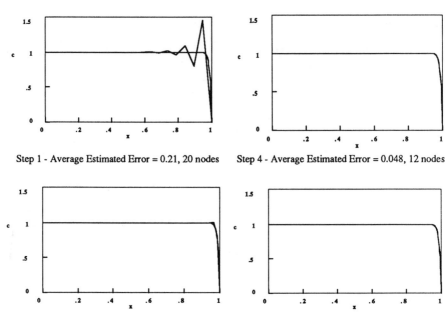

Step 1 - Average Estimated Error = 0.21, 20 nodes

Step 4 - Average Estimated Error = 0.048, 12 nodes

Step 2 - Average Estimated Error = 0.071, 17 nodes

Step 5 - Average Estimated Error = 0.018, 18 nodes

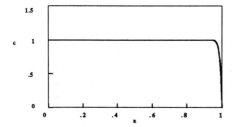

Step 3 - Average Estimated Error = 0.074, 12 nodes

Figure 7.10.
Adaptive Mesh Using the Estimated Error Criterion (smooth curve is the exact solution)

$$c_{j-1}\left[\frac{1}{\Delta x_{j-1}} + \frac{Pe}{2}(1+\alpha)\right] + c_j\left[-\frac{1}{\Delta x_{j-1}} - \frac{1}{\Delta x_j} - \alpha\, Pe\right] + c_{j+1}\left[\frac{1}{\Delta x_j} - \frac{Pe}{2}(1-\alpha)\right] = 0. \quad (7.35)$$

The same process is applied, using the strategy described in Eq. (7.23). Carey and Plover [1983] go beyond this method by using an upstream parameter that changes from element to element.

The results for this upstream method beginning with six nodes and using the estimated error criterion are shown in Figure 7.9. The solutions are smooth even for the coarse mesh, but the fine mesh solutions in Step 5 are not as good as when no upstream parameter is used. The same conclusion holds when the criteria is the residual or the gradient. Thus, the solution obtained on an adaptive mesh is better than the one derived using an upstream method and even the one

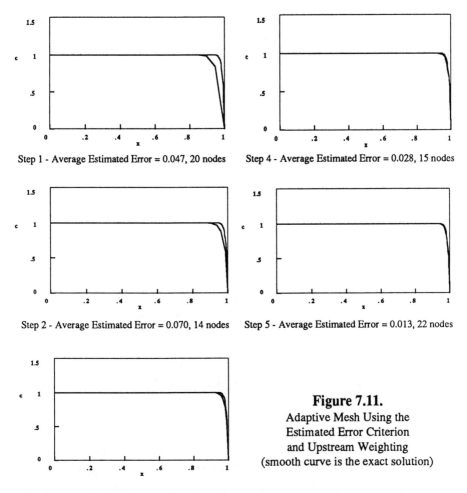

Step 1 - Average Estimated Error = 0.047, 20 nodes

Step 4 - Average Estimated Error = 0.028, 15 nodes

Step 2 - Average Estimated Error = 0.070, 14 nodes

Step 5 - Average Estimated Error = 0.013, 22 nodes

Step 3 - Average Estimated Error = 0.051, 13 nodes

Figure 7.11.
Adaptive Mesh Using the Estimated Error Criterion and Upstream Weighting (smooth curve is the exact solution)

derived using an adaptive upstream method. The same conclusion is reached when starting with twenty nodes. [See Figure 7.10 for the results for the standard method and Figure 7.11 for the results for the upstream method.] The final solution is best when no upstream parameter is used. By comparing the solution in Step 1 (uniform mesh) with that in Step 5 (non-uniform mesh), we immediately see the value of adaptive meshes.

If we compare Figure 7.10 (20 nodes) with Figure 7.5 (6 nodes) we see whether or not using more nodes improves the solution. These results suggest that it does not; the solution is just as good when starting with 6 nodes as with 20 nodes. Thus the added computation is wasted. If we use the upstream parameter, then this conclusion must be modified. Comparison of Figure 7.11 (20 nodes) with Figure 7.9 shows that the best solution is in Step 5 of Figure 7.11, which was obtained by starting with 20 nodes. Here the added nodes do improve the solution. It is clear, however, that the best solutions are obtained without the upstream parameter. The only reason for us to use the upstream parameter is if we want the intermediate solutions to be free of oscillations.

An adaptive mesh scheme can be used to put small elements in regions where they are needed. The criterion for placement of the elements can be either the residual, the estimated error, or the gradient. An upstream parameter can be used to keep the solution from oscillating. As the refinement of the mesh takes place, however, the dissipation errors in the upstream method become smaller. The best solutions are obtained by not using an upstream parameter and using the estimate error or gradient as the criterion for placement of the elements.

7.2. Moving Fronts with Analytical Placement of Nodes

The previous section discussed steady-state problems and used various criteria to move the nodes around until some desirable goal had been reached. Similar approaches can be applied to transient problems, except that now the mesh changes might occur continuously or at each time-step. The methods for treating transient problems fall into two classifications: those that add and subtract nodes and those that move the nodes around. Of course, combinations are also possible. In this section and several following sections we will consider a variety of methods for adding and subtracting nodes or moving them around. First we will consider cases that are simple enough that the required motion is obvious and can be specified by the analyst. These are cases in which the hyperbolic problem can be solved exactly or at least the movement of the front or shocks can be determined *a priori*. Then we will advance to more difficult cases in which some property of the solution is used to control the nodal movement. Next we will consider a so-called least-squares criterion to move the nodes and try to relate it to several Petrov-Galerkin formulations. We will conclude the chapter by discussing Euler-Lagrange methods, in which the hyperbolic part of the problem is solved exactly (or at least explicitly) and the diffusive part of the problem is appended later. These methods employ some of the features of using analytical solutions to move the nodes and also are related to the random choice method discussed in Section 6.1.

For many problems, such as the convective diffusion equation, we know how the front moves. In these cases we can use that information to provide a superior solution. There are two ways to view the moving nodes. One method involves transforming the domain into a moving coordinate system. The other method involves analyzing the finite element basis functions. Both methods turn out to be identical and it is valuable to see how they correspond.

There are several types of time derivatives possible, depending on what an observer is doing. For example, the notation

$$\partial c/\partial t \tag{7.36}$$

refers to a time derivative taken when an observer is stationary, keeping his or her position fixed. A convected derivative,

$$\frac{dc}{dt} \equiv \frac{\partial c}{\partial t} + \mathbf{v}^e \cdot \nabla\ c, \tag{7.37}$$

corresponds to a time derivative when an observer is moving with velocity $\mathbf{v}^e$. If the velocity is the fluid velocity, $\mathbf{v}$, then we have the usual convected derivative,

$$\frac{Dc}{Dt} \equiv \frac{\partial c}{\partial t} + \mathbf{v} \cdot \nabla\ c = \frac{\partial c}{\partial t} + v_x \frac{\partial c}{\partial x} + v_y \frac{\partial c}{\partial y} + v_z. \tag{7.38}$$

Any problem of the type

$$\frac{\partial c}{\partial t} + \mathbf{v} \cdot \nabla c = L c, \tag{7.39}$$

where the operator L did not involve time derivatives, would then become

$$\frac{dc}{dt} + (\mathbf{v} - \mathbf{v}^e) \cdot \nabla c = L c \tag{7.40}$$

in a coordinate system moving with velocity $\mathbf{v}^e$. By setting $\mathbf{v}^e = \mathbf{v}$ we can make the convective term disappear. Then the boundary conditions occur on surfaces that are generally moving. It may be possible to choose $\mathbf{v}^e = \mathbf{v}$ only near a sharp front; then the convective term disappears there; we have seen that this is the only place we need to worry about large convective terms. The convective term does not cause oscillations near the boundaries where the front is not steep.

Next let us consider a one-dimensional problem with x as the fixed spatial coordinate (fixed in time) and t the time. We transform the coordinates to new coordinates, ξ and τ, chosen such that the new coordinates move with velocity $v^e = U^e$.

$$\xi = x - \int_0^t U^e(x, t')\, dt' \qquad \tau = t \tag{7.41}$$

A function of x and t then becomes an implicit function of ξ and τ. Since we are going to solve for the solution in terms of ξ and τ, we need derivatives with respect to x and t. These are

$$\begin{aligned}\frac{\partial c}{\partial t} &= \frac{\partial \xi}{\partial t}\frac{\partial c}{\partial \xi} + \frac{\partial \tau}{\partial t}\frac{\partial c}{\partial \tau}, \\ \frac{\partial c}{\partial x} &= \frac{\partial \xi}{\partial x}\frac{\partial c}{\partial \xi} + \frac{\partial \tau}{\partial x}\frac{\partial c}{\partial \tau}.\end{aligned} \tag{7.42}$$

When we differentiate the new variables we get

$$\begin{aligned}\frac{\partial \tau}{\partial t} &= 1, & \frac{\partial \tau}{\partial x} &= 0 \\ \frac{\partial \xi}{\partial t} &= -U^e, & \frac{\partial \xi}{\partial x} &= 1 - \int_0^t \frac{\partial U^e}{\partial x}(x, t')\, dt'.\end{aligned} \tag{7.43}$$

The derivatives are then given by

$$\frac{\partial}{\partial t} = \frac{\partial}{\partial \tau} - U^e \frac{\partial}{\partial \xi}, \quad \frac{\partial}{\partial x} = \frac{\partial \xi}{\partial x} \frac{\partial}{\partial \xi} \qquad (7.44)$$

$$\frac{\partial^2}{\partial x^2} = \left(\frac{\partial \xi}{\partial x}\right)^2 \frac{\partial^2}{\partial \xi^2} + \frac{\partial^2 \xi}{\partial x^2} \frac{\partial}{\partial \xi}. \qquad (7.45)$$

If the original problem is

$$\frac{\partial u}{\partial t} = F\left(x, t, \frac{\partial u}{\partial x}, \frac{\partial^2 u}{\partial x^2}\right) \qquad (7.46)$$

and the right-hand side does not include any time derivatives, then the new problem (in transformed coordinates) is

$$\frac{\partial u}{\partial \tau} - U^e(\tau) \frac{\partial u}{\partial \xi} = F'\left(\xi + \int_0^\tau U^e(t')\, dt', \tau, \frac{\partial u}{\partial \xi}, \frac{\partial^2 u}{\partial \xi^2}\right). \qquad (7.47)$$

This can be rearranged as

$$\frac{\partial u}{\partial \tau} = F\left(\xi + \int_0^\tau U^e(t')\, dt', \tau, \frac{\partial u}{\partial \xi}, \frac{\partial^2 u}{\partial \xi^2}\right) + U^e(\tau) \frac{\partial u}{\partial \xi}. \qquad (7.48)$$

This transformation to a moving grid does not change the initial-value character of the problem. It does change the boundary conditions, however, since in the ξ-τ coordinate system, the boundaries occur at moving locations. Thus, initial or boundary conditions of the type

$$u(x, 0) = g(x)$$
$$u(x_1, t) = f_1(t) \qquad (7.49)$$
$$u(x_2, t) = f_2(t)$$

become

$$u(\xi, 0) = g(\xi)$$
$$u\left(\xi_1 + \int_0^\tau U^e(x, t')\, dt', \tau\right) = f_1(\tau) \qquad (7.50)$$
$$u\left(\xi_2 + \int_0^\tau U^e(x, t')\, dt', \tau\right) = f_2(\tau).$$

For example, if

$$F = -Pe \frac{\partial u}{\partial x} + \frac{\partial^2 u}{\partial x^2} \quad \text{and we use } U^e = Pe, \qquad (7.51)$$

then Eq. (7.48) is

$$\frac{\partial u}{\partial \tau} = -\text{Pe}\frac{\partial u}{\partial \xi} + \frac{\partial^2 u}{\partial \xi^2} + \text{Pe}\frac{\partial u}{\partial \xi} \quad (7.52)$$

or

$$\frac{\partial u}{\partial \tau} = \frac{\partial^2 u}{\partial \xi^2}. \quad (7.53)$$

Lynch [1982] shows that the weighted residual formulation in a moving coordinate system is the same as a finite element formulation with an added convective term that is evaluated on the moving grid. His arguments are summarized here in three dimensions, by taking

$$\begin{array}{ll} \varphi_i & \text{basis functions} \\ \mathbf{X}_i & \text{nodal coordinates} \\ \varphi_i = \varphi_i(\mathbf{x}, \mathbf{X}_1, \mathbf{X}_2, \ldots, \mathbf{X}_N). \end{array} \quad (7.54)$$

We write

$$\varphi_i = \varphi_i(\xi), \quad (7.55)$$

where ξ is a vector of a local coordinate system related by

$$\mathbf{x} = \sum_j \mathbf{X}_j \psi_j(\xi). \quad (7.56)$$

The differential of φ_i is obtained by using the chain rule

$$d\varphi_i = \nabla \varphi_i \cdot d\mathbf{x} + \sum_j \nabla_j \varphi_i \cdot d\mathbf{X}_j, \quad (7.57)$$

where the gradient, with respect to the nodes, is

$$\nabla_j = \left\{ \frac{\partial}{\partial x_j}, \frac{\partial}{\partial y_j}, \frac{\partial}{\partial z_j} \right\}. \quad (7.58)$$

The differential of $\mathbf{x}$ is obtained using the chain rule:

$$d\mathbf{x} = \sum_j (d\mathbf{X}_j \psi_j + \mathbf{X}_j d\psi_j). \quad (7.59)$$

Combining Eq. (7.36) with Eq. (7.59) gives

$$d\varphi_i = \sum_j (\psi_j \nabla \varphi_i + \nabla_j \varphi_i) \cdot d\mathbf{X}_j + \sum_j (\nabla \varphi_i \cdot \mathbf{X}_j) d\psi_j. \quad (7.60)$$

Both φ_i and ψ_i depend on ξ alone. As a result, when $d\xi = 0$, it follows that $d\varphi_i = 0$ and $d\psi_i = 0$. Thus Eq. (7.60) becomes

$$\sum_j (\psi_j \nabla \varphi_i + \nabla_j \varphi_i) \cdot d\mathbf{X}_j = 0. \tag{7.61}$$

Since $d\xi = 0$ can be achieved with an arbitrary displacement, $d\mathbf{X}_j$, each component must be zero.

$$\psi_j \nabla \varphi_i + \nabla_j \varphi_i = 0$$
$$\text{or} \quad \nabla_j \varphi_i = -\psi_j \nabla \varphi_i \tag{7.62}$$
$$\text{or} \quad \frac{\partial \varphi_i}{\partial x_j} = -\psi_j \frac{\partial \varphi_i}{\partial x}$$

Then we differentiate Eq. (7.54) while keeping **x** constant.

$$\left(\frac{\partial \varphi_i}{\partial t}\right)_x = \sum_j \nabla_j \varphi_i \cdot \frac{d\mathbf{X}_j}{dt} \tag{7.63}$$

We use Eq. (7.62) to change this to

$$\left(\frac{\partial \varphi_i}{\partial t}\right)_x = -\sum_j \psi_j \nabla \varphi_i \cdot \frac{d\mathbf{X}_j}{dt}. \tag{7.64}$$

If we define

$$\mathbf{V}^e \equiv \sum_j \psi_j \frac{d\mathbf{X}_j}{dt}, \tag{7.65}$$

then Eq. (7.64) becomes

$$\left(\frac{\partial \varphi_i}{\partial t}\right)_x = -\mathbf{V}^e \cdot \nabla \varphi_i. \tag{7.66}$$

Next let us consider the method of weighted residuals applied to the equation

$$\frac{\partial U}{\partial t} + LU = f \tag{7.67}$$

and expand the trial function in terms of the functions φ_i:

$$\hat{U}(\mathbf{X}, t) = \sum_j U_j(t) \varphi_j[\mathbf{X}, \mathbf{X}_1(t), \mathbf{X}_2(t), \ldots, \mathbf{X}_N(t)]. \tag{7.68}$$

Then, using Eq. (7.66),

$$\frac{\partial \hat{U}}{\partial t} = \sum_j \frac{dU_j}{dt} \varphi_j - \mathbf{V}^e \cdot \nabla \hat{U}. \tag{7.69}$$

The standard weighted residual

$$\int \left[\frac{\partial \hat{U}}{\partial t} + L\hat{U} - f \right] W_i \, dX = 0 \tag{7.70}$$

becomes

$$\int \left[\sum_j \frac{dU_j}{dt} \varphi_j - \mathbf{v}^e \cdot \nabla \hat{U} + L\hat{U} - f \right] W_i \, dX = 0. \tag{7.71}$$

The alternative interpretation involves making the transformation

$$\mathbf{X} = \mathbf{X}(\chi, t), \tag{7.72}$$

where χ is the position of a fluid particle at time zero (thus it identifies the same particle at all times).

$$U(\mathbf{X}, t) = U(\mathbf{X}(\chi, t), t) \tag{7.73}$$

$$\left.\frac{\partial U}{\partial t}\right|_\chi = \left.\frac{\partial U}{\partial t}\right|_X + \nabla U \cdot \left.\frac{\partial \mathbf{X}}{\partial t}\right|_\chi \tag{7.74}$$

The differential equation is

$$\left.\frac{\partial U}{\partial t}\right|_\chi - \left.\frac{\partial \mathbf{X}}{\partial t}\right|_\chi \cdot \nabla U + LU = f. \tag{7.75}$$

If we use Eq. (7.68) we get

$$\left(\frac{\partial \hat{U}}{\partial t}\right)_\chi = \sum_j \frac{dU_j}{dt} \varphi_i. \tag{7.76}$$

Now we identify χ with ξ so that χ = constant is the same as ξ = constant:

$$\left.\frac{\partial \mathbf{X}}{\partial t}\right|_\chi = \sum_j \frac{d\mathbf{X}_j}{dt} \psi_j = \mathbf{v}^e. \tag{7.77}$$

The weighted residual on the transformed domain is then

$$\int \left[\sum_j \frac{dU_j}{dt} \varphi_i - \mathbf{v}^e \cdot \nabla \hat{U} + L\hat{U} - f \right] W_i \, dX = 0, \tag{7.78}$$

which is the same as Eq. (7.71). Thus, the two approaches are the same. When we evaluate the terms in Eq. (7.78), we must include the determinant of the Jacobian since the integrals are evaluated on the transformed domain. An application of moving grid points to the solution of the Fokker-Planck equation is given by Harrison [1988].

These ideas are illustrated with the simplest application: the convective diffusion equation with a constant velocity. Examples of two alternatives, moving the nodes or adding and subtracting the nodes, will be given. The first alternative, moving the nodes, comes from work by Jensen and Finlayson [1980], although they used orthogonal collocation on finite elements rather than the Galerkin method. A more general formulation was later provided by Chawla, *et al.* [1987]. This formulation permitted each node to move independently, although the only application presented has the nodes moving with the same velocity, $v = Pe$. Chawla, *et al.* also use the collocation method with Hermite polynomials on finite elements (which is equivalent to the orthogonal collocation method). Sheintuch [1990] gives more examples involving the transformation of the coordinate system.

Here the original formulation of Jensen and Finlayson [1980] will be presented, but using the Galerkin finite element method. The convective diffusion equation is

$$\frac{\partial c}{\partial t} + Pe \frac{\partial c}{\partial x} = \frac{\partial^2 c}{\partial x^2} \qquad (7.79)$$

with the boundary and initial conditions

$$\begin{aligned} c(x,0) &= 0 & x &> 0 \\ c(0,t) &= 1 & t &> 0 \\ \frac{\partial c}{\partial x}(1,t) &= 0 & t &> 0. \end{aligned} \qquad (7.80)$$

We apply the transformation, Eq. (7.41), with the velocity $v = Pe$.

$$\frac{\partial c}{\partial \tau} = \frac{\partial^2 c}{\partial \xi^2} \qquad (7.81)$$

The boundary conditions become

$$\begin{aligned} c &= 0 & \xi &> 0, \tau = 0 \\ c &= 1 & \xi &= -Pe\,\tau \\ \frac{\partial c}{\partial \xi} &= 0 & \xi &= 1 - Pe\,\tau. \end{aligned} \qquad (7.82)$$

A grid in ξ coordinates is identified. This grid is maintained as it moves along with velocity Pe. In the original calculations, Jensen and Finlayson applied the boundary conditions at the closest node. In the calculations reported here we discretize the domain from $\xi = -1$ to $\xi = 1$. At time zero, the physical domain is from $\xi = 0$ to $\xi = 1$. At time 1/Pe, the physical domain is from $\xi = -1$ to $\xi = 0$. At intermediate times, the domain is in between (see Figure 7.12). Since the front is always near $\xi = 0$, we can use small elements there and large elements elsewhere.

The amount we expect the front to broaden can be found from the solution of the diffusion equation [Eq. (7.81)] in an infinite medium with an initial step

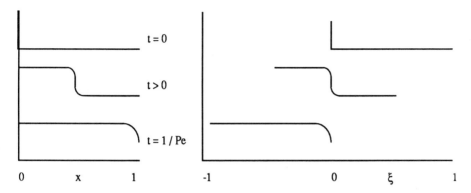

Figure 7.12. Fixed and Moving Coordinate Systems

change. The solution is

$$c = \text{erfc}\left(\frac{x}{\sqrt{4Dt}}\right). \tag{7.83}$$

For a change in c of 99% (c = 0.01), we get

$$\frac{x}{\sqrt{4Dt}} \approx 1.8 \tag{7.84}$$

so that the diffusion length is

$$L \cong 3.6\sqrt{Dt}. \tag{7.85}$$

Here $D = 1$ and $t_{max} = 1/Pe$, so that the maximum length of broadening will be

$$L \cong \frac{3.6}{\sqrt{Pe}}. \tag{7.86}$$

With Pe = 1000, this distance is about 0.1. Thus we use small elements in the region between $\xi = -0.1$ and $\xi = 0.1$. The elements within this region are graded with the smallest ones having $\Delta\xi = 0.01$; the actual element distribution is listed in Table 7.3. The boundary conditions are applied at $\xi = -1$ and $\xi = 1$. This is an approximation; it relies on the fact that the problem is convection-dominated. We are supposing that convection is so rapid that a fluid element is swept away from $x = 0$ before diffusion can influence it significantly. If this is true, applying the boundary condition at $\xi = -1$ or $x = 0$ should give the same result. Likewise, the last boundary condition is immaterial until the front reaches there. Thus we cannot continue the calculation indefinitely without adjusting the location of the boundary conditions.

We first solve Eq. (7.81) with an explicit method:

$$\frac{1}{2}(\Delta\xi_{j-1} + \Delta\xi_j)\frac{c_j^{n+1} - c_j^n}{\Delta\tau} = \frac{c_{j-1}^n - c_j^n}{\Delta\xi_{j-1}} - \frac{c_j^n - c_{j+1}^n}{\Delta\xi_j}. \tag{7.87}$$

Table 7.3. Element Sizes for a Moving Coordinate System

$\Delta\xi$	$\Sigma\Delta\xi$
0.01	0.01
0.02	0.03
0.04	0.07
0.08	0.15
0.16	0.31
0.32	0.63
0.37	1.00

As an approximate stability analysis, let us consider the stability limitation for this finite difference method (or lumped Galerkin finite element method) when all $\Delta\xi$ values are the same,

$$\Delta\tau \leq \frac{1}{2}\Delta\xi^2. \tag{7.88}$$

Since $\Delta\xi$ is small near $\xi = 0$, Eq. (7.88) must be satisfied for the smallest element. If we take $\Delta\xi = 0.01$, we need $\Delta\tau \leq 5 \times 10^{-5}$. However, if Pe = 1000, the front traverses a distance in x of 0.05 each time-step; thus, not many time-steps are needed to reach x = 0.5. Eq. (7.87) does not have a convective term, so that difficulty is bypassed. Typical results are shown in Figure 7.13; the profiles are quite steep with no oscillations and did not take long to obtain. Thus, the moving coordinate system is a viable method for this problem.

If we use an implicit method, there is no stability limitation. The calculations use a trapezoidal rule to integrate in time, giving

$$\frac{1}{2}(\Delta\xi_{j-1} + \Delta\xi_j)\frac{c_j^{n+1} - c_j^n}{\Delta\tau} = \frac{1}{2}\left(\frac{c_{j-1}^n - c_j^n}{\Delta\xi_{j-1}} - \frac{c_j^n - c_{j+1}^n}{\Delta\xi_j}\right) + \frac{1}{2}\left(\frac{c_{j-1}^{n+1} - c_j^{n+1}}{\Delta\xi_{j-1}} - \frac{c_j^{n+1} - c_{j+1}^{n+1}}{\Delta\xi_j}\right). \tag{7.89}$$

These can be rearranged in the form of

$$c_{j-1}^{n+1}\left(-\frac{\Delta\tau}{\Delta\xi_{j-1}}\right) + c_{j+1}^{n+1}\left(-\frac{\Delta\tau}{\Delta\xi_j}\right) + c_j^{n+1}\left[\Delta\xi_{j+1} + \Delta\xi_j + \Delta\tau\left(\frac{1}{\Delta\xi_{j-1}} + \frac{1}{\Delta\xi_j}\right)\right] =$$
$$= c_j^n(\Delta\xi_{j+1} + \Delta\xi_j) + \Delta\tau\left(\frac{c_{j-1}^n - c_j^n}{\Delta\xi_{j-1}} - \frac{c_j^n - c_{j+1}^n}{\Delta\xi_j}\right). \tag{7.90}$$

However, actual calculations showed that there was an accuracy limitation: if the time-steps were too big, a spike appeared at $\xi = 0$. Thus we still need $\Delta t \leq 5 \times 10^{-4}$; in one step of this size, the front moves all the way from x = 0 to x = 0.5. Calculations with $\Delta t = 1 \times 10^{-4}$ are shown in Figure 7.14. These results are also quite good, except for a small error in the middle of the first curve. Thus the moving coordinate system is a viable method when using either explicit or implicit integration. The inherent stability of the implicit method did not make a major impact because the size of the time-step was controlled by accuracy

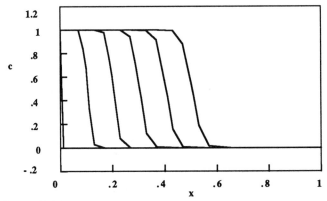

Figure 7.13. Convective Diffusion Equation, Moving Coordinate System
Explicit Method, Pe = 1000, $\Delta t = 0.00005$

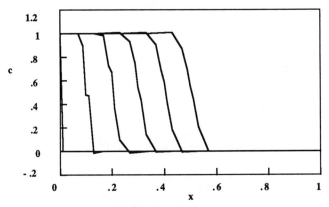

Figure 7.14. Convective Diffusion Equation, Moving Coordinate System
Implicit Method, Pe = 1000, $\Delta t = 0.0001$

requirements not to be much larger than the step-size for the explicit method.

Another alternative adaptive method keeps the mesh fixed in position but adds nodes before the front and deletes them after the front. In this manner, it is always possible to have small nodes around the front. Since the velocity is known, the locations to add and subtract nodes are known. An early example using this method was given by Price, et al. [1968]; they did not actually add and subtract nodes, but did keep a high-order element near the front. Yang, et al. [1991] also used a high-order polynomial at the location of the front. Let us consider a domain divided into uniform regions, with a linear finite element basis on each region. Then for the element on either side of the point where c = 0.5 (i.e., the mid-point of the front), the linear basis functions are replaced by quintic polynomials. This is similar to inserting four additional nodes in each of those elements, except that the high-order elements have a smaller truncation error. Here we employ the strategy of adding nodes near the front. We add nine additional nodes to the element with the front in it (i.e., the one with $x = t/Pe$), as

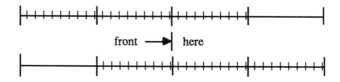

Figure 7.15. Movement of Micro Elements from Receding Macro Element to Forward Macro Element

well as to the element on either side of this one. When the front passes from one element to the next, we remove the extra nodes in the trailing refined element and add them to the element in front of the last refined element. An example of this change is illustrated in Figure 7.15. The interpolation is performed as follows. For the element at the left, the interior nodes are removed. The solution is kept at the original values at the ends of the elements. For the element at the right, the solution is interpolated (using a straight line) to give the values at the new interior points.

The method with moving nodes is run with the Galerkin method using Eq. (7.34) but with the time derivatives lumped.

$$\frac{1}{6}(\Delta x_{j-1} + \Delta x_j)\frac{dc_j}{dt} + \frac{Pe}{2}[(1 - \alpha)c_{j+1} + 2\alpha c_j - (1 + \alpha)c_{j-1}] +$$
$$+ \frac{1}{\Delta x_{j-1}}(-c_{j-1} + c_j) + \frac{1}{\Delta x_j}(c_j - c_{j+1}) = 0 \quad (7.91)$$

When Pe = 1000, there are some optimum choices. We can imagine covering the domain with macro elements of uniform size and subdividing the three elements near the front into micro elements. We have to decide how many macro elements are needed and how large the micro elements should be. For the micro element we use

$$\frac{Pe \, \Delta x}{2} = 1. \quad (7.92)$$

For Pe = 1000, this gives $\Delta x = 0.002$. We also use Eq. (7.86) to determine how broad the diffusion spreading is; in the time it takes for the front to move to $x = 0.5$, the front broadens by a distance of 0.08 (on each side). We thus need to have a few macro elements in a region of this size. We can make several choices about the size of the macro elements and this in turn determines the total number of nodes. Table 7.4 lists several possibilities; the size of the macro element is chosen arbitrarily and then the three macro elements near the front are filled with elements of size $\Delta x = 0.002$. Also shown in the table is the ratio of the number of nodes to the maximum number, 500, when the mesh has all elements of size 0.002. Results when the macro element sizes are 0.02 or 0.05 are also reported. Of course, more sophisticated schemes would use even larger elements farther away from the region of diffusion broadening; in that case, approximately half of

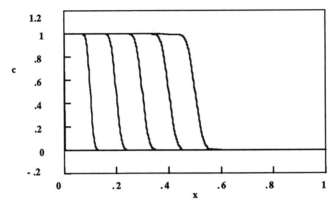

Figure 7.16. Convective Diffusion Equation, Moving Elements
Pe = 1000, Δx = 0.05 and 0.002, Δt = 9.09E-7

Table 7.4. Options for Macro Elements and Micro Elements

Δx_{macro}	No. of micro points per macro element	Total grid points	500/total points
0.1	49	160	3.1
0.05	24	93	5.4
0.0333	17	83	6.0
0.02	9	78	6.4
0.01	4	113	4.4

the nodes can be eliminated, making the schemes twice as fast to solve.

Calculations for Δx = 0.05 are quite good, as shown in Figure 7.16. The fronts are steep, there are no oscillations, and the results are relatively fast to obtain, improving over a fine mesh by a factor of 5.4 in the total number of points ($5.4^2 = 29$ in computer time because of the Courant number limitation). These solutions do, however, use more points than the moving coordinate system, which used only 15. Results when Δx = 0.02 are not as good, as shown in Figure 7.17. The reason is that while small elements are used precisely where the front is, the solution diffuses outside of the region where micro elements are used. The movement of the nodes is very important since if it is not done correctly, the solutions will not be improved. The next few sections will describe methods that predict nodal movement as well as the solution.

7.3 Moving Fronts with Placement of Nodes by Equidistribution Principles

The limitation of the methods in the previous section is that the movement of the front had to be known in advance; in some cases, all the nodes moved

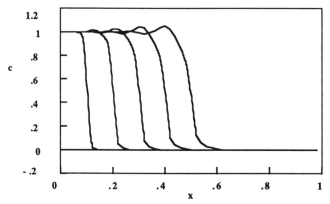

Figure 7.17. Convective Diffusion Equation, Moving Elements
Pe = 1000, Δx = 0.02 and 0.002, Δt = 9.09E-7

together. It would be desirable to allow each node to move independently and to have its movement determined by the solution to the problem without requiring intervention by the analyst. Here various techniques for doing this will be discussed.

Coyle, et al. [1986] studied the stability of one method for moving the nodes. They used an equidistribution principle to place the nodes, while attempting to keep some quantity uniform throughout the mesh.

$$\int_{x_{j-1}}^{x_j} w(x,t)\,dx = \frac{1}{N}\int_a^b w(x,t)\,dx \qquad (7.93)$$

The function w can be proportional to the absolute value of the gradient, the curvature, the local discretization error, or some combination of these. The points x_j are moved in such a way that Eq. (7.93) is satisfied. To do this, we differentiate Eq. (7.93) with respect to time:

$$\frac{dx_j}{dt}w(x_j,t) - \frac{dx_{j-1}}{dt}w(x_{j-1},t) + \int_{x_{j-1}}^{x_j}\frac{\partial w}{\partial t}(x,t)\,dx = \frac{1}{N}\int_a^b \frac{\partial w}{\partial t}(x,t)\,dx. \qquad (7.94)$$

This provides a differential equation in time for x_j that is to be solved together with the partial differential equation. Unfortunately, experience using the method has indicated that the mesh trajectories can leave the domain, cross each other, and occasionally oscillate wildly. Coyle, et al. [1986] did a stability analysis and showed that the method is stable to linear perturbations when

$$L(t) = \max_{0 < j < N} \frac{w(x_j(0),0)}{w(x_j(t),t)} < 1. \qquad (7.95)$$

Unfortunately, this does not hold for dissipative systems. Coyle, et al. then provide several adjustments to improve the results, but none of them are completely successful in providing a robust method. Dupont [1982] also provides a proof of the stability of methods with moving grids and finds that in some cases such methods are unstable. Thus, the movement of nodes numerically using an equidistribution principle is not straightforward. An excellent discussion is provided by Bieterman and Babuska [1986].

Smooke and Kozykowski [1983] used an equidistribution principle that proved successful. Their strategy was to try to ensure that both of the following equations hold for each element:

$$\int_{x_j}^{x_{j+1}} \left| \frac{dc}{dx} \right| dx \leq \delta \left| \max_{a \leq x \leq b} c - \min_{a \leq x \leq b} c \right| \tag{7.96}$$

$$\int_{x_j}^{x_{j+1}} \left| \frac{d^2c}{dx^2} \right| dx \leq \gamma \left| \max_{a \leq x \leq b} \frac{dc}{dx} - \min_{a \leq x \leq b} \frac{dc}{dx} \right|. \tag{7.97}$$

In addition, to prevent large changes of mesh size from element to element, they also require that (when $A \geq 1$)

$$\frac{1}{A} \leq \frac{\Delta x_j}{\Delta x_{j-1}} \leq A. \tag{7.98}$$

The strategy is as follows. At each time-step (or steady-state solution) we evaluate Eq. (7.96)-(7.97). If either criterion is not satisfied, we add a node in the middle of that element. We do this for all elements and then test Eq. (7.98). If this is not satisfied, we add a node in the larger element. We continue this until Eq. (7.98) holds for every element. To obtain the solution at the new points, we use a linear interpolation of the solution on the original mesh. This procedure adds nodes but does not remove them. However, in applications it was found that the solution soon reached a point where Eq. (7.96)-(7.97) was satisfied everywhere. When this happened over several time-steps, the points were allowed to move. They can be moved by extrapolating their past position in a linear fashion:

$$\frac{dx_j}{dt} = \frac{x_j^n - x_j^{n-1}}{t^n - t^{n-1}}, \quad x_j(t^{n-1}) = x_j^{n-1}, \tag{7.99}$$

$$\text{or } x_j = \left(\frac{x_j^n - x_j^{n-1}}{t^n - t^{n-1}} \right) t + x_j^{n-1}. \tag{7.100}$$

The paths of the points can cross or leave the domain, however. If the points cross, the new points are merely reordered. If a point leaves the domain, it is put back into the domain at a location halfway between the boundary and the first grid point. This scheme proved to be useful in several applications (Smooke and Kozykowski [1983]).

Another application of similar ideas results from work by Hu and Schiesser [1981]. They used a finite difference method to convert partial differential equations to ordinary differential equations; this method is sometimes called the method of lines. The criterion they used was the second derivative; if the second derivative was below a certain threshold value, they did not add a point. If it was above a certain threshold value, they added a point. Indeed, for difficult problems it proved necessary to have five thresholds, where up to 32 points were added. When points were added, the solution at the new point was obtained by interpolating the solution on the old mesh using a polynomial as accurate as needed to reproduce the finite difference equations. The Runge-Kutta method was used to integrate in time. Use of packages such as LSODE would be time-consuming because each time the mesh is changed the integrator would have to be restarted. Dissinger [1983] shows in his thesis that a fixed-mesh solution of Burger's equation was preferable to an adaptive technique using LSODE to integrate in time because each change of the mesh required that a new Jacobian be evaluated. Other early examples are by Dwyer, *et al.* [1980] and Davis and Flaherty [1982], while Tzanos [1989] gives a more recent application. Use of moving grids in conjunction with random choice-type methods is given by Harten and Hyman [1983].

An application to the convective diffusion equation is made here. For the first example, the initial conditions are

$$c_1(0) = 1,$$
$$c_i(0) = 0, \quad 2 \le i \le NT. \tag{7.101}$$

The initial conditions are taken on a mesh that is a combination of a fixed mesh with $\Delta x = 0.04$ (called the base grid) and added points at $x = 0, 0.003, 0.007, 0.012, 0.02$, and 0.03. The finite difference method is applied with centered first derivatives and a Crank-Nicolson treatment of the time derivatives. Thus we use

$$\frac{c_i^{n+1} - c_i^n}{\Delta t} + Pe \frac{\alpha(c_{i+1}^{n+1} - c_{i-1}^{n+1}) + (1-\alpha)(c_{i+1}^n - c_{i-1}^n)}{2(\Delta x_{i-1} + \Delta x_i)} =$$
$$= \frac{1}{\Delta x_{i-1} + \Delta x_i} \left[\alpha \left(\frac{c_{i-1}^{n+1} - c_i^{n+1}}{\Delta x_{i-1}} - \frac{c_i^{n+1} - c_{i+1}^{n+1}}{\Delta x_i} \right) + (1-\alpha) \left(\frac{c_{i-1}^n - c_i^n}{\Delta x_{i-1}} - \frac{c_i^n - c_{i+1}^n}{\Delta x_i} \right) \right]. \tag{7.102}$$

At each step of the calculation, Eq. (7.103) is applied.

$$\left(\frac{\Delta x_{j-1} + \Delta x_j}{2} \right)^2 \left| \frac{\partial^2 c}{\partial x^2} \right|_j \le \delta \tag{7.103}$$

The value of δ is usually taken as 0.1 and the second derivative is calculated using Eq. (7.15). If Eq. (7.103) is not satisfied, points can be added in accordance with Eq. (7.23). Points could also be eliminated, but the points on the base grid are always retained. The solution is shown in Figure 7.18. It is generally a good solution except for a few small (0.1%) overshoots behind the front. When the value of δ is reduced to 0.03, the solution is quite good, too, with slightly larger

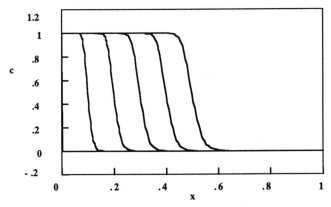

Figure 7.18. Convective Diffusion Equation, Equidistribution of Second Derivative Pe = 1000, 26 base nodes, $\delta = 0.1$, $\Delta t = 5E-6$

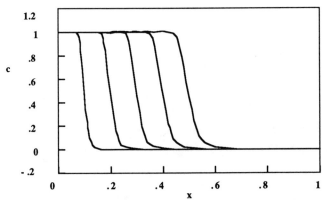

Figure 7.19. Convective Diffusion Equation, Equidistribution of Second Derivative Pe = 1000, 26 base nodes, $\delta = 0.03$, $\Delta t = 5E-6$

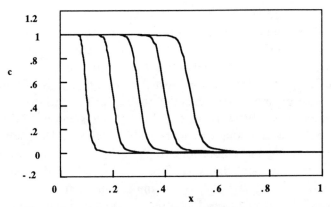

Figure 7.20. Convective Diffusion Equation, Equidistribution of Second Derivative Pe = 1000, all points moving, $\Delta t = 5E-6$

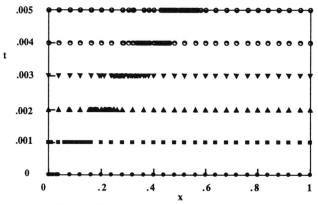

Figure 7.21. Nodal Locations for Figure 7.18

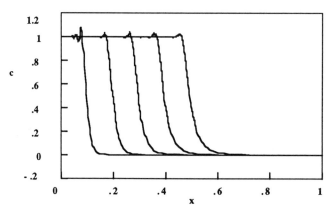

Figure 7.22. Advection Equation, Equidistribution of Second Derivative, $\alpha = 1$, $\delta = 0.01$

(0.7%) overshoots (see Figure 7.19). If all the points are allowed to move, the solution is as shown in Figure 7.20. In this figure, the initial condition had points at x = 0.0, 0.003, 0.007, 0.012, 0.02, 0.03, 0.04, 0.08, and 1.0. This solution is better than one calculated using a base grid that remains inviolate. The mesh is illustrated in Figure 7.21 for the case illustrated in Figure 7.18. The region of small elements clearly moves with the solution. Figure 7.22 shows the results for the advection equation. We also use $\alpha = 1$ to provide stability (if Δx is small enough, the upstream weighting will not be too deleterious). The solution is not very good despite using many more points (up to 170) than used elsewhere in this book. In contrast, Hu and Schiesser [1981] achieved good results for the advection equation in the case of a smooth initial condition.

Figure 7.23 shows the results for the second initial condition. Now the base grid has 20 elements of size $\Delta x = 0.1$ with the added points at 0.21, 0.22, 0.23, 0.238, 0.243, 0.247, and 0.25, the mirror image at approximately 0.25, and the same points centered on 0.75. The solution is a reasonable solution but again uses

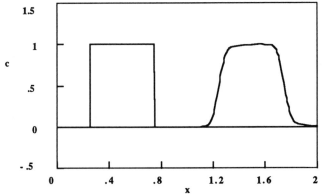

Figure 7.23. Convective Diffusion Equation, Equidistribution of Second Derivative
Pe = 1000, δ = 0.01, Δt = 5E-6

many grid points. It is clear that good results can be achieved with such equidistribution methods, provided that we are willing to adjust the tuning parameters (here δ and the base mesh).

7.4. Moving Fronts with Weighted Residual Placement of Nodes

In this section we will move nodes using principles that are derived from weighted residual criteria based on the differential equation. Different criteria are interpreted as least squares methods or the Petrov-Galerkin method; they are related to the equidistribution principles discussed in Section 7.3. The essential difference between these methods and those discussed previously is that here the nodes are moved continuously throughout a time-step. By contrast, the methods discussed in Section 7.3 involved placement of the nodes at each time-step, with possible addition and subtraction of nodes. However, the new nodes did not necessarily follow a continuous path from nodes at the previous time-step.

The first method that we will discuss was presented as a least squares method by the authors Miller and Miller [1981], Miller [1981], and Galinas, *et al.* [1981]. It will be presented here as a Petrov-Galerkin method and then we will see how the least squares interpretation can be used to good advantage (even though it is not a "least square"). Let us consider the differential equation

$$\frac{\partial c}{\partial t} = L(c), \qquad (7.104)$$

where the operator L(c) involves spatial gradients but not time derivatives. The trial function is taken as

$$c = \sum_{j=1}^{NT} c_j(t) N_j(x,t), \qquad (7.105)$$

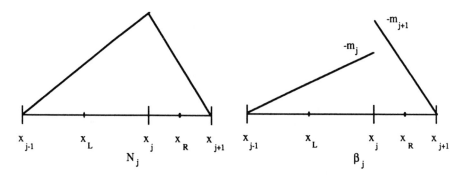

Figure 7.24. Basis Functions for Trial Functions and Time Derivatives

where the values of N_j are given by

$$N_j = \begin{cases} \dfrac{\xi+1}{2} & \xi = \dfrac{2(x-x_{j-1})}{x_j - x_{j-1}} - 1 \quad \text{for } x_{j-1} \leq x \leq x_j \\ \dfrac{1-\xi}{2} & \xi = \dfrac{2(x-x_j)}{x_{j+1} - x_j} - 1 \quad \text{for } x_j < x \leq x_{j+1} \\ 0 & \text{otherwise} \end{cases} \quad (7.106)$$

and are illustrated in Figure 7.24. The location of the nodes x_j can change with time. Thus the time derivative of c is

$$\frac{\partial c}{\partial t} = \sum_{j=1}^{NT} \frac{dc_j}{dt} N_j(x,t) + \sum_{j=1}^{NT} c_j \frac{\partial N_j}{\partial t}. \quad (7.107)$$

The derivative of the function N_j is

$$\frac{\partial N_j}{\partial t} = \frac{\partial N_j}{\partial \xi} \frac{\partial \xi}{\partial t} \quad (7.108)$$

$$\frac{\partial N_j}{\partial t} = \begin{cases} \dfrac{1}{2}\left[\dfrac{-2(x-x_{j-1})}{(x_j - x_{j-1})^2}(\dot{x}_j - \dot{x}_{j-1}) + \dfrac{-2\dot{x}_{j-1}}{(x_j - x_{j-1})} \right] & \text{for } x_{j-1} \leq x \leq x_j \\ -\dfrac{1}{2}\left[\dfrac{-2(x-x_j)}{(x_{j+1} - x_j)^2}(\dot{x}_{j+1} - \dot{x}_j) + \dfrac{-2\dot{x}_j}{(x_{j+1} - x_j)} \right] & \text{for } x_j < x \leq x_{j+1} \\ 0 & \text{otherwise} \end{cases} \quad (7.109)$$

or

$$\frac{\partial N_j}{\partial t} = \begin{cases} \left[-\left(\frac{1+\xi}{2}\right)\frac{\dot{\Delta x}_{j-1}}{\Delta x_{j-1}} - \frac{\dot{x}_{j-1}}{\Delta x_{j-1}}\right] & \text{for } x_{j-1} \leq x \leq x_j \\ \left[\left(\frac{1+\xi}{2}\right)\frac{\dot{\Delta x}_j}{\Delta x_j} + \frac{\dot{x}_j}{\Delta x_j}\right] & \text{for } x_j < x \leq x_{j+1} \\ 0 & \text{otherwise} \end{cases} \quad (7.110)$$

We next assemble the terms of the product, giving

$$\sum c_j \frac{\partial N_j}{\partial t}. \quad (7.111)$$

In the element to the left of x_j we have

$$c_{j-1}\frac{\partial N_{j-1}}{\partial t} + c_j\frac{\partial N_j}{\partial t} = \left[\left(\frac{1+\xi}{2}\right)\frac{\dot{\Delta x}_{j-1}}{\Delta x_{j-1}} + \frac{\dot{x}_{j-1}}{\Delta x_{j-1}}\right] + c_j\left[-\left(\frac{1+\xi}{2}\right)\frac{\dot{\Delta x}_{j-1}}{\Delta x_{j-1}} - \frac{\dot{x}_{j-1}}{\Delta x_{j-1}}\right], \quad (7.112)$$

for $x_{j-1} \leq x \leq x_j$

whereas in the element to the right of x_j we have

$$c_j\frac{\partial N_j}{\partial t} + c_{j+1}\frac{\partial N_{j+1}}{\partial t} = c_j\left[\left(\frac{1+\xi}{2}\right)\frac{\dot{\Delta x}_j}{\Delta x_j} + \frac{\dot{x}_j}{\Delta x_j}\right] + c_{j+1}\left[-\left(\frac{1+\xi}{2}\right)\frac{\dot{\Delta x}_j}{\Delta x_j} - \frac{\dot{x}_j}{\Delta x_j}\right], \quad (7.113)$$

for $x_j < x \leq x_{j+1}$

where

$$\dot{\Delta x}_{j-1} = \dot{x}_j - \dot{x}_{j-1}$$
$$\dot{\Delta x}_j = \dot{x}_{j+1} - \dot{x}_j. \quad (7.114)$$

It is convenient to rewrite the product, Eq. (7.111), in terms of the velocity of the nodes.

$$\sum \frac{\partial x_j}{\partial t} \beta_j \quad (7.115)$$

We collect the terms from Eq. (7.112) that include dx_j/dt:

$$c_{j-1}\left[\left(\frac{1+\xi}{2}\right)\frac{\dot{x}_j}{\Delta x_{j-1}}\right] + c_j\left[-\left(\frac{1+\xi}{2}\right)\frac{\dot{x}_j}{\Delta x_{j-1}}\right] \quad (7.116)$$

and those from Eq. (7.113):

$$c_j\left[-\left(\frac{1+\xi}{2}\right)\frac{\dot{x}_j}{\Delta x_j} + \frac{\dot{x}_j}{\Delta x_j}\right] + c_{j+1}\left[\left(\frac{1+\xi}{2}\right)\frac{\dot{x}_j}{\Delta x_j} - \frac{\dot{x}_j}{\Delta x_j}\right]. \quad (7.117)$$

In the left element, ξ goes from $\xi = -1$ to $\xi = 1$ and the coefficient of dx_j/dt goes from 0 to

$$-\frac{c_j - c_{j-1}}{x_j - x_{j-1}}. \tag{7.118}$$

In the right element, ξ goes from $\xi = -1$ to $\xi = 1$ and the coefficient of dx_j/dt goes from

$$-\frac{c_{j+1} - c_j}{x_{j+1} - x_j} \tag{7.119}$$

to 0. If we define the new functions m_j as

$$m_j = \frac{c_j - c_{j-1}}{x_j - x_{j-1}}, \quad m_{j+1} = \frac{c_{j+1} - c_j}{x_{j+1} - x_j}, \tag{7.120}$$

we can write the coefficient of dx_j/dt as

$$\beta_j = -m_j \left(\frac{1+\xi}{2}\right) = -m_j N_j \quad \text{on the left}$$

$$= m_{j+1} \left(\frac{\xi-1}{2}\right) = -m_{j+1} \left(\frac{1-\xi}{2}\right) = -m_{j+1} N_j \quad \text{on the right.} \tag{7.121}$$

The function β_j is illustrated in Figure 7.24. It is a discontinuous function at node j. The time derivative, Eq. (7.107), is then written as

$$\frac{\partial c}{\partial t} = \sum_{j=1}^{NT} \frac{dc_j}{dt} N_j(x,t) + \sum_{j=1}^{NT} \beta_j \frac{dx_j}{dt}, \tag{7.122}$$

with N_j given by Eq. (7.106) and β_j given by Eq. (7.121).
To summarize, the trial solution is

$$c = \sum_{j=1}^{NT} c_j(t) N_j(x,t) \tag{7.123}$$

and the time derivative is Eq. (7.122). Let us solve for the unknown variables by using a method of weighted residuals (actually a Petrov-Galerkin criterion):

$$\int \left(\frac{\partial c}{\partial t} - L(c)\right) N_i(x,t) \, dx = 0, \tag{7.124}$$

$$\int \left(\frac{\partial c}{\partial t} - L(c)\right) \beta_i(x,t) \, dx = 0. \tag{7.125}$$

Eq. (7.123) is substituted into Eq. (7.124)-(7.125) to give

$$\sum_j \frac{dc_j}{dt} \int N_j N_i \, dx + \sum_j \frac{dx_j}{dt} \int \beta_j N_i \, dx = \int L(c) N_i \, dx \tag{7.126}$$

$$\sum_j \frac{dc_j}{dt} \int N_j \beta_i \, dx + \sum_j \frac{dx_j}{dt} \int \dot{\beta}_j \beta_i \, dx = \int L(c) \beta_i \, dx. \tag{7.127}$$

These are the equations that must be solved for $c_j(t)$ and $x_j(t)$. The mesh, of course, will be irregular. Then the second derivative of c at x_j is given by Eq. (7.15) as

$$\frac{\partial^2 c}{\partial x^2} = \frac{m_{j+1} - m_j}{\frac{x_j + x_{j+1}}{2} - \frac{x_j + x_{j-1}}{2}} = \frac{2(m_{j+1} - m_j)}{\Delta x_{j-1} + \Delta x_j}. \tag{7.128}$$

If the solution is linear, it has no second derivative; in these cases, $m_j = m_{j+1}$. The trial function for $\dot{\beta}_j$ is no longer discontinuous, so the trial functions for $\dot{\beta}_j$ and for N_j are the same. Thus, some of the equations in Eq. (7.127) are redundant. This redundancy complicates the solution method. Another problem is that the nodes can get close to each other and the functions β_j become very large or undefined (if $x_j = x_{j+1}$). The nodes are free to move according to Eq. (7.126)-(7.127), which could lead to this situation.

The original derivation of the method employed a least squares criterion to solve this problem. Let us consider the functional

$$J(t) = \int \left[\frac{\partial c}{\partial t} - L(c) \right]^2 dx. \tag{7.129}$$

A functional is a mapping from a function space to the real line. In Eq. (7.129) the result is still a function because integration over time has not been performed. Thus Eq. (7.129) is not a proper functional and this will cloud the least squares interpretation. We add terms to the function that will retard the movement of the nodes or keep the nodes separated. Both of these empirical additions will be useful in practice. The form used by Miller [1981] is

$$J(t) = \int \left[\frac{\partial c}{\partial t} - L(c) \right]^2 dx + \sum_{j=1}^{NT} \left[c_1 \dot{\Delta x}_j - \frac{c_2}{\Delta x_j} \right]^2. \tag{7.130}$$

The constants c_1 and c_2 can be adjusted to obtain the best solution. When the trial functions are inserted, we obtain

$$J(t) = \int \left[\sum_{j=1}^{NT} \frac{dc_j}{dt} N_j(x,t) + \sum_{j=1}^{NT} \beta_j \frac{dx_j}{dt} - L(c) \right]^2 dx + \sum_{j=1}^{NT} \left[c_1 \dot{\Delta x}_j - \frac{c_2}{\Delta x_j} \right]^2. \tag{7.131}$$

Then we "minimize" Eq. (7.131) with respect to dc_j/dt and dx/dt, while keeping c_j and x_j fixed; this is the feature that prevents the method from being a least squares method. The result is

$$\sum_{j=i-1}^{i+1} \frac{dc_j}{dt} \int N_j N_i \, dx + \sum_{j=i-1}^{i+1} \frac{dx_j}{dt} \int \beta_j N_i \, dx = \int L(c) N_i \, dx \qquad (7.132)$$

$$\sum_{j=i-1}^{i+1} \frac{dc_j}{dt} \int N_j \beta_i \, dx + \sum_{j=i-1}^{i+1} \frac{dx_j}{dt} \int \beta_j \beta_i \, dx + c_1^2 (\Delta \dot{x}_j - \Delta \dot{x}_{j+1}) =$$

$$= \int L(c) \beta_i \, dx + c_2 c_1 \left(\frac{1}{x_j} - \frac{1}{x_{j+1}} \right). \qquad (7.133)$$

These equations are the same as Eq. (7.126)-(7.127), with additional terms that keep the nodes separated and prevent them from moving too quickly. The difficulty with the least squares interpretation of the Petrov-Galerkin method is that it is difficult to envisage variations of dc_j/dt at every instant in time while keeping $c_j(t)$ fixed. It is much more straightforward to interpret the method as a Petrov-Galerkin method and not confuse the interpretation with reference to *ad hoc* least squares ideas. Problem 7.7 explores this issue and compares the Galerkin and least squares methods, both as a function of time and when integrated from time zero to infinite time.

Lynch [1982] employs a slightly different way of moving the nodes. The simplification of his two-dimensional scheme is to solve for the nodal velocities using an equation for the elastic displacement of a solid. This prevents the nodes from crossing over each other.

$$\frac{d^2 \dot{x}_j}{dx^2} = 0 \qquad (7.134)$$

The Petrov-Galerkin method is used on this moving grid.

Another treatment of a weighted residual placement of nodes is discussed by Herbst, *et al.* [1982, 1983, 1984]. Their treatment, presented here for Burger's equation, allows us to relate the Petrov-Galerkin method presented above to the equidistribution principles of previous sections. It also provides a connection with upstream techniques used in other chapters. Let us consider the steady-state Burger equation in order to see the upstream nature of the results:

$$v \frac{d^2 u}{dx^2} - \frac{dV(u)}{dx} = 0 \qquad (7.135)$$

with

$$V(u) = \frac{1}{2} u^2. \qquad (7.136)$$

As long as V is used in the formulas, it is possible to change the differential equation to another equation by redefining the value of V; for example, using V = u gives the steady-state convective diffusion equation. We change the notation here to correspond with that of Herbst, *et al.* [1984]. The variable s goes from -1 to 1 as the x position goes from x_{j-1} to x_{j+1} (i.e., the local domain encompasses two elements). The trial function is

$$u = \sum_{i=1}^{NT} u_i N_i(x) \tag{7.137}$$

with

$$N_i(s) = \begin{cases} 1 - |s| & |s| \le 1 \\ 0 & |s| > 1. \end{cases} \tag{7.138}$$

The weighting function is

$$W_j = S_j(x) + T_j(x), \tag{7.139}$$

where the functions $S(x)$ and $T(x)$ are defined as

$$S_j(x) = S\left(\frac{x}{\Delta x} - j + 1\right), \quad T_j(x) = T\left(\frac{x}{\Delta x} - j + 1\right). \tag{7.140}$$

The weighted residual is then

$$\int_0^1 W_j \left[v \frac{d^2u}{dx^2} - \frac{dV(u)}{dx} \right] dx = 0. \tag{7.141}$$

Since the weighting function is not the same as the trial function, Eq. (7.138), the method is a Petrov-Galerkin method (rather than a Galerkin method alone).

We choose the weighting functions to satisfy certain properties. If we take $S(s)$ to be an even function of s and $T(s)$ to be an odd function of s, the following equations are satisfied:

$$\int_{-1}^{1} \frac{dS}{ds} ds = 0, \quad \int_{-1}^{1} T(s) ds = 0, \quad \text{and} \quad \int_{-1}^{1} \frac{dT}{ds} \text{sgn}(s) ds = 0. \tag{7.142}$$

In addition, we require that

$$\int_{-1}^{1} S(s) ds = 1, \quad S(-1) = S(1) = 0, \tag{7.143}$$

as well as requiring the consistency conditions

$$\int_{-1}^{1} \frac{dS}{ds} \text{sgn}(s) ds = -2, \quad \int_0^1 \frac{dT}{ds} ds = 0. \tag{7.144}$$

Several derived conditions follow from these; they are derived in the electronic text entitled "Special Petrov-Galerkin Weighting Functions." The result is

$$\nu\left(1 + \frac{u_j h \alpha}{2\nu}\right)(m_{j+1} - m_j) - \frac{u_j}{2}(u_{j+1} - u_{j-1}) =$$
$$= h^2 (m_{j+1}^2 - m_j^2) \int_0^1 s\, S(s)\, ds + h^2 (m_{j+1}^2 + m_j^2) \int_0^1 s\, T(s)\, ds, \tag{7.145}$$

where
$$m_j h = u_j - u_{j-1}. \tag{7.146}$$

In this equation
$$\alpha = -2 \int_0^1 T(s)\, ds. \tag{7.147}$$

The parameter α thus determines the amount of upstream weighting; if both α and u_j have the same sign, then including the function T_j increases the coefficient of the diffusion term. The presence of the function T_j also decreases the truncation error. This development shows us how to get an upstream method using the weighting function given in Eq. (7.139).

Next we apply the same ideas to the transient Burger equation. The equation is
$$\frac{\partial u}{\partial t} = \nu \frac{\partial^2 u}{\partial x^2} - \frac{\partial}{\partial x} V(u). \tag{7.148}$$

The same trial function and weighting functions are used as in Eq. (7.137)-(7.139), but now the x_j nodes are movable. The time derivative is given by
$$\frac{\partial u}{\partial t} = \sum_{i=1}^{NT} \left(\frac{du_i}{dt} - \frac{\partial u}{\partial x}\frac{dx_i}{dt}\right) N_i \tag{7.149}$$

and this can be expressed as
$$\frac{\partial u}{\partial t} = \sum_{i=1}^{NT} \left(\frac{du_i}{dt} N_i + \beta_i \frac{dx_i}{dt}\right), \tag{7.150}$$

where
$$\beta_i = -\frac{\partial u}{\partial x} N_i. \tag{7.151}$$

Eq. (7.151) is the analog to Eq. (7.122). We use the primary weighting function

$$S_j(x) = \begin{cases} S\left(\dfrac{x - x_j}{x_j - x_{j-1}}\right) & x_{j-1} \le x \le x_j \\ S\left(\dfrac{x - x_j}{x_{j+1} - x_j}\right) & x_j < x \le x_{j+1} \\ 0 & \text{elsewhere} \end{cases} \tag{7.152}$$

and the secondary weighting function

$$T_j(x) = \begin{cases} \Delta x_{j-1}^r \, T\left(\dfrac{x - x_j}{x_j - x_{j-1}}\right) & x_{j-1} \leq x \leq x_j \\ \Delta x_j^r \, T\left(\dfrac{x - x_j}{x_{j+1} - x_j}\right) & x_j < x \leq x_{j+1} \\ 0 & \text{elsewhere.} \end{cases} \qquad (7.153)$$

The power r is either 0 or 1 and the functions S and T satisfy the same conditions as given in Eq. (7.142)-(7.144). Since we have both the function and nodal velocities to determine, we use the weighted residual method with each weighting function:

$$\sum_{i=1}^{NT} \left[\frac{du_i}{dt} \int_0^1 N_i \, S_j \, dx + \frac{dx_i}{dt} \int_0^1 \beta_i \, S_j \, dx \right] = -\int_0^1 \frac{du}{dx} \frac{dS_j}{dx} dx + \int_0^1 V(u) \frac{dS_j}{dx} dx, \qquad (7.154)$$

$$\sum_{i=1}^{NT} \left[\frac{du_i}{dt} \int_0^1 N_i \, T_j \, dx + \frac{dx_i}{dt} \int_0^1 \beta_i \, T_j \, dx \right] = -\int_0^1 \frac{du}{dx} \frac{dT_j}{dx} dx + \int_0^1 V(u) \frac{dT_j}{dx} dx. \qquad (7.155)$$

The truncation error is

$$\text{truncation error} \propto O(\Delta x_{j+1} - \Delta x_j) + O(\Delta x^2)$$

$$\Delta x = \max_j \{\Delta x_j\} \qquad (7.156)$$

and Herbst, *et al.* [1982] show that, when r = 1, Eqs. (7.155-7.156) are equivalent to

$$\Delta x_{j-1}^{r+1} \, v \, \frac{\partial^2 u}{\partial x^2}\bigg|_{j-} = \Delta x_j^{r+1} \, v \, \frac{\partial^2 u}{\partial x^2}\bigg|_{j+} + O(\Delta x^{r+2}). \qquad (7.157)$$

This, of course, looks very similar to equidistribution principles. Herbst, *et al.* [1983] show that the Petrov-Galerkin scheme presented above, Eq. (7.127) (sometimes called a least squares method), is equivalent to Eq. (7.157) when r = 0. Thus we can regard the Petrov-Galerkin (least squares) method as one equidistribution principle [r = 0 in Eq. (7.157)]; other equidistribution principles can also be used (r = 1).

Implementation of the Petrov-Galerkin (least squares) method is not easy [Miller and Miller, 1981]; it is discussed in detail in two papers by Hrymak, *et al.* [1986] and Hrymak and Westerberg [1986]. The first difficulty is how to treat inner products involving the function β_j and the second derivative of the trial solution, since the function β_j is discontinuous and the second derivative of the trial function does not exist. The usual solution to this dilemma in finite element methods is to integrate by parts the term involving the second derivative. However, the weighting function is discontinuous here, so this is more complicated. Let us evaluate the inner product over regions:

$$\int_{x_{j-1}}^{x_{j+1}} \frac{\partial \beta}{\partial x} \frac{\partial u}{\partial x} dx = \int_{x_{j-1}}^{x_j-\varepsilon} \frac{\partial \beta}{\partial x} \frac{\partial u}{\partial x} dx + \int_{x_j+\varepsilon}^{x_{j+1}} \frac{\partial \beta}{\partial x} \frac{\partial u}{\partial x} dx + \int_{x_j-\varepsilon}^{x_j+\varepsilon} \frac{\partial \beta}{\partial x} \frac{\partial u}{\partial x} dx. \quad (7.158)$$

In the first region, from x_{j-1} to $x_j-\varepsilon$, the value of the β and u derivatives are [see Eq. (7.106) and Eq. (7.121)]

$$\frac{\partial \beta}{\partial x} = -\frac{m_j}{\Delta x_{j-1}}, \quad \frac{\partial u}{\partial x} = m_j. \quad (7.159)$$

In the last region, from $x_j+\varepsilon$ to x_{j+1}, the value of the β and u derivatives are [see Eq. (7.106) and Eq. (7.121)]

$$\frac{\partial \beta}{\partial x} = \frac{m_{j+1}}{\Delta x_j}, \quad \frac{\partial u}{\partial x} = m_{j+1}. \quad (7.160)$$

For the region at x_j, where neither the β nor u derivatives exist, we use the following approximations:

$$\frac{\partial \beta}{\partial x} = \frac{-m_{j+1} + m_j}{2\varepsilon}, \quad \frac{\partial u}{\partial x} = \frac{m_j + m_{j+1}}{2}. \quad (7.161)$$

Then

$$\int_{x_j-\varepsilon}^{x_j+\varepsilon} \frac{\partial \beta}{\partial x} \frac{\partial u}{\partial x} dx = \frac{m_j^2 - m_{j+1}^2}{2}. \quad (7.162)$$

The combined result is

$$\int_{x_{j-1}}^{x_{j+1}} \frac{\partial \beta}{\partial x} \frac{\partial u}{\partial x} dx = \frac{m_{j+1}^2 - m_j^2}{2}. \quad (7.163)$$

Another difficulty in implementing the Petrov-Galerkin (least squares) method is the choice of the constants in the node control terms of Eq. (7.130). Hrymak, *et al.* [1986] suggest

$$\sum_{j=1}^{NT} (X_j \Delta \dot{x}_j - \omega_j)^2, \quad X_j^2 = \frac{c_1^2}{\Delta x_j - \delta}, \quad X_j \omega_j = \frac{c_1^2}{(\Delta x_j - \delta)^2}. \quad (7.164)$$

Both c_1 and c_2 should take values close to the requested error from the differential equation solver. The term δ is the minimum approach distance; the nodes will not get closer than this. The nodes can also cross, which then makes the equations hard to solve (this is the same as when the internodal distances approach δ). To prevent this, Hrymak, *et al.* [1986] modified the differential equation solver by multiplying the corrector error by the fraction f.

$$f = \frac{|\Delta x_j^n|}{|\Delta x_j^n| + |\Delta x_j^{n+1}|} \quad (7.165)$$

The fraction is evaluated using the index j, for which the value of $x_{j+1} - x_j$ is the most negative at the new time. The effect of this change is to keep the time-step small enough so that the nodes do not cross. Finally, Hrymak and Westerberg [1986] found that when solving the convective diffusion equation with a step change at the boundary condition, they had to start with the step change within the domain rather than at the boundary. Convergence of the moving finite element method [similar to that shown in Eq. (7.126)-(7.127)] has been proven by DuPont [1982] for linear, smoothly varying problems.

Carey and Jiang [1988] and Jiang and Carey [1988] employ the least squares method on a discrete level. For the advection equation they take the functional

$$I(t^*) = \int_0^1 \left[\frac{\partial u}{\partial t}(x,t^*) + Pe \frac{\partial u}{\partial x}(x,t^*) \right]^2 dx. \quad (7.166)$$

The time derivative is written in terms of a difference expression,

$$\frac{\partial u}{\partial t} = \frac{u(x,t+\Delta t) - u(x,t)}{\Delta t} = \frac{u^{n+1} - u^n}{\Delta t}. \quad (7.167)$$

The convective term is evaluated implicitly to give

$$\frac{\partial u}{\partial x} = \theta \frac{\partial u}{\partial x}\bigg|^{n+1} + (1-\theta)\frac{\partial u}{\partial x}\bigg|^n. \quad (7.168)$$

The functional then depends on u^{n+1}:

$$I(u^{n+1}) = \int_0^1 \left\{ \frac{u^{n+1} - u^n}{\Delta t} + Pe \left[\theta \frac{\partial u}{\partial x}\bigg|^{n+1} + (1-\theta)\frac{\partial u}{\partial x}\bigg|^n \right] \right\}^2 dx. \quad (7.169)$$

Differentiation with respect to u^{n+1} gives

$$\delta I = 2 \int_0^1 \left\{ \frac{u^{n+1} - u^n}{\Delta t} + Pe \left[\theta \frac{\partial u}{\partial x}\bigg|^{n+1} + (1-\theta)\frac{\partial u}{\partial x}\bigg|^n \right] \right\} \left\{ \frac{\delta u^{n+1}}{\Delta t} + Pe \theta \frac{\partial \delta u^{n+1}}{\partial x} \right\} dx. \quad (7.170)$$

If $\theta = 0$, this is the standard Galerkin method. It is also a Petrov-Galerkin method with a weighting function of

$$W_j = N_j + Pe \, \theta \, \Delta t \frac{\partial N_j}{\partial x}. \quad (7.171)$$

If $\theta = 0.5$, it is related to a Taylor-Galerkin method. Carey and Jiang [1988] give comparisons that show that the method using θ = 0.5 is better than the method using θ = 0. This also proved true for the Galerkin method, since implicit methods caused the solutions to be smoothed out. When these methods were applied to nonlinear problems, Jiang and Carey [1988] found numerical instability when shocks formed and had to introduce additional numerical damping to correct this. Because of this instability, we will not consider the method further. See the paper by Park and Liggett [1990], who develop the method for linear problems.

To illustrate the results, solutions will be shown for Burger's equation, since that problem has been solved by all the authors referred to previously. We take ν = 0.001 and first consider a sinusoidal initial condition:

$$u(x,0) = \sin(2\pi x) + \frac{1}{2}\sin(\pi x). \tag{7.172}$$

Figure 7.25 shows the solution obtained by Gelinas, *et al.* [1981], which uses 21 nodes. Figure 7.26 shows the same problem using the method described above [Hrymak, *et al.*, 1986]. Both methods give sharp profiles; Figure 7.26 uses only 19 nodes. However, approximately twice as many equations must be integrated as there are nodes, since the nodes can move; thus Galinas, *et al.* integrated 40 equations while Hrymak, *et al.* integrated 36 equations.

Since this is not one of the standard problems solved in other chapters, we will solve it now. We use 51 nodes, which gives us only 50 equations to integrate since we are not using an equidistribution principle. The function u(x,t) starts out smooth but develops a shock. Figure 7.27a shows the solution obtained by using MacCormack's method. It is quite good, with only a small bump at the shock front. Figure 7.27b shows the results obtained with the MacCormack flux-corrected method; they are even better. If a smaller time-step is used, the results are excellent (see Figure 7.27c). The random choice method gives the results shown in Figure 7.27d; the solution is excellent with no oscillations, but the front is in slightly the wrong place. The Galerkin method requires a smaller step-size and gives a solution that oscillates wildly (see Figure 7.27e). The Taylor-Galerkin method with a finite element representation of the extra terms gives good results with small oscillations (see Figure 7.27f). If a larger time-step is used ($\Delta t = 0.01$), the results are better (see Figure 7.27g). We thus conclude for this example that some of the methods we have discussed in prior chapters solve this problem almost as well as the moving finite element methods.

The next comparison employs the function

$$f(\zeta) = \frac{[\mu + \alpha + (\mu - \alpha)\exp(\alpha\zeta/\varepsilon)]}{[1 + \exp(\alpha\zeta/\varepsilon)]} \tag{7.173}$$

and uses Burger's equation with the following initial and boundary conditions:

$$\begin{aligned} u(x,0) &= f(x-\beta) \\ u(0,t) &= f(-\mu t-\beta) \\ u(1,t) &= f(1-\beta-\mu t). \end{aligned} \tag{7.174}$$

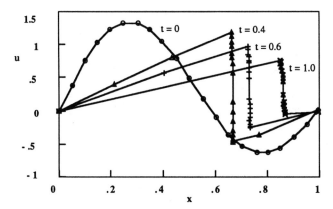

Figure 7.25. Burger's Equation, Moving Finite Elements, $\nu = 0.001$
Data from Galinas, *et al.* [1981]

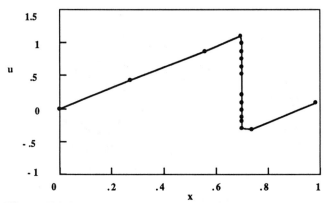

Figure 7.26. Burger's Equation, Moving Finite Elements, $\nu = 0.001$
Data from Hrymak, *et al.* [1986]

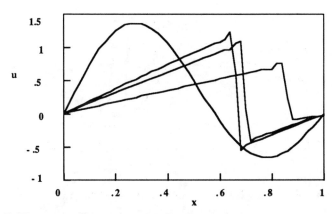

Figure 7.27a. Burger's Equation, MacCormack Method, $\nu = 0.001$, $\Delta x = 0.02$, $\Delta t = 0.01$

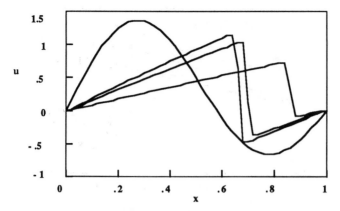

Figure 7.27b. Burger's Equation, MacCormack Method with Flux-correction
ν = 0.001, Δx = 0.02, Δt = 0.01

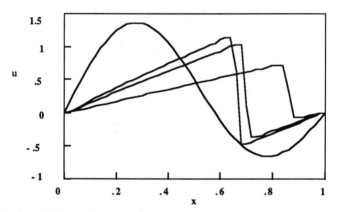

Figure 7.27c. Burger's Equation, MacCormack Method with Flux-correction
ν = 0.001, Δx = 0.02, Δt = 0.005

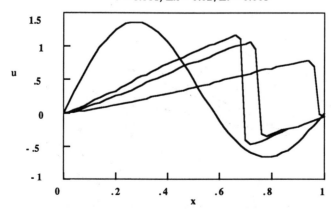

Figure 7.27d. Burger's Equation, Random Choice Method, ν = 0.001
Δx = 0.02, Δt = 0.01

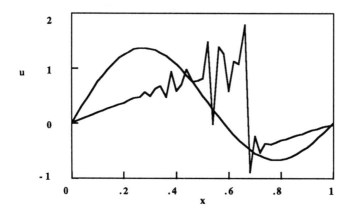

Figure 7.27e. Burger's Equation, Galerkin Method, $\nu = 0.001$, $\Delta x = 0.02$, $\Delta t = 0.005$

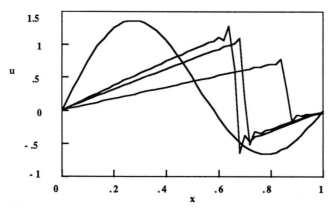

Figure 7.27f. Burger's Equation, Taylor-Galerkin Method, $\nu = 0.001$
$\Delta x = 0.02$, $\Delta t = 0.005$

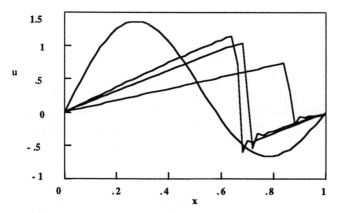

Figure 7.27g. Burger's Equation, Taylor-Galerkin Method, $\nu = 0.001$
$\Delta x = 0.02$, $\Delta t = 0.01$

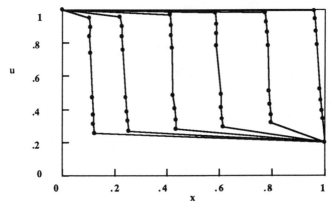

Figure 7.28a. Burger's Equation, Moving Finite Elements Petrov-Galerkin Method, r = 0, Data from Herbst, *et al.* [1983]

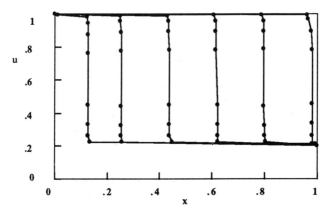

Figure 7.28b. Burger's Equation, Moving Finite Elements, Equidistribution Principle, r = 1, Data from Herbst, *et al.* [1983]

The exact solution is

$$u(x,t) = f(\zeta), \quad \zeta = x - \mu t - \beta. \tag{7.175}$$

It represents a travelling wave front that starts at $x = \beta$ and moves with a speed μ. The parameters are $\alpha = 0.4$, $\beta = 0.125$, $\varepsilon = 0.001$, and $\mu = 0.6$. The solutions using nine elements are shown in Figure 7.28a for the Petrov-Galerkin method [Eq. (7.157) and r = 0] and in Figure 7.28b for the equidistribution principle [Eq. (7.157) and r = 1]. Clearly the equidistribution principle with r = 1 gives better results.

7.5. Euler-Lagrange Methods

The concentration of a fluid can be characterized in several ways. If a Lagrangian method is used, the concentration is referred to a coordinate system that moves with the fluid velocity; we consider a fluid particle (or packet) and the coordinate system follows it. The time derivative is usually expressed as

$$\frac{Dc}{Dt} = \frac{\partial c}{\partial t} + \mathbf{v} \cdot \nabla c. \tag{7.176}$$

The convective diffusion equation in a Lagrangian system is

$$\frac{Dc}{Dt} = \frac{\partial^2 c}{\partial x^2}. \tag{7.177}$$

The convective term has seemingly disappeared! Unfortunately, the concentration is that of a fluid particle, which is constantly moving in time. Thus the equation is more complicated than it seems. By contrast, in a Eulerian system the coordinate system is fixed in space and the convective diffusion equation is

$$\frac{\partial c}{\partial t} + v \frac{\partial c}{\partial x} = \frac{\partial^2 c}{\partial x^2}. \tag{7.178}$$

Now the concentration is defined with respect to the fixed location and time, rather than the particle identification used for Eq. (7.177). This equation includes the convective term. It would be useful if we could combine these two approaches in some way: a convective movement and then a diffusion step. The Euler-Lagrange methods do just that.

The classical Euler-Lagrange method is the particle-in-cell finite difference method. In this method, the region of space is divided up into smaller cells and a number of particles are placed within each cell. The equations then determine the movement of the particles from cell to cell, while the concentration is determined by the number of particles within a cell at any given time. This approach was applied to the convective diffusion equation by Garder, *et al.* [1964]; more recently it has been applied to multi-dimensional problems by Farmer [1985]. The Euler-Lagrange methods are related to these approaches.

First, we decompose the problem into two parts, one that is predominately convective and another that is predominately diffusive with a forcing term. We take the convective diffusion equation,

$$\frac{\partial c}{\partial t} + Pe \frac{\partial c}{\partial x} = \frac{\partial^2 c}{\partial x^2}, \tag{7.179}$$

and write the solution as the sum of two parts

$$c = c_1 + c_2. \tag{7.180}$$

The function c_1 is taken as the solution to

$$\frac{\partial c_1}{\partial t} + Pe \frac{\partial c_1}{\partial x} = 0. \tag{7.181}$$

The equations that the function c_2 must satisfy are found by subtracting Eq. (7.181) from Eq. (7.179):

$$\frac{\partial c_2}{\partial t} + \text{Pe}\,\frac{\partial c_2}{\partial x} = \frac{\partial^2 c_2}{\partial x^2} + \frac{\partial^2 c_1}{\partial x^2}. \tag{7.182}$$

We have a convective diffusion equation for c_2 with a known forcing term involving c_1. If the function c_1 accounts for the convective nature of the solution, then the convective term for c_2 will be small and Eq. (7.182) will be primarily diffusive in nature. The function c_2 may be discontinuous, however, presenting new difficulties.

Let us consider a method of solving Eq. (7.179) that uses a time-splitting scheme. We split the calculation into two steps, one involving the convection operator and one involving the diffusion operator. We solve first the convective step:

$$\frac{c_i^{n+1/2} - c_i^n}{\Delta t} = -\text{Pe}\,\frac{c_i^n - c_{i-1}^n}{\Delta x} \tag{7.183}$$

and then the diffusion step

$$\frac{c_i^{n+1} - c_i^{n+1/2}}{\Delta t} = \frac{c_{i+1}^{n+1/2} - 2 c_i^{n+1/2} + c_{i-1}^{n+1/2}}{\Delta x^2}. \tag{7.184}$$

If we add these equations, we get

$$\frac{c_i^{n+1} - c_i^n}{\Delta t} = -\text{Pe}\,\frac{c_i^n - c_{i-1}^n}{\Delta x} + \frac{c_{i+1}^{n+1/2} - 2 c_i^{n+1/2} + c_{i-1}^{n+1/2}}{\Delta x^2}. \tag{7.185}$$

This is a finite difference formulation of Eq. (7.179) that is correct to the first-order in Δx and Δt. If we employ steps such as in the MacCormack method, we can make the convective terms second-order as well. In the Euler-Lagrange methods the convective step is solved much more accurately than to the first-order in Δx and Δt.

The first Euler-Lagrange method to be discussed follows the presentation by Ewing, Russell, and Wheeler [1983] and Russell [1983]; convergence proofs are in Douglas and Russell [1982]. We write the equation as

$$\frac{Dc}{Dt} = \frac{\partial^2 c}{\partial x^2}. \tag{7.186}$$

We write the difference approximation as

$$\frac{c_i^{n+1} - {}_i c^n}{\Delta t} = \frac{c_{i+1}^{n+1} - 2 c_i^{n+1} + c_{i-1}^{n+1}}{\Delta x^2}. \tag{7.187}$$

In this equation

$$_i c^n = c(x_{old} = x_i - u_i^n \Delta t, t = t^n) \tag{7.188}$$

represents the value of concentration at the time level n and a position x_{old}, such that movement takes it exactly to the position x_i at the time level n+1. The index that is appended before the concentration is the one that is irregular, i.e., the one that must be solved for by solving the convection equation. Here the concentration at $c(x_{old}, t^n)$ is convected along with the fluid and ends up at x_i, t^{n+1}. Then the concentration is allowed to diffuse at the time level t^{n+1}. This method has a grid that is always fixed and regular. The complication comes in solving the convective equation backwards and identifying the solution $c(x_{old}, t^n)$. This method is similar to the random choice method, except that there is no randomly moving Riemann solution.

The second Euler-Lagrange method solves the diffusion step explicitly [Thomaidis, et al. 1988]. Now the diffusion occurs at the n-th time level and the resulting solution is convected forward in time.

$$\frac{_i c^{n+1} - c_i^n}{\Delta t} = \frac{\frac{c_{i+1}^n - c_i^n}{\Delta x_i} - \frac{c_i^n - c_{i-1}^n}{\Delta x_{i-1}}}{\frac{1}{2}(\Delta x_i + \Delta x_{i-1})} \tag{7.189}$$

The index i is placed before the concentration symbol because the location of the i-th node changes with time and must be solved for. Since the convection need not end on a grid line, the grid at the new time, t^{n+1}, is irregular, which means that the grid is always irregular. This is why the diffusion equation is written for an irregular mesh. The positions are related by

$$x_i^{n+1} = x_i^n + u_i^n \Delta t. \tag{7.190}$$

For the convective diffusion equation, u_i^n is the Peclet number.

Neuman [1984] has used finite element methods that are similar to both of these approaches. Near a front, he uses forward tracking; in other words, he convects from time t^n to time t^{n+1} to a new time position and uses an implicit scheme for diffusion at the new time. Away from a front, however, he convects backwards and finds the concentration at a point that ends up at position x at time t^{n+1}. Neuman says that it is necessary to treat areas near the front and away from the front differently to avoid the dispersion that occurs if forward tracking is used everywhere. The forward tracking itself is done simply by introducing a cloud of particles and moving them according to the convection, as did Garner, et al. [1964].

The truncation error of the implicit scheme, Eq. (7.187)-(7.188), is found by setting

$$_i c^n = c^n(x_i - u\Delta t) = c_i^n + \frac{\partial c}{\partial x}(-u\Delta t) + \frac{\partial^2 c}{\partial x^2}\frac{(-u\Delta t)^2}{2} + \frac{\partial^3 c}{\partial x^3}\frac{(-u\Delta t)^3}{6} + \ldots \quad (7.191)$$

The term on the left of Eq. (7.187) is then

$$\frac{c_i^{n+1} - {}_i c^n}{\Delta t} = \frac{1}{\Delta t}\left\{ c_i^n + \frac{\partial c}{\partial t}\Delta t + \frac{\partial^2 c}{\partial t^2}\frac{\Delta t^2}{2} - c_i^n + \frac{\partial c}{\partial x}u\Delta t - \frac{\partial^2 c}{\partial x^2}\frac{(u\Delta t)^2}{2} + \frac{\partial^3 c}{\partial x^3}\frac{(u\Delta t)^3}{6}\right\}, \quad (7.192)$$

which becomes

$$\frac{c_i^{n+1} - {}_i c^n}{\Delta t} = \frac{\partial c}{\partial t} + u\frac{\partial c}{\partial x} + \frac{\partial^2 c}{\partial t^2}\frac{\Delta t}{2} - \frac{\partial^2 c}{\partial x^2}\frac{u^2 \Delta t}{2} + \frac{\partial^3 c}{\partial x^3}\frac{u^3 \Delta t^2}{6}. \quad (7.193)$$

If convection dominates, then

$$\frac{\partial c}{\partial t} = -u\frac{\partial c}{\partial x}, \quad \frac{\partial^2 c}{\partial t^2} = -u\frac{\partial^2 c}{\partial x \partial t}, \quad \frac{\partial^2 c}{\partial x \partial t} = -u\frac{\partial^2 c}{\partial x^2} \quad (7.194)$$

and this gives

$$\frac{\partial^2 c}{\partial t^2} = u^2 \frac{\partial^2 c}{\partial x^2}. \quad (7.195)$$

Thus the term in Eq. (7.193) that is first-order in Δt is small. The remaining terms are second-order in Δt and Δx. The truncation error of the right-hand side of Eq. (7.187) is proportional to Δt and Δx^2.

The truncation error of the explicit scheme, Eq. (7.189)-(7.190), is obtained by taking

$$_i c^{n+1} = c(x = x_i^{n+1}, t = t^{n+1}) = c(x = x_i^n + u_i^n \Delta t, t = t^{n+1}). \quad (7.196)$$

$$_i c^{n+1} = c_i^n + \frac{\partial c}{\partial t}\Delta t + u\Delta t\frac{\partial c}{\partial x} + \frac{\partial^2 c}{\partial t^2}\frac{\Delta t^2}{2} + \frac{\partial^2 c}{\partial x^2}\frac{u^2 \Delta t^2}{2} + 2\frac{\partial^2 c}{\partial t \partial x}\frac{u \Delta t^2}{2}. \quad (7.197)$$

Then

$$\frac{_i c^{n+1} - c_i^n}{\Delta t} = \frac{\partial c}{\partial t} + u\frac{\partial c}{\partial x} + \frac{\Delta t}{2}\left[\frac{\partial^2 c}{\partial t^2} + \frac{\partial^2 c}{\partial x^2}u^2 + 2\frac{\partial^2 c}{\partial t \partial x}u\right]. \quad (7.198)$$

To the same approximation as Eq. (7.194), we get

$$\frac{\partial^2 c}{\partial t^2} + \frac{\partial^2 c}{\partial x^2}u^2 + 2\frac{\partial^2 c}{\partial t \partial x}u = \frac{\partial^2 c}{\partial x^2}u^2 + \frac{\partial^2 c}{\partial x^2}u^2 - 2\frac{\partial^2 c}{\partial x^2}u^2 = 0, \quad (7.199)$$

an expression that is second-order in Δt. The truncation error of the diffusion term in Eq. (7.189) is Δt and Δx^2. Thus, for both implicit and explicit methods the

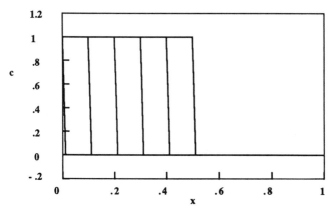

Figure 7.29. Advection Equation, Implicit or Explicit Euler-Lagrange Method
$\Delta t = 0.005$

overall truncation error is proportional to Δt and Δx^2.

Typical results for the advection equation are shown in Figure 7.29; these results are exact. The implicit and explicit methods are the same since there is no diffusion term and that is the only term handled using implicitly or explicitly. The initial mesh used points at x = 0, 0.01, 0.03, 0.07, 0.15, 0.31, 0.63, and 1. These points are convected with the velocity of the fluid. Additional points are introduced at the left; these mirror image points are introduced as if they were initially at x = -1, -0.63, -0.31, -0.15, -0.07, -0.03, and -0.01.

Results for the convective diffusion equation are shown in Figures 7.30a and 7.30b. The implicit Euler-Lagrange method gives excellent results. In this case the initial mesh was at x = 0, 0.01, 0.02, 0.03, 0.04, 0.05, 0.06, 0.07, 0.08, 0.1, 0.12, 0.14, 0.17, 0.2, 0.25, 0.3, 0.6, and 1. The mirror image points were also introduced at the left; at t = 0 they are at -1, -0.6, -0.3, -0.25, -0.17, -0.14, -0.12, -0.1, -0.08, -0.07, -0.06, -0.05, -0.04, -0.03, -0.02, and -0.01. The small oscillations in the explicit Euler-Lagrange method arise because of the interpolation necessary to find the solution at a point which is not a nodal point. Using a smaller Δt made little difference in the size of the oscillations.

Similar results occur for the other initial condition, as shown in Figure 7.31 for the advection equation (exact results) and in Figure 7.32 for the convective diffusion equation. For the advection equation the initial mesh was uniform on x = 0 to 1, with $\Delta x = 0.02$, and these points were convected with the velocity of the fluid without introducing any points at the left. For the convective diffusion equation the same points were used initially, but points were introduced at the left to keep $\Delta x \leq 0.05$ throughout the region. The solutions are excellent.

When the problem is nonlinear, both Douglas and Russell [1982] and Thomaidis, *et al.* [1988] suggest using an explicit scheme for the velocity. For Burger's equation,

$$\frac{\partial u}{\partial t} + u \frac{\partial u}{\partial x} = \nu \frac{\partial^2 u}{\partial x^2}, \qquad (7.200)$$

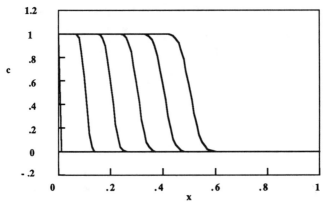

Figure 7.30a. Convective Diffusion Equation, Implicit Euler-Lagrange Method Pe = 1000, Δt = 1E-5, Special Mesh (see text)

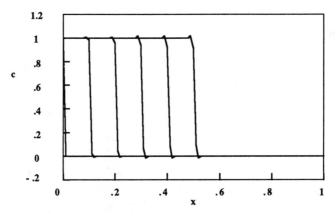

Figure 7.30b. Convective Diffusion Equation, Explicit Euler-Lagrange Method Pe = 1000, Δt = 2E-6, Special Mesh (see text)

the recommendation is

$$\frac{{}_i u_i^n - u_i^n}{\Delta t} = \nu \frac{\frac{u_{i+1}^n - u_i^n}{\Delta x_i} - \frac{u_i^n - u_{i-1}^n}{\Delta x_{i-1}}}{\frac{1}{2}(\Delta x_i + \Delta x_{i-1})} \tag{7.201}$$

along with

$$x_i^{n+1} = x_i^n + u_i^n \, \Delta t. \tag{7.202}$$

If a shock forms, the velocity in Eq. (7.202) would be the average of the solution before and after the shock, in accordance with the exact solution. A solution from Thomaidis, et al. [1988] is displayed in Figure 7.33. The front is quite sharp without oscillations. Thus the explicit Euler-Lagrange method provides a viable method for solving these convective diffusion problems.

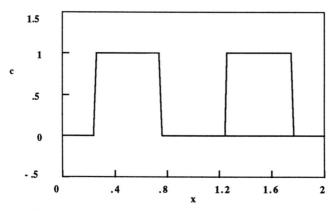

Figure 7.31. Advection Equation, Explicit Euler-Lagrange Method
$\Delta t = 1E\text{-}5$, Special Mesh (see text)

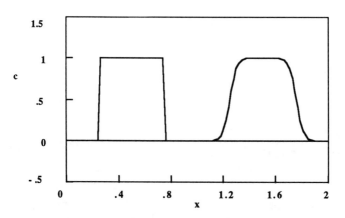

Figure 7.32. Convective Diffusion Equation, Explicit Euler-Lagrange Method
$Pe = 1000$, $\Delta t = 2E\text{-}6$, Special Mesh (see text)

Another approach is to use space-time elements, finite elements in both space and time. The elements in time are allowed to deform to follow the convective movement. Lynch [1982] shows how this method is similar to those expressed by Eq. (7.71) and Eq. (7.78). In this approach we write the trial function as

$$c = \sum_j \sum_l c_j^{\,l}\, \omega^{\,l}(t)\, N_j(x,t). \qquad (7.203)$$

If we rewrite Eq. (7.203) as

$$c = \sum_j c_j(t)\, N_j(x,t), \qquad (7.204)$$

the time derivative is

$$c_j(t) = c = \sum_l c_j^{\,l}\, \omega^{\,l}(t). \qquad (7.205)$$

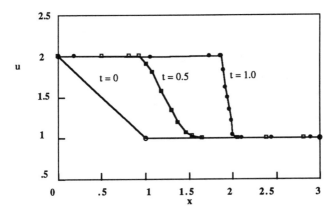

Figure 7.33. Burger's Equation, Explicit Euler-Lagrange Method, v = 0.001
Data from Thomaidis, *et al.* [1988]

When $X_j(t)$ moves, we have [as we did in (Eq. 7.69)]

$$\frac{\partial c}{\partial t} = \sum_j \left[\frac{dc_j}{dt} N_j + c_j \frac{\partial N_j}{\partial t} \right] = \sum_j \frac{dc_j}{dt} N_j - V^e \frac{\partial c}{\partial x}. \quad (7.206)$$

The weighted residual formulation is then

$$\int_{t^n}^{t^{n+1}} \int_0^1 \left[\sum_j \frac{dc_j}{dt} N_j + (Pe - V^e) \frac{\partial c}{\partial x} - \frac{\partial^2 c}{\partial x^2} \right] N_j \, dx \, \omega^l(t) \, dt = 0. \quad (7.207)$$

This is the same as Eq. (7.71) and Eq. (7.78). The difference lies in the approximation of time-dependence by using finite elements. The motion of the nodes must, of course, be specified. This idea has been applied to problems with melting and solidification and is discussed in more detail in Chapter 11.

7.6 Comparison of Methods

This chapter ends with a summary of the best methods presented in Chapters 6 and 7. We have treated linear problems in the form of the advection equation,

$$\frac{\partial c}{\partial t} + Pe \frac{\partial c}{\partial x} = 0, \quad (7.208)$$

where the concentration is convected along. We have also treated linear problems in the form of the convective diffusion equation,

$$\frac{\partial c}{\partial t} + Pe \frac{\partial c}{\partial x} = \frac{\partial^2 c}{\partial x^2}, \quad (7.209)$$

where the concentration is smeared as it is convected. The nonlinear problems we used were Burger's equation without viscosity,

$$\frac{\partial u}{\partial t} + u \frac{\partial u}{\partial x} = 0, \qquad (7.210)$$

and with viscosity,

$$\frac{\partial u}{\partial t} + u \frac{\partial u}{\partial x} = v \frac{\partial^2 u}{\partial x^2}. \qquad (7.211)$$

Eq. (7.210)-(7.211) were solved with initial conditions that led to self-sharpening fronts in one region and spreading fronts in another region.

The best solutions to the advection equation are shown in Figure 7.34 for the initial condition of a moving concentration wave. The three best methods are the random choice method, the MacCormack flux-corrected method, and the explicit Euler-Lagrange method. The explicit Euler-Lagrange method is exact, since the underlying convection problem is solved using the method of characteristics, which is easy for a linear advection problem. The random choice method is nearly exact, with the error in the location of the front (which on the average is in the correct place); the front itself is completely sharp without oscillations. The MacCormack flux-corrected method shows some added dispersion but still gives a sharp front without oscillations. The ENO method and Taylor-Galerkin method are almost as good as these three methods.

When diffusion is added to the advection equation, the best solutions are as shown in Figure 7.35. The explicit Euler-Lagrange method and the random choice method give excellent results. The MacCormack flux-corrected method, TVD and ENO methods are quite good. Of comparable accuracy are the QUICK method and the Taylor-Galerkin method, both of which have only a few wiggles. However, the QUICK method, using explicit integration in time, requires an extremely small step-size; for that reason, it would not be chosen. The Laumbach method is not suitable for the advection equation although it works well for the convective diffusion equation. The implicit ($\theta = 0.5$) Galerkin, Petrov-Galerkin, Taylor-Galerkin, and Morton-Parrott methods are as good as the explicit Taylor-Galerkin method. Another comparison of TVD with QUICK methods is by Lin and Chieng [1991].

The best solutions to Burger's equation without viscosity were found with the random choice method, the MacCormack flux-corrected method, and the TVD and ENO methods, as shown in Figure 7.36. A shock is propagated in both cases. The random choice method has a steep front at the shock and the solution has small wiggles where there is no shock. The MacCormack flux-corrected method, TVD and ENO methods are excellent. When viscosity is added, the same methods are again the best, as shown in Figure 7.37. With viscosity, however, the explicit Euler-Lagrange method is also an excellent method.

The method of moving finite elements is shown in Figure 7.38 for still another initial condition. This method provides excellent solutions, too. Shown

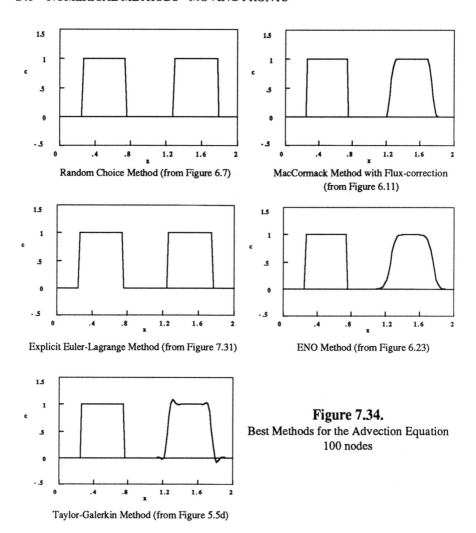

Figure 7.34.
Best Methods for the Advection Equation
100 nodes

in Figure 7.38 are the solutions derived with the random choice method, the MacCormack flux-corrected method, and the Taylor-Galerkin method. The results shown in Figure 7.38 indicate that all three of these methods are excellent.

Next let us compare the above results with those from Chapter 6. Since we are interested in general methods, applicable to a wide range of problems, we will eliminate special methods that can be applied only to certain problems. Thus we eliminate the Laumbach method and the Rosenberg method. The remaining methods are listed in Table 7.5; the ~ symbol indicates that very small oscillations are present in the results for that method. Methods with large oscillations are left off the list entirely. The MacCormack methods are not listed under linear problems because for pure advection they are the same as the Lax-Wendroff method, which oscillates some (although not a lot). The flux-corrected version is much better than the MacCormack method alone and thus is always recom-

ADAPTIVE AND MOVING MESHES

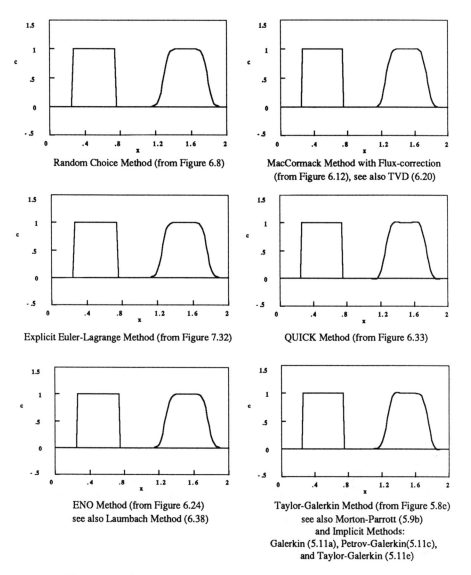

Figure 7.35. Best Methods for the Convective Diffusion Equation
Pe = 1000, 100 nodes

mended. The QUICK method is not listed under nonlinear problems because the results are not nearly as good as the other methods (see Figures 6.36-6.37). The filtered leapfrog and Morton-Parrott methods did not perform as well or nonlinear problems and the implicit Petrov-Galerkin method is complicated to apply to nonlinear problems; other methods achieve similar results and are easier to apply.

This summary indicates that there are several methods that give solutions with sharp fronts, without oscillations (or with very small oscillations), and are economical to use. Some of these are finite difference methods and some

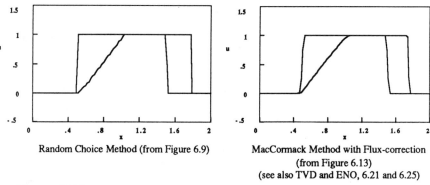

Random Choice Method (from Figure 6.9)

MacCormack Method with Flux-correction
(from Figure 6.13)
(see also TVD and ENO, 6.21 and 6.25)

Figure 7.36. Best Methods for Burger's Equation Without Viscosity, 100 nodes

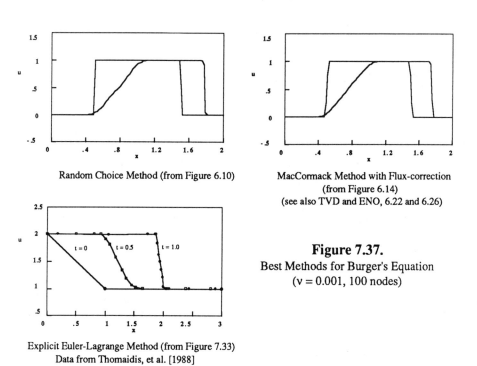

Random Choice Method (from Figure 6.10)

MacCormack Method with Flux-correction
(from Figure 6.14)
(see also TVD and ENO, 6.22 and 6.26)

Explicit Euler-Lagrange Method (from Figure 7.33)
Data from Thomaidis, et al. [1988]

Figure 7.37.
Best Methods for Burger's Equation
($\nu = 0.001$, 100 nodes)

are finite element methods. In the remainder of the book, these methods will be illustrated for specific applications to both one- and two-dimensional problems. When the problem is a two-dimensional situation with complicated geometry, some of the methods are easier to apply than others. Thus the next feature of interest is the extent to which the methods can be generalized to two-dimensional situations. Considerations that are important in two dimensional cases are discussed in Chapter 8.

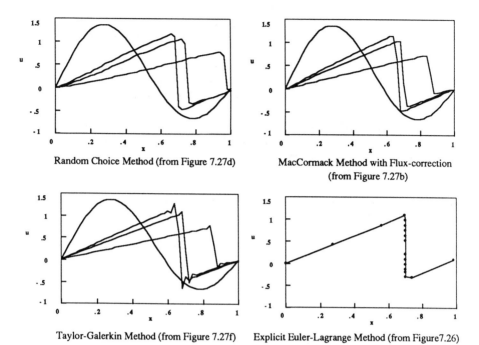

Figure 7.38. Best Methods for Burger's Equation, $\nu = 0.001$, 51 or 19 nodes

Table 7.5. Summary of Best Methods

Methods for Linear Problems	Methods for Nonlinear Problems
random choice	random choice
MacCormack flux-corrected	MacCormack flux-corrected
TVD and ENO	TVD and ENO
explicit Euler-Lagrange	explicit Euler-Lagrange
micro/macro moving elements	moving element methods
equidistribution with moving elements	~ MacCormack
~ Galerkin	~ Galerkin
~ Taylor-Galerkin	~ Taylor-Galerkin
~ QUICK	
~ filtered leapfrog	
~ Morton-Parrott	~ Indicates methods which are
~ implicit Petrov-Galerkin	not quite as qood as the other methods.

Problems

1_2. Derive the first and second derivatives, Eq. (7.14) and (7.15), for the nomenclature in Figure 7.3. Use a linear interpolation between c_L and c_R to derive the first derivative at point x_2. Evaluate the first derivative at c_L and c_R and then use the values obtained to evaluate the second derivative at the point x_2.

252 NUMERICAL METHODS - MOVING FRONTS

2_2. Derive Eq. (7.30). [Hint: Use integration by parts.]

3_3. Derive Eq. (7.34). [Hint: See your solution to Problems 2.16 and 2.17.]

4_2. Use the program CONVECT to solve the advection equation with a step change for initial conditions. Use the three-point upstream, QUICK, Laumbach, and Rosenberg methods. Compare your results with the results derived in Problems 4.5, 5.5, and 6.11.

5_2. Use the program CONVECT to solve the convective diffusion equation with Pe = 1000 and a step change for initial conditions. Use the three-point upstream, QUICK, Laumbach, and Rosenberg methods. Compare your results with the results derived in Problems 4.11, 5.13, and 6.12.

6_1. Propose an element distribution to solve the problem shown in Figure 7.34. Use a finite element method that adds and subtracts elements before and after the front.

7_3. To illustrate why Eq. (7.126)-(7.127) is not really a least squares method, consider (for notational simplicity) the following problem:

$$\frac{\partial c}{\partial t} = \frac{\partial^2 c}{\partial x^2}. \qquad (7.212)$$

$$c(x,0)=x(1-x), \quad c(0,t)=0, \quad c(1,t)=0. \qquad (7.213)$$

Try a solution of the form

$$c(x,t) = a(t)\, x\, (1-x). \qquad (7.214)$$

This function satisfies both the boundary conditions and the initial conditions with $a(0) = 1$. Put the trial function into the differential equation to form the residual

$$x(1-x)\frac{da}{dt} = -2a. \qquad (7.215)$$

Use the Galerkin method and the least squares method to solve for a.
(i) The Galerkin method solves for $a(t)$ by making the residual orthogonal to $x(1-x)$.

$$\int_0^1 x(1-x)\left[x(1-x)\frac{da}{dt} + 2a\right]dx = 0 \qquad (7.216)$$

Next write the integral

$$J\left[a, \frac{da}{dt}\right] = \int_0^1 \left[x(1-x)\frac{da}{dt} + 2a\right]^2 dx \qquad (7.217)$$

and differentiate it with respect to da/dt while keeping the value of a fixed. This step is analogous to the derivation of Eq. (7.132)-(7.133).

$$\frac{\partial J}{\partial \dot{a}} = 2\int_0^1 x(1-x)\left[x(1-x)\frac{da}{dt} + 2a\right]dx = 0 \qquad (7.218)$$

This is the same equation as derived with the Galerkin method.
(ii). Next consider the complete least squares method. You must now integrate Eq.

(7.217) over time so that you have a proper functional.

$$I[a] = \int_0^\infty \int_0^1 \left[x(1-x)\frac{da}{dt} + 2a \right]^2 dx\, dt \qquad (7.219)$$

Take the functional derivative (called a variation in the calculus of variations, δa) with respect to a(t) and integrate by parts the term involving $d\delta a/dt$. Specify a at $t = 0$ and $t = \infty$; thus, $\delta a = 0$ there. The variation is then

$$\delta I = 2 \int_0^\infty \int_0^1 \delta a \left\{ 2x(1-x)\frac{da}{dt} + 4a - x(1-x)\left[x(1-x)\frac{d^2a}{dt^2} + 2\frac{da}{dt} \right] \right\} dx\, dt = 0. \qquad (7.220)$$

If this is true for arbitrary δa functions, then the following equation must be satisfied (this is the Fundamental Lemma of the Calculus of Variations):

$$4a - \frac{1}{30}\frac{d^2a}{dt^2} = 0. \qquad (7.221)$$

Now evaluate the functions J and I using the two solutions. J depends on time. For what values of time is J smaller with the Galerkin solution? With the least squares solution? Which method has the smallest I?

References

Babuska, I. and Miller, A., "The Post-processing Approach in the Finite Element Method - Part 3: *A Posteriori* Error Estimates and Adaptive Mesh Selection," Int. J. Num. Methods Engn. 20 2311-2324 [1984].

Bieterman, M. and Babuska, I., "An Adaptive Method of Lines with Error Control for Parabolic Equations of the Reaction-diffusion Type," J. Comp. Phys. 63 33-66 [1986].

Blottner, F. G., "Numerical Solution of Diffusion-Convection Equations," Comp. Fluids 6 15-24 [1978].

Carey, G. F., "Adaptive Refinement and Nonlinear Fluid Problems," Comp. Methods Appl. Mech. Eng. 17/18 541-560 [1979].

Carey, G. F. and Finlayson, B. A., "Orthogonal Collocation on Finite Elements," Chem. Eng. Sci. 30 587-596 [1975].

Carey, G. F. and Humphrey, D., "Residuals, adaptive refinement and iteration for finite element computations," Int. J. Num. Methods Engn. 17 1717-1734 [1981].

Carey, G. F. and Plover, T., "Variable Upwinding and Adaptive Mesh Refinement in Convection-Diffusion," Int. J. Num. Methods Engn. 19 341-353 [1983].

Carey, G. F. and Jiang, B. N. "Least-Squares Finite Elements for First-Order Hyperbolic Systems," Int. J. Num. Methods Engn. 26 81-93 [1988].

Chawla, T. C., Leaf, G., and Minkowycz, W. J., "Collocation Method for Convection

Dominated Flows," Int. J. Num. Methods Fluids 4 271-281 [1984].

Christiansen, J. and Russell, R. D., "Error Analysis for Spline Collocation Methods With Application to Knot Selection," Math. Comp. 32 415-419 [1978].

Coyle, J. M., Flaherty, J. E., Ludwig, R., "On the Stability of Mesh Equidistribution Strategies for Time-Dependent Partial Differential Equations," J. Comp. Phys. 62 26-39 [1986].

Davis, S. F. and Flaherty, J. E., "An Adaptive Finite Element Method for Initial-Boundary Value Problems for Partial Differential Equations," SIAM J. Sci. Stat. Comput. 3 6-27 [1982].

Dissinger, G. R., "GRD1 - A New Implicit Integration Code for the Numerical Solution of Partial Differential Equations on Either Fixed or Adaptive Spatial Grids," Ph. D. Dissertation, Lehigh University [1983].

Douglas, J., Jr., and Russell, T. F., "Numerical Methods for Convection-Dominated Diffusion Problems Based on Combining the Method of Characteristics with Finite Element or Finite Difference Procedures," SIAM J. Numer. Anal. 19 871-885 [1982].

DuPont, T., "Mesh Modification for Evolution Equations," Math. Comp. 39 85-107 [1982].

Dwyer, H., A., Kee, R. J., and Sanders, B. R., "Adaptive Grid Method for Problems in Fluid Mechanics and Heat Transfer," AIAA J. 18 1205-1212 [1980].

Ewing, R. E., Russell, T. F. and Wheeler, M. F., "Simulation of Miscible Displacement Using Mixed Methods and a Modified Method of Characteristics," pp. 71-81 in Proceedings, Seventh SPE Symposium on Reservoir Simulation, San Francisco, CA, Nov. 16-18, 1983.

Farmer, C. L., "A Moving Point Method for Arbitrary Peclet Number Multi-Dimensional Convection-Diffusion Equations," IMA J. Num. Anal. 5 465-480 [1980].

Ferguson, N. B. and Finlayson, B. A., "Error Bounds for Approximate Solutions to Nonlinear Ordinary Differential Equations," A. I. Ch. E. J. 18 1053-1059 [1972].

Finlayson, B. A., *The Method of Weighted Residuals and Variational Principles,* Academic Press [1972].

Finlayson, B. A., *Nonlinear Analysis in Chemical Engineering,* McGraw-Hill [1980].

Galinas, R. J., Doss, S. K., Miller, K., "The Moving Finite Element Methods: Applications to General Partial Differential Equations with Multiple Large Gradients," J. Comp. Phys. 40 202-249 [1981].

Garder, A. O., Peaceman, D. W., and Pozzi, A. L., "Numerical Calculation of Multidimensional Miscible Displacement by the Method of Characteristics," Soc. Pet. Eng. J. 4 26-36 (1964).

Harrison, G. W., "Numerical Solution of the Fokker Planck Equation Using Moving Finite Elements," Num. Methods Part. Diff. Eqn. 4 219-232 [1988].

Harten, A., and Hyman, J. M., "Self-adjusting Grid Methods for One-dimensional Hyperbolic

Conservation Laws," J. Comp. Phys. 50 235-269 [1983].
Herbst, B. M., Schoombie, S. W., and Mitchell, A. R., "A Moving Petrov-Galerkin Method for Transport Equations," Int. J. Num. Methods Engn. 18 1321-1336 [1982].

Herbst, B. M., Mitchell, A. R.and Schoombie, S. W., "Equidistributing principles involved in two moving finite element methods, J. Comp. Appl. Mech. 9 377-389 [1983].

Herbst, B. M., Schoombie, S. W., Griffiths, D. F. and Mitchell, A. R., "Generalized Petrov-Galerkin Methods for the Numerical Solution of Burgers' Equation," Int. J. Num. Methods Engn. 20 1273-1289 [1984].

Hrymak, A. N., McRae, G. J. and Westerberg, A. W., "An Implementation of a Moving Finite Element Method," J. Comp. Phys. 63 168-190 [1986].

Hrymak, A. N. and Westerberg, A. W., "Computational Considerations for Moving Finite Element Methods," Chem. Eng. Sci. 41 1673-1680 [1986].

Hu, S. S. and Schiesser, W. E., "An Adaptive Grid Method in the Numerical Method of Lines," Adv. Comp. Methods Part. Diff. Eqn.-IV 305-311 [1981].

Jensen, O. K. and Finlayson, B. A., "Solution of the transport equations using a moving coordinate system," Adv. Water Res. 3 9-18 [1980].

Jiang, B. N., and Carey, G. F., "A Stable Least-squares Finite Element Method for Non-linear Hyperbolic Problems," Int. J. Num. Methods Fluids 8 933-942]1988].

Lin, H. and Chieng, C. C., "Characteristic-based Flux Limiters of an Essentially Third-order Flux-splitting Method for Hyperbolic Conservation Laws," Int. J. Num. Methods Fluids 13 287-307]1991].

Lynch, D. R., "Unified Approach to Simulation on Deforming Elements with Application to Phase Change Problems," J. Comp. Phys. 47 387-411 (1982).

Miller, K. and Miller, R. N., "Moving finite element methods, Part I," SIAM J. Numer. Anal. 18 1019-1032 [1981].

Miller, K. and Miller, R., "Moving finite element methods, Part II," SIAM J. Numer. Anal. 18 1033-1057 [1981].

Neuman, S. P., "Adaptive Eulerian-Lagrangian Finite Element Method for Advection-Dispersion," Int. J. Num. Methods Eng. 20 321-337 [1984].

Park, N. S. and Liggett, J. A., "Taylor-Least-Squares Finite Element for Two-dimensional Advection-dominated Unsteady Advection-diffusion Problems," Int. J. Num. Methods Fluids 11 21-38 [1990].

Pearson, C. E., "On a differential equation of boundary layer type," J. Math. Phys. 47 134-154 [1968].

Pereyra, V. and Sewell, E. G., "Mesh selection for discrete solution of boundary value problems in ordinary differential equations," Numer. Math 23 261-268 [1975].

Prenter, P. M., *Splines and Variational methods*, Wiley [1975].

Price, H. S., Cavendish, J. C., and Varga, R. S., "Numerical Methods of Higher-Order Accuracy for Diffusion-Convection Equations," Soc. Pet. Eng. J. 8 293-303 [1968].

Russell, R. D., "A comparison of collocation and finite differences for two-point boundary value problems," SIAM J. Numer. Anal. 14 19-39 [1977].

Russell, R. D. and Christiansen, J., "Adaptive Mesh Selection Strategies for Solving Boundary Value Problems," SIAM J. Numer. Anal. 15 59-80 [1978].

Russell, T. F., "Galerkin Time Stepping Along Characteristics for Burger's Equation," in *Scientific Computing*, R. S. Stepleman, *et al.* (ed.), IMACS, North Holland, NY [1983].

Sheintuch, M., "Numerical Approaches for Computation of Fronts," Num. Methods Part. Diff. Eqn. 6 43-58 [1990].

Smooke, M. D. and Koszykowski, M. L., "Fully Adaptive Solutions of One-Dimensional Mixed Initial-Boundary Value Problems with Applications to Unstable Problems in Combustion," SIAM J. Sci. Stat. Comp. 7 301-321 [1986].

Thomaidis, G., Zygourakis, K. and Wheeler, M. F., "An Explicit Finite Difference Scheme Based on the Modified Method of Characteristics for Solving Diffusion-Convection Problems in One Space Dimension," Num. Methods Partial Diff. Eqn. 4 119-138 (1988).

Tzanos, C. P., "A Method of Adaptive Nodes for Convective Heat Transfer Problems," Num. Heat Trans., Part B 15 153-169 [1989].

White, A. B., "On Selection of Equidistributing Meshes for Two-Point Boundary Value Problems," SIAM J. Num. Anal. 16 472-502 [1979].

White, A. B., "On the Numerical Solution of Initial/Boundary Value Problems in One Space Dimension," SIAM J. Num. Anal. 19 683-697 [1982].

Yang, G., Belleudy, P. and Temperville, A., "A Higher-order Eulerian Scheme for Coupled Advection-diffusion Transport," Int. J. Num. Methods Fluids 12 43-58 [1991].

CHAPTER
EIGHT

TWO-DIMENSIONAL CONSIDERATIONS

The equations solved in Chapters 1-7 had only one spatial dimension. Certainly a viable method has to be able to work in one dimension; it must also work in two or more dimensions. Some of the methods are easier to generalize to more dimensions than others. Here we will consider some of the important factors when changing from one to two dimensions. The calculations are reserved for other chapters on various areas of application.

8.1. Finite Difference Methods

To analyze the two-dimensional aspects of finite difference methods, let us consider the transient convective diffusion equation in two dimensions:

$$\frac{\partial c}{\partial t} + u \frac{\partial c}{\partial x} + v \frac{\partial c}{\partial y} = D \left(\frac{\partial^2 c}{\partial x^2} + \frac{\partial^2 c}{\partial y^2} \right). \tag{8.1}$$

In two dimensions, we use the subscripts i and j to represent the x and y locations; the i refers to an index in the x direction and the j refers to an index in the y direction. A superscript n still refers to the n-th time level. Thus

$$c_{i,j}^n = c(x_i, y_j, t^n)$$
for uniform grid $x_i = (i-1) \Delta x$, $y_j = (j-1) \Delta y$. \tag{8.2}

An explicit scheme with centered first derivatives gives

$$\frac{c_{i,j}^{n+1} - c_{i,j}^n}{\Delta t} + u_{ij} \frac{c_{i+1,j}^n - c_{i-1,j}^n}{2 \Delta x} + v_{ij} \frac{c_{i,j+1}^n - c_{i,j-1}^n}{2 \Delta y} = D \left(\frac{c_{i+1,j}^n - 2c_{i,j}^n + c_{i-1,j}^n}{\Delta x^2} + \frac{c_{i,j+1}^n - 2c_{i,j}^n + c_{i,j-1}^n}{\Delta y^2} \right). \tag{8.3}$$

This equation can be rearranged with all the known quantities on the right-hand side.

$$c_{ij}^{n+1} = c_{i+1,j}^n \left(\frac{D \Delta t}{\Delta x^2} - \frac{u \Delta t}{2 \Delta x} \right) + c_{i-1,j}^n \left(\frac{D \Delta t}{\Delta x^2} + \frac{u \Delta t}{2 \Delta x} \right) + c_{i,j+1}^n \left(\frac{D \Delta t}{\Delta y^2} - \frac{v \Delta t}{2 \Delta y} \right) +$$

$$+ c_{i,j-1}^n \left(\frac{D \Delta t}{\Delta y^2} + \frac{v \Delta t}{2 \Delta y} \right) + c_{ij}^n \left(1 - \frac{2 D \Delta t}{\Delta x^2} - \frac{2 D \Delta t}{\Delta y^2} \right) \tag{8.4}$$

The positivity rule states that if all of the coefficients on the right-hand side are positive and they add to 1.0, the numerical scheme is stable. This can be proved by bounding $|c_{ij}^{n+1}|$ in terms of the maximum (over all i and j values) of $|c_{ij}^n|$ (see Problem 8.1). In this case, the numerical scheme is stable if

$$\Delta t \left(\frac{2D}{\Delta x^2} + \frac{2D}{\Delta y^2} \right) \leq 1,$$
$$\frac{|u| \Delta x}{2 D} \leq 1 \text{ and } \frac{|v| \Delta y}{2 D} \leq 1. \tag{8.5}$$

If $\Delta x = \Delta y$, then this is

$$\Delta t \leq \frac{\Delta x^2}{4D}. \tag{8.6}$$

The time-step here is half as large as in the one-dimensional case. If $|u| > |v|$ we need

$$\frac{|u| \Delta x}{2 D} \leq 1, \tag{8.7}$$

which is the condition for no oscillations in a one-dimensional convection diffusion problem. These conditions are sufficient for stability; however, they may be more stringent than necessary. If both Eq. (8.6) and Eq. (8.7) hold, then we have

$$\frac{|u| \Delta t}{\Delta x} \leq \frac{1}{2}. \tag{8.8}$$

In other words, the Courant number is at most one-half.

There are other methods of employing a finite difference method. If a Crank-Nicolson method is used, we would write

$$\frac{c_{i,j}^{n+1} - c_{i,j}^n}{\Delta t} + \frac{1}{2} \left(u_{ij} \frac{c_{i+1,j}^{n+1} - c_{i-1,j}^{n+1}}{2 \Delta x} + v_{ij} \frac{c_{i,j+1}^{n+1} - c_{i,j-1}^{n+1}}{2\Delta y} + u_{ij} \frac{c_{i+1,j}^n - c_{i-1,j}^n}{2 \Delta x} + v_{ij} \frac{c_{i,j+1}^n - c_{i,j-1}^n}{2 \Delta y} \right) =$$
$$= \frac{D}{2} \left(\frac{c_{i+1,j}^{n+1} - 2 c_{i,j}^{n+1} + c_{i-1,j}^{n+1}}{\Delta x^2} + \frac{c_{i,j+1}^{n+1} - 2 c_{i,j}^{n+1} + c_{i,j-1}^{n+1}}{\Delta y^2} \right) +$$
$$+ \frac{D}{2} \left(\frac{c_{i+1,j}^n - 2 c_{i,j}^n + c_{i-1,j}^n}{\Delta x^2} + \frac{c_{i,j+1}^n - 2 c_{i,j}^n + c_{i,j-1}^n}{\Delta y^2} \right). \tag{8.9}$$

This system of equations can be written in the form

$$AA \, c = f. \tag{8.10}$$

If the numbering scheme is as shown in Figure 8.1, then the matrix has non-zero values in the pattern shown in Figure 8.2. During the solution of this system of equations, the areas between the three center diagonals and the two outlying diagonals get filled in. Although there are only a few non-zero entries in the beginning, most of the entries become non-zero as the solution proceeds. As a result, the matrix problem is large and time-consuming to solve. If the grid is an n x n grid, the matrix is $n^2 \times n^2$ in size. It generally takes n^4 multiplications to solve such a matrix problem. By contrast, in the one-dimensional case the grid is n long and it takes 5n multiplications to solve such a problem. (See Finlayson [1980] for the details of these calculations.) For a large n value (corresponding to small Δx and Δy values), the two-dimensional problem requires a significant increase in computation time over a one-dimensional problem.

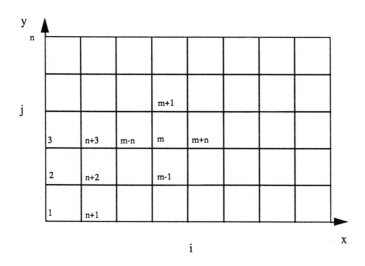

Figure 8.1. Finite Difference Numbering Scheme

In an effort to reduce computation time, the alternating direction method was introduced (see Peaceman [1977] for a recent exposition). This method is an operator splitting technique in which the two resulting subproblems are one-dimensional in character. First we solve the following one-dimensional subproblem in the x-direction for each index j.

$$\frac{c_{i,j}^{n+1/2} - c_{i,j}^{n}}{\Delta t / 2} + u_{ij} \frac{c_{i+1,j}^{n+1/2} - c_{i-1,j}^{n+1/2}}{2 \Delta x} + v_{ij} \frac{c_{i,j+1}^{n} - c_{i,j-1}^{n}}{2 \Delta y}$$

$$= D \left(\frac{c_{i+1,j}^{n+1/2} - 2 c_{i,j}^{n+1/2} + c_{i-1,j}^{n+1/2}}{\Delta x^2} + \frac{c_{i,j+1}^{n} - 2 c_{i,j}^{n} + c_{i,j-1}^{n}}{\Delta y^2} \right) \tag{8.11}$$

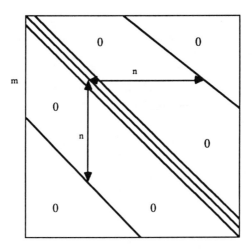

Figure 8.2. Location of Non-Zero Entries in Matrix with the Numbering Scheme of Figure 8.1

For the unknown concentrations (those with a superscript of n+1/2), the indices involved are i, i+1, and i-1, along with j. For the known concentrations (those with a superscript of n), the indices also include j+1 and j-1. Thus the problem is essentially one-dimensional (in x or i), with a different subproblem for each value of j. We thus solve n number of one-dimensional subproblems. Then we solve several one-dimensional subproblems in the y-direction, Eq. (8.12).

$$\frac{c_{i,j}^{n+1/2} - c_{i,j}^{n}}{\Delta t / 2} + u_{ij} \frac{c_{i+1,j}^{n+1/2} - c_{i-1,j}^{n+1/2}}{2 \Delta x} + v_{ij} \frac{c_{i,j+1}^{n} - c_{i,j-1}^{n}}{2 \Delta y}$$

$$= D \left(\frac{c_{i+1,j}^{n+1/2} - 2 c_{i,j}^{n+1/2} + c_{i-1,j}^{n+1/2}}{\Delta x^2} + \frac{c_{i,j+1}^{n} - 2 c_{i,j}^{n} + c_{i,j-1}^{n}}{\Delta y^2} \right) \quad (8.12)$$

For each of the subproblems the matrix problem is tri-diagonal, which makes it easy to solve.

There are other options available, such as the line successive overrelaxation method, the strongly implicit method, and the conjugate gradient method. (See Peaceman [1977] for the first two methods and Hestenes [1980], Eisenstat [1981], and McIntosh [1982] for the third method.) What should be appreciated here is that adding a second dimension creates additional difficulties, even if the Peclet number is small. Furthermore, these difficulties are magnified if the grid sizes (Δx and Δy) are small because of the large Peclet number. Presented here are applications of the Lax-Wendroff and MacCormack methods for two-dimensional problems.

If the Lax-Wendroff method is applied using fractional steps (as shown in Table 4.2 for the one-dimensional problem) in both x and y directions simultaneously, a difficulty arises. The intermediate solution is then at the i+1/2 and j+1/2 points. However, the boundary conditions are specified on grid lines, but not at the half points. Thus, the boundary conditions cannot be applied to the intermediate solution. This was not a serious difficulty for one-dimensional problems, since the boundary condition that extended outside the domain was the outflow condition (at x = 1) and the solution never reached there. One way to avoid difficulty in two-dimenisonal problems is to apply the Lax-Wendroff method in two steps to obtain the intermediate solution, according to work by Thommen [1966]. We will apply this method to the following problem:

$$\frac{\partial u}{\partial t} + \frac{\partial F(u)}{\partial x} + \frac{\partial G(u)}{\partial y} = 0. \qquad (8.13)$$

In the first intermediate step we use

$$u_{i+1/2,j}^{n+1/2} = \frac{1}{2}(u_{i+1,j}^n + u_{i,j}^n) - \frac{\Delta t}{2\Delta x}(F_{i+1,j}^n - F_{i,j}^n) - \frac{\Delta t}{4\Delta y}\left[\frac{1}{2}(G_{i,j+1}^n + G_{i+1,j+1}^n) - \frac{1}{2}(G_{i,j-1}^n + G_{i+1,j-1}^n)\right], \qquad (8.14)$$

while in the second intermediate step we use

$$u_{i,j+1/2}^{n+1/2} = \frac{1}{2}(u_{i,j+1}^n + u_{i,j}^n) - \frac{\Delta t}{2\Delta y}(G_{i,j+1}^n - G_{i,j}^n) - \frac{\Delta t}{4\Delta x}\left[\frac{1}{2}(F_{i+1,j}^n + F_{i+1,j+1}^n) - \frac{1}{2}(F_{i-1,j}^n + F_{i-1,j+1}^n)\right]. \qquad (8.15)$$

As before, we take

$$F_{i,j}^n = F(u_{i,j}^n) \text{ and } G_{i,j}^n = G(u_{i,j}^n). \qquad (8.16)$$

The solution is derived at the (i+1/2, j) points and these points will be on a boundary when j is on a boundary. The G terms in Eq. (8.14) are averaged to obtain the necessary values at the i+1/2 points. Then the final step is the usual one:

$$u_{i,j}^{n+1} = u_{i,j}^n - \frac{\Delta t}{\Delta x}(F_{i+1/2,j}^{n+1/2} - F_{i-1/2,j}^{n+1/2}) - \frac{\Delta t}{\Delta y}(G_{i,j+1/2}^{n+1/2} - G_{i,j-1/2}^{n+1/2}). \qquad (8.17)$$

For the equation with viscosity,

$$\frac{\partial u}{\partial t} + \frac{\partial F(u)}{\partial x} + \frac{\partial G(u)}{\partial y} = \nu\left(\frac{\partial^2 u}{\partial x^2} + \frac{\partial^2 u}{\partial y^2}\right), \qquad (8.18)$$

we use (again following work by Thommen)

$$u_{i+1/2,j}^{n+1/2} = \frac{1}{2}(u_{i+1,j}^n + u_{i,j}^n) - \frac{\Delta t}{2\Delta x}(F_{i+1,j}^n - F_{i,j}^n) - \frac{\Delta t}{4\Delta y}\left[\frac{1}{2}(G_{i,j+1}^n + G_{i+1,j+1}^n) - \frac{1}{2}(G_{i,j-1}^n + G_{i+1,j-1}^n)\right]$$
$$+ \frac{\nu\Delta t}{4\Delta x^2}[(u_{i+1,j}^n - 2u_{i,j}^n + u_{i-1,j}^n) + (u_{i+2,j}^n - 2u_{i+1,j}^n + u_{i,j}^n)]. \qquad (8.19)$$

$$u_{i,j+1}^{n+1/2} = \frac{1}{2}(u_{i,j+1}^n + u_{i,j}^n) - \frac{\Delta t}{2\Delta y}(G_{i,j+1}^n - G_{i,j}^n) - \frac{\Delta t}{4\Delta x}\left[\frac{1}{2}(F_{i+1,j}^n + F_{i+1,j+1}^n) - \frac{1}{2}(F_{i-1,j}^n + F_{i-1,j+1}^n)\right]$$

$$+ \frac{v\Delta t}{4\Delta y^2}[(u_{i,j+1}^n - 2u_{i,j}^n + u_{i,j-1}^n) + (u_{i,j+2}^n - 2u_{i,j+1}^n + u_{i,j}^n)] \tag{8.20}$$

$$u_{i,j}^{n+1} = u_{i,j}^n - \frac{\Delta t}{\Delta x}(F_{i+1/2,j}^{n+1/2} - F_{i-1/2,j}^{n+1/2}) - \frac{\Delta t}{\Delta y}(G_{i,j+1/2}^{n+1/2} - G_{i,j-1/2}^{n+1/2})$$

$$+ \frac{v\Delta t}{4\Delta x^2}(u_{i+1,j}^n - 2u_{i,j}^n + u_{i-1,j}^n) + \frac{v\Delta t}{4\Delta y^2}(u_{i,j+1}^n - 2u_{i,j}^n + u_{i,j-1}^n) \tag{8.21}$$

The two-dimensional MacCormack method is a straightforward generalization of the one-dimensional scheme. Peyret and Taylor [1983] indicate that there are several variations of MacCormack's method and that all of these have not been explored. They suggest the following variation:

$$u*_{i,j}^{n+1} = u_{i,j}^n - \frac{\Delta t}{\Delta x}(F_{i,j+1}^n - F_{i,j}^n) - \frac{\Delta t}{\Delta y}(G_{i,j+1}^n - G_{i,j}^n)$$
$$u_{i,j}^{n+1} = \frac{1}{2}(u_{i,j}^n + u*_{i,j}^{n+1}) - \frac{\Delta t}{2\Delta x}(F*_{i,j}^{n+1} - F*_{i-1,j}^{n+1}) - \frac{\Delta t}{2\Delta y}(G*_{i,j}^{n+1} - G*_{i,j-1}^{n+1}). \tag{8.22}$$

For the advection equation we take $F(u) = cu$ and $G(u) = cv$, where u and v are components of the velocity (supposedly known). This corresponds to writing Eq. (8.1) in conservative form as

$$\frac{\partial c}{\partial t} + \frac{\partial(cu)}{\partial x} + \frac{\partial(cv)}{\partial y} = 0 \text{ where } \frac{\partial u}{\partial x} + \frac{\partial v}{\partial y} = 0. \tag{8.23}$$

The MacCormack method is then

$$c*_{i,j}^{n+1} = c_{i,j}^n - \frac{\Delta t}{\Delta x}(u_{i+1,j} c_{i,j}^n - u_{i+1,j} c_{i,j}^n) - \frac{\Delta t}{\Delta y}(v_{i,j+1} c_{i,j+1}^n - v_{i,j} c_{i,j}^n)$$

$$c_{i,j}^{n+1} = \frac{1}{2}(c_{i,j}^n + c*_{i,j}^{n+1}) - \frac{\Delta t}{2\Delta x}(u_{i,j} c*_{i,j}^{n+1} - u_{i-1,j} c*_{i-1,j}^{n+1}) -$$
$$\frac{\Delta t}{2\Delta y}(v_{i,j} c*_{i,j}^{n+1} - v_{i,j-1} c*_{i,j-1}^{n+1}). \tag{8.24}$$

Eq. two equations of Eq. (8.24) can then be combined to give

$$c_{i,j}^{n+1} = c_{i,j}^n - \frac{\Delta t}{2\Delta x}(u_{i+1,j} c_{i+1,j}^n - u_{i-1,j} c_{i-1,j}^n) - \frac{\Delta t}{2\Delta y}(v_{i,j+1} c_{i,j+1}^n - v_{i,j-1} c_{i,j-1}^n)$$

$$+ \frac{\Delta t^2}{2\Delta x^2}[u_{i,j}(u_{i+1,j} c_{i+1,j}^n - u_{i,j} c_{i,j}^n) - u_{i-1,j}(u_{i,j} c_{i,j}^n - u_{i-1,j} c_{i-1,j}^n)]$$

$$+ \frac{\Delta t^2}{2\Delta y^2}[v_{i,j}(v_{i,j+1} c_{i,j+1}^n - v_{i,j} c_{i,j}^n) - v_{i,j-1}(v_{i,j} c_{i,j}^n - v_{i,j-1} c_{i,j-1}^n)] \tag{8.25}$$

$$+ \frac{\Delta t^2}{2\Delta x \Delta y}[u_{i,j}(v_{i,j+1} c_{i,j+1}^n - v_{i,j} c_{i,j}^n) - u_{i-1,j}(v_{i-1,j+1} c_{i-1,j+1}^n - v_{i-1,j} c_{i-1,j}^n) +$$

$$+ v_{i,j}(u_{i+1,j} c_{i+1,j}^n - u_{i,j} c_{i,j}^n) - v_{i,j-1}(u_{i+1,j-1} c_{i+1,j-1}^n - u_{i,j-1} c_{i,j-1}^n)]$$

When Taylor expansions are used, this may be rewritten as (see Problem 8.2)

$$\frac{\partial c}{\partial t} + \frac{\partial(cu)}{\partial x} + \frac{\partial(cv)}{\partial y} = \frac{\Delta t}{2}\left\{\frac{\partial}{\partial x}\left[u\left(u\frac{\partial c}{\partial x} + v\frac{\partial c}{\partial y}\right)\right] + \frac{\partial}{\partial y}\left[v\left(u\frac{\partial c}{\partial x} + v\frac{\partial c}{\partial y}\right)\right]\right\}. \tag{8.26}$$

This equation shows that there is an added dissipation term in the method, as in the one-dimensional case. Furthermore, the dissipation is anisotropic in that the coefficient depends on the velocity. This same result is obtained below using the Taylor-Galerkin method (Section 8.3). It was done using finite difference methods by Dukowicz and Ramshaw [1979].

For an equation with viscosity, Eq. (8.18), we expand Eq. (8.22) to give

$$u*_{i,j}^{n+1} = u_{i,j}^n - \frac{\Delta t}{\Delta x}(F_{i,j+1}^n - F_{i,j}^n) - \frac{\Delta t}{\Delta y}(G_{i,j+1}^n - G_{i,j}^n) +$$

$$+ \frac{\upsilon \Delta t}{\Delta x^2}(u_{i+1,j}^n - 2u_{i,j}^n + u_{i-1,j}^n) + \frac{\upsilon \Delta t}{\Delta x^2}(u_{i,j+1}^n - 2u_{i,j}^n + u_{i,j-1}^n) \tag{8.27}$$

$$u_{i,j}^{n+1} = \frac{1}{2}(u_{i,j}^n + u*_{i,j}^{n+1}) - \frac{\Delta t}{2\Delta x}(F*_{i,j}^{n+1} - F*_{i-1,j}^{n+1}) - \frac{\Delta t}{2\Delta y}(G*_{i,j}^{n+1} - G*_{i,j-1}^{n+1})$$

$$+ \frac{\upsilon \Delta t}{\Delta x^2}(u*_{i+1,j}^{n+1} - 2u*_{i,j}^{n+1} + u*_{i-1,j}^{n+1}) + \frac{\upsilon \Delta t}{\Delta x^2}(u*_{i,j+1}^{n+1} - 2u*_{i,j}^{n+1} + u*_{i,j-1}^{n+1}). \tag{8.28}$$

For the convective diffusion equation, Eq. (8.23) is replaced with

$$\frac{\partial c}{\partial t} + \frac{\partial(cu)}{\partial x} + \frac{\partial(cv)}{\partial y} = D\left(\frac{\partial^2 c}{\partial x^2} + \frac{\partial^2 c}{\partial y^2}\right). \tag{8.29}$$

We expand Eq. (8.24) to give

$$c*_{i,j}^{n+1} = c_{i,j}^n - \frac{\Delta t}{\Delta x}(u_{i+1,j} c_{i+1,j}^n - u_{i,j} c_{i,j}^n) - \frac{\Delta t}{\Delta y}(v_{i,j+1} c_{i,j+1}^n - v_{i,j} c_{i,j}^n) +$$

$$+ \frac{D\Delta t}{\Delta x^2}(c_{i+1,j}^n - 2c_{i,j}^n + c_{i-1,j}^n) + \frac{D\Delta t}{\Delta y^2}(c_{i,j+1}^n - 2c_{i,j}^n + c_{i,j-1}^n). \tag{8.30}$$

$$c_{i,j}^{n+1} = \frac{1}{2}(c_{i,j}^n + c^*{}_{i,j}^{n+1}) - \frac{\Delta t}{2\Delta x}(u_{i,j}c^*{}_{i,j}^{n+1} - u_{i-1,j}c^*{}_{i-1,j}^{n+1}) - \frac{\Delta t}{2\Delta y}(v_{i,j}c^*{}_{i,j}^{n+1} - v_{i,j-1}c^*{}_{i,j-1}^{n+1})$$
$$+ \frac{D\Delta t}{\Delta x^2}(c^*{}_{i+1,j}^{n+1} - 2c^*{}_{i,j}^{n+1} + c^*{}_{i-1,j}^{n+1}) + \frac{D\Delta t}{\Delta y^2}(c^*{}_{i,j+1}^{n+1} - 2c^*{}_{i,j}^{n+1} + c^*{}_{i,j-1}^{n+1}) \qquad (8.31)$$

Thus, the same techniques used in one-dimensional problems can be extended to two-dimensional problems. There are also ways of applying the MacCormack method as a time-splitting method by taking a half-step in y, a whole step in x, and another half-step in y. In this way, the severe stability limitations on the time-step are relaxed because each half-step can use a stability limit appropriate to the grid in that direction. Anderson, et al. [1984] demonstrate this technique and also give a split coefficient method in which the operators are divided into forward and backward operators so that upstream conditions can be used on some of them.

Finite difference methods have traditionally been applied to problems where the domain is square or composed of rectangular pieces. When the shape of the boundary is irregular, then body-fitted coordinates may be appropriate. We then need to apply a transformation,

$$\xi = \hat{\xi}(x,y), \quad \eta = \hat{\eta}(x,y), \qquad (8.32)$$

and transform the differential equations from ones that use x and y derivatives to ones that use ξ and η derivatives with variable coefficients. The functions $\hat{\xi}$ and $\hat{\eta}$ are constructed so that the ξ and η coordinate lines lie along the boundary; then the boundary conditions can be applied easily. (See Section 8.7 for further details.)

8.2. Galerkin Finite Element Method

The Petrov-Galerkin method applied to Eq. 8.1 gives

$$\int_\Omega W_j \frac{\partial c}{\partial t} d\Omega + \int_\Omega W_j \left(u \frac{\partial c}{\partial x} + v \frac{\partial c}{\partial y} \right) d\Omega = \int_\Omega W_j D \left(\frac{\partial^2 c}{\partial x^2} + \frac{\partial^2 c}{\partial y^2} \right) d\Omega. \qquad (8.33)$$

If the weighting function, W_j, is the same as the trial function, N_j, then the method is a Galerkin method. The boundary conditions can be three types: the Dirichlet, Neumann, and Robin boundary conditions or boundary conditions of the first, second, and third kinds. Here we will use all three types and identify the part of the boundary on which they apply as $\partial \Omega_i$.

$$\begin{aligned} c &= c_1 & \text{on } \partial\Omega_1 \\ -D\frac{\partial c}{\partial n} &= m_2 & \text{on } \partial\Omega_2 \\ -D\frac{\partial c}{\partial n} &= k_m(c - c_3) & \text{on } \partial\Omega_3 \end{aligned} \qquad (8.34)$$

We integrate by parts the last term in Eq. (8.33), giving

$$\int_\Omega W_j D\left(\frac{\partial^2 c}{\partial x^2} + \frac{\partial^2 c}{\partial y^2}\right) d\Omega = -\int_\Omega D\left(\frac{\partial W_j}{\partial x}\frac{\partial c}{\partial x} + \frac{\partial W_j}{\partial y}\frac{\partial c}{\partial y}\right) d\Omega +$$
$$+ \int_\Omega D\left[\frac{\partial}{\partial x}\left(W_j \frac{\partial c}{\partial x}\right) + \frac{\partial}{\partial y}\left(W_j \frac{\partial c}{\partial y}\right)\right] d\Omega. \qquad (8.35)$$

The last term is evaluated using the divergence theorem.

$$\int_\Omega D\left[\frac{\partial}{\partial x}\left(W_j \frac{\partial c}{\partial x}\right) + \frac{\partial}{\partial y}\left(W_j \frac{\partial c}{\partial y}\right)\right] d\Omega = \int_{\partial\Omega} D\left(n_x W_j \frac{\partial c}{\partial x} + n_y W_j \frac{\partial c}{\partial y}\right) ds = \int_{\partial\Omega} D W_j \frac{\partial c}{\partial n} ds \qquad (8.36)$$

The variable s refers to a boundary arc length. Next we apply the boundary conditions according to Eq. (8.34):

$$\int_{\partial\Omega} D W_j \frac{\partial c}{\partial n} ds = \int_{\partial\Omega_1} D W_j \frac{\partial c}{\partial n} ds + \int_{\partial\Omega_2} D W_j \frac{\partial c}{\partial n} ds + \int_{\partial\Omega_3} D W_j \frac{\partial c}{\partial n} ds$$
$$= -\int_{\partial\Omega_2} W_j m_2 \, ds - \int_{\partial\Omega_3} W_j k_m (c - c_3) \, ds. \qquad (8.37)$$

We ignore the evaluation of the integrals on $\partial\Omega_1$ because we choose the weighting function to equal zero there. (Actually we employ the boundary condition there and never write the weighting function corresponding to the trial function on the boundary.) Combining Eq. (8.37) with Eq. (8.33)-(8.36) gives us the complete Petrov-Galerkin statement,

$$\int_\Omega W_j \frac{\partial c}{\partial t} d\Omega + \int_\Omega W_j \left(u \frac{\partial c}{\partial x} + v \frac{\partial c}{\partial y}\right) d\Omega = -\int_\Omega D\left(\frac{\partial W_j}{\partial x}\frac{\partial c}{\partial x} + \frac{\partial W_j}{\partial y}\frac{\partial c}{\partial y}\right) d\Omega$$
$$-\int_{\partial\Omega_2} W_j m_2 \, ds - \int_{\partial\Omega_3} W_j k_m (c - c_3) \, ds. \qquad (8.38)$$

Next we use a trial function,

$$c = \sum_{i=1}^{NT} c_i N_i(x,y), \qquad (8.39)$$

which turns Eq. (8.38) into

$$\sum_{i=1}^{NT}\left\{\int_\Omega W_j N_i \, d\Omega\right\} \frac{dc_i}{dt} + \sum_{i=1}^{NT}\left\{\int_\Omega W_j \left(u \frac{\partial N_i}{\partial x} + v \frac{\partial N_i}{\partial y}\right) d\Omega\right\} c_i =$$
$$-\sum_{i=1}^{NT}\left\{\int_\Omega D\left(\frac{\partial W_j}{\partial x}\frac{\partial N_i}{\partial x} + \frac{\partial W_j}{\partial y}\frac{\partial N_i}{\partial y}\right) d\Omega\right\} c_i$$
$$-\int_{\partial\Omega_2} W_j m_2 \, ds - \sum_{i=1}^{NT}\left\{\int_{\partial\Omega_3} W_j k_m N_i \, ds\right\} c_i + \int_{\partial\Omega_3} W_j k_m c_3 \, ds. \qquad (8.40)$$

We can define the element matrices as

$$M_{JI}^e = \int_{\Omega^e} W_J N_I \, d\Omega, \quad A_{JI}^e = \int_{\Omega^e} W_J \left(u \frac{\partial N_I}{\partial x} + v \frac{\partial N_I}{\partial y} \right) d\Omega \qquad (8.41)$$

$$B_{JI}^e = \int_{\Omega^e} D \left(\frac{\partial W_J}{\partial x} \frac{\partial N_I}{\partial x} + \frac{\partial W_J}{\partial y} \frac{\partial N_I}{\partial y} \right) d\Omega, \quad \partial B_{JI}^e = \int_{\partial \Omega_3^e} W_J k_m N_I \, ds \qquad (8.42)$$

$$f_J^e = -\int_{\partial \Omega_2^e} W_J m_2 \, ds + \int_{\partial \Omega_3^e} W_J k_m c_3 \, ds \qquad (8.43)$$

and write Eq. (8.40) as

$$\sum_e M_{JI}^e \frac{dc_I^e}{dt} + \sum_e A_{JI}^e c_I^e = -\sum_e B_{JI}^e c_I^e - \sum_e \partial B_{JI}^e c_I^e + \sum_e f_J^e. \qquad (8.44)$$

In addition to Eq. (8.44), we need the boundary conditions

$$c_i = c_1(x_i, y_i) \text{ and } x_i, y_i \text{ on } \partial \Omega_1. \qquad (8.45)$$

Next let us consider how to evaluate the element matrices. Presented here are the details for only those elements that are triangular in shape and have linear basis functions defined on them. A typical element is shown in Figure 8.3.

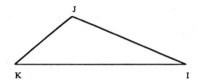

Figure 8.3. General Triangular Element; Plan View

The trial function on the element is taken as

$$c = N_I(x,y) c_I + N_J(x,y) c_J + N_K(x,y) c_K. \qquad (8.46)$$

The linear basis functions are

$$N_I = \frac{a_I + b_I x + d_I y}{2\Delta}, \qquad (8.47)$$

where

$$a_I = x_J y_K - x_K y_J$$
$$b_I = y_J - y_K \qquad \text{plus permutations on I, K, and J.} \qquad (8.48)$$
$$d_I = x_K - x_J$$

$$2\Delta = \det \begin{bmatrix} 1 & x_I & y_I \\ 1 & x_J & y_J \\ 1 & x_K & y_K \end{bmatrix} = 2 \text{ (area of triangle)} \qquad (8.49)$$

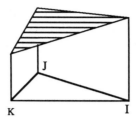

Figure 8.4. General Triangular Element; Perspective View

A typical trial function is shown in Figure 8.4. These parameters obey the following restrictions:

$$a_I + a_J + a_K = 1$$
$$b_I + b_J + b_K = 0 \quad (8.50)$$
$$d_I + d_J + d_K = 0.$$

One of the element terms in Eq. (8.42) is then (with D constant)

$$B^e_{JI} = D \frac{b_J b_I + d_J d_I}{4\Delta}. \quad (8.51)$$

The other element terms are more complicated and are best derived using triangular coordinates. The centroids of the triangle are given by

$$\bar{x} = \frac{1}{3}(x_I + x_J + x_K)$$
$$\bar{y} = \frac{1}{3}(y_I + y_J + y_K) \quad (8.52)$$
$$a_I + b_I \bar{x} + d_I \bar{y} = \frac{2}{3}\Delta.$$

If the domain is covered by such triangles, we can evaluate Eq. (8.51) for each one and place the element matrix in the proper position of a larger matrix. The complete problem can be written as

$$\sum_i MM_{ji} \frac{dc_i}{dt} = \sum_i AA_{ji} c_i + f_j. \quad (8.53)$$

Most of the entries of the matrices MM_{ji} and AA_{ji} are zero. This is the set of equations that must be integrated to provide the solution. The integration of Eq. (8.53) seems to require implicit methods of integration. However, if the step-size and mesh do not change from one time-step to another, we can solve the problem once using an LU decomposition. Then for successive time-steps only a fore-aft sweep is made, which is much faster [see Eq. (2.87)-(2.89) for an example for a one-dimensional problem]. For large finite element problems, the author has found that the fore-aft sweeps take only 2-5% of the computer time taken by the original LU decomposition. Thus, Eq. (8.53) can be solved even with explicit schemes after the first time-step. If the mass matrix is lumped, we get

$$\left(\sum_i MM_{ji}\right) \frac{dc_j}{dt} = \sum_i AA_{ji} c_i + f_j \tag{8.54}$$

and the normal explicit methods can be used. Additional details of finite element methods as applied to two-dimensional problems can be found in books by Becker, et al. [1981] and Akin [1986].

Thus, solution techniques exist for finite difference methods and the Galerkin finite element method applied to the two-dimensional convective diffusion equation. The methods will not be any better than they were in one dimension, however, so we want to employ special techniques for two-dimensional problems. The next few sections discuss how these techniques are applied to two-dimensional problems.

8.3. Taylor-Galerkin Methods

Some of the specialized techniques require more complicated treatment for two-dimensional problems. Here we will discuss complications for the Taylor-Galerkin method. Let us consider a flow that is two-dimensional, incompressible, and steady. It then satisfies the following condition:

$$\nabla \cdot \mathbf{u} \equiv \frac{\partial u}{\partial x} + \frac{\partial v}{\partial y} = 0. \tag{8.55}$$

(The development is made in terms of planar geometry but a similar development holds for cylindrical geometry as well.) To derive the added terms for the Taylor-Galerkin addition, we consider the advection equation in two dimensions:

$$\frac{\partial c}{\partial t} + \mathbf{u} \cdot \nabla c \equiv \frac{\partial c}{\partial t} + u \frac{\partial c}{\partial x} + v \frac{\partial c}{\partial y} = 0. \tag{8.56}$$

Now the convective term can be rewritten using differentiation by parts:

$$\mathbf{u} \cdot \nabla c = \nabla \cdot (\mathbf{u} c) - c \nabla \cdot \mathbf{u}. \tag{8.57}$$

Since the fluid is incompressible, Eq. (8.55) causes the last term to disappear. Then the advection equation is

$$\frac{\partial c}{\partial t} + \nabla \cdot (\mathbf{u} c) \equiv \frac{\partial c}{\partial t} + \frac{\partial}{\partial x}(u c) + \frac{\partial}{\partial y}(v c) = 0. \tag{8.58}$$

We differentiate this equation once with respect to time, giving

$$\frac{\partial^2 c}{\partial t^2} + \nabla \cdot \left(\mathbf{u} \frac{\partial c}{\partial t}\right) \equiv \frac{\partial^2 c}{\partial t^2} + \frac{\partial}{\partial x}\left(u \frac{\partial c}{\partial t}\right) + \frac{\partial}{\partial y}\left(v \frac{\partial c}{\partial t}\right) = 0. \tag{8.59}$$

Then we substitute Eq. (8.56) into Eq. (8.59):

$$\frac{\partial^2 c}{\partial t^2} = -\frac{\partial}{\partial x}\left(u\frac{\partial c}{\partial t}\right) - \frac{\partial}{\partial y}\left(v\frac{\partial c}{\partial t}\right) = +\frac{\partial}{\partial x}\left[u\left(u\frac{\partial c}{\partial x} + v\frac{\partial c}{\partial y}\right)\right] + \frac{\partial}{\partial y}\left[v\left(u\frac{\partial c}{\partial x} + v\frac{\partial c}{\partial y}\right)\right].$$

(8.60)

Next we use the Taylor expansion

$$\frac{\partial c}{\partial t} = \frac{c^{n+1} - c^n}{\Delta t} - \frac{\Delta t}{2}\frac{\partial^2 c}{\partial t^2} \qquad (8.61)$$

and substitute for $\partial c/\partial t$ in Eq. (8.56).

$$\frac{c^{n+1} - c^n}{\Delta t} + \mathbf{u} \cdot \nabla\, c = \frac{\Delta t}{2}\left\{\frac{\partial}{\partial x}\left[u\left(u\frac{\partial c}{\partial x} + v\frac{\partial c}{\partial y}\right)\right] + \frac{\partial}{\partial y}\left[v\left(u\frac{\partial c}{\partial x} + v\frac{\partial c}{\partial y}\right)\right]\right\} \quad (8.62)$$

Eq. (8.62) is the Taylor form and it is solved in the Taylor-Galerkin method. We notice that the added diffusion term is anisotropic in the direction of the velocity. It can be written as

$$\frac{c^{n+1} - c^n}{\Delta t} + \mathbf{u} \cdot \nabla\, c = \frac{\Delta t}{2}\left\{\frac{\partial}{\partial x}\left[D_{xx}\frac{\partial c}{\partial x} + D_{xy}\frac{\partial c}{\partial y}\right] + \frac{\partial}{\partial y}\left[D_{yx}\frac{\partial c}{\partial x} + D_{yy}\frac{\partial c}{\partial y}\right]\right\}, \quad (8.63)$$

where

$$\mathbf{D} = u\,u\,\mathbf{e}_x\mathbf{e}_x + u\,v\,\mathbf{e}_x\mathbf{e}_y + v\,u\,\mathbf{e}_y\mathbf{e}_x + v\,v\,\mathbf{e}_y\mathbf{e}_y. \qquad (8.64)$$

The Galerkin formulation of this equation is

$$\sum_{i=1}^{NT}\int_\Omega W_j N_i\, d\Omega\, \frac{c_i^{n+1} - c_i^n}{\Delta t} + \sum_{i=1}^{NT}\int_\Omega W_j\left(u\frac{\partial N_i}{\partial x} + v\frac{\partial N_i}{\partial y}\right)d\Omega\, c_i =$$

$$-\frac{\Delta t}{2}\sum_{i=1}^{NT}\int_\Omega\left\{\frac{\partial W_j}{\partial x}\left[D_{xx}\frac{\partial N_i}{\partial x} + D_{xy}\frac{\partial N_i}{\partial y}\right] + \frac{\partial W_j}{\partial y}\left[D_{yx}\frac{\partial N_i}{\partial x} + D_{yy}\frac{\partial N_i}{\partial y}\right]\right\}d\Omega\, c_i. \quad (8.65)$$

The last term was integrated by parts and the boundary term was ignored since it is zero for the case without diffusion. On the boundary either the concentration is specified (in which case the weighting function W_j is not used) or the applied flux is zero (for a solid boundary or outflow boundary).

It is instructive to compare the added diffusion term with that used in the streamwise upstream weighting method developed by Hughes and Brooks [1979] [see Eq. (6.141)]. In that case the equation is

$$\frac{\partial c}{\partial t} + \mathbf{u} \cdot \nabla\, c = \frac{\partial}{\partial x}\left[D_{xx}\frac{\partial c}{\partial x} + D_{xy}\frac{\partial c}{\partial y}\right] + \frac{\partial}{\partial y}\left[D_{yx}\frac{\partial c}{\partial x} + D_{yy}\frac{\partial c}{\partial y}\right], \quad (8.66)$$

where

$$D_{ij} = \tilde{D}\,\hat{u}_i\,\hat{u}_j\,, \quad \hat{u}_i = \frac{u_i}{\|\mathbf{u}\|}. \tag{8.67}$$

To see why this is called streamwise upstream weighting, we note that $\hat{u}$ is a unit vector along the direction of the velocity (i.e., along the streamline). For simplicity, let us consider a point and identify two directions, $\mathbf{t} = \hat{\mathbf{u}}$, a unit vector in the direction of flow, and $\mathbf{n}$, a unit vector normal to the direction of flow. The diffusion term is

$$\mathbf{D} \cdot \nabla c. \tag{8.68}$$

In the direction of flow the diffusion is

$$\mathbf{t} \cdot \mathbf{D} \cdot \nabla c = \tilde{D}\,\frac{1}{\|\mathbf{u}\|}\,\mathbf{u} \cdot \nabla c, \tag{8.69}$$

which can be non-zero. In the direction perpendicular to flow, the diffusion is equal to zero:

$$\mathbf{n} \cdot \mathbf{D} \cdot \nabla c = 0. \tag{8.70}$$

Thus the added numerical dispersion is only in the direction of flow.

The xx-coefficient and xy-coefficient in the Taylor-Galerkin method are

$$\frac{\Delta t}{2}\,u\,u \quad \text{and} \quad \frac{\Delta t}{2}\,u\,v, \quad \text{ratio} = \frac{u}{v}. \tag{8.71}$$

These same coefficients in the streamwise upstream Petrov-Galerkin method are

$$\tilde{D}\,\frac{1}{\|\mathbf{u}\|^2}\,u\,u \quad \text{and} \quad \tilde{D}\,\frac{1}{\|\mathbf{u}\|^2}\,u\,v, \quad \text{ratio} = \frac{u}{v}. \tag{8.72}$$

The <u>ratio</u> of the xx- to xy-coefficients is the same for the Taylor-Galerkin and Petrov-Galerkin methods. This ratio is also true for the other coefficients. This means that the <u>direction</u> of added diffusion is the same in both cases; however, the magnitude of the coefficients is different. In the streamwise upstream Petrov-Galerkin method, the xx-coefficient is

$$\tilde{D}\,\frac{1}{\|\mathbf{u}\|^2}\,u\,u \quad \text{where} \quad \tilde{D} = \frac{\tilde{\xi}\,u\,\Delta x + \tilde{\eta}\,v\,\Delta y}{2}. \tag{8.73}$$

The values of $\tilde{\xi}$ and $\tilde{\eta}$ are given by

$$\tilde{\xi} = \coth(\alpha_\xi) - \frac{1}{\alpha_\xi}, \quad \alpha_\xi = \frac{u\,\Delta x}{2} \quad \text{and} \quad \tilde{\eta} = \coth(\alpha_\eta) - \frac{1}{\alpha_\eta}, \quad \alpha_\eta = \frac{v\,\Delta y}{2}. \tag{8.74}$$

When α_ξ and α_η are large (greater than 3.0), however, $\tilde{\xi}$ and $\tilde{\eta}$ are close to 1.0. Let us suppose that u is large, v is small, and $\alpha_\xi = 1$. Then the xx-coefficient in the streamwise upstream Petrov-Galerkin method is approximately

$$\frac{u \Delta x}{2}. \tag{8.75}$$

We also take the Courant number as 1.0. Then the xx-coefficient in the Taylor-Galerkin method is

$$\frac{\Delta t}{2} u^2 = \frac{u \Delta x}{2} \quad \text{when} \quad \frac{u \Delta t}{\Delta x} = 1. \tag{8.76}$$

In that case, the Taylor-Galerkin and Petrov-Galerkin methods are the same. Of course, u is not always much larger than v and the Courant number is not always 1.0; if these conditions are not met, the methods differ. However, both methods have an anisotropy that is identical (except in magnitude) in all cases. The streamwise upstream method was introduced as a way to ensure that added diffusion, necessary in the streamwise direction, did not also introduce unnecessary added diffusion in the cross-stream direction. This same property is shared by the Taylor-Galerkin method.

Next let us consider an application to the two-dimensional Burger's equation.

$$\frac{\partial u}{\partial t} = -\frac{\partial F(u)}{\partial x} - \frac{\partial G(u)}{\partial y} \tag{8.77}$$

To apply the Taylor-Galerkin method, we differentiate this equation with respect to time.

$$\begin{aligned}\frac{\partial^2 u}{\partial t^2} &= -\frac{\partial^2 F}{\partial t \partial x} - \frac{\partial^2 G}{\partial t \partial y} \\ &= -\frac{\partial}{\partial x}\left(\frac{dF}{du}\frac{\partial u}{\partial t}\right) - \frac{\partial}{\partial y}\left(\frac{dG}{du}\frac{\partial u}{\partial t}\right)\end{aligned} \tag{8.78}$$

We insert Eq. (8.77), giving

$$\frac{\partial^2 u}{\partial t^2} = -\frac{\partial}{\partial x}\left[\frac{dF}{du}\left(-\frac{\partial F}{\partial x} - \frac{\partial G}{\partial y}\right)\right] - \frac{\partial}{\partial y}\left[\frac{dG}{du}\left(-\frac{\partial F}{\partial x} - \frac{\partial G}{\partial y}\right)\right]. \tag{8.79}$$

Using the Taylor expansion of the time difference expression, we have

$$\frac{\partial u}{\partial t} = \frac{u^{n+1} - u^n}{\Delta t} - \frac{\Delta t}{2}\frac{\partial^2 u}{\partial t^2}. \tag{8.80}$$

Combining all these terms gives

$$\frac{u^{n+1} - u^n}{\Delta t} = -\frac{\partial F}{\partial x} - \frac{\partial G}{\partial y} + \frac{\Delta t}{2}\left\{\frac{\partial}{\partial x}\left[\frac{dF}{du}\left(\frac{\partial F}{\partial x} + \frac{\partial G}{\partial y}\right)\right] + \frac{\partial}{\partial y}\left[\frac{dG}{du}\left(\frac{\partial F}{\partial x} + \frac{\partial G}{\partial y}\right)\right]\right\}. \tag{8.81}$$

This can be written as

$$\frac{u^{n+1} - u^n}{\Delta t} = -\frac{\partial F}{\partial x} - \frac{\partial G}{\partial y} + \frac{\Delta t}{2}\left\{\frac{\partial}{\partial x}\left[\left(\frac{dF}{du}\right)^2\frac{\partial u}{\partial x} + \frac{dF}{du}\frac{dG}{du}\frac{\partial u}{\partial y}\right]\right.$$
$$\left. + \frac{\partial}{\partial y}\left[\frac{dG}{du}\frac{dF}{du}\frac{\partial u}{\partial x} + \left(\frac{dG}{du}\right)^2\frac{\partial u}{\partial y}\right]\right\} \quad (8.82)$$

or as

$$\frac{u^{n+1} - u^n}{\Delta t} = -\frac{\partial F}{\partial x} - \frac{\partial G}{\partial y} + \frac{\Delta t}{2}\left\{\frac{\partial}{\partial x}\left[D_{xx}\frac{\partial u}{\partial x} + D_{xy}\frac{\partial u}{\partial y}\right] + \frac{\partial}{\partial y}\left[D_{yx}\frac{\partial u}{\partial x} + D_{yy}\frac{\partial u}{\partial y}\right]\right\}, \quad (8.83)$$

where

$$D_{xx} = \left(\frac{dF}{du}\right)^2, \quad D_{xy} = \frac{dF}{du}\frac{dG}{du} = D_{yx}, \quad D_{yy} = \left(\frac{dG}{du}\right)^2. \quad (8.84)$$

Thus, we have additional non-isotropic dispersion added in the two-dimensional case. This is the Taylor version of the two-dimensional Burger's equation.

Next we apply the Galerkin method. We expand the trial function

$$u = \sum_{i=1}^{NT} u_i N_i(x,y) \quad (8.85)$$

and use a weighting function, W_j. The time derivative then becomes

$$\frac{\partial u}{\partial t} \rightarrow \sum_{i=1}^{NT}\int_\Omega W_j N_i \, d\Omega \, \frac{d u_i}{dt}. \quad (8.86)$$

The convective term can be handled by writing it as a divergence:

$$\text{convective term} \rightarrow -\frac{\partial F}{\partial x} - \frac{\partial G}{\partial y} = -\nabla \cdot \mathbf{v}, \quad v_x = F, \quad v_y = G. \quad (8.87)$$

The Galerkin terms from the convective term are treated using integration by parts.

$$-\int_\Omega W_j \nabla \cdot \mathbf{v} \, d\Omega = -\int_\Omega \nabla \cdot (W_j \mathbf{v}) \, d\Omega + \int_\Omega \mathbf{v} \cdot \nabla W_j \, d\Omega \quad (8.88)$$

Use of the divergence theorem gives

$$-\int_\Omega W_j \nabla \cdot \mathbf{v} \, d\Omega = -\int_{\partial\Omega} W_j \mathbf{n} \cdot \mathbf{v} \, ds + \int_\Omega \mathbf{v} \cdot \nabla W_j \, d\Omega. \quad (8.89)$$

This is

$$-\int_\Omega W_j \left(\frac{\partial F}{\partial x} - \frac{\partial G}{\partial y}\right) d\Omega = -\int_{\partial\Omega} W_j \mathbf{n} \cdot (F \mathbf{e}_x + G \mathbf{e}_y) \, ds + \int_\Omega \left(F\frac{\partial W_j}{\partial x} + G\frac{\partial W_j}{\partial y}\right) d\Omega. \quad (8.90)$$

On an inflow boundary, the value of u is specified and the weighting function W_j

is not used. On other boundaries, the boundary term may be needed.

The best method of treating the nonlinear qualities of one-dimensional problems is to expand the function F in terms of the trial function. Thus, here we use

$$F = \sum_{i=1}^{NT} F_i N_i(x,y), \quad G = \sum_{i=1}^{NT} G_i N_i(x,y). \tag{8.91}$$

The convective term becomes

$$-\int_\Omega W_j \left(\frac{\partial F}{\partial x} - \frac{\partial G}{\partial y}\right) d\Omega = -\sum_{i=1}^{NT} \int_{\partial\Omega} W_j \, \mathbf{n} \cdot (N_i F_i \mathbf{e}_x + N_i G_i \mathbf{e}_y) \, ds + \\ + \sum_{i=1}^{NT} \int_\Omega \left(\frac{\partial W_j}{\partial x} N_i\right) d\Omega \, F_i + \sum_{i=1}^{NT} \int_\Omega \left(\frac{\partial W_j}{\partial y} N_i\right) d\Omega \, G_i. \tag{8.92}$$

In order to derive the equation for the added dispersion term in Eq. (8.83), we write

$$\begin{aligned} v_x &= D_{xx} \frac{\partial u}{\partial x} + D_{xy} \frac{\partial u}{\partial y} \\ v_y &= D_{yx} \frac{\partial u}{\partial x} + D_{yy} \frac{\partial u}{\partial y}, \end{aligned} \tag{8.93}$$

where the D's are given by Eq. (8.84). The dispersion terms can then be integrated by parts, giving

$$\text{dispersion term} \rightarrow \frac{\Delta t}{2} \sum_{i=1}^{NT} \int_{\partial\Omega} W_j \, \mathbf{n} \cdot \left[\left(D_{xx} \frac{\partial N_i}{\partial x} + D_{xy} \frac{\partial N_i}{\partial y} \right) \mathbf{e}_x + \left(D_{yx} \frac{\partial N_i}{\partial x} + D_{yy} \frac{\partial N_i}{\partial y} \right) \mathbf{e}_y \right] ds \, u_i + \\ - \frac{\Delta t}{2} \sum_{i=1}^{NT} \int_\Omega \frac{\partial W_j}{\partial x} \left(D_{xx} \frac{\partial N_i}{\partial x} + D_{xy} \frac{\partial N_i}{\partial y} \right) d\Omega \, u_i - \frac{\Delta t}{2} \sum_{i=1}^{NT} \int_\Omega \frac{\partial W_j}{\partial y} \left(D_{yx} \frac{\partial N_i}{\partial x} + D_{yy} \frac{\partial N_i}{\partial y} \right) d\Omega \, u_i.$$

The complete Taylor-Galerkin method is then

$$\sum_{i=1}^{NT} \int_\Omega W_j N_i \, d\Omega \frac{d u_i}{dt} = -\sum_{i=1}^{NT} \int_{\partial\Omega} W_j \, \mathbf{n} \cdot (N_i F_i \mathbf{e}_x + N_i G_i \mathbf{e}_y) \, ds + \\ + \sum_{i=1}^{NT} \int_\Omega \left(\frac{\partial W_j}{\partial x} N_i\right) d\Omega \, F_i + \sum_{i=1}^{NT} \int_\Omega \left(\frac{\partial W_j}{\partial y} N_i\right) d\Omega \, G_i + \\ + \frac{\Delta t}{2} \sum_{i=1}^{NT} \int_{\partial\Omega} W_j \, \mathbf{n} \cdot \left[\left(D_{xx} \frac{\partial N_i}{\partial x} + D_{xy} \frac{\partial N_i}{\partial y} \right) \mathbf{e}_x + \left(D_{yx} \frac{\partial N_i}{\partial x} + D_{yy} \frac{\partial N_i}{\partial y} \right) \mathbf{e}_y \right] ds \, u_i + \\ - \frac{\Delta t}{2} \sum_{i=1}^{NT} \int_\Omega \frac{\partial W_j}{\partial x} \left(D_{xx} \frac{\partial N_i}{\partial x} + D_{xy} \frac{\partial N_i}{\partial y} \right) d\Omega \, u_i - \frac{\Delta t}{2} \sum_{i=1}^{NT} \int_\Omega \frac{\partial W_j}{\partial y} \left(D_{yx} \frac{\partial N_i}{\partial x} + D_{yy} \frac{\partial N_i}{\partial y} \right) d\Omega \, u_i. \tag{8.95}$$

274 NUMERICAL METHODS - MOVING FRONTS

If viscosity is added to Eq. (8.77), the equation is

$$\frac{\partial u}{\partial t} = -\frac{\partial F(u)}{\partial x} - \frac{\partial G(u)}{\partial y} + v\left(\frac{\partial^2 u}{\partial x^2} + \frac{\partial^2 u}{\partial y^2}\right). \tag{8.96}$$

Possible boundary conditions now include

$$u = u_1 \text{ on } \partial\Omega_1$$
$$-v\frac{\partial u}{\partial n} = m_2 \text{ on } \partial\Omega_2. \tag{8.97}$$

The Galerkin method adds a term:

$$\text{diffusion term} \to v \sum_{i=1}^{NT} \int_\Omega W_j \left(\frac{\partial^2 N_i}{\partial x^2} + \frac{\partial^2 N_i}{\partial y^2}\right) d\Omega \, u_i. \tag{8.98}$$

The viscous term is evaluated according to Eq. (8.35)-(8.37). The result is

$$v \sum_{i=1}^{NT} \int_\Omega W_j \left(\frac{\partial^2 N_i}{\partial x^2} + \frac{\partial^2 N_i}{\partial y^2}\right) d\Omega \, u_i = -v \sum_{i=1}^{NT} \int_\Omega \left(\frac{\partial W_j}{\partial x}\frac{\partial N_i}{\partial x} + \frac{\partial W_j}{\partial x}\frac{\partial N_i}{\partial y}\right) d\Omega \, u_i$$
$$- \int_{\partial\Omega_2} W_j \, m_2 \, ds. \tag{8.99}$$

The Taylor-Galerkin method applied to Burger's equation in two dimensions is thus

$$\sum_{i=1}^{NT} \int_\Omega W_j N_i \, d\Omega \, \frac{d u_i}{dt} = -\sum_{i=1}^{NT} \int_{\partial\Omega} W_j \, \mathbf{n} \cdot (N_i F_i \mathbf{e}_x + N_i G_i \mathbf{e}_y) \, ds +$$
$$+ \sum_{i=1}^{NT} \int_\Omega \left(\frac{\partial W_j}{\partial x} N_i\right) d\Omega \, F_i + \sum_{i=1}^{NT} \int_\Omega \left(\frac{\partial W_j}{\partial y} N_i\right) d\Omega \, G_i$$
$$- v \sum_{i=1}^{NT} \int_\Omega \left(\frac{\partial W_j}{\partial x}\frac{\partial N_i}{\partial x} + \frac{\partial W_j}{\partial x}\frac{\partial N_i}{\partial y}\right) d\Omega \, u_i - \int_{\partial\Omega_2} W_j \, m_2 \, ds \tag{8.100}$$
$$+ \frac{\Delta t}{2} \sum_{i=1}^{NT} \int_{\partial\Omega} W_j \, \mathbf{n} \cdot \left[\left(D_{xx}\frac{\partial N_i}{\partial x} + D_{xy}\frac{\partial N_i}{\partial y}\right)\mathbf{e}_x + \left(D_{yx}\frac{\partial N_i}{\partial x} + D_{yy}\frac{\partial N_i}{\partial y}\right)\mathbf{e}_y\right] ds \, u_i +$$
$$- \frac{\Delta t}{2} \sum_{i=1}^{NT} \int_\Omega \frac{\partial W_j}{\partial x}\left(D_{xx}\frac{\partial N_i}{\partial x} + D_{xy}\frac{\partial N_i}{\partial y}\right) d\Omega \, u_i - \frac{\Delta t}{2} \sum_{i=1}^{NT} \int_\Omega \frac{\partial W_j}{\partial y}\left(D_{yx}\frac{\partial N_i}{\partial x} + D_{yy}\frac{\partial N_i}{\partial y}\right) d\Omega \, u_i.$$

This approach can be compared with the artificial viscosity term used by Löhner, et al. [1985]. They solve the equation

$$\frac{\partial u}{\partial t} + \frac{\partial F(u)}{\partial x} = 0 \qquad (8.101)$$

using a Taylor-Galerkin method.

$$\begin{aligned} \frac{u^{n+1/2} - u^n}{\Delta t/2} &= -\frac{\partial F}{\partial x}\bigg|_i^n \\ \frac{u^{n+1} - u^n}{\Delta t} &= -\frac{\partial F}{\partial x}\bigg|_i^{n+1/2} \end{aligned} \qquad (8.102)$$

Then they add an artificial viscosity term to the right-hand side:

$$\int_\Omega \left[\frac{\partial}{\partial x}\left(\mu_x \frac{\partial u}{\partial x}\right) + \frac{\partial}{\partial y}\left(\mu_y \frac{\partial u}{\partial y}\right) \right] W_j \, d\Omega. \qquad (8.103)$$

The artificial viscosity is

$$\mu_x = c_v \, h_e^2 \left|\frac{\partial u}{\partial x}\right|, \quad \mu_y = c_v \, h_e^2 \left|\frac{\partial u}{\partial y}\right|. \qquad (8.104)$$

This is *ad hoc*, but does provide additional viscosity in regions where the gradient is large.

8.4. Flux-corrected Methods

The TVD and ENO methods are particularly easy to apply to two dimensional cases. The calculations are first performed in one direction, as a one-dimensional problem, and then in the other direction. Thus one solves a succession of one-dimensional problems. Referring to Figure 8.1, TVD would apply the flux correction, Eq. (6.92)-(6.97) along the line given by points 1, n+1, 2n+1,..., and then along the line with points 2, n+2, 2n+2,... After all calculations are done in the x-direction, the fluxes are calculated along the y-direction; first along points 1, 2, 3,... then n+1, n+2, n+3,...,etc. This completes one sub-cycle. Then the process is repeated using Eq. (6.98)-(6.101) to complete the next sub-cycle. ENO is applied similarly except that there are 3 sub-cycles; all calculations are done in the x-direction, then all calculations are done in the y-direction in each sub-cycle.

Flux-corrected steps can be applied to both finite difference and finite element methods, too. First we write equations for the flux-corrected finite difference method in two dimensions, according to work by Zalesak [1979]. We need to evaluate flows using both a high-order method and a low-order method. We write the standard equation in the form

$$u_{i,j}^{n+1} = u_{i,j}^n - \Delta t \, (\mathcal{F}_{i+1/2,j} - \mathcal{F}_{i-1/2,j} + \mathcal{G}_{i,j+1/2} - \mathcal{G}_{i,j-1/2}). \qquad (8.105)$$

When we use a low-order upstream method, we write this as

$$u_{i,j}^L = u_{i,j}^n - \Delta t \, (\mathcal{F}_{i+1/2,j}^L - \mathcal{F}_{i-1/2,j}^L + \mathcal{G}_{i,j+1/2}^L - \mathcal{G}_{i,j-1/2}^L). \qquad (8.106)$$

Since the upstream method is

$$u_{i,j}^{n+1} = u_{i,j}^n - \Delta t \left[\frac{1}{\Delta x} (F_{i,j}^n - F_{i-1,j}^n) + \frac{1}{\Delta y} (G_{i,j}^n - G_{i,j-1}^n) \right], \qquad (8.107)$$

we make the identifications

$$\begin{aligned} \mathcal{F}_{i+1/2,j}^L &= F_{i,j}^n / \Delta x, \text{ velocity gives movement from node } i{-}1 \text{ to node } i, \\ \mathcal{G}_{i,j+1/2}^L &= G_{i,j}^n / \Delta y, \text{ velocity gives movement from node } j{-}1 \text{ to node } j. \end{aligned} \qquad (8.108)$$

For two dimensional problems, the flow can be in various directions. Thus the upstream method is written as Eq. (8.107) when the velocity is from node i-1 to node i and from node j-1 to node j. When the flow is from node i to node i-1, the function F is evaluated at the nodes i+1 and i; when the flow is from node j to node j-1, the function G is evaluated at the nodes j+1 and j.

$$\begin{aligned} \mathcal{F}_{i+1/2,j}^L &= F_{i+1,j}^n / \Delta x, \text{ velocity gives movement from node } i \text{ to node } i{-}1 \\ \mathcal{G}_{i,j+1/2}^L &= G_{i,j+1}^n / \Delta y, \text{ velocity gives movement from node } j \text{ to node } j{-}1 \end{aligned} \qquad (8.109)$$

The advection problem for a rotating cone will be solved in Section 9.4; in this problem, the velocity is a specified function. The sign of the velocity terms then determines whether to use Eq. (8.108) or Eq. (8.109); an upstream derivative is always used.

When we use a high-order method of evaluating flows, we write Eq. (8.105) as

$$u_{i,j}^H = u_{i,j}^n - \Delta t \, (\mathcal{F}_{i+1/2,j}^H - \mathcal{F}_{i-1/2,j}^H + \mathcal{G}_{i,j+1/2}^H - \mathcal{G}_{i,j-1/2}^H). \qquad (8.110)$$

Since the centered difference scheme is

$$u_{i,j}^{n+1} = u_{i,j}^n - \frac{\Delta t}{\Delta x} (F_{i+1/2,j}^n - F_{i-1/2,j}^n) - \frac{\Delta t}{\Delta y} (G_{i,j+1/2}^n - G_{i,j-1/2}^n), \qquad (8.111)$$

we make the identifications

$$\begin{aligned} \mathcal{F}_{i+1/2,j}^H &= (F_{i+1,j}^n + F_{i,j}^n)/2\Delta x, \\ \mathcal{G}_{i,j+1/2}^H &= (G_{i,j+1}^n + G_{i,j}^n)/2\Delta y. \end{aligned} \qquad (8.112)$$

The flux-corrected method involves the following six steps:

Step 1. We evaluate the low-order flows using Eq. (8.108).

Step 2. We evaluate the high-order flows using Eq. (8.112).

Step 3. We define the following quantities:

$$A_{i+1/2,j} = F^H_{i+1/2,j} - F^L_{i+1/2,j}$$
$$A_{i,j+1/2} = G^H_{i,j+1/2} - G^L_{i,j+1/2}. \tag{8.113}$$

These terms represent the difference between the flows evaluated using high-order methods and those evaluated using low-order methods.

Step 4. We calculate the low-order solution using Eq. (8.106).

Step 5. We perform the flux-limiting step using the following algorithm. The final solution involves flows with a superscript c. These are given by

$$A^c_{i+1/2,j} = C_{i+1/2,j} A_{i+1/2,j}, \quad 0 \le C_{i+1/2,j} \le 1, \text{ and}$$
$$A^c_{i,j+1/2} = C_{i,j+1/2} A_{i,j+1/2}, \quad 0 \le C_{i,j+1/2} \le 1. \tag{8.114}$$

The purpose of the flux-correction step is to define the C values such that the final result, Eq. (8.119), will not exceed a certain maximum value nor go below a certain minimum value. The algorithm is as follows. We calculate the following six quantities for all points:

$$P^+_{i,j} = \max(0, A_{i-1/2,j}) - \min(0, A_{i+1/2,j}) + \max(0, A_{i,j-1/2}) - \min(0, A_{i,j+1/2})$$
$$Q^+_{i,j} = (u^{max}_{i,j} - u^L_{i,j})/\Delta t$$
$$R^+_{i,j} = \begin{cases} \min(1, Q^+_{i,j}/P^+_{i,j}) & \text{if } P^+_{i,j} > 0 \\ 0 & \text{if } P^+_{i,j} = 0 \end{cases} \tag{8.115}$$

$$P^-_{i,j} = \max(0, A_{i+1/2,j}) - \min(0, A_{i-1/2,j}) + \max(0, A_{i,j+1/2}) - \min(0, A_{i,j-1/2})$$
$$Q^-_{i,j} = (u^L_{i,j} - u^{min}_{i,j})/\Delta t$$
$$R^-_{i,j} = \begin{cases} \min(1, Q^-_{i,j}/P^-_{i,j}) & \text{if } P^-_{i,j} > 0 \\ 0 & \text{if } P^-_{i,j} = 0, \end{cases} \tag{8.116}$$

where

$$u^a_{i,j} = \max(u^n_{i,j}, u^L_{i,j})$$
$$u^{max}_{i,j} = \max(u^a_{i-1,j}, u^a_{i+1,j}, u^a_{i,j}, u^a_{i,j-1}, u^a_{i,j+1})$$
$$u^b_{i,j} = \min(u^n_{i,j}, u^L_{i,j}) \tag{8.117}$$
$$u^{min}_{i,j} = \min(u^b_{i-1,j}, u^b_{i+1,j}, u^b_{i,j}, u^b_{i,j-1}, u^b_{i,j+1}).$$

We note that $P^+_{i,j} \ge 0$ since each term, including the sign in front of it, is positive. Then the C values are defined by

$$C_{i+1/2,j} = \begin{cases} \min(R^-_{i+1,j}, R^+_{i,j}) & A_{i+1/2,j} < 0 \\ \min(R^-_{i,j}, R^+_{i+1,j}) & A_{i+1/2,j} \geq 0 \end{cases}$$

$$C_{i,j+1/2} = \begin{cases} \min(R^-_{i,j+1}, R^+_{i,j}) & A_{i,j+1/2} < 0 \\ \min(R^-_{i,j}, R^+_{i,j+1}) & A_{i,j+1/2} \geq 0. \end{cases} \tag{8.118}$$

Step 6. We obtain the final solution as a correction or limitation to the low-order solution:

$$u_{i,j}^{n+1} = u_{i,j}^L - \Delta t\, (A^c_{i+1/2,j} - A^c_{i-1/2,j} + A^c_{i,j+1/2} - A^c_{i,j-1/2}). \tag{8.119}$$

There are additional steps that can be used prior to Eq. (8.115), but Book [1981, p. 45] says that the effect is minimal. When viscosity is included, the fluxes must be defined with the viscous terms included. When source terms are used, they are included in Eq. (8.106) [Book, 1981, p. 165], but not in the calculation of the anti-diffusive fluxes [Oran and Boris, 1987, p. 283].

The use of flux-correction with finite element methods is a five-step process that follows work by Parrott and Christie [1986]. This is an alternative to the finite element treatment presented in Chapter 6 that is based on the work of Löhner et al. [1987]. Let us consider the advection equation,

$$\frac{\partial c}{\partial t} + \nabla \cdot (\mathbf{u}\, c) = 0. \tag{8.120}$$

Step 1. We evaluate the low-order solution, which must not oscillate. Parrott and Christie [1986] use a method in which the time derivative is lumped and the velocity terms are evaluated in an upstream fashion. Finite element methods applied to Eq. (8.120) give

$$\sum_i \int_\Omega N_j N_i\, d\Omega\, \frac{dc_i}{dt} = -\sum_i \int_\Omega N_j \nabla \cdot (\mathbf{u}\, N_i)\, d\Omega\, c_i. \tag{8.121}$$

The time derivatives are lumped by defining

$$\Omega_j = \sum_i \int_\Omega N_j N_i\, d\Omega \tag{8.122}$$

and taking

$$\sum_i \int_\Omega N_j N_i\, d\Omega\, \frac{dc_i}{dt} \approx \Omega_j \frac{dc_j}{dt}. \tag{8.123}$$

The right-hand side of Eq. (8.121) is integrated by parts:

$$-\sum_i \int_\Omega N_j \nabla \cdot (\mathbf{u} N_i) \, d\Omega \, c_i = \int_\Omega \sum_i N_i c_i \mathbf{u} \cdot \nabla N_j \, d\Omega - \sum_i \int_{\partial\Omega} \mathbf{n} \cdot \mathbf{u} N_j N_i \, dS \, c_i$$

$$= \sum_e \int_{\Omega_e} c^e \mathbf{u}^e \cdot \nabla N_J \, d\Omega - \sum_i \int_{\partial\Omega} \mathbf{n} \cdot \mathbf{u} N_j N_i \, dS \, c_i, \quad (8.124)$$

where dS is an incremental surface area on the boundary $\partial\Omega$. The first term on the right-hand side is written as

$$\sum_e \int_{\Omega_e} c^e \mathbf{u}^e \cdot \nabla N_J \, d\Omega \equiv \sum_e f_j^e \quad \text{where } f_j^e = \int_{\Omega_e} c^e \mathbf{u}^e \cdot \nabla N_J \, d\Omega. \quad (8.125)$$

Now the equation is

$$\Omega_j \frac{dc_j}{dt} = \sum_e f_j^e - \sum_i \int_{\partial\Omega} \mathbf{n} \cdot \mathbf{u} N_j N_i \, dS \, c_i. \quad (8.126)$$

To define upstream evaluation for a finite element method, we use an algorithm from Parrott and Christie [1986] for triangular finite elements. The upstream approximation is

$$(f_j^e)^{ups} = (c_j^e)^{ups} \int_{\Omega_e} \mathbf{u}^e \cdot \nabla N_J \, d\Omega = (c_j^e)^{ups} v_j^e \quad \text{where } v_j^e \equiv \int_{\Omega_e} \mathbf{u}^e \cdot \nabla N_J \, d\Omega, \quad (8.127)$$

where the upstream concentrations are defined with respect to node j by

$$(c_j^e)^{ups} = \begin{cases} c_j & \text{for } v_j^e \geq 0 \\ \dfrac{w_k c_k + w_i c_i}{w_k + w_i} & \text{for } v_j^e < 0. \end{cases} \quad (8.128)$$

Here

$$w_j = \max(0, v_j^e) \quad \text{for all nodes j within element e.} \quad (8.129)$$

The nodes of element e are i, j, and k. The quantity v_j^e is the flow vector $\mathbf{u}$ resolved perpendicular to the side opposite node j in element e (i.e., $\mathbf{n}_j \cdot \mathbf{u}$, where $\mathbf{n}_j$ is the normal to the side opposite j) (see Figure 8.5). Next we define

$$F_j^L \equiv \sum_e (f_j^e)^{ups}. \quad (8.130)$$

Now the equation can be written as

$$\Omega_j \frac{c_j^L - c_j^n}{\Delta t} = F_j^L - \sum_i \int_{\partial\Omega} \mathbf{n} \cdot \mathbf{u} N_j N_i \, dS \, c_i. \quad (8.131)$$

This gives the low-order solution, c_j^L, and the low-order flows, F_j^L.

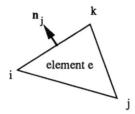

Figure 8.5. Element Notation for Flux-corrected Finite Elements

Step 2. We evaluate the high-order solution. Parrott and Christie [1986] use a lumped version (i.e., a Taylor finite difference method on triangles). Presented here is an improvement using a Taylor-Galerkin method, following work by Löhner, et al. [1987].

$$\sum_i \int_\Omega N_j N_i \, d\Omega \, \frac{dc_i}{dt} = \sum_e \int_{\Omega_e} c^e \mathbf{u}^e \cdot \nabla N_J \, d\Omega - \sum_i \int_{\partial\Omega} \mathbf{n} \cdot \mathbf{u} \, N_j N_i \, dS \, c_i +$$

$$- \frac{\Delta t}{2} \sum_i \int_\Omega \left\{ \frac{\partial N_j}{\partial x} \left[u u \frac{\partial N_i}{\partial x} + u v \frac{\partial N_i}{\partial y} \right] + \frac{\partial N_j}{\partial y} \left[v u \frac{\partial N_i}{\partial x} + v v \frac{\partial N_i}{\partial y} \right] \right\} d\Omega \, c_i \quad (8.132)$$

Eq. (8.132) is written in matrix notation as

$$(M_c)_{ji} \frac{c_i^H - c_i^n}{\Delta t} = \sum_e f_j^e - \sum_i \int_{\partial\Omega} \mathbf{n} \cdot \mathbf{u} \, N_j N_i \, dS \, c_i +$$

$$- \frac{\Delta t}{2} \sum_i \int_\Omega \left\{ \frac{\partial N_j}{\partial x} \left[u u \frac{\partial N_i}{\partial x} + u v \frac{\partial N_i}{\partial y} \right] + \frac{\partial N_j}{\partial y} \left[v u \frac{\partial N_i}{\partial x} + v v \frac{\partial N_i}{\partial y} \right] \right\} d\Omega \, c_i. \quad (8.133)$$

We rewrite it as

$$(M_L)_{ji} \frac{c_i^H - c_i^n}{\Delta t} = (M_L - M_c)_{ji} \frac{c_i^H - c_i^n}{\Delta t} + \sum_e f_j^e - \sum_i \int_{\partial\Omega} \mathbf{n} \cdot \mathbf{u} \, N_j N_i \, dS \, c_i +$$

$$- \frac{\Delta t}{2} \sum_i \int_\Omega \left\{ \frac{\partial N_j}{\partial x} \left[u u \frac{\partial N_i}{\partial x} + u v \frac{\partial N_i}{\partial y} \right] + \frac{\partial N_j}{\partial y} \left[v u \frac{\partial N_i}{\partial x} + v v \frac{\partial N_i}{\partial y} \right] \right\} d\Omega \, c_i \quad (8.134)$$

and define the high-order flows as

$$F_j^H \equiv (M_L - M_c)_{ji} \frac{c_i^H - c_i^n}{\Delta t} + \sum_e f_j^e +$$

$$- \frac{\Delta t}{2} \sum_i \int_\Omega \left\{ \frac{\partial N_j}{\partial x} \left[u u \frac{\partial N_i}{\partial x} + u v \frac{\partial N_i}{\partial y} \right] + \frac{\partial N_j}{\partial y} \left[v u \frac{\partial N_i}{\partial x} + v v \frac{\partial N_i}{\partial y} \right] \right\} d\Omega \, c_i. \quad (8.135)$$

Eq. (8.134) represents a set of linear equations. Rather than using a direct solution technique, we solve it using successive substitution.

$$(M_L)_{ji} \frac{c_i^{H,k+1} - c_i^n}{\Delta t} = (M_L - M_c)_{ji} \frac{c_i^{H,k} - c_i^n}{\Delta t} + \sum_e f_j^e - \sum_i \int_{\partial \Omega} \mathbf{n} \cdot \mathbf{u}\, N_j\, N_i\, dS\, c_i +$$

$$- \frac{\Delta t}{2} \sum_i \int_\Omega \left\{ \frac{\partial N_j}{\partial x} \left[uu \frac{\partial N_i}{\partial x} + uv \frac{\partial N_i}{\partial y} \right] + \frac{\partial N_j}{\partial y} \left[vu \frac{\partial N_i}{\partial x} + vv \frac{\partial N_i}{\partial y} \right] \right\} d\Omega\, c_i \qquad (8.136)$$

This gives the high-order solution that is used to find the high-order fluxes in Eq. (8.135).

Step 3. We define the anti-diffusion contributions as

$$AD_j^e \equiv f_j^H - f_j^L. \qquad (8.137)$$

Step 4. We perform the flux-limiting step to find the anti-diffusion correction factors, ADC_j^e, according to Eq. (6.79)-(6.83). Here the algorithm is

$$\begin{aligned}
&\text{For each node: } c_i^a = \max\{c_i^L, c_i^n\} \\
&\text{For all nodes in an element: } c_e^* = \max\nolimits_{\text{all nodes in element } e} \{c_I^a\} \\
&\text{For all elements touching node i: } c_i^{max} = \max\nolimits_{\text{all elements touching node i}} \{c_e^*\} \\
&\text{For each node: } c_i^b = \min\{c_i^L, c_i^n\} \\
&\text{For all nodes in an element: } c_e^\dagger = \min\nolimits_{\text{all nodes in element } e} \{c_I^b\} \\
&\text{For all elements touching node i: } c_i^{min} = \min\nolimits_{\text{all elements touching node i}} \{c_e^\dagger\}
\end{aligned} \qquad (8.138)$$

For elements contributing to node I:

$$P_i^+ = \sum_e \max\{0, AD_I^e\}, \quad Q_i^+ = (M_L)_i (c_i^{max} - c_i^L) / \Delta t,$$
$$R_i^+ = \min\{1, Q_i^+ / P_i^+\} \text{ if } P_i^+ > 0; \quad 0 \text{ if } P_i^+ = 0. \qquad (8.139)$$

$$P_i^- = -\sum_e \min\{0, AD_I^e\}, \quad Q_i^- = (M_L)_i (c_i^L - c_i^{min}) / \Delta t,$$
$$R_i^- = \min\{1, Q_i^- / P_i^-\} \text{ if } P_i^- > 0; \quad 0 \text{ if } P_i^- = 0. \qquad (8.140)$$

We compute the correction factors

$$R_i^e = \begin{cases} R_i^+ & \text{if } AD_i^e > 0 \\ \\ R_i^- & \text{if } AD_i^e < 0, \end{cases} \qquad (8.141)$$

$$C_e = \min_{\text{all nodes in element}} (R_I^e). \tag{8.142}$$

$$ADC_j^e = C_e \, AD_j^e \qquad 0 \le C_e \le 1. \tag{8.143}$$

Step 5. We obtain the final solution using the flux-limited corrections:

$$\Omega_j \frac{c_j^{n+1} - c_j^L}{\Delta t} = \sum_e ADC_j^e. \tag{8.144}$$

8.5. Grid and Mesh Refinement

Grid generation refers to the process of specifying the finite difference grid points that, when connected, cover a domain. Mesh generation refers to specifying the finite elements that cover a domain. Grid and mesh refinement are the processes of changing the grid to meet some goal. Both grid and mesh generation and grid and mesh refinement are more difficult to do in two dimensions than in one dimension. For finite difference methods, the grid points are usually organized in some regular fashion. For finite element methods, the domain is covered with contiguous elements, but these are often arranged in an irregular fashion. Once the grid or mesh is generated, it can be moved. In this section techniques involving grid and mesh refinement on fixed grids and meshes will be described; moving grids and meshes are described in Section 8.6. Grid and mesh generation techniques will not be discussed.

There are several approaches to grid and mesh refinement. In one approach, a coarse mesh is established and maintained at all times. A fine mesh is added over parts of the region, as needed, and the regions overlaid can change as the solution changes. This approach is usually the only one used with finite difference methods. With finite element methods there are more options. The domain can be covered with quadrilateral or triangular elements (or both) and these elements need not be arranged in a regular manner. How the elements are joined together, however, is an important detail.

Let us first consider grid refinement for finite difference methods, as used by Smooke and Koszykowski [1984]. Suppose we wish to refine a region in the upper right-hand corner of Figure 8.6(a). The finite difference algorithm would not work with the grid shown. However, if we extend the grid lines to the boundaries of the domain, as shown in Figure 8.6(b), then we can apply a finite difference method to any grid point in a straightforward extension of the techniques applied to one-dimensional problems. For example, the convective diffusion equation would replace Eq. (8.3) with Eq. (8.145). Smooke and Koszykowski [1984] achieved good results when using this technique, even when it was coupled with moving grid lines. This is probably the easiest way to refine the grid for a finite difference method. It does, however, introduce some

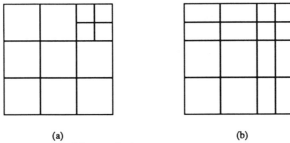

(a) (b)

Figure 8.6. Grid Refinement

$$\frac{c_{i,j}^{n+1} - c_{i,j}^n}{\Delta t} + u_{i,j} \frac{c_{i+1,j}^n - c_{i-1,j}^n}{\Delta x_i + \Delta x_{i-1}} + v_{i,j} \frac{c_{i,j+1}^n - c_{i,j-1}^n}{\Delta y_j + \Delta y_{j-1}} =$$

$$+ D \frac{\frac{c_{i+1,j}^n - c_{i,j}^n}{\Delta x_i} - \frac{c_{i,j}^n - c_{i-1,j}^n}{\Delta x_{i-1}}}{\frac{1}{2}(\Delta x_i + \Delta x_{i-1})} + D \frac{\frac{c_{i,j+1}^n - c_{i,j}^n}{\Delta y_j} - \frac{c_{i,j}^n - c_{i,j-1}^n}{\Delta y_{j-1}}}{\frac{1}{2}(\Delta y_j + \Delta y_{j-1})}. \quad (8.145)$$

inefficiencies, since some of the grid points are needed by the finite difference method rather than the solution.

Another method of grid refinement is one developed by Gropp [1980]. Let us suppose we have the solution at time t on the coarse grid shown in Figure 8.7(a). We integrate the equation on this grid from time t to time t+Δt; for example, we could use Eq. (8.3). Then we determine which regions need to be refined; these are marked with a $\sqrt{}$ in Figure 8.7(b). We also mark the regions that form a boundary between the coarse mesh and the refined mesh; these are marked with a ∂ symbol in Figure 8.7(b). The purpose of this boundary layer is to provide a region for oscillations that develop in the fine region to be damped out before reaching the coarse mesh (where they would be propagated very far). We then construct a fine mesh in regions marked with a $\sqrt{}$ and ∂ in Figure 8.7(c). We integrate from time t to time t+Δt on this mesh. For the grid points on the bounding line, noted by dots in Figure 8.7(c), we require the concentration to be the average of the solution on the coarse grid at time t and time t+Δt. Whenever a fine mesh solution is not available at time t, the coarse mesh solution is interpolated linearly to provide it. Sometimes the fine mesh solution is available because it was generated at the previous time-step. Finally, the solution at the new time t+Δt is taken as the fine mesh solution where it exists and as the coarse mesh solution elsewhere. As shown in Figure 8.7(d), when the coarse mesh and fine mesh overlap on an interior region, the solution from the fine mesh is transferred to the coarse mesh before integrating on the coarse mesh for the next time-step. Berger and Oliger [1984] use this type of mesh refinement and give a good discussion of data structures.

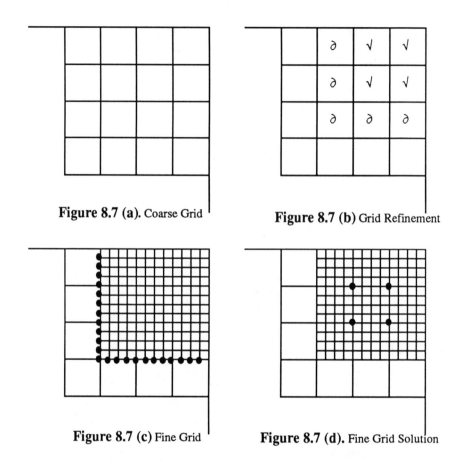

Figure 8.7 (a). Coarse Grid

Figure 8.7 (b) Grid Refinement

Figure 8.7 (c) Fine Grid

Figure 8.7 (d). Fine Grid Solution

For the interpolation process it is sometimes necessary to obtain the value of the solution at a point inside the coarse mesh. This can be done using a finite element interpolation. Let us suppose that the points 1 through 4 in Figure 8.8 represent the nodes of a coarse mesh and that we need the value of the solution at point x. We can represent the solution on the rectangle using a bilinear trial function,

$$N_1 = (1-u)(1-v), \quad N_2 = u(1-v)$$
$$N_3 = u v, \quad N_4 = v(1-u), \tag{8.146}$$

where

$$u = \frac{x - x_{i-1}}{\Delta x_{i-1}}, \quad v = \frac{y - y_{j-1}}{\Delta y_{j-1}}. \tag{8.147}$$

The solution at any point is then

$$c(u,v) = c_1 N_1(u,v) + c_2 N_2(u,v) + c_3 N_3(u,v) + c_4 N_4(u,v), \tag{8.148}$$

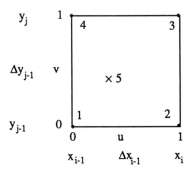

Figure 8.8. Finite Element Representation for Interpolation

This can be rearranged at evaluated at the point (u_5, v_5) to give

$$c_5 = c_1 + u_5(-c_1 + c_2) + v_5(-c_1 + c_4) + u_5 v_5(c_1 - c_2 + c_3 - c_4). \quad (8.149)$$

Of course, interpolation can be used along a grid line or a straight line.

Next let us consider how a finite element method can be applied to the refinement shown in Figure 8.6(a). On a square, the shape functions for a bilinear element are

$$N_1 = \tfrac{1}{4}(-1+\xi)(-1+\eta), \quad N_2 = -\tfrac{1}{4}(1+\xi)(-1+\eta)$$
$$N_3 = \tfrac{1}{4}(1+\xi)(1+\eta), \quad N_4 = -\tfrac{1}{4}(-1+\xi)(1+\eta). \quad (8.150)$$

These are the same as those in Eq. (8.146), except that the domain extends from -1 to 1 in both ξ and η. The solution within the element is then

$$c(\xi,\eta) = c_1 N_1(\xi,\eta) + c_2 N_2(\xi,\eta) + c_3 N_3(\xi,\eta) + c_4 N_4(\xi,\eta). \quad (8.151)$$

Figure 8.9 shows the node locations in the different elements. Some of the square elements have only four nodes; the shape functions can then be represented as in Eq. (8.150). Other elements, however, have five nodes and the shape functions given by Eq. (8.150) are not correct. Within the larger element, Eq. (8.150) assumes the shape function is linear along a side; however, this is true only if the solution at the mid-side node is the average of the solutions at the ends of the side.

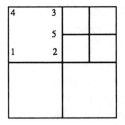

Figure 8.9. Nodal Arrangement

There are three approaches to handling this difficulty. In the first approach, a new shape function is defined that replaces Eq. (8.150). In this new trial function there is a fifth shape function; each function again is 1 at one node and 0 at the other nodes. One example (where the element goes from -1 to 1 in both ξ and η) is

$$N_1 = \tfrac{1}{4}(-1+\xi)(-1+\eta), \quad N_2 = \tfrac{1}{4}(1+\xi)(-1+\eta)\eta$$
$$N_3 = \tfrac{1}{4}(1+\xi)(1+\eta)\eta, \quad N_4 = -\tfrac{1}{4}(-1+\xi)(1+\eta) \quad (8.152)$$
$$N_5 = \tfrac{1}{2}(1+\xi)(1-\eta^2).$$

The second approach to handling this difficulty is to use a linear constraint. In this case, the equation governing Node 5 is changed to

$$c_5 = \tfrac{1}{2}(c_2 + c_3). \quad (8.153)$$

The third approach is to grade the mesh so that it moves smoothly from the fine region to the coarse region. Ang and Valliappan [1986] give examples of this approach, which are shown in Figure 8.10.

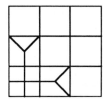
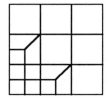

Figure 8.10 (a). Using Triangular and Quadrilateral Elements

Figure 8.10 (b). Using Only Quadrilateral Elements

When triangular elements are used, similar difficulties arise with mesh refinement, but they are more easily solved. Let us suppose that we have two triangles, one of which is to be refined, as shown in Figure 8.11(a). Two ways of doing this are demonstrated in Figure 8.11(b) and 8.11(d). In the first case [Figure 8.11(b)], the refinement makes no changes in the second element. However, on successive refinements the elements become long and thin, as shown in Figure 8.11(c). Furthermore, there is no refinement along the boundary. The other approach, as shown in Figure 8.11(d), makes a change in the adjacent element but this change does not propagate any further than the next element. Thus it is a good way to refine the elements. We must decide which side is to be split or if several sides are to be split, as shown in Figure 8.11(e).

Triangular regions can also be refined into quadrilaterals, as shown in Figure 8.12.

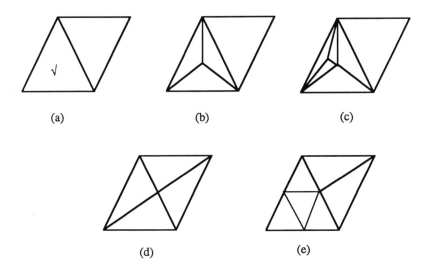

Figure 8.11. Grid Refinement Using Triangles

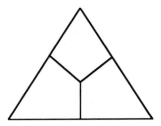

Figure 8.12. Grid Refinement of Triangle Using Quadrilateral Elements
(according to Zienkiewicz, private communication, 1988)

The refinement shown above is called h-refinement, because the sizes of the elements (noted by h = Δx) are changed in certain regions. There is also p-refinement, in which the degree of polynomial of the trial function is raised but the element size is retained. Both schemes are used together by Oden [1989].

Some sort of criteria must be used to decide whether or not the element is to be refined. The criteria can be the value of the first or second derivative, some variable that relates to the error, or the residual. Babuska and Miller [1984a,b,c] derive criteria that are designed to give the best accuracy to some desired property of the solution. All of these criteria have been used in the past and are currently being applied in the literature. General articles are available by Demkowicz and Oden [1986], Kikuchi [1986], Frey [1987], Peraire, *et al*. [1987], and Zhu and Zienkiewicz [1988].

8.6. Moving Grids and Finite Elements

In this section, three methods for solving problems with sharp fronts in two dimensions will be discussed. The generalization from one to two dimensions is not straightforward, but it has been accomplished for several methods. We will discuss here the random choice method, the method of moving finite elements, and the Euler-Lagrange methods. In each case the distinguishing feature is how the grid is represented and moved.

Front tracking and the random choice method. For one-dimensional problems, the random choice method provides an excellent solution method for solving problems that have moving fronts. For two-dimensional problems, it is necessary to advance the front (which can be done with the random choice method), but the front must be advanced along a direction normal to the front. In the method of front tracking (as the random choice method is called for two-dimensional problems), the fronts are located explicitly and propagated dynamically. Riemann problems are used to find the one-dimensional wave speed in the direction normal to the front. Before and after the front the smooth part of the solution is found using any method, but most previous work has used finite difference methods. (See Glimm, et al. [1986] for a finite element application.) The problem of colliding fronts is also important in two dimensions, since the Riemann problem must be solved in such circumstances. In this case, the Riemann problem is essentially two-dimensional; it has been solved in some special cases [Courant and Friedrichs, 1948]. The front tracking method applied to such cases is given by Glimm, et al. [1987, 1988]. Roe [1991] applies discontinuous solution methods when there are discontinuities oblique to the grid.

The work by Glimm and co-workers on front tracking will be summarized here [Glimm, et al. 1980, 1981, 1985; Chern, et al. 1986]. Since movement of a front is characterized by a Lagrangian treatment, the location of the front must be solved for and described at each instant in time. This requires at least two grids: a hyperbolic grid, on which the state variables are defined at each grid point, and a hyperbolic interface grid, which defines the location of the front. The hyperbolic grid is generally fixed in time while the hyperbolic interface grid moves through the region with the solution. If the interface is defined by piecewise linear shapes, then the hyperbolic interface grid can be defined locally by locating the intersection of these lines with the hyperbolic grid, as shown in Figure 8.13. Thus, the normal to the interface is clearly defined (except at the nodes where discontinuities can exist in the slope).

If there is an elliptic part of the problem, then the third grid would be the elliptic grid. If the coefficients were discontinuous across the front, this grid might have to be realigned so that a boundary lies along the discontinuity. This can be done by sliding nodes until they intersect the front, and then triangulating locally. A fourth grid might be necessary to define the front on the elliptic grid.

TWO-DIMENSIONS   289

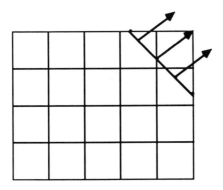

Figure 8.13. Hyperbolic Grid and Hyperbolic Interface

Let us examine the details of the movement of a front, according to work by Glimm, *et al.* [1986]. We will let z_0 be a point on the front at time t_0 and establish a new coordinate system that is imbedded in the front, as shown in Figure 8.14. The coordinate lines with s = constant are perpendicular to the front, while those with r = constant are parallel to the front. The unit vectors $\hat{n}(r,s)$ and $\hat{s}(r,s)$ are the vectors normal to and parallel to the front at z_0.

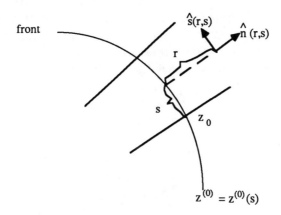

Figure 8.14. Geometry of the Front

The problem

$$\frac{\partial u}{\partial t} + \nabla \cdot F(u) = 0 \qquad (8.154)$$

can be written as

$$\frac{\partial u}{\partial t} + \hat{n} \cdot [\hat{n} \cdot \nabla \ F(u)] + \hat{s} \cdot [\hat{s} \cdot \nabla \ F(u)] = 0. \qquad (8.155)$$

In order to advance the front, we wish to solve this equation subject to the initial condition

$$u(t=t_0) = u^{(0)}, \qquad (8.156)$$

where $u^{(0)}$ is smooth except for possible discontinuities across the front. The computation uses operator splitting. The first step is

$$\frac{\partial u}{\partial t} + \hat{n} \cdot [\hat{n} \cdot \nabla \; F(u)] = 0, \quad u(t=t_0) = u^{(0)} \qquad (8.157)$$

and the solution is called u^{cn}. Then the tangential equations are solved to obtain the solution, u^c.

$$\frac{\partial u}{\partial t} + \hat{s} \cdot [\hat{s} \cdot \nabla \; F(u)] = 0, \quad u(t=t_0) = u^{cn} \qquad (8.158)$$

The error in the entire process can be shown to be $O(\Delta t)$. The first step, Eq. (8.157), uses the random choice method, as outlined in Section 6.1. The second step, Eq. (8.158), uses the Lax-Wendroff method; in this step the mesh is held fixed, since movement of grid points tangential to the front is merely a remeshing of the front. If diffusion or viscosity is included in the model, then presumably we would do an additional calculation, as done in Eq. (6.21); this step would be done on the elliptic grid.

The above calculations describe the basis of the method. The most important part of the algorithm is the solution of the Riemann problem (especially when oblique waves can interact) and the identification and interaction of the different meshes. The first problem is discussed in this book in conjunction with specific applications while the second is described by Glimm and McBryan [1985].

Moving finite elements. Let us consider the differential equation

$$\frac{\partial c}{\partial t} = L(c), \qquad (8.159)$$

where $L(c)$ is a spatial operator in two dimensions. The trial function is taken as

$$c = \sum_{i=1}^{NT} c_j(t) N_j(x,y,t). \qquad (8.160)$$

We define

$$\begin{aligned} x_j &= \text{x-location of } N_j \\ y_j &= \text{y-location of } N_j, \text{ such that } N_j(x_j, y_j, t) = 1 \end{aligned} \qquad (8.161)$$

and allow the points $\{x_j, y_j\}$ to move with time, giving

$$\frac{dx_j}{dt} \text{ and } \frac{dy_j}{dt}. \qquad (8.162)$$

The time derivative of the trial function is then

$$\frac{\partial c}{\partial t} = \frac{dc_j}{dt}\frac{\partial c}{\partial c_j} + \frac{dx_j}{dt}\frac{\partial c}{\partial x_j} + \frac{dy_j}{dt}\frac{\partial c}{\partial y_j} \equiv \frac{dc_j}{dt}\alpha_j + \frac{dx_j}{dt}\beta_j + \frac{dy_j}{dt}\gamma_j. \quad (8.163)$$

The three trial functions (α_j, β_j, γ_j) are defined by Alexander, et al. [1980] as follows. We suppose the triangular elements are arranged such that six of them form a hexagon, as illustrated in Figure 8.15. The variable α_j takes the value 1.0 at the center, j-th node and the value zero on the boundary nodes; it is thus N_j. Within an element, β_j is a plane taking the value zero on the boundary and the value $-u_x$ at the center. The function γ_j is similar to β_j, except that it takes the value $-u_y$ at the center. If a velocity is identified for each element, then β_j and γ_j are discontinuous functions along the boundaries between elements but they connect with point j.

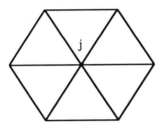

Figure 8.15. Triangular Elements for Moving Finite Elements

The weighted residual formulation is then

$$\sum_{j=1}^{NT} \frac{dc_j}{dt}\int_\Omega N_j N_i\, d\Omega + \sum_{j=1}^{NT} \frac{dx_j}{dt}\int_\Omega \beta_j N_i\, d\Omega + \sum_{j=1}^{NT} \frac{dy_j}{dt}\int_\Omega \gamma_j N_i\, d\Omega = \int_\Omega L(c)\, N_i\, d\Omega. \quad (8.164)$$

$$\sum_{j=1}^{NT} \frac{dc_j}{dt}\int_\Omega N_j \beta_i\, d\Omega + \sum_{j=1}^{NT} \frac{dx_j}{dt}\int_\Omega \beta_j \beta_i\, d\Omega + \sum_{j=1}^{NT} \frac{dy_j}{dt}\int_\Omega \gamma_j \beta_i\, d\Omega = \int_\Omega L(c)\, \beta_i\, d\Omega. \quad (8.165)$$

$$\sum_{j=1}^{NT} \frac{dc_j}{dt}\int_\Omega N_j \gamma_i\, d\Omega + \sum_{j=1}^{NT} \frac{dx_j}{dt}\int_\Omega \beta_j \gamma_i\, d\Omega + \sum_{j=1}^{NT} \frac{dy_j}{dt}\int_\Omega \gamma_j \gamma_i\, d\Omega = \int_\Omega L(c)\, \gamma_i\, d\Omega. \quad (8.166)$$

As was done in Eq. (7.130), empirical additions are used to keep the mesh regular; the constants thus introduced must be determined by trial and error.

Euler-Lagrange methods. The Euler-Lagrange methods presented in Section 7.5 can be generalized to two dimensions. Let us consider the convective diffusion equation and write the equation as

$$\frac{Dc}{Dt} = \frac{\partial^2 c}{\partial x^2} + \frac{\partial^2 c}{\partial y^2}. \quad (8.167)$$

In the implicit form, we provide a finite difference method as

$$\frac{c_{i,j}^{n+1} - {}_{i,j}c^n}{\Delta t} = \frac{1}{\Delta x^2}(c_{i+1,j}^{n+1} - 2c_{i,j}^{n+1} + c_{i-1,j}^{n+1}) + \frac{1}{\Delta y^2}(c_{i,j+1}^{n+1} - 2c_{i,j}^{n+1} + c_{i,j-1}^{n+1}), \quad (8.168)$$

where

$$_{i,j}c^n = c(x = x_i - u_{i,j}\Delta t,\ y = y_j - v_{i,j}\Delta t,\ t = t^n) \quad (8.169)$$

is the solution at the position (x,y) and time t that ends up at position (x_{ij}, y_{ij}) at time t+Δt. The grid is fixed, so the most difficult part of the method is solving Eq. (8.169). Of course the implicit step, Eq. (8.168), is also slow and cumbersome if a direct solution technique is used.

There are two important difficulties when using the explicit version. First, the grid deforms into irregular patterns, so a finite element method is nearly the only method that can be used. Finite element methods, unfortunately, are not suited to explicit methods, since a matrix problem must be solved at each time-step. In addition, the movement of the nodes is essentially at the mercy of the solution and numerical artifacts; the nodes can become very close to each other or even cross, and the mesh can become tangled. Of course, these are the same difficulties treated in the front tracking methods, so they can be overcome with sufficient effort. Neuman [1984] gives results obtained for simple two-dimensional cases.

8.7. Spine Representation of Mesh

Since many of the methods employed in the previous section moved the mesh according to some physical principle, we consider here a representation of the mesh that allows this mesh movement. First let us consider a heat transfer problem on the domain shown in Figure 8.16. The heat conduction equation is

$$\frac{\partial^2 T}{\partial x^2} + \frac{\partial^2 T}{\partial y^2} = 0. \quad (8.170)$$

Let us suppose the boundary conditions have temperature specified along the boundary. Here we take them as

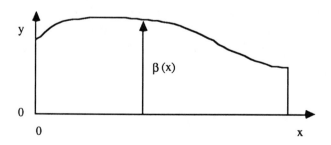

Figure 8.16. Domain for a Heat Transfer Problem

$$T(x,0) = g_1(x), \quad T(0,y) = g_2(y)$$
$$T(x,\beta(x)) = g_3(x), \quad T(x_1,y) = g_4(y), \tag{8.171}$$

where the functions g_1 through g_4 are specified, known functions. We can transform the domain using the new variables

$$\xi = \frac{x}{x_1}, \quad \eta = \frac{y}{\beta(x)}. \tag{8.172}$$

Under this transformation we have

$$\frac{\partial T}{\partial x} = \frac{\partial T}{\partial \xi}\frac{\partial \xi}{\partial x} + \frac{\partial T}{\partial \eta}\frac{\partial \eta}{\partial x},$$
$$\frac{\partial T}{\partial y} = \frac{\partial T}{\partial \xi}\frac{\partial \xi}{\partial y} + \frac{\partial T}{\partial \eta}\frac{\partial \eta}{\partial y},$$
$$\frac{\partial \xi}{\partial x} = \frac{1}{x_1}, \quad \frac{\partial \xi}{\partial y} = 0, \quad \frac{\partial \eta}{\partial x} = -\frac{\eta}{\beta}\frac{d\beta}{dx}, \quad \frac{\partial \eta}{\partial y} = \frac{1}{\beta}. \tag{8.173}$$

$$\frac{\partial^2 \xi}{\partial x^2} = 0, \quad \frac{\partial^2 \xi}{\partial y^2} = 0, \quad \frac{\partial^2 \eta}{\partial x^2} = \frac{2\eta}{\beta^2}\left(\frac{d\beta}{dx}\right)^2 - \frac{\eta}{\beta}\frac{d^2\beta}{dx^2}, \quad \frac{\partial^2 \eta}{\partial y^2} = 0 \tag{8.174}$$

$$\frac{\partial^2 T}{\partial x^2} = \frac{\partial^2 T}{\partial \xi^2}\left(\frac{\partial \xi}{\partial x}\right)^2 + 2\frac{\partial^2 T}{\partial \xi \partial \eta}\left(\frac{\partial \xi}{\partial x}\right)\left(\frac{\partial \eta}{\partial x}\right) + \frac{\partial^2 T}{\partial \eta^2}\left(\frac{\partial \eta}{\partial x}\right)^2 + \frac{\partial T}{\partial \xi}\frac{\partial^2 \xi}{\partial x^2} + \frac{\partial T}{\partial \eta}\frac{\partial^2 \eta}{\partial x^2}$$

$$\frac{\partial^2 T}{\partial y^2} = \frac{\partial^2 T}{\partial \xi^2}\left(\frac{\partial \xi}{\partial y}\right)^2 + 2\frac{\partial^2 T}{\partial \xi \partial \eta}\left(\frac{\partial \xi}{\partial y}\right)\left(\frac{\partial \eta}{\partial y}\right) + \frac{\partial^2 T}{\partial \eta^2}\left(\frac{\partial \eta}{\partial y}\right)^2 + \frac{\partial T}{\partial \xi}\frac{\partial^2 \xi}{\partial y^2} + \frac{\partial T}{\partial \eta}\frac{\partial^2 \eta}{\partial y^2} \tag{8.175}$$

The differential equation is then

$$\frac{1}{x_1^2}\frac{\partial^2 T}{\partial \xi^2} - \frac{2\eta}{\beta x_1}\frac{d\beta}{dx}\frac{\partial^2 T}{\partial \xi \partial \eta} + \left[\frac{1}{\beta^2} + \frac{\eta^2}{\beta^2}\left(\frac{d\beta}{dx}\right)^2\right]\frac{\partial^2 T}{\partial \eta^2} + \frac{\partial T}{\partial \eta}\left[\frac{2\eta}{\beta^2}\left(\frac{d\beta}{dx}\right)^2 - \frac{\eta}{\beta}\frac{d^2\beta}{dx^2}\right] = 0. \tag{8.176}$$

The boundary conditions are now applied as

$$T(\xi,0) = g^*_1(\xi) = g_1(x=x_1\xi), \quad T(\xi,0) = g^*_2(\eta) = g_2(y=\eta\beta(0)),$$
$$T(\xi,1) = g^*_3(\xi) = g_3(x=x_1\xi), \quad T(1,\eta) = g^*_4(\eta) = g_4(y=\eta\beta(x_1)). \tag{8.177}$$

The boundary conditions are applied on a square, so the problem can be solved on the square region:

$$0 \leq \xi \leq 1, \, 0 \leq \eta \leq 1. \tag{8.178}$$

We have thus changed from a simple differential equation on an irregular domain to a complicated differential equation on a regular domain. A finite difference method can easily be applied on the revised domain, since the new domain is regular. We then say the method is solved using using body-fitted coordinates

[Thompson, *et al.* 1974]. Similar ideas were used by Young and Finlayson [1976] when they applied the method of orthogonal collocation to irregular domains. Recently Ramanathan and Kumar [1988] have compared the use of body-fitted coordinates for finite difference methods and finite element methods. This approach opens up a wide class of problems that can be solved using a finite difference method; however, sometimes we need to solve problems on even more complicated geometries. The approach given here, using a spine representation of the boundary, allows for these geometries.

To appreciate the finite element approach as applied to problems with irregular geometries, we first outline how the geometric transformation affects the terms appearing in the Galerkin formulation. Let us consider the terms given in Eq. (8.41)-(8.43) and identify the following two representative terms:

$$I_1^e = \int_{\Omega_e} W_J \, u \, \frac{\partial N_I}{\partial x} \, d\Omega,$$

$$I_2^e = \int_{\Omega_e} \frac{\partial N_J}{\partial x} \frac{\partial N_I}{\partial x} \, d\Omega. \quad (8.179)$$

These integrals are to be evaluated on an element. It is convenient to transform the element to a regular domain such as a square or triangle (see Figure 8.17). Thus we write

$$x = x(\xi,\eta), \; y = y(\xi,\eta). \quad (8.180)$$

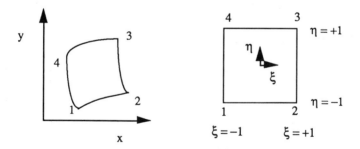

Figure 8.17. Transformation of Quadrilateral Element to Standard Form

The relationship between terms in the x-y coordinate system and the ξ–η coordinate system is

$$dxdy = \det J \, d\xi \, d\eta, \quad (8.181)$$

where the determinant of J (det J) is given by

$$\det J = \frac{\partial x}{\partial \xi} \frac{\partial y}{\partial \eta} - \frac{\partial x}{\partial \eta} \frac{\partial y}{\partial \xi}. \quad (8.182)$$

The Jacobian is

$$[J] = \begin{bmatrix} \dfrac{\partial x}{\partial \xi} & \dfrac{\partial y}{\partial \xi} \\ \dfrac{\partial x}{\partial \eta} & \dfrac{\partial y}{\partial \eta} \end{bmatrix}. \quad (8.183)$$

First derivatives are given by

$$\frac{\partial f}{\partial x} = \frac{\partial f}{\partial \xi} \frac{\partial \xi}{\partial x} + \frac{\partial f}{\partial \eta} \frac{\partial \eta}{\partial x}. \quad (8.184)$$

To obtain the needed derivatives, we use

$$\begin{bmatrix} \dfrac{\partial \xi}{\partial x} & \dfrac{\partial \eta}{\partial x} \\ \dfrac{\partial \xi}{\partial y} & \dfrac{\partial \eta}{\partial y} \end{bmatrix} = [J]^{-1} = \frac{1}{\det J} \begin{bmatrix} \dfrac{\partial y}{\partial \eta} & -\dfrac{\partial y}{\partial \xi} \\ -\dfrac{\partial x}{\partial \eta} & \dfrac{\partial x}{\partial \xi} \end{bmatrix}. \quad (8.185)$$

Thus we can write

$$I_1^e = \int_{\Omega_e} W_j^e u \frac{\partial N_i^e}{\partial x} dx\, dy = \int_{-1}^{1}\int_{-1}^{1} W_J u^e \left\{ [J]_{11}^{-1} \frac{\partial N_I}{\partial \xi} + [J]_{12}^{-1} \frac{\partial N_I}{\partial \xi} \right\} \det J\, d\xi\, d\eta \quad (8.186)$$

$$= \int_{-1}^{1}\int_{-1}^{1} W_J u^e \left\{ \frac{\partial y}{\partial \eta} \frac{\partial N_I}{\partial \xi} - \frac{\partial y}{\partial \xi} \frac{\partial N_I}{\partial \xi} \right\} d\xi\, d\eta$$

and

$$I_2^e = \int_{\Omega_e} \frac{\partial N_j^e}{\partial x} \frac{\partial N_i^e}{\partial x} dx\, dy = \int_{-1}^{1}\int_{-1}^{1} \left\{ [J]_{11}^{-1} \frac{\partial N_J}{\partial \xi} + [J]_{12}^{-1} \frac{\partial N_J}{\partial \eta} \right\}$$

$$\left\{ [J]_{11}^{-1} \frac{\partial N_I}{\partial \xi} + [J]_{12}^{-1} \frac{\partial N_I}{\partial \eta} \right\} \det J\, d\xi\, d\eta \quad (8.187)$$

$$= \int_{-1}^{1}\int_{-1}^{1} \left\{ \frac{\partial y}{\partial \eta} \frac{\partial N_J}{\partial \xi} - \frac{\partial y}{\partial \xi} \frac{\partial N_J}{\partial \eta} \right\} \left\{ \frac{\partial y}{\partial \eta} \frac{\partial N_I}{\partial \xi} - \frac{\partial y}{\partial \xi} \frac{\partial N_I}{\partial \eta} \right\} \frac{1}{\det J} d\xi\, d\eta.$$

The integrals, Eq. (8.186) and Eq. (8.187), are evaluated using Gaussian quadrature. Thus, Eq. (8.186) becomes

$$I_1^e = \sum_{IG=1}^{NG} W_{IG} W_J(\xi_{IG},\eta_{IG}) u^e(\xi_{IG},\eta_{IG}) \left\{ [J]_{11}^{-1} \frac{\partial N_I}{\partial \xi} + [J]_{12}^{-1} \frac{\partial N_I}{\partial \xi} \right\}\bigg|_{\xi_{IG},\eta_{IG}} \det J(\xi_{IG},\eta_{IG}), \quad (8.188)$$

while Eq. (8.187) becomes

$$I_2^e = \sum_{IG=1}^{NG} W_{IG} \left\{ [J]_{11}^{-1} \frac{\partial N_J}{\partial \xi} + [J]_{12}^{-1} \frac{\partial N_J}{\partial \eta} \right\} \left\{ [J]_{11}^{-1} \frac{\partial N_I}{\partial \xi} + [J]_{12}^{-1} \frac{\partial N_I}{\partial \eta} \right\} \bigg|_{\xi_{IG}, \eta_{IG}} \det J\, (\xi_{IG}, \eta_{IG}).$$

(8.189)

In many codes, the values of $[J^{-1}]_{11}$, $[J^{-1}]_{12}$, $[J^{-1}]_{21}$, $[J^{-1}]_{22}$, and det J are computed at every Gauss point of each element and are stored in an array to be used as needed. The derivatives of $\partial N_J/\partial \xi$, for example, depend only on ξ and η and are the same for all elements (since all elements are the same in the ξ-η coordinate system).

Next we look at the functions x and y. In isoparametric elements, any position can be expressed using the same trial functions as are used for the solution. Thus we take

$$x = \sum x_i^e N_i^e(\xi,\eta) = \sum x_I N_I(\xi,\eta)$$

$$y = \sum y_i^e N_i^e(\xi,\eta) = \sum y_I N_I(\xi,\eta) \quad \text{in e-th element.}$$

(8.190)

Now let us suppose we have an equation that gives the location of each point in the domain in terms of certain parameters. A typical parameter might be the function $\beta(x)$ at the same x location, as shown in Figure 8.16. The function $\beta(x)$ is then characterized by the values of β_i at specific x-locations at the edges of the elements (see Figure 8.18). The locations are given by

$$x = \sum x_i^e(\beta) N_i^e(\xi,\eta) = \sum x_I(\beta) N_I(\xi,\eta)$$

$$y = \sum y_i^e(\beta) N_i^e(\xi,\eta) = \sum y_I(\beta) N_I(\xi,\eta),$$

(8.191)

where $\beta = (\beta_1, \beta_2, ..., \beta_{NB})$ represents the nodal locations defining the boundary. In the equations below we also need the derivative of the equation with respect to the parameters, β.

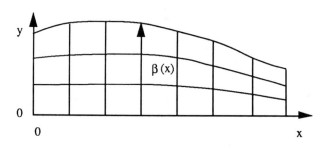

Figure 8.18. Finite Elements on an Irregular Domain

We can differentiate Eq. (8.191), giving

$$\frac{\partial x}{\partial \beta_k} = \sum \frac{\partial x_I}{\partial \beta_k} N_I(\xi,\eta) \quad \text{where} \quad \frac{\partial x_I}{\partial \beta_k} = \frac{\partial x_i^e}{\partial \beta_k} \text{ in e-th element}$$

$$\frac{\partial y}{\partial \beta_k} = \sum \frac{\partial y_I}{\partial \beta_k} N_I(\xi,\eta) \quad \text{where} \quad \frac{\partial y_I}{\partial \beta_k} = \frac{\partial y_i^e}{\partial \beta_k} \text{ in e-th element,} \qquad (8.192)$$

since the N_I values depend only on ξ and η within the element. In addition,

$$\frac{\partial}{\partial \beta_k}\left(\frac{\partial x}{\partial \xi}\right) = \sum \frac{\partial x_I}{\partial \beta_k} \frac{\partial N_I}{\partial \xi}, \quad \frac{\partial}{\partial \beta_k}\left(\frac{\partial x}{\partial \eta}\right) = \sum \frac{\partial x_I}{\partial \beta_k} \frac{\partial N_I}{\partial \eta}$$

$$\frac{\partial}{\partial \beta_k}\left(\frac{\partial y}{\partial \xi}\right) = \sum \frac{\partial y_I}{\partial \beta_k} \frac{\partial N_I}{\partial \xi}, \quad \frac{\partial}{\partial \beta_k}\left(\frac{\partial y}{\partial \eta}\right) = \sum \frac{\partial y_I}{\partial \beta_k} \frac{\partial N_I}{\partial \eta}. \qquad (8.193)$$

Finally,

$$\frac{\partial (\det J)}{\partial \beta_k} = \sum \frac{\partial x_I}{\partial \beta_k} \frac{\partial N_I}{\partial \xi} \frac{\partial y}{\partial \eta} + \frac{\partial x}{\partial \xi} \sum \frac{\partial y_I}{\partial \beta_k} \frac{\partial N_I}{\partial \eta} - \sum \frac{\partial x_I}{\partial \beta_k} \frac{\partial N_I}{\partial \eta} \frac{\partial y}{\partial \xi} - \frac{\partial x}{\partial \eta} \sum \frac{\partial y_I}{\partial \beta_k} \frac{\partial N_I}{\partial \xi}. \qquad (8.194)$$

Suppose we write the problem as

$$F(T,\beta) = 0. \qquad (8.195)$$

T represents the vector of temperature values at the nodes in the finite element mesh. For now, we assume that the β parameters are specified. The Newton-Raphson method applied to Eq. (8.195) gives

$$0 = F(T^n, \beta) + \sum_{k=1}^{NT} \frac{\partial F}{\partial T_k}\bigg|_{T^n} (T_k^{n+1} - T_k^n). \qquad (8.196)$$

We write this using the Jacobian:

$$J^{(n)}(T^{n+1} - T^n) = -F(T^n, \beta), \quad J_{ij}^n = \frac{\partial F_i}{\partial T_j}\bigg|_{T^n}. \qquad (8.197)$$

Eq. (8.197) represents NT simultaneous equations to be solved. Now let us suppose that the β parameters are not specified, but that they are the solutions to a set of equations, written as

$$G(T,\beta) = 0. \qquad (8.198)$$

Eq. (8.198) represents NB equations. Now we apply the Newton-Raphson method to both T and β variables.

$$0 = F(T^n, \beta) + \sum_{k=1}^{NT} \frac{\partial F}{\partial T_k}\bigg|_{T^n,\beta^n} (T_k^{n+1} - T_k^n) + \sum_{k=1}^{NB} \frac{\partial F}{\partial \beta_k}\bigg|_{T^n,\beta^n} (\beta_k^{n+1} - \beta_k^n) \qquad (8.199)$$

$$0 = G(T^n, \beta) + \sum_{k=1}^{NT} \frac{\partial G}{\partial T_k}\bigg|_{T^n,\beta^n}(T_k^{n+1} - T_k^n) + \sum_{k=1}^{NB} \frac{\partial G}{\partial \beta_k}\bigg|_{T^n,\beta^n}(\beta_k^{n+1} - \beta_k^n) \quad (8.200)$$

We need to calculate the derivative of terms such as I_1 and I_2 with respect to β_k. These calculations are done according to work by Saito and Scriven [1981] and Kistler and Scriven [1983]. An integral evaluated on an element can be written in the form

$$I(T, \beta) = \int_\Omega F(x,y,T,\beta)\, dx\, dy \quad (8.201)$$

When the transformation is made to the local coordinate system, this is

$$I(T, \beta) = \int_{-1}^{1}\int_{-1}^{1} F[x(\xi,\eta,\beta),y(\xi,\eta,\beta),T,\beta] \det J\, d\xi\, d\eta. \quad (8.202)$$

Thus, derivatives with respect to β_k are

$$\frac{\partial I}{\partial \beta_k} = \int_{-1}^{1}\int_{-1}^{1} \frac{\partial}{\partial \beta_k}\{F[x(\xi,\eta,\beta),y(\xi,\eta,\beta),T,\beta] \det J\}\, d\xi\, d\eta. \quad (8.203)$$

For calculating the derivatives with respect to β_k, it is convenient to use the second form of both Eq. (8.186) and Eq. (8.187).

$$\frac{\partial I_1^e}{\partial \beta_k} = \int_{-1}^{1}\int_{-1}^{1} W_J\, u^e \left\{ \frac{\partial}{\partial \beta_k}\left(\frac{\partial y}{\partial \eta}\right)\frac{\partial N_I}{\partial \xi} - \frac{\partial}{\partial \beta_k}\left(\frac{\partial y}{\partial \xi}\right)\frac{\partial N_I}{\partial \xi} \right\} d\xi\, d\eta \quad (8.204)$$

$$\begin{aligned}\frac{\partial I_2^e}{\partial \beta_k} =& \int_{-1}^{1}\int_{-1}^{1} \left\{ \frac{\partial}{\partial \beta_k}\left(\frac{\partial y}{\partial \eta}\right)\frac{\partial N_J}{\partial \xi} - \frac{\partial}{\partial \beta_k}\left(\frac{\partial y}{\partial \xi}\right)\frac{\partial N_J}{\partial \eta} \right\}\left\{\frac{\partial y}{\partial \eta}\frac{\partial N_I}{\partial \xi} - \frac{\partial y}{\partial \xi}\frac{\partial N_I}{\partial \eta}\right\}\frac{1}{\det J} d\xi\, d\eta \\ &+ \int_{-1}^{1}\int_{-1}^{1} \left\{ \frac{\partial y}{\partial \eta}\frac{\partial N_J}{\partial \xi} - \frac{\partial y}{\partial \xi}\frac{\partial N_J}{\partial \eta} \right\}\left\{ \frac{\partial}{\partial \beta_k}\left(\frac{\partial y}{\partial \eta}\right)\frac{\partial N_I}{\partial \xi} - \frac{\partial}{\partial \beta_k}\left(\frac{\partial y}{\partial \xi}\right)\frac{\partial N_I}{\partial \eta}\right\} \frac{1}{\det J} d\xi\, d\eta \\ &+ \int_{-1}^{1}\int_{-1}^{1} \left\{ \frac{\partial y}{\partial \eta}\frac{\partial N_J}{\partial \xi} - \frac{\partial y}{\partial \xi}\frac{\partial N_J}{\partial \eta} \right\}\left\{ \frac{\partial y}{\partial \eta}\frac{\partial N_I}{\partial \xi} - \frac{\partial y}{\partial \xi}\frac{\partial N_I}{\partial \eta}\right\} \frac{\partial}{\partial \beta_k}\left(\frac{1}{\det J}\right) d\xi\, d\eta\end{aligned} \quad (8.205)$$

The various derivatives are found in Eq. (8.192)-(8.194) along with

$$\frac{\partial}{\partial \beta_k}\left(\frac{1}{\det J}\right) = -\frac{1}{(\det J)^2}\frac{\partial}{\partial \beta_k}(\det J). \quad (8.206)$$

These derivatives also need to be evaluated at each Gauss point for every element.
To show how these equations can be solved, we organize them in the following format:

$$\begin{bmatrix} A & B \\ C & D \end{bmatrix}\begin{bmatrix} u \\ v \end{bmatrix} = \begin{bmatrix} f \\ g \end{bmatrix}. \quad (8.207)$$

where

$$A = J^{(n)} = \frac{\partial F}{\partial T}, \ B = \frac{\partial F}{\partial \beta}, \ C = \frac{\partial G}{\partial T}, \ D = \frac{\partial G}{\partial \beta}$$
$$u = T^{n+1} - T^n, \ v = \beta^{n+1} - \beta^n, \ f = -F, \ g = -G.$$
(8.208)

The matrix **A** is an NT by NT matrix, **B** is a NT by NB matrix, **C** is a NB by NT matrix, and **D** is a NB by NB matrix. The vectors **u** and **f** have NT entries, while the vectors **v** and **g** have NB entries. We note that

$$\mathbf{A u = f} \tag{8.209}$$

is the fixed mesh problem, with β specified. The form of Eq. (8.207) is called an arrow matrix, since the non-zero values take the pattern shown in Figure 8.19. The matrix **A** has a significant number of zeros in it, whereas all the other matrices are dense; in other words, most entries are non-zero. There is an efficient way to solve such problems when the size of the bands is small (i.e., NB<<NT).

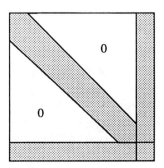

Figure 8.19. Arrow Matrix

We can solve Eq. (8.209) in several steps, as follows. The numbers above the matrix represent the dimensions of the matrix. First we solve the two problems

$$\begin{array}{ccc} \text{NTxNT} & \text{NTx1} & \text{NTx1} \\ \mathbf{A} & \mathbf{x} = & \mathbf{f} \end{array} \tag{8.210}$$

$$\begin{array}{ccc} \text{NTxNT} & \text{NTxNB} & \text{NTxNB} \\ \mathbf{A} & \mathbf{y} = & \mathbf{B}. \end{array} \tag{8.211}$$

Since the matrix **A** is the same in both equations, one LU decomposition of **A** is all that is necessary. The matrix **A** is the one with a significant number of zeros in it; presumably the LU decomposition is done using an efficient method for a matrix with such a structure. Next we solve the dense matrix problem

$$\underset{NBxNB}{(D} - \underset{NBxNT}{C} \underset{NTxNB}{y}) \underset{NBx1}{v} = - \underset{NBxNT}{C} \underset{NTx1}{x} + \underset{NBx1}{g} \tag{8.212}$$

and evaluate

$$\underset{NTx1}{u} = \underset{NTx1}{x} - \underset{NTxNB}{y} \underset{NBx1}{v}. \tag{8.213}$$

To verify the solution, we calculate the terms in Eq. (8.207) as

$$\begin{aligned} A u + B v &= A(x - y v) + B v = A x - (A y) v + B v \\ &= f - B v + B v = f. \end{aligned} \tag{8.214}$$

From Eq. (8.212) we have

$$D v = C y v - C x + g. \tag{8.215}$$

From Eq. (8.213) we have

$$C u = C x - C y v. \tag{8.216}$$

Then

$$C u + D v = g. \tag{8.217}$$

Thus, the correct equations are solved.

This verifies the solution of the arrow matrix problem. This solution method allows the original problem to be solved without change, Eq. (8.210). Then a fore-aft sweep is performed [Eq. (8.211)] and a smaller dense matrix problem is solved [Eq. (8.212)]. The additional details required to handle the spine representation of the mesh are thus carried out separately without changing the original problem. This feature is convenient when adding a spine representation to an existing, verified computer code.

8.8. Natural Boundary Conditions

One of the advantageous features of finite element methods is the ability to easily use natural boundary conditions. These conditions have been used throughout the book in one-dimensional problems by using one-sided difference expressions at the boundary (i.e., processing only the element to the inside of the boundary point, not the missing element outside the boundary point). This was useful for outflow conditions. To see the impact of natural boundary conditions in two-dimensional problems, let us consider the following heat transfer problem:

$$k \nabla^2 T = Q \tag{8.218}$$

$$\begin{aligned} T &= T_1 \text{ on } \partial \Omega_1 \\ -k \, \mathbf{n} \cdot \nabla T &= q_2 \text{ on } \partial \Omega_2 \\ -k \, \mathbf{n} \cdot \nabla T &= h (T - T_3) \text{ on } \partial \Omega_3. \end{aligned} \tag{8.219}$$

On Region 1 of the boundary, the temperature is a specified function; this is called an essential boundary condition in the calculus of variations. On Region 2 or 3 the boundary conditions are called natural boundary conditions in the calculus of variations. One rule of thumb is that if the differential equation involves second-order derivatives, then any boundary condition involving first-order derivatives is natural while any condition involving no derivatives is essential. Another rule of thumb is that the natural boundary conditions can be combined with the differential equation, in the manner illustrated below. These rules come from the calculus of variations, where the variational principle determines whether the boundary conditions are essential or natural. Essential boundary conditions are ones that must be satisfied by the trial function because the variational principle does not force them to be satisfied. By contrast, a variational principle will force natural boundary conditions to be satisfied as the number of unknowns is increased. When no variational principle exists, the Galerkin test can still be used: if the residual is zero, will the Galerkin criterion force the boundary conditions to be satisfied? If so, they are natural.

We look at the boundary conditions, Eq. (8.219), and decide that Region 1 is an essential boundary condition; the trial function for temperature must satisfy this condition. The other regions involve natural boundary conditions so the trial function need not satisfy these conditions. Instead, we take the Galerkin integral and combine it with a weighted boundary residual. Applying the Galerkin method to the differential equation gives

$$\int_\Omega k W_j \nabla^2 T \, dx \, dy - \int_\Omega W_j Q \, dx \, dy = 0$$
$$\int_\Omega k W_j \nabla \cdot \nabla T \, dx \, dy - \int_\Omega W_j Q \, dx \, dy = 0. \qquad (8.220)$$

Applying the Galerkin method to the natural boundary conditions gives

$$-\int_{\partial\Omega_2} \mathbf{n} \cdot (k W_j \nabla T) \, ds = \int_{\partial\Omega_2} W_j q_2 \, ds$$
$$-\int_{\partial\Omega_3} \mathbf{n} \cdot (k W_j \nabla T) \, ds = \int_{\partial\Omega_3} W_j h (T - T_3) \, ds. \qquad (8.221)$$

Rather than satisfying each of these separately (we cannot do this anyway since we do not have enough unknowns), let us combine them. We integrate by parts the term with the Laplacian,

$$\int_\Omega k W_j \nabla \cdot \nabla T \, dx \, dy = \int_\Omega \nabla \cdot (k W_j \nabla T) \, dx \, dy - \int_\Omega k \nabla T \cdot \nabla W_j \, dx \, dy. \qquad (8.222)$$

Then we use the divergence theorem to give

$$\int_\Omega kW_j \nabla \cdot \nabla T \, dx \, dy = \int_{\partial\Omega} \mathbf{n} \cdot (k\, W_j \nabla T) \, ds - \int_\Omega k \nabla T \cdot \nabla W_j \, dx \, dy. \qquad (8.223)$$

The boundary integral is evaluated over each of the Regions 1, 2, and 3.

$$\int_\Omega kW_j \nabla \cdot \nabla T \, dx \, dy = \int_{\partial\Omega_1} \mathbf{n} \cdot (k\, W_j \nabla T) \, ds + \int_{\partial\Omega_2} \mathbf{n} \cdot (k\, W_j \nabla T) \, ds +$$
$$\int_{\partial\Omega_3} \mathbf{n} \cdot (k\, W_j \nabla T) \, ds - \int_\Omega k \nabla T \cdot \nabla W_j \, dx \, dy. \qquad (8.224)$$

If the node j is on Region 1, we use the boundary condition $T_j = T_1$ and do not write Eq. (8.224). For all other nodes, $W_j = 0$ in Region 1. Thus the boundary integral on Region 1 vanishes. We next insert the weighted natural boundary conditions, Eq. (8.221), to give

$$\int_\Omega kW_j \nabla \cdot \nabla T \, dx \, dy = -\int_{\partial\Omega_2} W_j\, q_2 \, ds +$$
$$-\int_{\partial\Omega_3} W_j\, h\, (T - T_3) \, ds - \int_\Omega k \nabla T \cdot \nabla W_j \, dx \, dy. \qquad (8.225)$$

The Galerkin finite element equations are then

$$\int_\Omega k \nabla T \cdot \nabla W_j \, dx \, dy + \int_{\partial\Omega_3} W_j\, h\, T \, ds =$$
$$= -\int_\Omega W_j\, Q \, dx \, dy - \int_{\partial\Omega_2} W_j\, q_2 \, ds + \int_{\partial\Omega_3} W_j\, T_3 \, ds. \qquad (8.226)$$

In essence, for an element near the boundary we replace the integrated residual with one that includes the integrated boundary conditions.

What is the advantage of this treatment? When we solve a problem and have an outflow boundary, or a boundary that is at infinity, we say there are no applied heat fluxes or applied forces at that boundary (because we cannot enforce them physically). Thus we set $q_2 = 0$ in Eq. (8.225) rather than applying any boundary condition. We can truncate the domain, but the result is still good. The reason is that a finite element method uses integrals evaluated on elements. A node in the center of the domain is determined by equations derived from elements surrounding it. A node on the boundary, however, is determined by equations derived partly from the differential equations (inside the domain) and partly from the boundary conditions. The Galerkin finite element integral correctly accounts for these influences.

To see how this happens, we take a Graetz problem, which is heat transfer to a fluid flowing steadily down a tube. The fluid enters at temperature zero, but the walls of the tube are maintained at temperature 1.0. The flow is fully developed throughout. We wish to solve for the temperature distribution in the

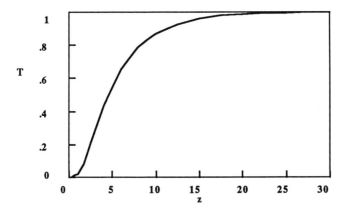

Figure 8.20a. Center Temperature for the Graetz Problem, Pe = 60

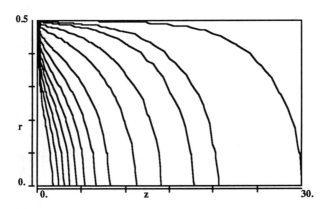

Figure 8.20b. Temperature Contours for the Graetz Problem, Pe = 60

fluid inside the tube. The equations are

$$\text{Pe}\, u(r) \frac{\partial T}{\partial z} = \frac{1}{r}\frac{\partial}{\partial r}\left(r\frac{\partial T}{\partial r}\right) + \frac{\partial^2 T}{\partial z^2}, \quad u = 2(1-r^2) \qquad (8.227)$$

and the boundary conditions are

$$T(r, 0) = 0, \; T(1, z) = 1, \; \frac{\partial T}{\partial r} = 0 \text{ at } r = 0. \qquad (8.228)$$

The expected solution when Pe = 60 is illustrated in Figure 8.20.

We will solve this Graetz problem four times, on the three meshes shown in Figure 8.21. In all four cases we will use the following boundary conditions:

$$T(r, 0) = 0, \; T(1, z) = 1, \; q_2(0, z) = 0. \qquad (8.229)$$

304 NUMERICAL METHODS - MOVING FRONTS

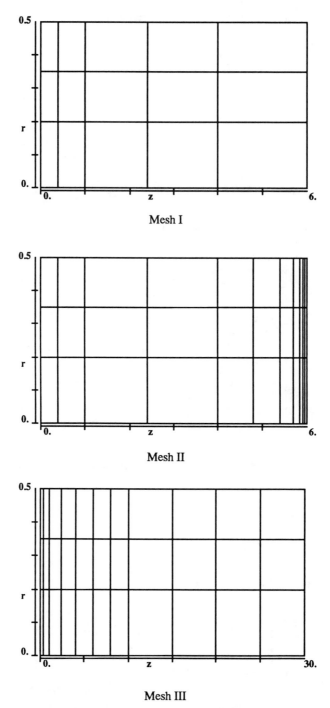

Figure 8.21. Meshes for the Graetz Problem

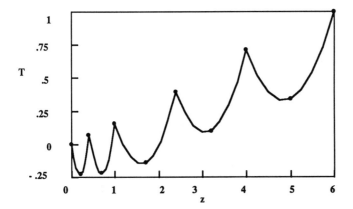

Figure 8.22a. Center Temperature Solved on Mesh I with Eq. (8.230)

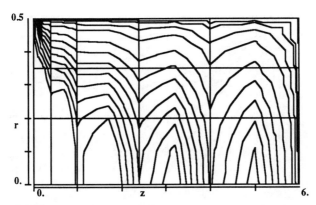

Figure 8.22b. Temperature Contours Solved on Mesh I with Eq. (8.230)

Along the centerline we use the natural boundary condition: we take q = 0 there. For the first solution, we solve on Mesh I under the following boundary conditions:

$$T(z = 6, r) = 1. \tag{8.230}$$

The solution is highly oscillatory, as shown in Figure 8.22. To see why this happens, we re-solve the problem under the same boundary condtions, but using Mesh II. The solution has a sharp boundary layer near the exit (see Figure 8.23). We know that the boundary layer requires a small Δz value to avoid oscillations; the solutions on Meshes I and II confirm that. Of course, the real problem does not have this boundary layer. The boundary layer developed because we specified the temperature to be 1.0 at the exit and we are physically unable to enforce that condition. If we replace this boundary condition by the natural condition

$$q = 0 \text{ at the outflow, } z = 6 \tag{8.231}$$

and solve on Mesh I again, we get the solution shown in Figure 8.24. This is clearly a good solution without oscillation. The temperature derivative normal

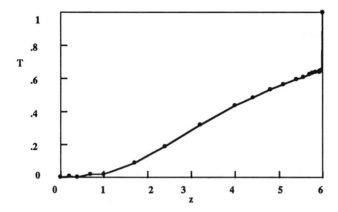

Figure 8.23a. Center Temperature Solved on Mesh II with Eq. (8.230)

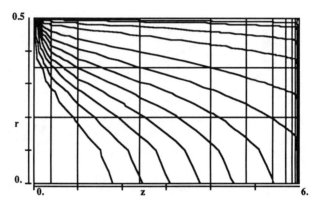

Figure 8.23b. Temperature Contours Solved on Mesh II with Eq. (8.230)

to the exit boundary is not zero, which means that the natural boundary condition is not satisfied pointwise. To see that this is the correct solution, we solve the problem again on Mesh III. This time the mesh is longer; indeed, it is long enough that the temperature should approach 1.0 at z = 30. If we solve it with the boundary condition

$$T(z = 30, r) = 1, \qquad (8.232)$$

we get the solution shown in Figure 8.25. If we use the natural boundary condition

$$q = 0 \text{ at the outflow, } z = 30, \qquad (8.233)$$

we get essentially the same solution (it is indistinguishable on the scale of the graph). Furthermore, if we compare the solution for the long domain in the region z = 0 to z = 6, with the solution in the short domain (z = 0 to z = 6) for the case with natural boundary conditions, we find the two solutions are virtually identical. These solutions are plotted on the same graph in Figure 8.26 and are indistinguishable. Thus we have proved that we can solve the problem on the shorter domain, Mesh I, as long as the outlet boundary condition is the natural one,

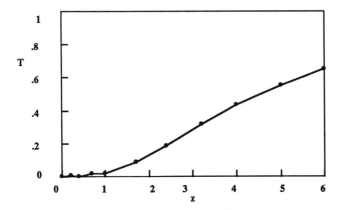

Figure 8.24a. Center Temperature Solved on Mesh I with Eq. (8.231)

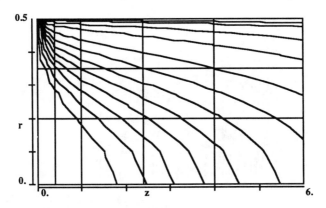

Figure 8.24b. Temperature Contours Solved on Mesh I with Eq. (8.231)

Eq. (8.231).

Let us contrast this solution method with that for a finite difference method. The boundary conditions at the inlet or on the wall or the centerline are standard. At the exit, the condition would be $\partial T/\partial z = 0$ at some length L. Since this means that the temperature profile is fully developed by $z = L$, we must use a grid long enough for that to be true. Finite element methods give us the option of using a shorter domain and still obtaining the correct solution within that domain.

8.9. Group Velocity

Propagation effects in two dimensions are best explained using the concept of group velocity. A summary of the treatment by Trefethen [1982] is presented here, which first considers one-dimensional problems and then two-dimensional problems. Group velocity is used when there is dispersion (that is,

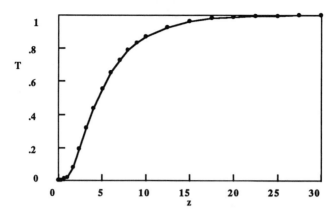

Figure 8.25a. Center Temperature Solved on Mesh III with Eq. (8.232)

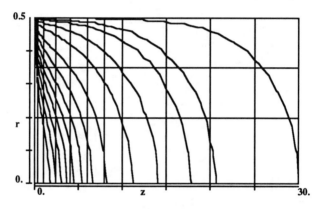

Figure 8.25b. Temperature Contours Solved on Mesh III with Eq. (8.232)

a change of phase) but no dissipation (or attenuation). Thus, we do the analysis for only the advection equation and a few numerical methods. We see that a wave packet moves with the group velocity rather than the phase speed.

Let us consider the one-dimensional advection equation,

$$\frac{\partial u}{\partial t} + \frac{\partial u}{\partial x} = 0. \tag{8.234}$$

We take a solution in the form

$$u(x,t) = e^{i(\omega t - \xi x)}. \tag{8.235}$$

(The sign of ξ is opposite to that used in Chapter 4.) For every ξ, there is a ω such that u is a solution. Substituting Eq. (8.235) into Eq. (8.234) yields

$$i\omega\, e^{i(\omega t - \xi x)} - i\xi\, e^{i(\omega t - \xi x)} = 0. \tag{8.236}$$

Thus if the frequency, ω, satisfies

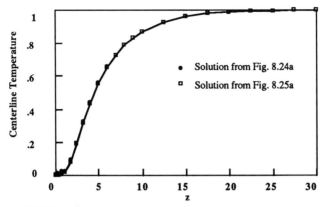

Figure 8.26a. Center Temperature, Plots of Figures 8.24a and 8.25a on the Scale of Fig. 8.25a

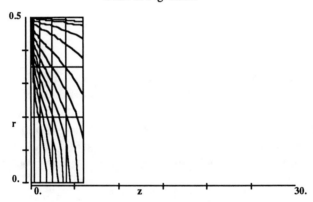

Figure 8.26b. Temperature Contours, Plot of Figure 8.24b on the Same Scale as Figure 8.25b

$$\omega = \xi, \tag{8.237}$$

then Eq. (8.235) is a solution. An equation relating ω and ξ is called the dispersion relation. If the Peclet number is set to 1.0 in Eq. (4.13), it differs from Eq. (8.237) only in the different sign. If

$$\omega t = \xi x, \text{ then } x = \frac{\omega}{\xi} t \tag{8.238}$$

and the velocity is

$$\text{velocity} = \frac{\omega}{\xi} = c(\xi) = \text{phase speed}, \tag{8.239}$$

The function $c(\xi)$ is the phase speed.

Next let us consider the following interpretation by Trefethen [1982]. We let an initial distribution [f(x,0)] have the Fourier transform $\hat{f}(\xi)$. Since there is no dissipation, the solution at a later time obeys

$$f(x,t) = \int_{-\infty}^{\infty} \hat{f}(\xi) \, e^{i(\omega(\xi)t - \xi x)} \, d\xi = \int_{-\infty}^{\infty} \hat{f}(\xi) \, e^{it[\omega(\xi) - \xi x/t]} \, d\xi. \tag{8.240}$$

Suppose we view the solution such that the ratio x/t is held fixed (i.e., we are in a moving reference frame). Then as $t \to \infty$, the exponential oscillates more and more rapidly with ξ. Thus the integral tends towards zero, except in regions where

$$\frac{d}{d\xi}\left[\omega(\xi) - \frac{\xi x}{t}\right] = 0, \text{ i.e. } \frac{d\omega}{d\xi} = \frac{x}{t}. \tag{8.241}$$

Therefore, we define the group velocity as

$$C(\xi) = \frac{d\omega}{d\xi}(\xi). \tag{8.242}$$

For the advection equation, Eq. (8.234), the dispersion relation gives

$$\omega(\xi) = \xi, \quad c(\xi) = 1, \quad C(\xi) = 1. \tag{8.243}$$

Let us suppose we use a numerical method to solve Eq. (8.234). Then we have a group velocity (and phase velocity) for the numerical method that may be different from those of the exact solution. Trevethen [1982] analyzes the second-order leapfrog method, the implicit trapezoid rule, and a fourth-order leapfrog method. Let us consider here the implicit trapezoid rule. Using centered differences for the convection term and a trapezoid rule in time gives us

$$u_i^{n+1} - u_i^n = -\frac{\Delta t}{2 \Delta x} [u_{i+1}^{n+1} - u_{i-1}^{n+1} + u_{i+1}^n - u_{i-1}^n]. \tag{8.244}$$

Substituting in

$$u(x,t) = e^{i(\omega t - \xi x)}, \quad t = n \Delta t, \, x = j \Delta x \tag{8.245}$$

gives

$$e^{i\omega\Delta t} - 1 = \frac{2\Delta t}{\Delta x} [e^{-i\xi\Delta x} - e^{i\xi\Delta x}][e^{i\omega\Delta t} + 1]$$

Real part: $\cos \omega \Delta t - 1 = -\frac{\Delta t}{\Delta x} [\sin \xi \Delta x][\sin \omega \Delta t]$

$$\tag{8.246}$$

Imaginary part: $\sin \omega \Delta t = \frac{\Delta t}{\Delta x} [\sin \xi \Delta x][\cos \omega \Delta t + 1].$

Apparently the group velocity comes from the imaginary portion of this equation, since that affects the phase; also, the only schemes treated are conservative ones with no dissipation. Thus we have

$$\sin \omega \Delta t = \frac{\Delta t}{\Delta x} [\sin \xi \Delta x][\cos \omega \Delta t + 1]. \tag{8.247}$$

Using the trignometric identities

$$\frac{\cos 2\theta + 1}{2} = \cos^2 \theta, \quad \sin \theta = 2 \sin \frac{\theta}{2} \cos \frac{\theta}{2} \tag{8.248}$$

this can be written as

$$2 \sin \frac{\omega \Delta t}{2} \cos \frac{\omega \Delta t}{2} = \frac{\Delta t}{\Delta x}[\sin \xi \Delta x] \left[2 \cos^2 \frac{\omega \Delta t}{2} \right]$$

$$2 \tan \frac{\omega \Delta t}{2} = \frac{\Delta t}{\Delta x} \sin \xi \Delta x. \tag{8.249}$$

Eq. (8.249) provides the dispersion relation for this numerical method. For small values of ξ, we get

$$\omega = \xi. \tag{8.250}$$

For other values of ξ, however, the relation is nonlinear and is even multivalued for some methods. Shown in Figure 8.27 is a sketch of the solution. The slope on this diagram, $d\omega/d\xi$, is the group velocity and ω/ξ is the phase speed. The results for the leapfrog method are (see Figure 8.27)

$$\sin \omega \Delta t = \frac{\Delta t}{\Delta x} \sin \xi \Delta x. \tag{8.251}$$

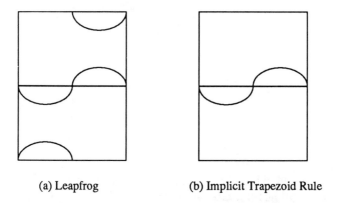

(a) Leapfrog (b) Implicit Trapezoid Rule

Figure 8.27. Dispersion Relations for Numerical Methods, Co = 0.5 (shown is a plot in ξ–η space, with ξ from $-\pi/\Delta x$ to $\pi/\Delta x$ and η from $-\pi/\Delta t$ to $\pi/\Delta t$) Data from Trefethen [1982]

Trefethen [1982] does a calculation in the region $0 \leq x \leq 3$ using $\Delta x = 1/160$ and Co = 0.4. In that case,

$$\frac{\text{Pe } \Delta x}{2} = \frac{1}{2} \frac{1}{160} = 0.003125. \tag{8.252}$$

An initial condition is taken as

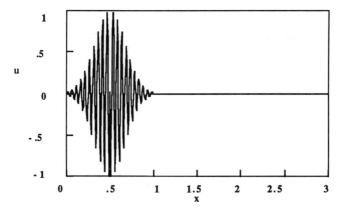

Figure 8.28a. Wave Packet at t = 0

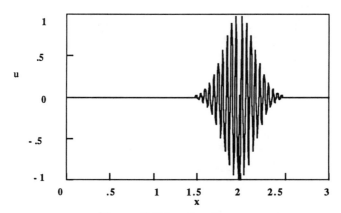

Figure 8.28b. Wave Packet at t = 2
Data from Trefethen [1982]

$$u(x,0) = e^{-16(x-0.5)^2} \sin \xi x \qquad (8.253)$$

with $\xi \Delta x = 2\pi/8 = 0.79$. Thus, the initial condition was a wave packet centered about $x = 0.5$ that had an oscillating profile (see Figure 8.28). The calculation proceeded from $t = 0$ to $t = 2$ in Trefethen's example using the leapfrog method. During the time from $t = 0$ to $t = 2$, the wave packet moved from $x = 0.5$ to 1.97, for a velocity of 0.74. The phase speed is 0.91 and the group velocity is 0.74. Thus the wave packet definitely moved with the velocity of the group velocity rather than the phase speed.

For two-dimensional problems, we have a wave number that is a vector; the dispersion relation is

$$\omega = \omega(\xi), \quad \xi = (\xi, \eta). \qquad (8.254)$$

The group velocity is then

Figure 8.29. Dispersion Plot for the Leapfrog method

$$C = e_x \frac{\partial \omega}{\partial \xi} + e_y \frac{\partial \omega}{\partial \eta}. \tag{8.255}$$

An example in Trefethen [1982] is for the second-order wave equation

$$\frac{\partial^2 u}{\partial t^2} = \frac{\partial^2 u}{\partial x^2} + \frac{\partial^2 u}{\partial y^2} \tag{8.256}$$

and the leapfrog method. The dispersion relation for the wave equation is

$$\omega^2 = \xi^2 + \eta^2, \tag{8.257}$$

but for the leapfrog method it is

$$\sin^2 \frac{\omega \Delta t}{2} = \left(\frac{\Delta t}{\Delta x}\right)^2 \left[\sin^2 \frac{\xi \Delta x}{2} + \sin^2 \frac{\eta \Delta y}{2} \right]. \tag{8.258}$$

Contour lines of constant ω in the ξ–η space should be a circle, according to Eq. (8.257). For the leapfrog method, they are as shown in Figure 8.29. The group velocity has the components

$$C_x = \frac{\Delta t}{\Delta x} \frac{\sin \xi \Delta x}{\sin \omega \Delta t}, \quad C_y = \frac{\Delta t}{\Delta x} \frac{\sin \eta \Delta y}{\sin \omega \Delta t}, \quad \Delta x = \Delta y. \tag{8.259}$$

The waves travel at a speed that is slower than the correct one, but the velocity is not the same in all directions. In particular, it is slower (by a factor of 2) in the direction $\theta = 0$ than in the direction $\theta = 45$. The net effect of this is that the numerical method propagates errors in different directions with different velocities, so that there are additional complications in two dimensions that are not present in one dimension.

An extension to finite element methods is given by Vichnevetsky

Problems

1_2. Prove the positivity rule. When

$$c_{i,j}^{n+1} = a_1 c_{i+1,j}^n + a_2 c_{i,j}^n + a_3 c_{i-1,j}^n + \ldots \qquad (8.260)$$

and

$$a_i \geq 0, \quad \sum a_i = 1, \qquad (8.261)$$

then

$$\max_{i,j} |c_{i,j}^{n+1}| \leq \max_{i,j} |c_{i,j}^n|. \qquad (8.262)$$

[Hint: Take absolute values of both sides of the equation and use the fact that a sum of absolute values is greater than the absolute value of the sum (see p. 216, Finlayson, [1980]).]

2_2. Derive the truncation error of Eq. (8.25).

3_1. List the conditions under which the Petrov-Galerkin method is the same as the Taylor-Galerkin method for a two-dimensional convective diffusion equation.

4_2. Verify Eq. (8.94).

5_2. Consider Figure 8.9. Suppose you knew the solutions at Points 1 through 5 and needed to refine the mesh in the upper-left hand corner by turning it into four rectangles (like the one in the upper right-hand corner). What values would the solution take at the new grid points, in terms of the solution at the Points 1 through 5?

6_3. Prove that Eq. (8.155) is the same as Eq. (8.154). [Hint: Write $\hat{n} = n_x \mathbf{e}_x + n_y \mathbf{e}_y$. Use $\hat{s} = \mathbf{e}_z \times \hat{n}$. Compute $\hat{n} \cdot \nabla\ F$ and $\hat{s} \cdot \nabla\ F$.]

7_1. Verify Eq. (8.173), Eq. (8.175), and Eq. (8.176).

8_1. Verify that Eq. (8.185) gives the inverse of [J]. Do so by calculation to see whether or not $[J][J]^{-1} = I$.

9_2. Verify Eq. (8.249). Rationalize Figure 8.27(b).

10_3. Verify Eq. (8.251). Rationalize Figure 8.27(a).

References

Akin, J. E., *Finite Element Analysis for Undergraduates,* Academic Press, New York [1986].

Alexander, R., Manselli, P. and Miller, K., "Moving Finite Elements for the Stefan Problem in Two Dimensions," Rendicont. Acc. Naz. Lincei, Rome, [1980].

Anderson, D. A., Tannehill, J. C., Pletcher, R. H., *Computational Fluid Mechanics and Heat Transfer,* McGraw-Hill [1984].

Ang, K. K. and Valliappan, S., "Mesh Grading Technique Using Modified Isoparametric Shape Functions and Its Application to Wave Propagation Problems," Int. J. Num. Methods Engn 23 331-348 [1986]..

Babuska, I. and Miller, A.,"The Post-Processing Approach in the Finite Element Method-Part 1: Calculation of Displacements, Stresses and other Higher Derivatives of the Displacements," Int. J. Num. Methods Engn 20 1085-1109 [1984a].

Babuska, I. and Miller, A.," The Post-Processing Approach in the Finite Element Method-Part 2:, The Calculation of Stress Intensity Factors" Int. J. Num. Methods Engn 20 1111-1129 [1984b].

Babuska, I. and Miller, A.," The Post-Processing Approach in the Finite Element Method-Part 3: *A Posteriori* Error Estimates and Adaptive Mesh Selection," Int. J. Num. Methods Engn 20 2311-2324 [1984c].

Becker, E. B, Carey, G. F., and Oden, J. T., *Finite Elements: An Introduction*, Prentice-Hall [1981].

Berger, M. J. and Oliger, J., "Adaptive Mesh Refinement for Hyperbolic Partial Differential Equations," J. Comp. Phys. 53 484-512 [1984].

Book, D. L. (ed.), *Finite-Difference Techniques for Vectorized Fluid Dynamics Calculations*, Springer-Verlag [1981].

Chern, I. L., Glimm, J., McBryan, O., Plohr, B. and Yaniv, S., "Front Tracking for Gas Dynamics," *J. Comp. Phys.* 62 83-110 [1986].

Courant, R. and Friedrichs, K., *Supersonic Flow and Shock Waves*, Interscience, New York [1948].

Demkowicz, L. and Oden, J. T., "An Adaptive Characteristic Petrov-Galerkin Finite Element Method for Convection-dominated Linear and Nonlinear Parabolic Problems in Two Space Variables," Comp. Meth. Appl. Mech. Eng. 55 63-87 [1986].

Dukowicz, J. K. and Ramshaw, J. D., "Tensor Viscosity Method for Convection in Numerical Fluid Dynamics," J. Comp. Phys. 32 71-79 [1979].

Eisenstat, S. C., "Efficient Implementation of a Class of Preconditioned Conjugate Gradient Methods," SIAM J. Sci. Stat. Comput. 2 1-4 [1981].

Finlayson, B. A., *Nonlinear Analysis in Chemical Enginering*, McGraw-Hill (1980).

Frey, W. H., "Selective Refinement: A New Strategy for Automatic Node Placement in Graded Triangular Meshes," Int. J. Num. Method. Engn. 24 2183-2200 [1987].

Glimm, J., Marchesin, D. and McBryan, O., "Subgrid Resolution of Fluid Discontinuities II," *J. Comp. Phys.* 37 336-354 [1980].

Glimm, J., Isaacson, E., Marchesin, D., and McBryan, O., "Front Tracking for Hyperbolic Systems," *Adv. Appl. Math.* 2 91-119 [1981].

Glimm, J. and McBryan, O., "A Computational Model for Interfaces," DOE/ER/03077-265 [1985].

Glimm, J., Klingenberg, C, McBryan, O., Plohr, B., Sharp, D., and Yaniv, S., "Front Tracking and Two Dimensional Riemann Problems," *Adv. Appl. Math.* 6 259-290 [1985].

Glimm, J., McBryan, O., Menikoff, R. and Sharp, D. H., "Front Tracking Applied to Rayleigh-Taylor Instability," SIAM J. Sci. Stat. Comp. 7 230-251 [1986].

Glimm, J., Lindquist, B., McBryan, O. A., and Tryggvason, G., "Sharp and Diffuse Fronts in Oil Reservoirs: Front Tracking and Capillarity," *Mathematical and Computational Methods in Seismic Exploration and Reservoir Modeling*, Fitzgibbon, W. E. (ed.) 68-84, SIAM, Philadelphia [1987].

Glimm, J., Grove, J., Lindquist, B., McBryan, O. A., and Tryggvason, G., "The Bifurcation of Tracked Scalar Waves," *SIAM J. Sci. Stat. Comp.* 9 61-79 [1988].

Gropp, W. D., "A Test of Moving Mesh Refinement for 2-D Scalar Hyperbolic Problems," SIAM J. Sci. Stat. Comp. 1 191-197 [1980].

Hestenes, M. R., *Conjugate Gradient Methods in Optimization*, Springer-Verlag [1980].

Hughes, T. J. R. and Brooks, A., "A Multi-dimensional Upwind Scheme with no Crosswind Diffusion," pp. 19-35 in *Finite Element Methods for Convection Dominated Flows*, AMD, Vol. 34, ASME [1979].

Kikuchi, N., "Adaptive Grid-Design Methods for Finite Element Analysis," Comp. Meth. Appl. Mech. Engn. 55 129-160 [1986].

Kistler, S. F. and Scriven, L. E., "Coating Flows," pp. 243-299 in *Computational Analysis of Polymer Processing* (ed. J. R. A. Pearson, S. M. Richardson), Appl. Sci. Publ., London [1983].

Löhner, R., Morgan, K. and Zienkiewicz, O. C., "An Adaptive Finite Element Procedure for Compressible High Speed Flows," Comp. Methods Appl. Mech. Eng. 51 441-465 [1985].

Löhner, R., Morgan, K., Peraire, J. and Vahdati, M., "Finite Element Flux-corrected Transport (FEM-FCT) for the Euler and Navier-Stokes Equations," Int. J. Num. Methods Fluids 7 1093-1109 [1987].

McIntosh, A., *Fitting Linear Models: an Application of Conjugate Gradient Algorithms*,, Springer-Verlag, New York [1982].

Neuman, S. P., "Adaptive Eulerian-Lagrangian Finite Element Method for Advection-Dispersion," Int. J. Num. Methods Eng. 20 321-337 [1984].

Oden, T., "Lecture at Seventh Int. Conf. Finite Elements Fluids,", Huntsville, Al. [1989].

Oran, E. S. and Boris, J. P., *Numerical Simulation of Reactive Flow*, Elsevier, New York [1987].

Parrott, A. K. and Christie, M. A., "FCT Applied to the 2-D Finite Element Solution of Tracer Transport by Single Phase Flow in a Porous Medium," pp. 609-619 in *Numerical Methods for*

Fluid Dynamics, Oxford University Press, (K. W. Morton and M. J. Baines, eds.) [1986].

Peaceman, D. W., *Fundamentals of Numerical Reservoir Simulation*, Elsevier (1977).

Peraire, J., Vahdati, M., Morgan, K. and Zienkiewicz, O. C., "Adaptive Remeshing for Compressible Flow Computation," J. Comp. Phys. 72 449-466 [1987].

Peyret, R. and Taylor, T. D., *Computational Methods for Fluid Flow*, Springer-Verlag [1983].

Ramanathan, S. and Kumar, R., "Comparison of Boundary-fitted Coordinates with Finite-Element Approach for Solution of Conduction Problems," Num. Heat Transfer 14 187-211 [1988].

Roe, P. L., "Discontinuous Soilutions to Hyperbolic Systems Under Operator Splitting", Num. Methods Part. Diff. Eqn. 7 277-297 [1991].

Saito, H. and Scriven, L. E., "Study of Coating Flow by the Finite Element Method," J. Comp. Phys. 42 53-76 [1981].

Smooke, M. D. and Koszykowski, M. L., "Two-dimensional Fully Adaptive Solutions of Solid-Solid Alloying Reactions," Sandia Report SAND83-8909 [Feb.,1984].

Trefethen, L. N., "Group Velocity in Finite Difference Schemes," SIAM Review 24 113-136 [1982].

Thommen, H. U., "Numerical Integration of the Navier-Stokes Equations," *ZAMP* 17 369-384 [1966].

Thompson, J. F., Thames, F. C., Mastin, C. W., "Automatic Numerical Generation of Body-Fitted Curvilinear Coordinate System for Field Containing Any Number of Arbitrary Two-Dimensional Bodies," J. Comp. Phys. 15 299-319 [1974].

Vichnevetsky, R. and Bowles, J. B., *Fourier Analysis of Numerical Approximations of Hyperbolic Equations*, SIAM, Philadelphia [1982].

Vichnevetsky, R., "Propagation and Spurious Reflection in Finite-Element Approximations of Hyperbolic Equations,", Comp. & Maths with Appls. 11 733-746 [1985].

Young, L. C. and Finlayson, B. A., "Mathematical Models of the Monolith Catalytic Converter. Part I. Development of Model and Application of Orthogonal Collocation," A. I. Ch. E. J. 22 331-343 [1976].

Zalesak, S. T., "Fully Multidimensional Flux-Corrected Transport Algorithms for Fluids," J. Comp. Phys. 31 335-362 [1979].

Zhu, J. Z. and Zienkiewicz, O. C."Adaptive Techniques in the Finite Element Method," Comm. Appl. Num. Meth. 4 197-204 [1988].

CHAPTER
NINE

THE CONVECTIVE DIFFUSION EQUATION WITH ADSORPTION

In this chapter we will extend the convective diffusion equation to include adsorption, thus adding a second phenomenon to the problem. For some chromatography problems and pressure swing adsorption problems the changes in initial conditions are very rapid, resulting in steep concentration profiles. Furthermore, different species may move at different velocities. Thus there potentially are some new complications when adsorption is allowed. We will also look at the convective diffusion equation in two dimensions: first for steady, uniform flow diagonal to the grid and second for a rotating cone problem. The rotating cone problem, in particular, clearly differentiates between methods that are excellent and those that are merely adequate.

9.1. Linear Adsorption

The discussion here of chromatography problems is based on an excellent book by Rhee, Aris, and Amundson [1986]. This book provides a detailed derivation of the equations and a very incisive description of the mathematics applicable to these problems. No numerical methods are discussed, but the book gives the theory necessary to appreciate the features of the numerical solution.

Let us consider a cylindrical tube filled with adsorbent. The adsorbent is a porous media on which chemicals can adsorb. Usually models are formulated in terms of an average concentration, defined over a volume large enough to average out the small-scale fluctuations in void space and pore space. This creates continuous phases (possibly more than one) co-existing in the same space. The problem is represented by two differential equations: one for c, the concentration in the fluid phase, and one for n, the concentration on the stationary phase. The differential equations are

$$\phi \frac{\partial c}{\partial t} + \phi V \frac{\partial c}{\partial z} + (1 - \phi) \frac{\partial n}{\partial t} = 0 \qquad (9.1)$$

and

$$\frac{\partial n}{\partial t} = k(\gamma c - n). \tag{9.2}$$

The fluid velocity is V and the porosity is ϕ. The initial conditions at time zero are

$$c(z,0) = f(z), \quad n(z,0) = h(z), \tag{9.3}$$

while the boundary condition is

$$c(0,t) = g(t). \tag{9.4}$$

We make these equations non-dimensional by taking

$$\xi = \frac{zk}{V}, \quad \eta = kt. \tag{9.5}$$

The result is

$$\frac{\partial c}{\partial \eta} + \frac{\partial c}{\partial \xi} + \frac{1-\phi}{\phi}(\gamma c - n) = 0, \tag{9.6}$$

$$\frac{\partial n}{\partial \eta} = \gamma c - n. \tag{9.7}$$

If we define the variable $u = \gamma c$, then the only dimensionless group is $\gamma(1-\phi)/\phi$; however, we will leave the equation in the form shown since concentration is the most important variable.

The theory of linear and semi-linear equations is described in detail in Rhee, et al. [1986]. Only a summary is presented here. The general form of a semi-linear equation is

$$P\frac{\partial z}{\partial x} + Q\frac{\partial z}{\partial y} = R, \tag{9.8}$$

where the functions P and Q depend on only x and y (to make the equation linear), while the function R can depend on only x and y (for the linear case) or on the solution, z, as well (for the semi-linear case). If P and Q are constant and R is linear, as is the case in Eq. (9.6)-(9.7), then discontinuities in the derivatives of the initial data and discontinuities in the initial data itself [Eq. (9.3)] are propagated along characteristics. If the function R depends on z, the discontinuity is still propagated along characteristics, but the size of the discontinuity does not remain constant. If P and Q are not constant in x and y, then discontinuities in slope and function are propagated along characteristics, but the characteristics are curved. Discontinuities induced by the initial conditions are called contact discontinuities.

The method of characteristics applied to linear or semi-linear equations [e.g., Eq. (9.8)] would solve the following equations:

$$\frac{dx}{ds} = P(x,y), \quad \frac{dy}{ds} = Q(x,y), \quad \frac{dz}{ds} = R(x,y,z). \tag{9.9}$$

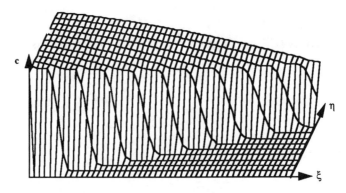

Figure 9.1. Exact Solution of Eq. 9.6-9.7

The relationship between Eq. (9.9) and Eq. (9.8) is similar to the relationship between the Lagrangian and Eulerian coordinate systems. The case we are solving, Eq. (9.6)-(9.7), is more complicated because it involves two unknowns rather than one, but the idea is the same. In any case, we see that it is possible to have steep fronts that follow characteristics in time.

We return to the specific problem, Eq. (9.6)-(9.7). Rhee, *et al.* [1986] derive the solution using Laplace transforms. When the initial concentrations of both c and n are zero and the inlet condition is c = 1, the solution is

$$c(\xi,\eta) = e^{-\beta\xi - (\eta - \xi)} I_0 [\, 2\sqrt{\beta\xi(\eta - \xi)} \,] +$$

$$+ e^{-\beta\xi - (\eta - \xi)} e^{\eta - \xi} \int_0^{\eta - \xi} e^{-\eta''} I_0 [\, 2\sqrt{\beta\xi\eta''} \,] \, d\eta'', \quad (9.10)$$

$$\beta = \gamma \frac{1 - \phi}{\phi}.$$

This solution is plotted in Figure 9.1. A steep front is obtained at small time values; this front becomes less steep as time proceeds. What are the requirements to model this problem numerically?

Let us consider the numerical solution of the transient problem. We have no diffusion in this equation, but the adsorption term seems to act like diffusion (see Figure 9.1). We apply the upstream finite element method, the random choice method, the Galerkin method, the Taylor-Galerkin method, the Taylor-finite difference method, the Lax-Wendroff method, the MacCormack method, the MacCormack method with flux-correction, and the Euler-Lagrange method. For some of the methods, we use Eq. (9.6)-(9.7) with dispersion added; the model is then a two-region model of the concentration of chemicals in plasma and the surrounding tissue for bioengineering situations {Bassingthwaighte [1974] and Lenhoff and Lightfoot [1984]}.

$$\frac{\partial c}{\partial \eta} + \frac{\partial c}{\partial \xi} + \frac{1-\phi}{\phi}[\gamma c - n] = \frac{1}{Pe_1}\frac{\partial^2 c}{\partial \xi^2} \qquad (9.11)$$

$$\frac{\partial n}{\partial \eta} - [\gamma c - n] = \frac{1}{Pe_2}\frac{\partial^2 n}{\partial \xi^2} \qquad (9.12)$$

Upstream weighting method. A finite difference method with upstream weighting of the convection term is

$$\frac{c_i^{n+1} - c_i^n}{\Delta \eta} = -\frac{c_i^n - c_{i-1}^n}{\Delta \xi} - \frac{1-\phi}{\phi}(\gamma c_i^n - n_i^n) + \frac{1}{Pe_1}\left(\frac{c_{i+1}^n - 2c_i^n + c_{i-1}^n}{\Delta \xi^2}\right), \qquad (9.13)$$

$$\frac{n_i^{n+1} - n_i^n}{\Delta \eta} = (\gamma c_i^n - n_i^n) + \frac{1}{Pe_2}\left(\frac{n_{i+1}^n - 2n_i^n + n_{i-1}^n}{\Delta \xi^2}\right). \qquad (9.14)$$

See Sun and Meunier [1991] for an alternative formulation using an adaptive upstream method to evaluate the convection term in a more complicated model.

Random choice method. The convection part of the problem is merely the advection equation:

$$\frac{\partial c}{\partial \eta} + \frac{\partial c}{\partial \xi} = 0. \qquad (9.15)$$

This equation is first solved using the random choice method to obtain $c*_i^{n+1}$. Then we calculate the solution from

$$\frac{c_i^{n+1} - c*_i^{n+1}}{\Delta \eta} = \frac{1-\phi}{\phi}[n_i^n - \gamma c*_i^{n+1}] + \frac{1}{Pe_1}\frac{c*_{i+1}^{n+1} - 2c*_i^{n+1} + c*_{i-1}^{n+1}}{\Delta \xi^2}, \qquad (9.16)$$

$$\frac{n_i^{n+1} - n_i^n}{\Delta \eta} = [\gamma c*_i^{n+1} - n_i^n] + \frac{1}{Pe_2}\frac{n_{i+1}^n - 2n_i^n + n_{i-1}^n}{\Delta \xi^2}. \qquad (9.17)$$

Galerkin method. The Galerkin method applied to Eq. (9.6)-(9.7) is

$$\frac{1}{6}\frac{dc_{i+1}}{d\eta} + \frac{2}{3}\frac{dc_i}{d\eta} + \frac{1}{6}\frac{dc_{i-1}}{d\eta} + \frac{1}{2\Delta \xi}[c_{i+1} - c_{i-1}] +$$
$$+ \frac{1-\phi}{\phi}\left\{\gamma\left[\frac{1}{6}c_{i+1} + \frac{2}{3}c_i + \frac{1}{6}c_{i-1}\right] - \left[\frac{1}{6}n_{i+1} + \frac{2}{3}n_i + \frac{1}{6}n_{i-1}\right]\right\} = 0, \qquad (9.18)$$

$$\frac{1}{6}\frac{dn_{i+1}}{d\eta} + \frac{2}{3}\frac{dn_i}{d\eta} + \frac{1}{6}\frac{dn_{i-1}}{d\eta} =$$
$$\left\{\gamma\left[\frac{1}{6}c_{i+1} + \frac{2}{3}c_i + \frac{1}{6}c_{i-1}\right] - \left[\frac{1}{6}n_{i+1} + \frac{2}{3}n_i + \frac{1}{6}n_{i-1}\right]\right\}. \qquad (9.19)$$

The equations for the last element are

$$\frac{1}{3}\frac{dc_{NT}}{d\eta} + \frac{1}{6}\frac{dc_{NT-1}}{d\eta} + \frac{1}{2\Delta\xi}[\,c_{NT} - c_{NT-1}\,] +$$
$$+ \frac{1-\phi}{\phi}\left\{\gamma\left[\frac{1}{3}c_{NT} + \frac{1}{6}c_{NT-1}\right] - \left[\frac{1}{3}n_{NT} + \frac{1}{6}n_{NT-1}\right]\right\} = 0, \quad (9.20)$$

$$\frac{1}{3}\frac{dn_{NT}}{d\eta} + \frac{1}{6}\frac{dn_{NT-1}}{d\eta} = \left\{\gamma\left[\frac{1}{3}c_{NT} + \frac{1}{6}c_{NT-1}\right] - \left[\frac{1}{3}n_{NT} + \frac{1}{6}n_{NT-1}\right]\right\}, \quad (9.21)$$

while those for the first element are

$$c_1 = 1,$$

$$\frac{1}{6}\frac{dn_2}{dt} + \frac{1}{3}\frac{dn_1}{dt} = \gamma\left(\frac{1}{6}c_2 + \frac{1}{3}c_1\right) - \left(\frac{1}{6}n_2 + \frac{1}{3}n_1\right). \quad (9.22)$$

Taylor-Galerkin method and Taylor-finite difference method. Next we will derive the Taylor version of Eq. (9.6)-(9.7). The Taylor-Galerkin method was first applied to the convective diffusion equation with a source term by Donea, et al. [1984]. To derive the equation here, we first differentiate with respect to η, giving

$$\frac{\partial^2 c}{\partial \eta^2} + \frac{\partial^2 c}{\partial \eta \partial \xi} + \frac{1-\phi}{\phi}\left(\gamma\frac{\partial c}{\partial \eta} - \frac{\partial n}{\partial \eta}\right) = 0, \quad (9.23)$$

$$\frac{\partial^2 n}{\partial \eta^2} = \gamma\frac{\partial c}{\partial \eta} - \frac{\partial n}{\partial \eta}. \quad (9.24)$$

For the first derivatives, we substitute from Eq. (9.6)-(9.7).

$$\frac{\partial^2 c}{\partial \eta^2} + \frac{\partial^2 c}{\partial \eta \partial \xi} + \frac{1-\phi}{\phi}\left[\gamma\left(-\frac{\partial c}{\partial \xi} - \frac{1-\phi}{\phi}(\gamma c - n)\right) - (\gamma c - n)\right] = 0 \quad (9.25)$$

$$\frac{\partial^2 n}{\partial \eta^2} = \gamma\left(-\frac{\partial c}{\partial \xi} - \frac{1-\phi}{\phi}(\gamma c - n)\right) - (\gamma c - n) \quad (9.26)$$

To obtain the cross derivative, we take the derivative of Eq. (9.6) with respect to ξ.

$$\frac{\partial^2 c}{\partial \eta \partial \xi} + \frac{\partial^2 c}{\partial \xi^2} + \frac{1-\phi}{\phi}\left[\gamma\frac{\partial c}{\partial \xi} - \frac{\partial n}{\partial \xi}\right] = 0 \quad (9.27)$$

We combine this result in Eq. (9.25) to get

$$\frac{\partial^2 c}{\partial \eta^2} - \frac{\partial^2 c}{\partial \xi^2} - \frac{1-\phi}{\phi}\left(\gamma\frac{\partial c}{\partial \xi} - \frac{\partial n}{\partial \xi}\right) +$$
$$+ \frac{1-\phi}{\phi}\left[\gamma\left(-\frac{\partial c}{\partial \xi} - \frac{1-\phi}{\phi}(\gamma c - n)\right) - (\gamma c - n)\right] = 0. \quad (9.28)$$

Now we use the Taylor series,

ADSORPTION PROBLEMS   323

$$\frac{\partial c}{\partial \eta} = \frac{c^{n+1} - c^n}{\Delta \eta} - \frac{\Delta \eta}{2} \frac{\partial^2 c}{\partial \eta^2} \tag{9.29}$$

$$\frac{\partial n}{\partial \eta} = \frac{n^{n+1} - n^n}{\Delta \eta} - \frac{\Delta \eta}{2} \frac{\partial^2 n}{\partial \eta^2}. \tag{9.30}$$

Next we substitute Eq. (9.28) into Eq. (9.29) and the result into Eq. (9.6). Then we substitute Eq. (9.26) into Eq. (9.30) and the result into Eq. (9.7). These steps give

$$\frac{c^{n+1} - c^n}{\Delta \eta} = -\frac{\partial c}{\partial \xi} - \frac{1 - \phi}{\phi}(\gamma c - n) + \frac{\Delta \eta}{2}\left\{\frac{\partial^2 c}{\partial \xi^2} + \frac{1 - \phi}{\phi}\left(\gamma \frac{\partial c}{\partial \xi} - \frac{\partial n}{\partial \xi}\right)\right.$$
$$\left. + \frac{1 - \phi}{\phi}\left[\gamma\left(\frac{\partial c}{\partial \xi} + \frac{1 - \phi}{\phi}(\gamma c - n)\right) + (\gamma c - n)\right]\right\}, \tag{9.31}$$

$$\frac{n^{n+1} - n^n}{\Delta \eta} = (\gamma c - n) - \frac{\Delta \eta}{2}\left[\gamma\left(\frac{\partial c}{\partial \xi} + \frac{1 - \phi}{\phi}(\gamma c - n)\right) + (\gamma c - n)\right]. \tag{9.32}$$

This is the set of equations to be solved when the Taylor terms are included. The Taylor-finite difference method is

$$\frac{c_i^{n+1} - c_i^n}{\Delta \eta} = -\frac{c_{i+1}^n - c_{i-1}^n}{2\Delta \xi} - \frac{1 - \phi}{\phi}(\gamma c_i^n - n_i^n) +$$
$$+ \frac{\Delta \eta}{2}\left\{\frac{c_{i+1}^n - 2c_i^n + c_{i-1}^n}{\Delta \xi^2} + \frac{1 - \phi}{\phi}\left(\gamma \frac{c_{i+1}^n - c_{i-1}^n}{2\Delta \xi} - \frac{n_{i+1}^n - n_{i-1}^n}{2\Delta \xi}\right) + \right.$$
$$\left. + \frac{1 - \phi}{\phi}\left[\gamma\left(\frac{c_{i+1}^n - c_{i-1}^n}{2\Delta \xi} + \frac{1 - \phi}{\phi}(\gamma c_i^n - n_i^n)\right) + (\gamma c_i^n - n_i^n)\right]\right\}, \tag{9.33}$$

$$\frac{n_i^{n+1} - n_i^n}{\Delta \eta} = (\gamma c_i^n - n_i^n) - \frac{\Delta \eta}{2}\left[\gamma \frac{c_{i+1}^n - c_{i-1}^n}{2\Delta \xi} + \frac{1 - \phi}{\phi}(\gamma c_i^n - n_i^n)\right) + (\gamma c_i^n - n_i^n)\right]. \tag{9.34}$$

The Taylor-Galerkin method is

$$\frac{1}{6}\frac{c_{i+1}^{n+1} - c_{i+1}^n}{\Delta \eta} + \frac{2}{3}\frac{c_i^{n+1} - c_i^n}{\Delta \eta} + \frac{1}{6}\frac{c_{i-1}^{n+1} - c_{i-1}^n}{\Delta \eta} = -\frac{c_{i+1}^n - c_{i-1}^n}{2\Delta \xi} - \frac{1 - \phi}{\phi}TR_i^n +$$
$$+ \frac{\Delta \eta}{2}\left\{\frac{c_{i+1}^n - 2c_i^n + c_{i-1}^n}{\Delta \xi^2} + \frac{1 - \phi}{\phi}\left(\gamma \frac{c_{i+1}^n - c_{i-1}^n}{2\Delta \xi} - \frac{n_{i+1}^n - n_{i-1}^n}{2\Delta \xi}\right) + \right.$$
$$\left. + \frac{1 - \phi}{\phi}\left[\gamma\left(\frac{c_{i+1}^n - c_{i-1}^n}{2\Delta \xi} + \frac{1 - \phi}{\phi}TR_i^n\right) - TR_i^n\right]\right\}, \tag{9.35}$$

$$\frac{1}{6}\frac{n_{i+1}^{n+1}-n_{i+1}^{n}}{\Delta\eta}+\frac{2}{3}\frac{n_{i}^{n+1}-n_{i}^{n}}{\Delta\eta}+\frac{1}{6}\frac{n_{i-1}^{n+1}-n_{i-1}^{n}}{\Delta\eta}=$$
$$TR_{i}^{n}-\frac{\Delta\eta}{2}\left[\gamma\left(\frac{c_{i+1}^{n}-c_{i-1}^{n}}{2\Delta\xi}+\frac{1-\phi}{\phi}TR_{i}^{n}\right)+TR_{i}^{n}\right], \quad (9.36)$$

where

$$TR_{i}^{n}=\gamma\left[\frac{1}{6}c_{i+1}+\frac{2}{3}c_{i}+\frac{1}{6}c_{i-1}\right]-\left[\frac{1}{6}n_{i+1}+\frac{2}{3}n_{i}+\frac{1}{6}n_{i-1}\right]. \quad (9.37)$$

Lax-Wendroff method. We next compare the Lax-Wendroff method and the MacCormack methods with the Taylor-finite difference method. The Lax-Wendroff method given in Table 4.5 generalizes to

$$\frac{c_{i+1/2}^{n+1/2}-\frac{1}{2}(c_{i}^{n}+c_{i+1}^{n})}{\Delta\eta}=-\frac{c_{i+1}^{n}-c_{i}^{n}}{2\Delta\xi}-\frac{1-\phi}{4\phi}[\gamma(c_{i+1}^{n}+c_{i}^{n})-(n_{i}^{n}+n_{i+1}^{n})] \quad (9.38)$$

$$\frac{n_{i+1/2}^{n+1/2}-\frac{1}{2}(n_{i}^{n}+n_{i+1}^{n})}{\Delta\eta}=[\gamma(c_{i+1}^{n}+c_{i}^{n})-(n_{i}^{n}+n_{i+1}^{n})]/2 \quad (9.39)$$

for the predictor and

$$\frac{c_{i}^{n+1}-c_{i}^{n}}{\Delta\eta}=-\frac{c_{i+1/2}^{n+1/2}-c_{i-1/2}^{n+1/2}}{\Delta\xi}-\frac{1-\phi}{\phi}(\gamma c_{i}^{n}-n_{i}^{n})$$
$$\frac{n_{i}^{n+1}-n_{i}^{n}}{\Delta\eta}=\gamma c_{i}^{n}-n_{i}^{n} \quad (9.40)$$

for the corrector. Here the source terms have been evaluated in the same way that the diffusion terms were evaluated in Table 4.5. The second equation, Eq. (9.39), is not actually necessary in this case. If these equations are combined, we do not get Eq. (9.33)-(9.34) so the truncation error is only $O(\Delta\eta)$ (see Problem 9.1).

MacCormack method. In constrast to the Lax-Wendroff method, the MacCormack method is

$$\frac{c_{i}^{*,n+1}-c_{i}^{n}}{\Delta\eta}=-\frac{c_{i+1}^{n}-c_{i}^{n}}{\Delta\xi}-\frac{1-\phi}{\phi}(\gamma c_{i}^{n}-n_{i}^{n}) \quad (9.41)$$

$$\frac{n_{i}^{*,n+1}-n_{i}^{n}}{\Delta\eta}=(\gamma c_{i}^{n}-n_{i}^{n}) \quad (9.42)$$

followed by the corrector

$$\frac{c_{i}^{n+1}-\frac{1}{2}(c_{i}^{n}+c_{i}^{*,n+1})}{\Delta\eta}=-\frac{c_{i}^{*,n+1}-c_{i-1}^{*,n+1}}{2\Delta\xi}-\frac{1-\phi}{2\phi}(\gamma c_{i}^{*,n+1}-n_{i}^{*,n+1}) \quad (9.43)$$

$$\frac{n_i^{n+1} - \frac{1}{2}(n_i^n + n*_i^{n+1})}{\Delta\eta} = \frac{1}{2}(\gamma c*_i^{n+1} - n*_i^{n+1}). \tag{9.44}$$

When these equations are combined, we get Eq. (9.33)-(9.34). Thus the Taylor-finite difference method and the MacCormack method give identical equations for this linear problem with a source term. The truncation error for the MacCormack method is also smaller than the truncation error for the Lax-Wendroff method, so we choose to apply a flux-correction step to the MacCormack method.

MacCormack method with flux-correction. We apply the MacCormack method with flux-correction to Eq. (9.11)-(9.12); in the first step we derive a predicted value,

$$\frac{c*_i^{n+1} - c_i^n}{\Delta\eta} = -\frac{c_{i+1}^n - c_i^n}{\Delta\xi} - \frac{1-\phi}{\phi}(\gamma c_i^n - n_i^n) + \frac{1}{Pe_1}\left(\frac{c_{i+1}^n - 2c_i^n + c_{i-1}^n}{\Delta\xi^2}\right) \tag{9.45}$$

$$\frac{n*_i^{n+1} - n_i^n}{\Delta\eta} = (\gamma c_i^n - n_i^n) + \frac{1}{Pe_2}\left(\frac{n_{i+1}^n - 2n_i^n + n_{i-1}^n}{\Delta\xi^2}\right), \tag{9.46}$$

while in the second step we use it to evaluate the derivatives:

$$\frac{\tilde{c}_i^{n+1} - \frac{1}{2}(c_i^n + c*_i^{n+1})}{\Delta\eta} = -\frac{c*_i^{n+1} - c*_{i-1}^{n+1}}{2\Delta\xi} -$$
$$\frac{1-\phi}{2\phi}(\gamma c*_i^{n+1} - n*_i^{n+1}) + \frac{1}{Pe_1}\left(\frac{c*_{i+1}^{n+1} - 2c*_i^{n+1} + c*_{i-1}^{n+1}}{2\Delta\xi^2}\right) \tag{9.47}$$

$$\frac{\tilde{n}_i^{n+1} - \frac{1}{2}(n_i^n + n*_i^{n+1})}{\Delta\eta} = \frac{1}{2}(\gamma c*_i^{n+1} - n*_i^{n+1}) + \frac{1}{Pe_2}\left(\frac{n*_{i+1}^{n+1} - 2n*_i^{n+1} + n*_{i-1}^{n+1}}{2\Delta\xi^2}\right). \tag{9.48}$$

The earlier treatment of flux-correction (Section 6.2) needs to be expanded here to include the source terms. Let us consider the equation

$$\frac{\partial u}{\partial t} = -\frac{\partial F(u)}{\partial x} + S. \tag{9.49}$$

According to work by Book [1981, p. 165], we use the algorithm in Section 6.2 but add the source terms to Eq. (6.23)-(6.24). The equations considered here, Eq. (9.6)-(9.7), are slightly more complicated since the equation for n has no flux term. The algorithm used here is Eq. (9.45)-(9.48), followed by

$$\hat{\Delta}^c_{i+1/2} = \eta\,(\tilde{c}^{n+1}_{i+1} - \tilde{c}^{n+1}_i)$$
$$\tilde{c}^{n+1}_i = \hat{c}^{n+1}_i + \eta\,(c^n_{i+1} - 2c^n_i + c^n_{i-1}) \tag{9.50}$$
$$\Delta^c_{i+1/2} = \hat{c}^{n+1}_{i+1} - \hat{c}^{n+1}_i$$

to obtain a correction to the fluxes:

$$f^c_{i+1/2} = \text{sign}(\hat{\Delta}^c_{i+1/2})\,\max\left\{0,\ \min\left\{\begin{array}{l}\text{sign}(\hat{\Delta}^c_{i+1/2})\Delta^c_{i-1/2}\\ |\hat{\Delta}^c_{i+1/2}|\\ \text{sign}(\hat{\Delta}^c_{i+1/2})\Delta^c_{i+3/2}\end{array}\right\}\right\} \tag{9.51}$$

The final step is

$$c^{n+1}_i = \hat{c}^{n+1}_i - (f^c_{i+1/2} - f^c_{i-1/2}). \tag{9.52}$$

This algorithm does not employ a correction to the n variable since n does not have fluxes in the equation.

Euler-Lagrange method. In the Euler-Lagrange method we allow the nodes to move along characteristics. Thus there is a moving mesh, which is denoted with the subscript j. The solid phase remains fixed so that it is represented on a fixed mesh, which is denoted with the subscript i. The equations for the Euler-Lagrange method are

$$\frac{c^{n+1}_j - c^n_j}{\Delta\eta} = \frac{1-\phi}{\phi}[n^n_j - \gamma c^n_j] + \frac{1}{\text{Pe}_1}\frac{\dfrac{c^n_{j+1}-c^n_j}{\Delta x_j} - \dfrac{c^n_j - c^n_{j-1}}{\Delta x_{j-1}}}{\tfrac{1}{2}(\Delta x_j + \Delta x_{j-1})}, \tag{9.53}$$

$$\frac{n^{n+1}_i - n^n_i}{\Delta\eta} = [\gamma c^n_i - n^n_i] + \frac{1}{\text{Pe}_2}\frac{\dfrac{n^n_{i+1}-n^n_i}{\Delta x_i} - \dfrac{n^n_i - n^n_{i-1}}{\Delta x_{i-1}}}{\tfrac{1}{2}(\Delta x_i + \Delta x_{i-1})}, \tag{9.54}$$

and

$$x^{n+1}_j = x^n_j + \Delta t,\quad x^{n+1}_i = x^n_i. \tag{9.55}$$

After this step, it is necessary to interpolate the values of n onto the j-mesh so that the mass transfer term can be evaluated on the moving mesh. This is done using a cubic spline. At the end of the step, both c and n are known values on the j-mesh.

The problem is solved first using $\phi = 0.4$ and $\gamma = 2$. The initial conditions for both c and n are zero and there is a boundary condition of $c = 1$ at $x = 0$. A total of 51 nodes are distributed uniformly from $x = 0$ to $x = 1$ with a spacing of 0.02; the size of the time-step is 0.01. The solution is plotted at $t = 0.2$, $t = 0.4$, and $t = 0.6$. The solution for c when using the upstream finite difference method is

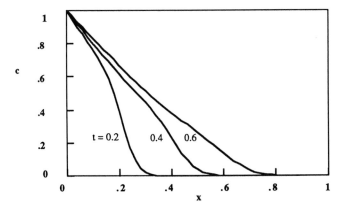

Figure 9.2a. Fluid Concentration, Upstream Method

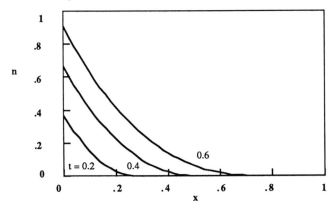

Figure 9.2b. Solid Concentration, Upstream Method, 51 nodes, $\Delta t = 0.01$

shown in Figure 9.2a. The corresponding solution for n is shown in Figure 9.2b. Both solutions are smooth, creating little difficulty in solving the equations. At equilibrium, the value of n is $\gamma c = 2$. This indicates that the adsorptive capacity of the bed is very large, which results in a rapid reduction of c with distance; all the material gets adsorbed. Figures 9.3 a and 9.3b show the results obtained with the random choice method; the profiles for c are very steep while the profiles for n are not. The Galerkin method gave a solution that oscillated wildly. When the Taylor-Galerkin method was employed, there were some oscillations for $\Delta t = 0.01$, but only small oscillations when the time-step was reduced to 0.005 (see Figure 9.4). The MacCormack method with flux-correction gave the results shown in Figure 9.5. These results are quite good, but the profile for c is not as steep as that obtained with the random choice method. The Euler-Lagrange method gave the results shown in Figure 9.6. In summary, all methods except the upstream and Galerkin methods gave good results.

Next we change the value of γ to 0.1. The adsorptive capacity of the bed is less, so the problem resembles the advection equation. The results are also

328 NUMERICAL METHODS - MOVING FRONTS

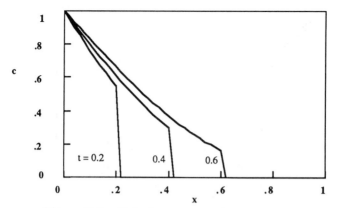

Figure 9.3a. Fluid Concentration, Random Choice Method

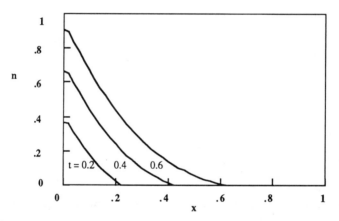

Figure 9.3b. Solid Concentration, Random Choice Method, 51 nodes, $\Delta t = 0.01$

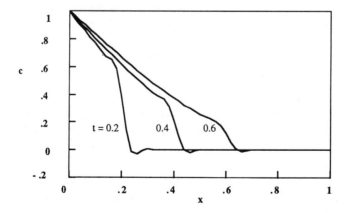

Figure 9.4a. Fluid Concentration, Taylor-Galerkin Method

ADSORPTION PROBLEMS   329

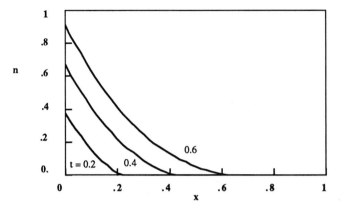

Figure 9.4b. Solid Concentration, Taylor-Galerkin Method, 51 nodes, $\Delta t = 0.005$

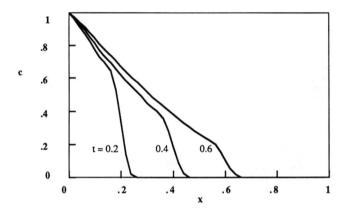

Figure 9.5a. Fluid Concentration, MacCormack Method with Flux-correction

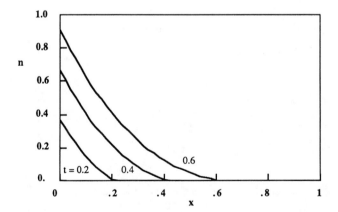

Figure 9.5b. Solid Concentration, MacCormack Method with Flux-correction 51 nodes, $\Delta t = 0.01$

330 NUMERICAL METHODS - MOVING FRONTS

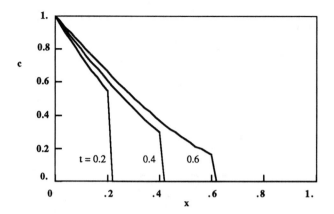

Figure 9.6a. Fluid Concentration, Euler Lagrange Method

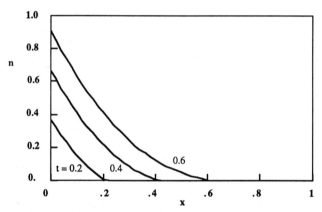

Figure 9.6b. Solid Concentration, Euler Lagrange Method, 51 nodes, $\Delta t = 0.01$

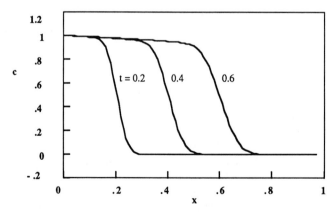

Figure 9.7a. Fluid Concentration, Upstream Method, 100 nodes, $\Delta t = 0.005$

ADSORPTION PROBLEMS 331

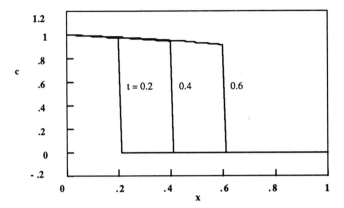

Figure 9.7b. Fluid Concentration, Random Choice Method, 100 nodes, $\Delta t = 0.005$

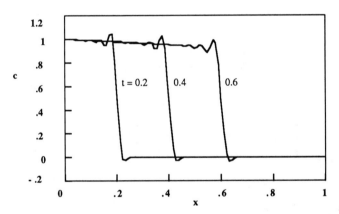

Figure 9.7c. Fluid Concentration, Taylor-Galerkin Method, 100 nodes, $\Delta t = 0.00125$

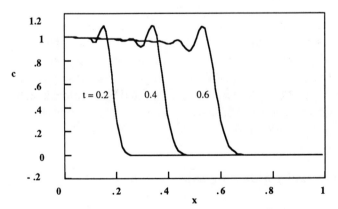

Figure 9.7d. Fluid Concentration, MacCormack Method, 100 nodes, $\Delta t = 0.005$

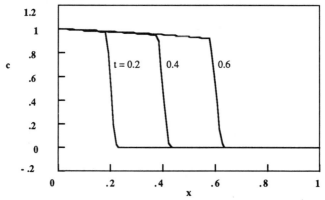

Figure 9.7e. Fluid Concentration, MacCormack Method with Flux-correction 100 nodes, $\Delta t = 0.005$

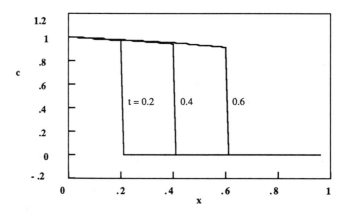

Figure 9.7f. Fluid Concentration, Euler-Lagrange Method, 100 nodes, $\Delta t = 0.005$

comparable with the results for the advection equation, as illustrated in Figures 9.7a-9.7f for 100 nodes. Figure 9.7a shows that the results using the upstream, explicit method are smoothed out. The random choice method, MacCormack method with flux-correction, and Euler-Lagrange method all give good solutions with steep profiles. The Taylor-Galerkin method and MacCormack methods give acceptable results with small oscillations. These comments parallel exactly the results obtained for the advection equation (see Problem 9.5).

9.2. Chromatography with Langmuir Adsorption

Under some circumstances, the moving and stationary phases are in equilibrium. We will consider that case next, according to work by Rhee, *et al.* [1986]. For equilibrium, the problem posed in Eq. (9.6)-(9.7) can be simplified further. Let us rewrite it as

$$\phi \frac{\partial c}{\partial t} + \phi V \frac{\partial c}{\partial x} + (1 - \phi) \frac{\partial n}{\partial t} = 0. \qquad (9.56)$$

Now, however, we will suppose that the concentration c is always in equilibrium with the concentration n. In Section 9.1 we used $\gamma c = n$. Here we use a more general relation:

$$n = f(c). \qquad (9.57)$$

Then

$$\frac{\partial n}{\partial t} = \frac{df}{dc} \frac{\partial c}{\partial t}. \qquad (9.58)$$

The problem can be reduced to

$$\phi \frac{\partial c}{\partial t} + \phi V \frac{\partial c}{\partial x} + (1 - \phi) \frac{df}{dc} \frac{\partial c}{\partial t} = 0. \qquad (9.59)$$

The non-dimensional form is

$$\left(1 + \frac{1-\phi}{\phi} \frac{df}{dc}\right) \frac{\partial c}{\partial \eta} + \frac{\partial c}{\partial \xi} = 0. \qquad (9.60)$$

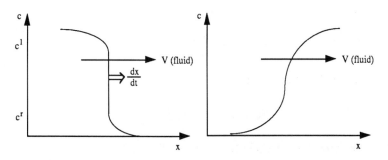

Figure 9.8a. Equation (9.71) Satisfied **Figure 9.8b.** Equation (9.71) Not Satisfied

The form of Eq. (9.60) is equivalent to that of a quasi-linear equation (when P and Q in Eq. (9.8) depend on c). The characteristic direction in Eq. (9.8) is given by

$$\frac{dy}{dx} = \frac{Q}{P}. \qquad (9.61)$$

Here the characteristic direction is

$$\frac{d\eta}{d\xi} = \left(1 + \frac{1-\phi}{\phi} \frac{df}{dc}\right). \qquad (9.62)$$

Let us consider Figure 9.8a and suppose the discontinuity moves with a velocity of dx/dt. Then the rate of transfer of material through the mobile phase is

$$\phi A c^l \left(V - \frac{dx}{dt} \right) - \phi A c^r \left(V - \frac{dx}{dt} \right), \qquad (9.63)$$

whereas the rate of transfer of adsorbed material in the reverse direction is

$$(1 - \phi) A \frac{dx}{dt} (n^r - n^l). \qquad (9.64)$$

The net rate of transfer must be zero since mass is conserved at the discontinuity; setting the sum of Eq. (9.63) and Eq. (9.64) to zero gives

$$V \frac{dt}{dx} = 1 + \frac{1 - \phi}{\phi} \frac{n^l - n^r}{c^l - c^r}. \qquad (9.65)$$

The step change in n is given by

$$[n] = f(c^l) - f(c^r) = [f]. \qquad (9.66)$$

In non-dimensional form, Eq. (9.65) [with Eq. (9.66)] is

$$\frac{d\eta}{d\xi} = 1 + \frac{1 - \phi}{\phi} \frac{[f]}{[c]}. \qquad (9.67)$$

The characteristic direction is given by Eq. (9.61) as

$$\sigma(c) \equiv 1 + \frac{1 - \phi}{\phi} \frac{df}{dc}. \qquad (9.68)$$

We now define the quantity $\tilde{\sigma}$ as

$$\tilde{\sigma}(c^l, c^r) \equiv 1 + \frac{1 - \phi}{\phi} \frac{[f]}{[c]}. \qquad (9.69)$$

Rhee, *et al.* [1986] show that the shock condition is

$$\tilde{\sigma}(c^l, c^r) \geq \tilde{\sigma}(c^l, c) \text{ for every c between } c^l \text{ and } c^r. \qquad (9.70)$$

Essentially the speed of the material behind the shock is faster than that in front of the shock, so the shock stays sharp. If the function f(c) is convex or sigmoid with only one inflection point, Eq. (9.70) can be replaced by [Rhee, *et al.*, 1986]

$$\sigma(c^l) \leq \sigma(c^r). \qquad (9.71)$$

Next let us consider a special case when the adsorption follows a Langmuir isotherm:

$$f = \frac{\gamma c}{1 + K c}. \qquad (9.72)$$

For such an isotherm, Eq. (9.71) is appropriate and the derivative is

$$\frac{df}{dc} = \frac{\gamma}{(1+Kc)^2}. \tag{9.73}$$

The characteristic direction is

$$\sigma(c) = 1 + \frac{1-\phi}{\phi}\frac{\gamma}{(1+Kc)^2}. \tag{9.74}$$

Let us consider the problem of flow through a packed bed with adsorption onto a solid. If we begin with a column loaded and change the inlet concentration to zero, then $c^l = 0$ and $c^r > 0$. Then the functions are

$$\sigma(c^l) = 1 + \frac{1-\phi}{\phi}\gamma,$$

$$\sigma(c^r) = 1 + \frac{1-\phi}{\phi}\frac{\gamma}{(1+Kc^r)^2}. \tag{9.75}$$

These equations satisfy

$$\sigma(c^l) > \sigma(c^r), \tag{9.76}$$

which means that a shock is not formed. We expect the solution to be broadening and the numerical solution to become easier as time proceeds. This broadening is shown in Figure 9.8b. If we start with an empty column and increase the inlet concentration, then $c^r = 0$ and $c^l > 0$. In that case we obtain

$$\sigma(c^l) < \sigma(c^r) \tag{9.77}$$

and the condition for a shock is satisfied. Since a shock will form even if one does not exist initially, the numerical method must be able to solve problems with shocks.

For this flow problem we will use the following methods: the upstream finite difference method, the random choice method, the MacCormack method with and without flux-correction, the Taylor-Galerkin method, and the Galerkin method. An example of the use of the orthogonal collocation method is given by Farooq and Ruthven [1991]. The Langmuir isotherm, Eq. (9.72), is used with the parameters $\phi = 0.4$, $\gamma = 0.1$, and $K = 2$. The initial concentration takes the value 1.0 between $\xi = 0.25$ and $\xi = 0.75$ and zero elsewhere. The injected fluid has zero concentration. The leading edge satisfies the conditions for a shock [Eq. (9.77)] while the trailing edge does not. For convenience, Eq. (9.60) is written as

$$P(c)\frac{\partial c}{\partial \eta} + \frac{\partial c}{\partial \xi} = 0, \quad P(c) = 1 + \frac{1-\phi}{\phi}\frac{df}{dc} = 1 + \frac{1-\phi}{\phi}\frac{\gamma}{(1+Kc)^2}. \tag{9.78}$$

Upstream finite difference method. The explicit, upstream finite difference method applied to Eq. (9.78) gives

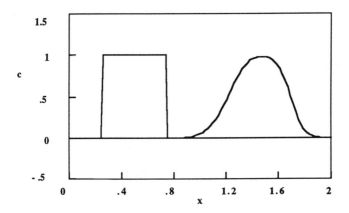

Figure 9.9a. Adsorption with Eq. (9.72), Upstream Method, 100 nodes, $\Delta t = 0.01$

$$P(c_i^n) \frac{c_i^{n+1} - c_i^n}{\Delta \eta} = - \frac{c_i^n - c_{i-1}^n}{\Delta \xi}. \quad (9.79)$$

Typical results are shown in Figure 9.9a; the concentration profile is smeared out over the front, as expected.

Random Choice. The random choice method requires a solution to the Riemann problem. This solution is patterned after the approximate solution to the Riemann problem for Burger's equation [Sod, 1985].

$$v(\xi/\eta, v_L, v_R) = \begin{cases} v_L & \xi/\eta < \text{shock speed} \\ v_R & \xi/\eta \geq \text{shock speed} \end{cases} \quad \text{for } v_L \geq v_R \quad (9.80)$$

$$v(\xi/\eta, v_L, v_R) = \begin{cases} v_L & \xi/\eta \leq S_L \\ \frac{1}{2}(v_L + v_R) & \text{otherwise} \\ v_R & \xi/\eta > S_R \end{cases} \quad \text{when } v_L < v_R \quad (9.81)$$

The solution at the half-time step is taken as

$$c_{i+1/2}^{n+1/2} = v\left(\xi_{i+1/2} + \theta_n \Delta\xi, \eta^n + \frac{\Delta\eta}{2}\right), \quad (9.82)$$

where v is the solution to the Riemann problem. The solution at the next half-time step is then

$$c_i^{n+1} = v(\xi_{i+1} + \theta_n \Delta\xi, \eta^n + \Delta\eta). \quad (9.83)$$

The shock speed is given by [Rhee, et al., 1986]

$$\text{shock speed} = \frac{1}{1 + \frac{1-\phi}{\phi} \frac{f(c_L) - f(c_R)}{c_L - c_R}}. \quad (9.84)$$

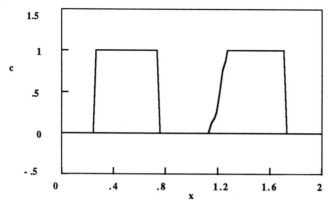

Figure 9.9b. Adsorption with Eq. (9.72), Random Choice Method, 100 nodes, $\Delta t = 0.01$

The speeds of the left-hand Riemann solution and right-hand Riemann solution are given by

$$S_L = \frac{1}{1 + \frac{1-\phi}{\phi}\frac{df}{dc}\big|_{c_L}}, \quad S_R = \frac{1}{1 + \frac{1-\phi}{\phi}\frac{df}{dc}\big|_{c_R}}. \tag{9.85}$$

Typical results are shown in Figure 9.9b; the concentration profile is quite sharp, retaining a shock on one side with a sloping, trailing edge, as exhibited by the exact solution.

MacCormack method. To apply the MacCormack method (with and without flux-correction), we rearrange Eq. (9.78) as

$$\frac{\partial c}{\partial \eta} + \frac{1}{P(c)}\frac{\partial c}{\partial \xi} = 0. \tag{9.86}$$

Then we define the quantity

$$M(c) = \int_0^c \frac{dc}{P(c)}. \tag{9.87}$$

The first derivative is then

$$\frac{\partial M(c)}{\partial \xi} = \frac{1}{P(c)}\frac{\partial c}{\partial \xi} \tag{9.88}$$

and the problem can be written as

$$\frac{\partial c}{\partial \eta} + \frac{\partial M(c)}{\partial \xi} = 0. \tag{9.89}$$

For the Langmuir adsorption we have

$$P(c) = 1 + \frac{1-\phi}{\phi}\frac{df}{dc}, \tag{9.90}$$

338 NUMERICAL METHODS - MOVING FRONTS

$$M(c) = \int_0^c \frac{dc}{1 + \frac{1-\phi}{\phi}\frac{df}{dc}}. \tag{9.91}$$

Expansion of this gives

$$M(c) = \int_0^c \frac{(1 + 2Kc + K^2 c^2)\, dc}{1 + \frac{1-\phi}{\phi}\gamma + 2Kc + K^2 c^2}. \tag{9.92}$$

This is cumbersome to use, but fortunately we do not have to. The MacCormack method without flux-correction applied to Eq. (9.86) is

$$\frac{c^{*n+1}_i - c^n_i}{\Delta\eta} = -\frac{M(c^n_{i+1}) - M(c^n_i)}{\Delta\xi}, \tag{9.93}$$

$$\frac{c^{n+1}_i - \frac{1}{2}(c^n_i + c^{*n+1}_i)}{\Delta\eta} = -\frac{M(c^{*n+1}_i) - M(c^{*n+1}_{i-1})}{2\Delta\xi}, \tag{9.94}$$

To evaluate the difference on the right-hand side, we write it as

$$M(c^n_{i+1}) - M(c^n_i) = \int_{c^n_i}^{c^n_{i+1}} \frac{dc}{P(c)}. \tag{9.95}$$

Next we evaluate the integral, using the trapezoid rule, as

$$\int_{c^n_i}^{c^{n+1}_i} \frac{dc}{P(c)} = \frac{1}{2}\left(\frac{1}{P(c^n_{i+1})} + \frac{1}{P(c^n_i)}\right)(c^n_{i+1} - c^n_i). \tag{9.96}$$

The MacCormack method is then

$$\frac{c^{*n+1}_i - c^n_i}{\Delta\eta} = -\frac{1}{2\Delta\xi}\left(\frac{1}{P(c^n_{i+1})} + \frac{1}{P(c^n_i)}\right)(c^n_{i+1} - c^n_i), \tag{9.97}$$

$$\frac{c^{n+1}_i - \frac{1}{2}(c^{*n+1}_i + c^n_i)}{\Delta\eta} = -\frac{1}{4\Delta\xi}\left(\frac{1}{P(c^{*n+1}_i)} + \frac{1}{P(c^{*n+1}_{i-1})}\right)(c^{*n+1}_i - c^{*n+1}_{i-1}). \tag{9.98}$$

Oscillations in the solution would be disastrous if they cause the coefficient P(c) to be evaluated for a negative c value because this could give P a value of zero. In this case that will not happen since df/dc ≥ 0 for all c values, whether positive or negative.

Results for the MacCormack method without flux-correction are shown in Figure 9.9c. The solution contains significant oscillations. Indeed, the solution looks similar to Figure 4.9b for the advection equation, except that the trailing

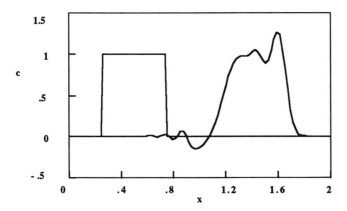

Figure 9.9c. Adsorption with Eq. (9.72), MacCormack Method, 100 nodes, $\Delta t = 0.01$

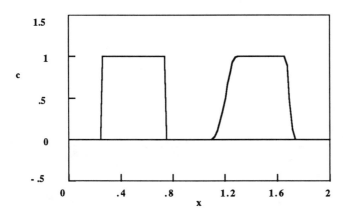

Figure 9.9d. Adsorption with Eq. (9.72), MacCormack Method with Flux-correction 100 nodes, $\Delta t = 0.01$

edge is more smoothed. The method is improved significantly when the flux-correction step is added: the oscillations disappear and the leading edge is quite sharp (see Figure 9.9d). The trailing edge is more diffuse than the leading edge, as expected.

Taylor-Galerkin method. The Taylor-Galerkin method can be obtained most easily by using Eq. (9.89) rather than Eq. (9.78). This is the form used in Chapter 5 when the Taylor-Galerkin method was applied to Burger's equation and the general flux equation, Eq. (5.25). The Taylor-Galerkin method is obtained by inspection from Eq. (5.53):

$$\frac{1}{6}\frac{c_{i+1}^{n+1} - c_{i+1}^{n}}{\Delta \eta} + \frac{2}{3}\frac{c_{i}^{n+1} - c_{i}^{n}}{\Delta \eta} + \frac{1}{6}\frac{c_{i-1}^{n+1} - c_{i-1}^{n}}{\Delta \eta} = -\frac{M_{i+1}^{n} - M_{i-1}^{n}}{2\Delta \xi} +$$
$$+ \frac{\Delta \eta}{4\Delta \xi^2}\left\{\left[\frac{1}{P^2(c_{i+1}^{n})} + \frac{1}{P^2(c_{i}^{n})}\right][c_{i+1}^{n} - c_{i}^{n}] - \left[\frac{1}{P^2(c_{i}^{n})} + \frac{1}{P^2(c_{i-1}^{n})}\right][c_{i}^{n} - c_{i-1}^{n}]\right\}. \quad (9.99)$$

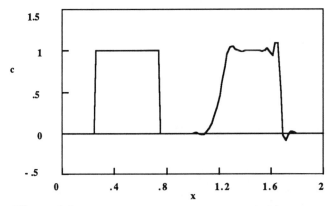

Figure 9.9e. Adsorption with Eq. (9.72), Taylor-Galerkin Method
100 nodes, $\Delta t = 0.005$

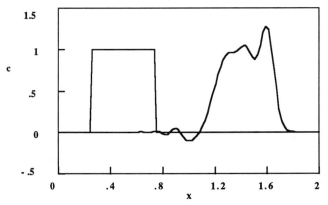

Figure 9.9f. Adsorption with Eq. (9.72), Taylor-Galerkin Method, Lumped in Time
100 nodes, $\Delta t = 0.01$

Again we use the trapezoid rule to evaluate the first term on the right-hand side; we use Eq. (9.95) and Eq. (9.96) to give

$$\frac{1}{6}\frac{c_{i+1}^{n+1} - c_{i+1}^{n}}{\Delta \eta} + \frac{2}{3}\frac{c_{i}^{n+1} - c_{i}^{n}}{\Delta \eta} + \frac{1}{6}\frac{c_{i-1}^{n+1} - c_{i-1}^{n}}{\Delta \eta} = -\frac{1}{4\Delta \xi}\left(\frac{1}{P(c_{i+1}^{n})} + \frac{1}{P(c_{i-1}^{n})}\right)(c_{i+1}^{n} - c_{i}^{n}) +$$

$$+ \frac{\Delta \eta}{4\Delta \xi^2}\left\{\left[\frac{1}{P^2(c_{i+1}^{n})} + \frac{1}{P^2(c_{i}^{n})}\right][c_{i+1}^{n} - c_{i}^{n}] - \left[\frac{1}{P^2(c_{i}^{n})} + \frac{1}{P^2(c_{i-1}^{n})}\right][c_{i}^{n} - c_{i-1}^{n}]\right\}.$$

(9.100)

Results for the Taylor-Galerkin method are shown in Figure 9.9e. These results have oscillations that are worse than those obtained for the advection equation (see Figure 5.8e). The leading edge has a steep front while the trailing edge is smoothed out. If the time-dependent terms are lumped, the method should be similar to the MacCormack method (and identical to it for a linear problem). Figure 9.9f, obtained with the Taylor-Galerkin method and mass

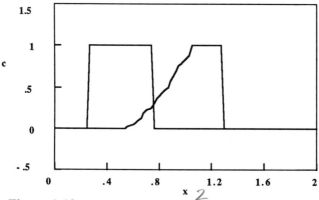

Figure 9.10a. Adsorption with $\gamma = 0.1$, Random Choice Method
100 nodes, $\Delta t = 0.01$, $t = 1.0$

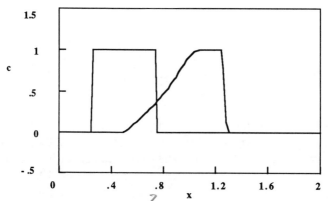

Figure 9.10b. Adsorption with $\gamma = 0.1$, MacCormack Method with Flux-correction
100 nodes, $\Delta t = 0.01$, $t = 1.0$

lumping, looks very similar to Figure 9.9c, obtained with the MacCormack method.

Galerkin method. The Galerkin method can be obtained without rearranging the equation. If the nonlinear coefficient is expanded in terms of the trial function, the result is (see Problem 9.7):

$$\frac{1}{12}[P(c_{i-1}) + P(c_i)]\frac{dc_{i-1}}{dt} + \left[\frac{1}{12}P(c_{i-1}) + \frac{1}{2}P(c_i) + \frac{1}{12}P(c_{i+1})\right]\frac{dc_i}{dt} +$$
$$+ \frac{1}{12}[P(c_i) + P(c_{i+1})]\frac{dc_{i+1}}{dt} = -\frac{c_{i+1} - c_{i-1}}{2\Delta\xi}. \quad (9.101)$$

Since the advection equation solved with the Galerkin method is unstable, we will not try Eq. (9.101). The theory is presented merely to show the process of handling a nonlinear coefficient that multiplies the time derivative.

The effect of changing parameter values is illustrated in Figures 9.10a

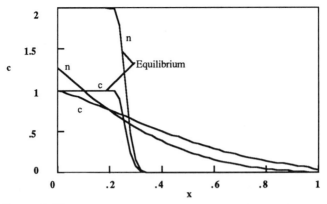

Figure 9.11. Comparison of Models With and Without Equilibrium, K = 0
MacCormack Method with Flux-correction, 51 nodes, $\Delta t = 0.01$

and 9.10b. Here $\gamma = 2$ rather than $\gamma = 0.1$. There is significantly more adsorption and when the pulse passes through, the material desorbs slowly. Figure 9.11 shows the solution for a linear equilibrium line (K = 0) and $\gamma = 2$. The concentration distribution in the fluid retains its shape as a slug; the corresponding concentration on the solid has the same shape but takes values twice as large. Also shown in Figure 9.11 are the curves for c and n, obtained using the equations in Section 9.1. In that case, equilibrium is not assumed and the solution for n changes much more slowly than the model presented above; furthermore, there is no shock in the c curve, which is spread out dramatically. The comparisons shown in Figure 9.11 demonstrate the effect of making the equilibrium assumption, or equivalently, the effect of having an adsorption rate that is very much faster than transport in the flowing medium.

For a final example, let us consider an adsorption problem solved by Hu, et al. [1981]. The equations are similar to those discussed here, except that there is more than one concentration to be followed. Because the changes in inlet conditions happen so fast, the profiles are extremely steep. Typical initial conditions are shown in Figure 9.12. The method of solution was the one discussed in Section 7.3. Additional nodes were added between two existing nodes whenever the second derivative became too large. Several thresholds were used: 2, 4, 8, 16, or 32 nodes were added as the second derivative became larger. The solution at later times is shown in Figure 9.13. The solutions were obtained on a length that went to 1200; however, only the first 240 units are shown in the figures. We can see that the fronts are modeled extremely well using this equidistribution technique. An example using moving finite elements and the method of orthogonal collocation on finite elements can be found in work by Yu and Wang [1986].

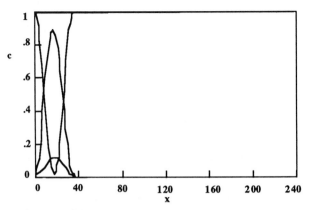

Figure 9.12. Initial Conditions for Adsorption Problem Data from Hu, *et al.* [1981]

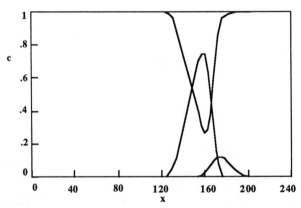

Figure 9.13. Solution to Adsorption Problem, Data from Hu, *et al.* [1981]

9.3. Two-dimensional Cross Flow

We next consider a two-dimensional example of convection and diffusion in a square region where the velocity is a constant flowing along the diagonal (see Figure 9.14). For any velocity, the differential equation is

$$u \frac{\partial c}{\partial x} + v \frac{\partial c}{\partial y} = \frac{\partial}{\partial x}\left(D_{xx} \frac{\partial c}{\partial x} + D_{xy} \frac{\partial c}{\partial y} \right) + \frac{\partial}{\partial y}\left(D_{yx} \frac{\partial c}{\partial x} + D_{yy} \frac{\partial c}{\partial y} \right) + f$$

$$c = g \quad \text{on } \Omega_1$$

$$n_x \left(D_{xx} \frac{\partial c}{\partial x} + D_{xy} \frac{\partial c}{\partial y} \right) + n_y \left(D_{yx} \frac{\partial c}{\partial x} + D_{yy} \frac{\partial c}{\partial y} \right) = h \quad \text{on } \Omega_2.$$

(9.102)

This form of the equation allows the diffusivity tensor to be non-isotropic.

This problem is a <u>steady-state</u> problem so the Taylor-Galerkin method, Lax-Wendroff method, and MacCormack method with and without flux-correction are not applicable. However, grid refinement is possible; this is illustrated

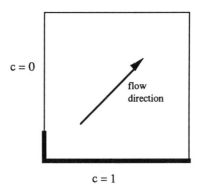

Figure 9.14. Steady-State Convection Problem

below by Demkowicz and Oden [1986] for a more complicated, time-dependent problem. Here the Petrov-Galerkin method is presented.

Petrov-Galerkin method. The Petrov-Galerkin method gives [see Eq. (8.65)]

$$\int_V W_j \left(u \frac{\partial c}{\partial x} + v \frac{\partial c}{\partial y} \right) dx\, dy + \int_V \frac{\partial W_j}{\partial x} \left(D_{xx} \frac{\partial c}{\partial x} + D_{xy} \frac{\partial c}{\partial y} \right) dx\, dy +$$

$$\int_V \frac{\partial W_j}{\partial y} \left(D_{yx} \frac{\partial c}{\partial x} + D_{yy} \frac{\partial c}{\partial y} \right) dx\, dy = \int_V W_j\, f\, dx\, dy + \int_S W_j\, h\, ds. \quad (9.103)$$

Hughes and Brooks [1979] suggested the following form for the artificial diffusion term; the form is a two-dimensional generalization of the quadrature upstream weighting method. The convective term in Eq. (9.103) is evaluated at the quadrature point

$$\tilde{\xi} = (\tilde{\xi}, \tilde{\eta}), \quad (9.104)$$

where

$$\tilde{\xi} = \coth \alpha_\xi - \frac{1}{\alpha_\xi}, \quad \tilde{\eta} = \coth \alpha_\eta - \frac{1}{\alpha_\eta}, \quad (9.105)$$

$$\alpha_\xi = \frac{u_\xi h_\xi}{2D}, \quad \alpha_\eta = \frac{u_\eta h_\eta}{2D}, \quad u_\xi = e_\xi \cdot u, \quad u_\eta = e_\eta \cdot u. \quad (9.106)$$

The values of h_ξ and h_η are determined by the size of the element along the lines $\eta = 0$ and $\xi = 0$, respectively. This scheme, however, has excessive "crosswind diffusion." We want extra diffusion in the direction of the velocity since that is where the strong gradients occur, but we do not want extra diffusion perpendicular to the streamlines since the gradients are generally not steep in that direction.

To avoid crosswind diffusion, Hughes and Brooks [1979] propose adding artificial diffusion to the problem to obtain

$$\int_V W_j \left(u \frac{\partial c}{\partial x} + v \frac{\partial c}{\partial y} \right) dx\, dy + \int_V \frac{\partial W_j}{\partial x} \left[(D_{xx}+\tilde{D}_{xx}) \frac{\partial c}{\partial x} + (D_{xy}+\tilde{D}_{xy}) \frac{\partial c}{\partial y} \right] dx\, dy +$$
$$\int_V \frac{\partial W_j}{\partial y} \left[(D_{yx}+\tilde{D}_{yx}) \frac{\partial c}{\partial x} + (D_{yy}+\tilde{D}_{yy}) \frac{\partial c}{\partial y} \right] dx\, dy = \int_V W_j f\, dx\, dy + \int_S W_j h\, ds. \quad (9.107)$$

The form of the added diffusion is

$$\tilde{D}_{ij} = \tilde{D}\, \hat{u}_i\, \hat{u}_j\,, \quad \hat{u}_i = \frac{u_i}{\|u\|}, \quad \|u\|^2 = \sum_{i=1}^{2} u_i u_i. \quad (9.108)$$

The quantity $\tilde{D}$ is chosen as

$$\tilde{D} = \frac{\tilde{\xi}\, u_\xi\, h_\xi + \tilde{\eta}\, u_\eta\, h_\eta}{2}. \quad (9.109)$$

In practice, when Eq. (9.108) is inserted into Eq. (9.107), we obtain

$$\int_V W_j \left(u \frac{\partial c}{\partial x} + v \frac{\partial c}{\partial y} \right) dx\, dy + \int_V \frac{\partial W_j}{\partial x} \left[\left(D_{xx}+\tilde{D} \frac{uu}{\|u\|^2} \right) \frac{\partial c}{\partial x} + \left(D_{xy}+\tilde{D} \frac{uv}{\|u\|^2} \right) \frac{\partial c}{\partial y} \right] dx\, dy +$$
$$\int_V \frac{\partial W_j}{\partial y} \left[\left(D_{yx}+\tilde{D} \frac{vu}{\|u\|^2} \right) \frac{\partial c}{\partial x} + \left(D_{yy}+\tilde{D} \frac{vv}{\|u\|^2} \right) \frac{\partial c}{\partial y} \right] dx\, dy = \int_V W_j f\, dx\, dy + \int_S W_j h\, ds.$$
$$(9.110)$$

The term involving $\tilde{D}$ can be combined with the convective term to obtain

$$\int_V \tilde{W}_j \left(u \frac{\partial c}{\partial x} + v \frac{\partial c}{\partial y} \right) dx\, dy + \int_V \frac{\partial W_j}{\partial x} \left[D_{xx} \frac{\partial c}{\partial x} + D_{xy} \frac{\partial c}{\partial y} \right] dx\, dy +$$
$$\int_V \frac{\partial W_j}{\partial y} \left[D_{yx} \frac{\partial c}{\partial x} + D_{yy} \frac{\partial c}{\partial y} \right] dx\, dy = \int_V W_j f\, dx\, dy + \int_S W_j h\, ds, \quad (9.111)$$

where

$$\tilde{W}_j = W_j + \tilde{D}\, \frac{\hat{u}\, \frac{\partial W_j}{\partial x} + \hat{v}\, \frac{\partial W_j}{\partial y}}{\|u\|}. \quad (9.112)$$

Two implementations of Eq. (9.111)-(9.112) are presented: one in which one-point quadrature is used on the convection term (SU1) and one in which the regular two-by-two Gaussian quadrature is used (SU2). Typical results are shown in Figure 9.15. None of the results are completely satisfactory, although SU2 is an improvement over the straight Galerkin method. Mizukami and Hughes [1985] tried a method in which diffusion was added in the direction of the gradient rather than the velocity, while Hughes et al. [1986] used discontinuous weighting functions. This method is considerably better than SU1 and SU2 since it has no

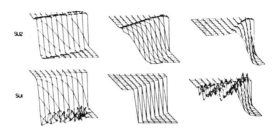

Figure 9.15. Petrov-Galerkin Solution to Steady Flow Problem (Figure 9.14) From Hughes and Brooks [1979]; copyright 1979; reprinted by permission of ASME.

Figure 9.16. Petrov-Galerkin Method with Discontinuous Weighting Functions From Mizukami and Hughes [1985]; copyright 1985; reprinted by permission of Elsevier Science Publishers.

oscillations, keeps a steep front, and is economical to use (see Figure 9.16).

Huyakorn [1977] solved a similar problem using the Petrov-Galerkin method, but for a case with a different velocity function. Smith and Hutton [1982] provide a summary of the solutions obtained using twenty different methods for a steady-state problem. Tzanos [1990] also presents comparisons for the steady, two-dimensional convective diffusion equation, including one method based on an adaptive grid determined by the second derivative of the solution.

9.4. A Moving Cone Problem

Similar techniques have been applied to transient two-dimensional convection problems. They are illustrated here for a problem without adsorption. To include adsorption we would need to generalize the two-dimensional methods, as we did for one-dimensional methods in Sections 9.1 and 9.2. Alternatively, we could generalize the one-dimensional methods with adsorption to include two dimensions, as we did for the cases without adsorption in Chapter 8.

Demkowicz and Oden [1986] used mesh refinement to solve a transient problem with a velocity that was constant in space and time. Their initial condition was a concentration profile in the shape of a Gaussian pulse. Since the velocity is constant in some direction, the pulse should move unchanged in that direction. They solved the problem first on a coarse mesh. If they found that the mesh needed to be refined in some region of the domain, they did so. Then a local problem was solved in this small region of the domain using the solution on the coarse mesh as the boundary conditions. The results achieved using this method are excellent, as shown in Figure 9.17.

Most of the calculations in this section are for a rotating cone problem. We solve the advection equation

$$\frac{\partial c}{\partial t} + u\frac{\partial c}{\partial x} + v\frac{\partial c}{\partial y} = 0, \qquad (9.113)$$

where the velocity is a vortex and satisfies

$$\nabla \cdot \mathbf{u} = 0. \qquad (9.114)$$

The velocity function is obtained by use of the stream function, defined by

$$u = \frac{\partial \psi}{\partial y}, \quad v = -\frac{\partial \psi}{\partial x}. \qquad (9.115)$$

Then the velocity equation becomes

$$\frac{\partial^2 \psi}{\partial x^2} + \frac{\partial^2 \psi}{\partial y^2} = 0. \qquad (9.116)$$

For the vortex, the stream function is

$$\psi = -\frac{1}{2}(x^2 + y^2). \qquad (9.117)$$

The flow field then consists of a rotary motion about the center, $x = y = 0$. The initial condition is taken in the shape of an inverted cone centered at the point (x_0, y_0) with radius r.

$$c(x, y, 0) = \begin{cases} 1 - \frac{z}{r} & z \le r \\ 0, & z > r \end{cases} \quad z^2 = (x-x_0)^2 + y^2 \qquad (9.118)$$

This inverted cone is centered on a point other than the center ($x = x_0$, $y = 0$) (see Figure 9.18). The calculation proceeds long enough for the center of the cone to revolve once completely and return to the initial point. The concentration plot has the shape of an upside-down ice cream cone turning on a record turntable.

The first method we apply is a second-order finite difference method known as Arakawa's method [Orszag, 1971]; this method uses the leapfrog method in time. The results show significant oscillations, as shown in Figure 9.19a. Gresho, et al [1978] indicate that this method is equivalent in accuracy to a lumped finite element method with bilinear trial functions. Orszag [1971] also gives results for a fourth-order Arakawa method (Figure 9.19b). These results are much better; Gresho et al [1978] indicate that this method is equivalent in

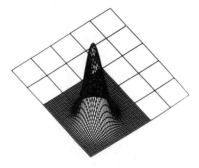

Initial Conditions

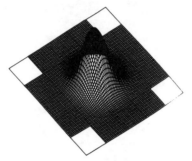

Solution at t = 0.5

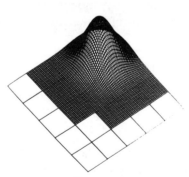

Figure 9.17. Solution to the Moving Gaussian Profile, t = 1.0
From Demkowicz and Oden [1986]; copyright 1986;
reprinted with permission of Elsevier Science Publishers.

accuracy to a finite element method using Lagrangian bilinear functions but with a consistent mass matrix (i.e. not lumping the time derivatives). Using a spectral method (Orszag [1971]) gives the best results (see Figure 9.19c). The consistent mass finite element method with bilinear functions (which is also Arakawa's fourth-order finite difference method) is a suitable method.

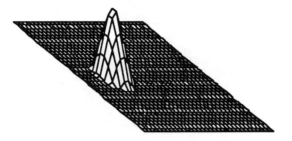

Figure 9.18. Initial Conditions for Inverted Cone Problem

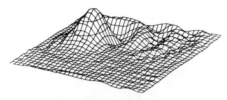

Figure 9.19a. Second-order Finite Difference Method, 32 x 32 grid
From Orszag [1971]; copyright 1971; reprinted with permission of
Cambridge University Press.

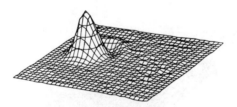

Figure 9.19b. Fourth-order Finite Difference Method, 32 x 32 grid
From Orszag [1971]; copyright 1971; reprinted with permission of
Cambridge University Press.

The upstream method gives the results shown in Figure 9.19d; clearly the cone is damped too much. The MacCormack method presented in Section 8.1 gives the results shown in Figure 9.19e, which have oscillations behind the cone. The Taylor-finite difference method presented in Section 8.1 gives results which are slightly better (see Figure 9.19f). The flux-corrected Taylor-finite difference method presented in Section 8.4 gives results that are slightly better than that (see Figure 9.19g). The ENO results are excellent (see Figure 9.19h).

Methods with moving nodes or nodes that are added and subtracted have also been applied to this problem. Gropp [1980, 1987] used a mesh refinement strategy, as described in Section 8.5. The solution was advanced one

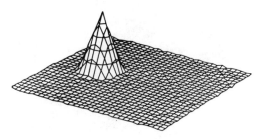

Figure 9.19c. Inverted Cone Problem, Galerkin Spectral Method, 32 x 32 grid
From Orszag [1971]; copyright 1971; reprinted with permission of
Cambridge University Press.

Figure 9.19d. Inverted Cone Problem, Upstream Method, 31 x 31 grid

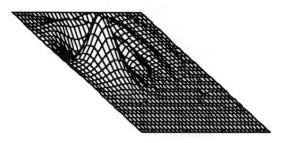

Figure 9.19e. Inverted Cone Problem, MacCormack Method, 31 x 31 grid

time-step, from t to t + Δt. If the gradient on any cell was too big, the cell was subdivided. The solution was restarted from time t and moved to time t + Δt on the refined mesh. The solution is excellent, as shown in Figure 9.20.

The final example is a flux-limited transport method described by Book, *et al.* [1981]. These authors solve a slightly different problem. They begin with an initial condition as shown in Figure 9.21a. The rotary motion causes a solid body rotation. After one revolution the results shown in Figure 9.21b are achieved; these results are excellent.

To obtain excellent results for the rotating cone problem, it is necessary to use the ENO method, refine the mesh, or move the points. The other methods are not satisfactory.

Figure 9.19f. Inverted Cone Problem, Taylor-Finite Difference Method, 31 x 31 grid

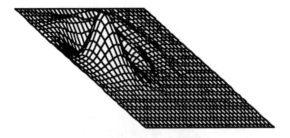

Figure 9.19g. Inverted Cone Problem, Taylor-Finite Difference Method with Flux-correction, 31 x 31 grid

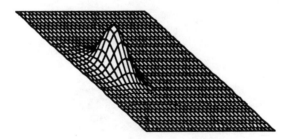

Figure 9.19h. Inverted Cone Problem, ENO Method, 31 x 31 grid

Problems

1_1. Combine Eq. (9.38)-(9.40) to show that they do not give Eq. (9.33)-(9.34).

2_1. Combine Eq. (9.41)-(9.44) to show that they do give Eq. (9.33)-(9.34).

3_2. Write down the Galerkin equations applied to Eq. (9.11)-(9.12), the problem with dispersion.

4_2. Using the option ADSORPTION in the program APPLICATIONS, use the Galerkin method to solve the same problem as is shown in Figure 9.4.

5_1. Write an essay comparing the results in Figures 9.7 with the results for the advection

Figure 9.20. Inverted Cone Problem, Adaptive Mesh, from Gropp [1987] Reprinted with permission from SIAM Journal on Scientific aand Statistical Computing, pp. 292-304. Copyright 1987 by the Society for Industrial and Applied Mathematics. All rights reserved.

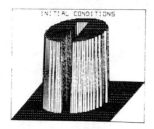

Figure 9.21a. Initial Conditions

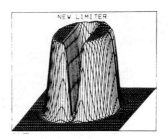

Figure 9.21b. Solid Body Rotation, Flux-limited Transport Method
From Book, *et al.* [1981]; copyright 1981; reprinted by permission of Springer-Verlag.

equation derived earlier using the same methods.

6_1. Derive the linear case of Eq. (9.59) from Eq. (9.1)-(9.2) by assuming the mass transfer coefficient goes to infinity. Divide Eq. (9.2) by k and let $k \to \infty$. This gives the equation relating c and n.

$$\frac{1}{k}\frac{\partial n}{\partial t} = \gamma c - n, \quad \lim_{k \to \infty} \frac{1}{k}\frac{\partial n}{\partial t} = \gamma c - n = 0 \qquad (9.119)$$

Differentiate the last equation to obtain the time derivative of n with respect to t and substitute the result into Eq. (9.1) to get Eq. (9.59) for the linear case. Note that c and n are linearly related at all times; their time derivatives are non-zero.

$$\frac{\partial c}{\partial t} \neq 0, \quad \frac{\partial n}{\partial t} \neq 0 \qquad (9.120)$$

However, because k is large, we have $\frac{1}{k}\frac{\partial n}{\partial t} = 0$. \hfill (9.121)

7₃. Derive Eq. (9.101). Expand P(c) in terms of the trial function and calculate the finite element integrals. Also give the simplification of Eq. (9.101) when the time derivatives are lumped.

References

Bassingthwaighte, J. B., "A Concurrent Flow Model for Extraction during Transcapillary Passage," Circ. Res. 35 483-503 [1974].

Book, D. L. (ed.), *Finite-Difference Techniques for Vectorized Fluid Dynamics Calculations*, Springer-Verlag [1981].

Book, D. L., Boris, J. P., and Zalesak, S. T., "Flux-Corrected Transport," pp. 29-55 in *Finite-Difference Techniques for Vectorized Fluid Dynamics Calculations*, Book D. L. (ed.), Springer-Verlag [1981].

Demkowicz, L. and Oden, J. T., "An Adaptive Characteristic Petrov-Galerkin Finite Element Method for Convection-dominated Linear and Nonlinear Parabolic Problems in Two Space Variables," Comp. Meth. Appl. Mech. Eng. 55 63-87 [1986].

Donea, J., Giuliani, S., Laval, H., Quartapelle, L, "Time-Accurate Solution of Advection-Diffusion Problems by Finite Elements," Comp. Meth. Appl. Mech. Eng. 45 123-145 [1984].

Farooq, S. and Ruthven, D. M., "Dynamics of Kinetically Controlled Binary Adsorption in a Fixed Bed," A. I. Ch. E. J. 37 299-301 [1991].

Gresho, P. M., Lee, R. L., Sani, R. L., "Advection-Dominated Flows, with Emphasis on the Consquences of Mass Lumping," pp.335-350 in *Finite Elements in Fluids*, Vol. 3, R. H. Gallagher, *et al.* (ed.) [1976].

Gropp, W. D., "A Test of Moving Mesh Refinement for 2-D Scalar Hyperbolic Problems," SIAM J. Sci. Stat. Comp. 1 191-197 [1980].

Gropp W. D., "Local Uniform Mesh Refinement with Moving Grids," SIAM J. Sci. Stat. Comp. 8 292-304 [1987]

Hu, S. S., Didwania, A. K., May, W. G., Pirkle, J. C., Jr. and Schiesser, W. E., "An Adaptive Grid for the Computer Simulation of Separation Systems," private communication [1987].

Hughes, T. J. R. and Brooks, A., "A Multi-dimensional Upwind Scheme with no Crosswind Diffusion," pp. 19-35 in *Finite Element Methods for Convection Dominated Flows*, AMD, Vol. 34, ASME [1979].

Hughes, T. J. R., Mallet, M. and Mizukami, A., "A New Finite Element Formulation for Computational Fluid Dynamics: II. Beyond SUPG", Comp. Meth. Appl. Mech. Eng. 54 341-355 [1986].

Huyakorn, P. S., "Solution of steady-state, convective transport equation using an upwind finite element scheme," Appl. Math. Modeling 1 187-195 [1977].

Lenhoff, A. M. and Lightfoot, E. N., "The Effects of Axial Diffusion and Permeability Barriers on the Transient Response of Tissue Cylinders. II. Solution in Time Domain," J. Theor. Biol. 106 207-238 [1984].

Mizukami, A. and Hughes, T. J. R., "A Petrov-Galerkin Finite Element Method for Convection-dominated Flows: An Accurate Upwinding Technique for Satisfying the Maximum Principle," Comp. Meth. Appl. Mech. Eng. 50 181-193 [1985].

Orszag, S. A., "Numerical Simulation of Incompressible Flows within Simple Boundaries: Accuracy," J. Fluid Mech. 49 75-112 [1971].

Rhee, H. K., Aris, R. and Amundson, N. R., *First-Order Partial Differential Equations, Vol. I, Theory and Application of Single Equations*, Prentice-Hall [1986].

Smith, R. M., and Hutton, A. G., "The Numerical Treatment of Advection: A Performance Comparison of Current Methods," Num. Heat Trans. 5 439-461 [1982].

Sod, G. A., *Numerical Methods in Fluid Dynamics*, Cambridge University Press [1985].

Sun, L. M. and Meunier, F., "An Improved Finite Difference method for Fixed-Bed Multicomponent Sorption," A. I. Ch. E. J. 37 244-254 [1991].

Tzanos, C. P., "Central Difference-Like Approximation for the Solution of the Convection-Diffusion Equation," Num. Heat Trans., Part B: Fund. 18 97-112 [1990].

Yu, Q., and Wang, N. H. L., "Computer Simulation of the Dynamics of Multicomponent Ion Exchange and Adsorption in Fixed Beds - Gradient-directed Moving Finite Element Method," Comp. Chem. Eng. 13 915-926 [1986].

CHAPTER
TEN

CONVECTION AND DIFFUSION WITH REACTION

In previous chapters we discussed moving fronts that were caused by convection. In this chapter we will add the complication of a chemical reaction and discuss applications to several problems. Moving fronts are caused by the convection that accompanies reaction; moving fronts can also be caused by reaction and diffusion alone (without convection). The moving fronts may be sharp (and frequently are) because the reactant gets used up. It is possible for the reaction front to move at some velocity other than the velocity of the fluid; numerical methods must be capable of resolving this movement. The first problem to be discussed is a model problem with an exact solution; for comparison, several methods will be applied to this problem. The second problem involves finding the flame speed for combustion problems (a steady-state one-dimensional problem). The third problem involves a transient case with a moving combustion zone. For all three of these problems the techniques of equidistribution or moving nodes are useful. The fourth problem is a transient model of a catalytic converter. Finally, we will discuss reaction diffusion problems without the convection term and show how the moving coordinate system can be easily used to give good results.

10.1. A Model Problem With Convection

Let us consider a cylindrical tube that is loosely packed with a catalyst. A gas flows around the catalyst particles, mixing as it goes. The chemicals in the gas can react, however, and this reaction is fast in the presence of the catalyst. Thus, when the chemicals are on the catalyst they react but when they are in the gas they do not. It is common to model this phenomenon using a packed bed model with two phases: a gas phase and a solid phase. The concentrations in the gas phase are averaged over a small region, as are the concentrations on the solid phase. Because of this averaging, we ignore the geometrical complications of the irregular arrangement of the catalyst particles. We then need to solve for the gas concentration and the solid concentration of each chemical species at every point in the packed bed. For this model problem we assume that the reaction is irreversible, involves only two components, and has one component available in great excess. In this case, the concentration of the component that is in great excess remains essentially constant, although a small part of it reacts.

The differential equation is

$$\frac{\partial c}{\partial t} + u \frac{\partial c}{\partial x} + k_m a_v (c - c_s) = 0, \qquad (10.1)$$

where c is the gas phase concentration, c_s is the solid phase concentration, u is the velocity of the reaction mixture, k_m is the mass transfer coefficient, a_v is the ratio of catalyst surface area to volume of the tube, and k is the first-order kinetic coefficient. The first term in Eq. (10.1) represents the accumulation or time rate of change of concentration; the second term represents convection down the tube; and the third term represents mass transfer between the gas phase and the solid phase. Because there is a mass transfer resistance, this transfer is not instantaneous. We thus need an equation that relates the rate of mass transfer to the rate of chemical reaction; we assume they are equal for this problem.

$$k_m a_v (c - c_s) = k c_s \qquad (10.2)$$

Initially the gas has some concentration,

$$c(x,0) = f(x), \qquad (10.3)$$

and the inlet concentration takes the value

$$c(0,t) = g(t). \qquad (10.4)$$

We can rearrange Eq. (10.2) to give

$$c_s = \frac{k_m a_v}{k_m a_v + k} c. \qquad (10.5)$$

Then

$$c - c_s = c \left[1 - \frac{k_m a_v}{k_m a_v + k} \right] = \frac{k c}{k_m a_v + k}. \qquad (10.6)$$

The original equation, Eq. (10.1), is then

$$\frac{\partial c}{\partial t} + u \frac{\partial c}{\partial x} + \frac{k k_m a_v}{k_m a_v + k} c = 0. \qquad (10.7)$$

It is instructive to look at the time constants in this equation. They are

$$t_{conv} = \frac{L}{u}, \quad t_{mass\, tr.} = \frac{1}{k_m a_v}, \quad t_{rxn} = \frac{1}{k}, \quad \text{or } t_{rxn-mass} = \frac{k_m a_v + k}{k k_m a_v} = \frac{1}{k} + \frac{1}{k_m a_v}. \qquad (10.8)$$

We can anticipate that the problem will be especially difficult when these time constants have widely differing magnitudes. Eq. (10.7) is made non-dimensional using

$$t' = \frac{tu}{L}, \quad x' = \frac{x}{L}, \quad c' = \frac{c}{c_0}, \tag{10.9}$$

where L is the length of the tube. The non-dimensional equation is then

$$\frac{\partial c'}{\partial t'} + \frac{\partial c'}{\partial x'} + R c' = 0, \tag{10.10}$$

where

$$R = \frac{k \, k_m \, a_v}{k_m \, a_v + k} \frac{L}{u}. \tag{10.11}$$

Eq. (10.11) is merely a ratio of the pertinent time constants, so $R \gg 1$ or $R \ll 1$ are expected to be important ranges of R in which the problem is especially difficult. We could absorb the parameter R into t' and x' if the tube is of indefinite length. Because of the form of Eq. (10.10), the length of the tube is irrelevant; we simply integrate until we reach the end, but the location of the end does not influence what happens upstream. It is convenient to retain the parameter R, however, since it occurs in the more general equations used below. We drop the primes and state the complete problem as

$$\frac{\partial c}{\partial t} + \frac{\partial c}{\partial x} + R c = 0.$$

$$c(x, 0) = f(x). \tag{10.12}$$

$$c(0, t) = g(t).$$

The solution to this problem is provided by Rhee, et al. [1986]. Using the method of characteristics, we compare it to the standard semi-linear equation

$$P \frac{\partial c}{\partial t} + Q \frac{\partial c}{\partial x} = R(x, t, c). \tag{10.13}$$

The method of characteristics solves Eq. (10.13) by solving the following ordinary differential equations:

$$\frac{dt}{ds} = P, \quad \frac{dx}{ds} = Q, \quad \frac{dc}{ds} = R(x, t, c). \tag{10.14}$$

In the case of Eq. (10.12) we get

$$\frac{dt}{ds} = 1, \quad \frac{dx}{ds} = 1, \quad \frac{dc}{ds} = -R c. \tag{10.15}$$

The characteristics are straight lines since the right-hand sides of both x- and t-derivatives are constants. Figure 10.1 shows these characteristics. One of the characteristics, the one going through the origin, is an important one. The solution below this dividing characteristic is influenced by the initial concentration, $f(x)$, but not by the boundary conditions. This region represents the fluid that is in the tube initially and is pushed out of the tube before any changes at the inlet can have

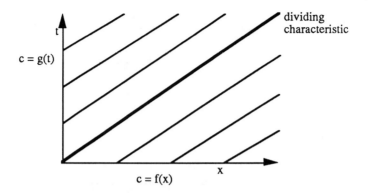

Figure 10.1. Characteristics for the Adsorption Problem

an impact. The region above the dividing characteristic is the region that is unaffected by the initial conditions; this region is fully determined by the boundary condition at the inlet to the tube. Thus we can separate the problem into two subproblems that are independent of each other. We first solve the subproblem within the region below the dividing characteristic. The functions $x(s)$, $t(s)$, and $c(s)$ are all obtained by solving Eq. (10.15). The initial conditions are taken as some point along the x-axis at time zero:

$$t(0) = 0, \quad x(0) = \xi, \quad c(0) = f(\xi). \tag{10.16}$$

Integrating Eq. (10.15) with these initial conditions gives

$$t = s, \quad x = \xi + s, \quad c = A e^{-Rs}, \quad A = f(\xi) \text{ so that } c(s,\xi) = f(\xi) e^{-Rs}. \tag{10.17}$$

By rearranging Eq. (10.15) we can write

$$s = t, \quad \xi = x - t, \quad c(x,t) = f(x-t) e^{-Rt}. \tag{10.18}$$

To examine the solution, let us consider the point where $x = t$ (i. e., the point moving with the velocity of the fluid). The solution there is

$$c(x = t) = f(0) e^{-Rt}, \tag{10.19}$$

which is a decreasing function of time. These points are plotted in Figure 10.2.

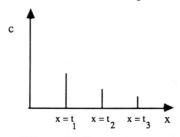

Figure 10.2. Solution at $x = t$

If the initial condition is zero, then the function f(x) is

$$f(x) = \begin{cases} 1 & \text{for } x \leq 0 \\ 0 & \text{for } x > 0. \end{cases} \qquad (10.20)$$

Alternatively,

$$f(\xi) = \begin{cases} 1 & \text{for } \xi = 0 \\ 0 & \text{for } \xi > 0 \end{cases} \qquad (10.21)$$

or

$$f(x-t) = \begin{cases} 1 & \text{for } x - t = 0 \\ 0 & \text{for } x - t > 0. \end{cases} \qquad (10.22)$$

For the subproblem in the region above the dividing characteristic, we solve Eq. (10.15) with the initial conditions

$$t(0) = \eta, \quad x(0) = 0, \quad c(0) = g(\eta). \qquad (10.23)$$

The solution is

$$t = \eta + s, \quad x = s, \quad c(s, \eta) = g(\eta) e^{-Rs}. \qquad (10.24)$$

We rearrange the variables to give

$$\eta = t - s = t - x, \quad s = x \qquad (10.25)$$

and the solution is

$$c(x, t) = g(t-x) e^{-Rx}. \qquad (10.26)$$

This solution is valid for $\eta > 0$ (or $t > x$) and gives the profile behind the front. If the inlet conditions are constant ($g = 1$), we have

$$c(x, t) = e^{-Rx}. \qquad (10.27)$$

This solution is plotted in Figure 10.3.

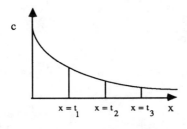

Figure 10.3. Solution to the Reaction Problem in a Packed Bed

The same type of analysis can be applied to a nonlinear reaction:

$$\frac{\partial c}{\partial t} + \frac{\partial c}{\partial x} + R(c) = 0. \tag{10.28}$$

The characteristics are determined by

$$\frac{dt}{ds} = 1, \quad \frac{dx}{ds} = 1, \quad \frac{dc}{ds} = -R(c). \tag{10.29}$$

Since the characteristic for c does not depend on x and t, the same type of analysis can be done, although the equation for c may be more difficult to solve analytically.

Next let us consider using numerical methods to solve such equations. The method of characteristics can be applied by simply integrating Eq. (10.15) directly. Then the solution c is known at certain unspecified values of $\{x,t\}$. Such methods are difficult to apply in two dimensions, however, so we will consider methods on a fixed grid. The method of characteristics also cannot be applied when diffusion is included, even if the amount of diffusion is very small. The applicable methods are the same ones that were used in Chapter 9. We will consider an explicit, upstream method, the MacCormack method with and without flux-correction, the Taylor-finite difference method, and the Taylor-Galerkin method. In order to be as general as possible, the general nonlinear convection term will be included as well. The equation is thus taken in the form

$$\frac{\partial u}{\partial t} + \frac{\partial F}{\partial x} = G(u). \tag{10.30}$$

The **explicit, upstream method** is simply

$$\frac{u_i^{n+1} - u_i^n}{\Delta t} = -\frac{F_i^n - F_{i-1}^n}{\Delta x} + G_i^n. \tag{10.31}$$

The **MacCormack method** is derived according to Eq. (9.41)-(9.43) to give

$$\frac{u*_i^{n+1} - u_i^n}{\Delta t} = -\frac{F_{i+1}^n - F_i^n}{\Delta x} + G_i^n$$

$$\frac{u_i^n - \frac{1}{2}(u_i^n + u*_i^{n+1})}{\Delta t} = -\frac{F*_i^{n+1} - F*_{i-1}^{n+1}}{2\Delta x} + \frac{1}{2} G*_i^{n+1}. \tag{10.32}$$

Taylor-finite difference method. The Taylor terms for Eq. (10.30) are derived in Problem 10.1 to give

$$\frac{u^{n+1} - u^n}{\Delta t} = -\frac{\partial F}{\partial x} + G + \frac{\Delta t}{2}\left\{\frac{\partial}{\partial x}\left[\left(\frac{dF}{du}\right)^2\frac{\partial u}{\partial x}\right] - \frac{\partial}{\partial x}\left[\frac{dF}{du}G\right] + \frac{dG}{du}\left(G(u) - \frac{\partial F}{\partial x}\right)\right\}. \tag{10.33}$$

The Taylor-finite difference method gives

$$\frac{u_i^{n+1} - u_i^n}{\Delta t} = -\frac{F_{i+1}^n - F_{i-1}^n}{2\Delta x} + G_i^n +$$

$$+ \frac{\Delta t}{4\Delta x^2}\left\{\left[\left(\frac{dF}{du}\right)_{i+1}^2 + \left(\frac{dF}{du}\right)_i^2\right][u_{i+1}^n - u_i^n] - \left[\left(\frac{dF}{du}\right)_i^2 + \left(\frac{dF}{du}\right)_{i-1}^2\right][u_i^n - u_{i-1}^n]\right\} \quad (10.34)$$

$$- \frac{\Delta t}{4\Delta x}\left\{\left[\frac{dF}{du}G\right]_{i+1} - \left[\frac{dF}{du}G\right]_{i-1}\right\} + \frac{\Delta t}{2}\left(\frac{dG}{du}G\right)_i - \frac{\Delta t}{4}\left(\frac{dG}{du}\right)_i \frac{F_{i+1} - F_{i-1}}{\Delta x}.$$

Taylor-Galerkin method. The Taylor-Galerkin method is more complicated; the convection terms are the same as derived in Eq. (5.53). The terms involving only the F function are

$$\frac{1}{6}\frac{u_{i+1}^{n+1} - u_{i+1}^n}{\Delta t} + \frac{2}{3}\frac{u_i^{n+1} - u_i^n}{\Delta t} + \frac{1}{6}\frac{u_{i-1}^{n+1} - u_{i-1}^n}{\Delta t} = -\frac{F_{i+1}^n - F_{i-1}^n}{2\Delta x} +$$

$$+ \frac{\Delta t}{4\Delta x^2}\left\{\left[\left(\frac{dF}{du}\right)_{i+1}^2 + \left(\frac{dF}{du}\right)_i^2\right][u_{i+1}^n - u_i^n] - \left[\left(\frac{dF}{du}\right)_i^2 + \left(\frac{dF}{du}\right)_{i-1}^2\right][u_i^n - u_{i-1}^n]\right\} \quad (10.35)$$

+ extra terms.

The terms involving the reaction rate term can be evaulated several ways. If Gaussian quadrature is used, we get

$$\int_0^1 N_i G(u) dx = \sum_e \frac{\Delta x_e}{2} \sum_{IG} WG_{IG}[N_I G(u^e)]|_{\xi_{IG}}. \quad (10.36)$$

If we interpolate the reaction rate term onto the finite element trial functions,

$$G = \sum_K G_K N_K(\xi), \quad (10.37)$$

then we get (where $\Delta x_e = \Delta x$ for all elements)

$$\sum_e \sum_K G_K \frac{\Delta x_e}{2} \int_{-1}^1 N_I(\xi) N_K(\xi) d\xi = \frac{\Delta x}{2}\left(\frac{1}{3}G_{i+1} + \frac{4}{3}G_i + \frac{1}{3}G_{i-1}\right). \quad (10.38)$$

Finally, we could lump this term, giving

$$\Delta x\, G_i, \quad (10.39)$$

which is the finite difference form. When the complete equation is developed, it is divided by Δx; this Δx value is the same in every element, so the contribution to Eq. (10.35) is

Gaussian quadrature, extra terms $= \sum_e \frac{1}{2} \sum_{IG} WG_{IG} [N_I G(u^e)]|_{\xi_{IG}}$, (10.40)

finite element interpolation, extra terms $= \left(\frac{1}{6} G_{i+1} + \frac{2}{3} G_i + \frac{1}{6} G_{i-1}\right)$, (10.41)

mass lumping, extra terms $= G_i$. (10.42)

Next consider another term needed in the Taylor-Galerkin method:

$$\int_0^1 N_i \frac{\partial}{\partial x}\left[\frac{dF}{du} G\right] dx = \int_0^1 \frac{\partial}{\partial x}\left[N_i \frac{dF}{du} G\right] dx - \int_0^1 \frac{\partial N_i}{\partial x}\left[\frac{dF}{du} G\right] dx. \quad (10.43)$$

We integrate this by parts and interpolate the nonlinear function onto a finite element basis set,

$$\frac{dF}{du} G = \sum_K \left[\frac{dF}{du} G\right]_K N_K. \quad (10.44)$$

When the finite element integrals are evaluated and divided by Δx we get (Problem 10.2):

$$\frac{\left[\frac{dF}{du} G\right]_{i+1} - \left[\frac{dF}{du} G\right]_{i-1}}{2 \Delta x}. \quad (10.45)$$

The term that involves $G \, dG/du$ can be handled in the same way that the term involving G was handled. Interpolating this function onto the finite element basis set gives

$$G \frac{dG}{du} = \sum_K \left(G \frac{dG}{du}\right)_K N_K(\xi). \quad (10.46)$$

This gives the Galerkin terms (see Problem 10.3)

$$\int_0^1 N_i G \frac{dG}{du} dx = \Delta x \left[\frac{1}{6}\left(G \frac{dG}{du}\right)_{i+1} + \frac{2}{3}\left(G \frac{dG}{du}\right)_i + \frac{1}{6}\left(G \frac{dG}{du}\right)_{i-1}\right]. \quad (10.47)$$

The last term to be evaluated is

$$\int_0^1 N_i \frac{dG}{du} \frac{\partial F}{\partial x} dx. \quad (10.48)$$

We write this as

$$\int_0^1 N_i \frac{dG}{du} \frac{dF}{du} \frac{\partial u}{\partial x} dx \text{ with } \frac{dG}{du} \frac{dF}{du} = \sum_K \left(\frac{dG}{du} \frac{dF}{du}\right)_K N_K(\xi) \quad (10.49)$$

and get

$$\sum_L \sum_K \left(\frac{dG}{du}\frac{dF}{du}\right)_K \int_{-1}^{1} N_K N_J \frac{dN_L}{d\xi} d\xi\, u_L. \tag{10.50}$$

The derivative is constant within an element ($\pm 1/2$) and the remaining integral is

$$\int_{-1}^{1} N_K N_J \, d\xi = \begin{bmatrix} \frac{2}{3} & \frac{1}{3} \\ \frac{1}{3} & \frac{2}{3} \end{bmatrix} \begin{matrix} K=1 \\ K=2 \end{matrix}. \tag{10.51}$$

$$J=1 \quad J=2$$

Careful accounting gives the result

$$\frac{1}{6}\left(\frac{dG}{du}\frac{dF}{du}\right)_{i-1}(u_i - u_{i-1}) + \frac{1}{3}\left(\frac{dG}{du}\frac{dF}{du}\right)_i (u_{i+1} - u_{i-1}) + \frac{1}{6}\left(\frac{dG}{du}\frac{dF}{du}\right)_{i+1}(u_{i+1} - u_i). \tag{10.52}$$

Combining all results then gives the Taylor-Galerkin method, with all nonlinear terms evaluated on the finite element basis set:

$$\frac{1}{6}\frac{u_{i+1}^{n+1} - u_{i+1}^n}{\Delta t} + \frac{2}{3}\frac{u_i^{n+1} - u_i^n}{\Delta t} + \frac{1}{6}\frac{u_{i-1}^{n+1} - u_{i-1}^n}{\Delta t} = -\frac{F_{i+1}^n - F_{i-1}^n}{2\Delta x} +$$

$$+ \frac{\Delta t}{4\Delta x^2}\left\{\left[\left(\frac{dF}{du}\right)_{i+1}^2 + \left(\frac{dF}{du}\right)_i^2\right][u_{i+1}^n - u_i^n] - \left[\left(\frac{dF}{du}\right)_i^2 + \left(\frac{dF}{du}\right)_{i-1}^2\right][u_i^n - u_{i-1}^n]\right\} +$$

$$+ \left(\frac{1}{6}G_{i+1} + \frac{2}{3}G_i + \frac{1}{6}G_{i-1}\right) - \frac{\Delta t}{4\Delta x}\left\{\left[\left(\frac{dF}{du}\right)^2 G\right]_{i+1} - \left[\left(\frac{dF}{du}\right)^2 G\right]_{i-1}\right\} +$$

$$+ \frac{\Delta t}{2}\left[\frac{1}{6}\left(G\frac{dG}{du}\right)_{i+1} + \frac{2}{3}\left(G\frac{dG}{du}\right)_i + \frac{1}{6}\left(G\frac{dG}{du}\right)_{i-1}\right] +$$

$$- \frac{\Delta t}{2\Delta x}\left[\frac{1}{6}\left(\frac{dG}{du}\frac{dF}{du}\right)_{i-1}(u_i - u_{i-1}) + \frac{1}{3}\left(\frac{dG}{du}\frac{dF}{du}\right)_i (u_{i+1} - u_{i-1}) + \right.$$

$$\left. + \frac{1}{6}\left(\frac{dG}{du}\frac{dF}{du}\right)_{i+1}(u_{i+1} - u_i)\right]. \tag{10.53}$$

If the reaction terms are evaluated using Gaussian quadrature or mass lumping, then one set of terms is changed, as indicated in Eq. (10.40)-(10.42).

Equations without convection. If there is no convection term but there are reaction terms, Eq. (10.33) reduces to

$$\frac{u^{n+1} - u^n}{\Delta t} = G + \frac{\Delta t}{2}\frac{dG}{du} G(u). \tag{10.54}$$

There are no spatial derivatives to create difficulties. The values of u at adjacent nodes do not interact, so the fronts can be as sharp as can be represented by the grid; no spurious oscillations are expected from this source. Any sharp fronts that appear in space must arise from the initial condition or the final condition, both of which thereby determine the maximum possible value of Δx. The Taylor-Galerkin method applied to Eq. (10.54) is

$$\frac{1}{6}\frac{u_{i+1}^{n+1}-u_{i+1}^n}{\Delta t}+\frac{2}{3}\frac{u_i^{n+1}-u_i^n}{\Delta t}+\frac{1}{6}\frac{u_{i-1}^{n+1}-u_{i-1}^n}{\Delta t}=\left(\frac{1}{6}G_{i+1}+\frac{2}{3}G_i+\frac{1}{6}G_{i-1}\right)+$$
$$+\frac{\Delta t}{2}\left[\frac{1}{6}\left(G\frac{dG}{du}\right)_{i+1}+\frac{2}{3}\left(G\frac{dG}{du}\right)_i+\frac{1}{6}\left(G\frac{dG}{du}\right)_{i-1}\right]. \quad (10.55)$$

If the reaction is first-order ($G = -k u$), then the problem is

$$\frac{du}{dt}=-k u, \quad u(0)=u_0. \quad (10.56)$$

The solution is

$$u = u_0 e^{-kt}. \quad (10.57)$$

The explicit, upstream method gives the following for Eq. (10.54):

$$u^{n+1}=u^n\left[1-k\Delta t+\frac{k^2\Delta t^2}{2}\right]. \quad (10.58)$$

The MacCormack method becomes a predictor-corrector method,

$$\frac{u*_i^{n+1}-u_i^n}{\Delta t}=-k u_i^n, \quad (10.59)$$

$$\frac{u_i^{n+1}-\frac{1}{2}(u_i^n+u*_i^{n+1})}{\Delta t}=-\frac{1}{2}k u*_i^{n+1}. \quad (10.60)$$

If Eq. (10.59)-(10.60) are combined, they give Eq. (10.58) exactly. The Taylor-Galerkin method gives

$$\frac{1}{6}\frac{u_{i+1}^{n+1}-u_{i+1}^n}{\Delta t}+\frac{2}{3}\frac{u_i^{n+1}-u_i^n}{\Delta t}+\frac{1}{6}\frac{u_{i-1}^{n+1}-u_{i-1}^n}{\Delta t}=-k\left(\frac{1}{6}u_{i+1}^n+\frac{2}{3}u_i^n+\frac{1}{6}u_{i-1}^n\right)+$$
$$+\frac{\Delta t}{2}k^2\left[\frac{1}{6}u_{i+1}^n+\frac{2}{3}u_i^n+\frac{1}{6}u_{i-1}^n\right]. \quad (10.61)$$

This equation is correct to $O(\Delta t^2)$. The Taylor-Galerkin method also has a term of $O(\Delta x^2)$.

The Taylor-Galerkin method has been applied to combustion problems by Chung, et al. [1987]. A streamwise, upstream Petrov-Galerkin method has been developed for reaction-convection-diffusion problems by Tezduyar, et al. [1987], including applications to two-dimensional problems. The Petrov-

Galerkin method will not be applied here.

Example calculations. For the example calculations, we set $u = c$ and take the convection term as $F = c$ and the reaction rate for a reaction with a Langmuir isotherm. Here Da_I is the Damköhler number, a dimensionless reaction rate.

$$G(c) = \frac{Da_I\, c}{1 + \alpha\, c} \tag{10.62}$$

The differential equation to be solved is

$$\frac{\partial c}{\partial t} + \frac{\partial c}{\partial x} = \frac{Da_I\, c}{1 + \alpha\, c}. \tag{10.63}$$

The standard problem has zero concentration in the packed bed initially but the inlet concentration jumps suddenly to some value (taken as 1.0 in non-dimensional form). The solution should then exhibit a moving front down the packed bed. Here we wish the reactant to be used up, so we use a negative Damköhler (Da_I) value. Calculations are presented below for $Da_I = -0.1$ and $\alpha = 2$.

Figure 10.4a shows the results obtained using the upstream finite difference method; the solution is smoothed out. Figure 10.4b gives the solution for the MacCormack method and 100 nodes: the solution oscillates. The effect of flux-correction is seen in Figure 10.4c, which has no oscillations. The Taylor-Galerkin method gives the results shown in Figure 10.4d, which are not as good. Clearly the best method is the MacCormack method with flux-correction.

If the reaction rate is increased, the steep front is missing. We change the Damköhler number to -10 (100 times larger). Figure 10.5 shows a solution obtained using the MacCormack method (without flux-correction) and only 26 nodes. This solution is smooth; during the time shown it reaches a steady-state condition. If the problem represented a chemical reactor, the reactant would have all reacted before $x = 0.6$. Thus, in this problem a larger reaction rate term makes the problem easier to solve. If the reaction rate term is small, the problem resembles the advection equation and is more difficult to solve. Indeed, the behavior shown in Figures 10.4a-d is similar to the results obtained with the advection equation.

10.2. A Model Problem with Convection and Diffusion

If diffusion or dispersion is allowed in the x-direction, a diffusion term can be added to Eq. (10.30), giving

$$\frac{\partial u}{\partial t} + \frac{\partial F(u)}{\partial x} = \frac{\partial^2 u}{\partial x^2} + G(u). \tag{10.64}$$

The explicit, upstream method gives

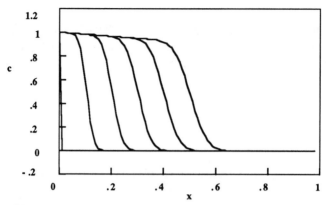

Figure 10.4a. Reaction with Convection, Upstream Method, $\Delta x = 0.01$, $\Delta t = 0.005$ Reaction Rate in Eq. (10.62), $Da_I = -0.1$, $\alpha = 2$

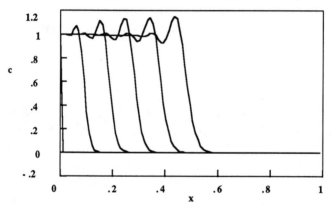

Figure 10.4b. Reaction with Convection, MacCormack Method, $\Delta x = 0.01$, $\Delta t = 0.005$ Reaction Rate in Eq. (10.62), $Da_I = -0.1$, $\alpha = 2$

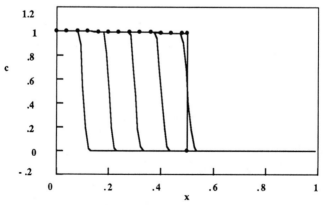

Figure 10.4c. Reaction with Convection, MacCormack Method with Flux-correction $\Delta x = 0.01$, $\Delta t = 0.005$, Reaction Rate in Eq. (10.62), $Da_I = -0.1$, $\alpha = 2$

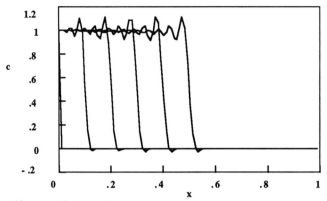

Figure 10.4d. Reaction with Convection, Taylor-Galerkin Method Reaction Rate in Eq. (10.62), $Da_I = -0.1$, $\alpha = 2$, $\Delta x = 0.01$, $\Delta t = 0.005$

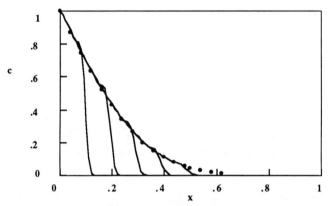

Figure 10.5. Reaction with Convection, MacCormack Method with Flux-correction $\Delta x = 0.01$, $\Delta t = 0.005$, Reaction Rate in Eq. (10.62), $Da_I = -10$, $\alpha = 2$

$$\frac{u_i^{n+1} - u_i^n}{\Delta t} = -\frac{F_i^n - F_{i-1}^n}{\Delta x} + G_i^n + \frac{u_{i+1}^n - 2u_i^n + u_{i-1}^n}{\Delta x^2}. \quad (10.65)$$

The MacCormack method gives

$$\frac{u*_i^{n+1} - u_i^n}{\Delta t} = -\frac{F_{i+1}^n - F_i^n}{\Delta x} + G_i^n + \frac{u_{i+1}^n - 2u_i^n + u_{i-1}^n}{\Delta x^2}$$

$$\frac{u_i^{n+1} - \frac{1}{2}(u_i^n + u*_i^{n+1})}{\Delta t} = -\frac{F*_i^{n+1} - F*_{i-1}^{n+1}}{2\Delta x} + \frac{1}{2}G*_i^{n+1} + \frac{u*_{i+1}^{n+1} - 2u*_i^{n+1} + u*_{i-1}^{n+1}}{2\Delta x^2}. \quad (10.66)$$

The Taylor form of the equation is

$$\frac{u^{n+1} - u^n}{\Delta t} = -\frac{\partial F}{\partial x} + G + \frac{\partial^2 u}{\partial x^2}$$

$$+ \frac{\Delta t}{2} \left\{ \frac{\partial}{\partial x} \left[\left(\frac{dF}{du}\right)^2 \frac{\partial u}{\partial x} \right] - \frac{\partial}{\partial x}\left[\frac{dF}{du} G\right] + \frac{dG}{du}\left(G(u) - \frac{\partial F}{\partial x}\right) \right\}. \tag{10.67}$$

The finite difference representation of the Taylor form is

$$\frac{u_i^{n+1} - u_i^n}{\Delta t} = -\frac{F_{i+1}^n - F_{i-1}^n}{2\Delta x} + G_i^n + \frac{u_{i+1}^n - 2u_i^n + u_{i-1}^n}{\Delta x^2}$$

$$+ \frac{\Delta t}{4\Delta x^2} \left\{ \left[\left(\frac{dF}{du}\right)^2_{i+1} + \left(\frac{dF}{du}\right)^2_i\right][u_{i+1}^n - u_i^n] - \left[\left(\frac{dF}{du}\right)^2_i + \left(\frac{dF}{du}\right)^2_{i-1}\right][u_i^n - u_{i-1}^n] \right\} \tag{10.68}$$

$$- \frac{\Delta t}{4\Delta x}\left\{ \left[\frac{dF}{du} G\right]_{i+1} - \left[\frac{dF}{du} G\right]_{i-1} \right\} + \frac{\Delta t}{2}\left(\frac{dG}{du} G\right)_i - \frac{\Delta t}{4}\left(\frac{dG}{du}\right)_i \frac{F_{i+1} - F_{i-1}}{\Delta x}.$$

The Taylor-Galerkin form of the equation is

$$\frac{1}{6}\frac{u_{i+1}^{n+1} - u_{i+1}^n}{\Delta t} + \frac{2}{3}\frac{u_i^{n+1} - u_i^n}{\Delta t} + \frac{1}{6}\frac{u_{i-1}^{n+1} - u_{i-1}^n}{\Delta t} = -\frac{F_{i+1}^n - F_{i-1}^n}{2\Delta x} + \frac{u_{i+1}^n - 2u_i^n + u_{i-1}^n}{\Delta x^2}$$

$$+ \frac{\Delta t}{4\Delta x^2}\left\{\left[\left(\frac{dF}{du}\right)^2_{i+1} + \left(\frac{dF}{du}\right)^2_i\right][u_{i+1}^n - u_i^n] - \left[\left(\frac{dF}{du}\right)^2_i + \left(\frac{dF}{du}\right)^2_{i-1}\right][u_i^n - u_{i-1}^n]\right\} + \tag{10.69}$$

$$+ \left(\frac{1}{6}G_{i+1} + \frac{2}{3}G_i + \frac{1}{6}G_{i-1}\right) - \frac{\Delta t}{4\Delta x}\left\{\left[\frac{dF}{du} G\right]_{i+1} - \left[\frac{dF}{du} G\right]_{i-1}\right\} +$$

$$+ \frac{\Delta t}{2}\left[\frac{1}{6}\left(G\frac{dG}{du}\right)_{i+1} + \frac{2}{3}\left(G\frac{dG}{du}\right)_i + \frac{1}{6}\left(G\frac{dG}{du}\right)_{i-1}\right] +$$

$$- \frac{\Delta t}{2\Delta x}\left[\frac{1}{6}\left(\frac{dG}{du}\frac{dF}{du}\right)_{i-1}(u_i - u_{i-1}) + \frac{1}{3}\left(\frac{dG}{du}\frac{dF}{du}\right)_i (u_{i+1} - u_{i-1}) + \right.$$

$$\left. + \frac{1}{6}\left(\frac{dG}{du}\frac{dF}{du}\right)_{i+1}(u_{i+1} - u_i)\right].$$

In addition to the methods discussed above, moving elements can be used. Gatica, et al. [1987] used the orthogonal collocation method on finite elements on a fixed mesh for a more complicated problem. They also used a mesh with seven to nine moving elements. Here we will solve the standard problem using the methods given above.

Eq. (10.64) now has all the major types of terms: accumulation, convection, diffusion, and reaction. Let us see what time-step limitations exist for each term. If we have only convection, then we use $F = Pe\, u$ and $G = 0$.

$$\frac{\partial u}{\partial t} + \text{Pe} \frac{\partial u}{\partial x} = 0 \text{ requires } \text{Co} = \frac{\text{Pe } \Delta t}{\Delta x} \le 1 \qquad (10.70)$$

If we have only diffusion, we take $F = G = 0$.

$$\frac{\partial u}{\partial t} = \frac{\partial^2 u}{\partial x^2} \text{ requires } \frac{\Delta t}{\Delta x^2} \le \frac{1}{2} \text{ for FD, } \frac{1}{6} \text{ for FEM} \qquad (10.71)$$

If we have only reaction terms, we take $F = 0$ and no diffusion terms.

$$\frac{du}{dt} = G(u) \text{ requires } \left|\frac{dG}{du}\right| \Delta t \le 2 \qquad (10.72)$$

This gives the stable size for a predictor-corrector method (such as the MacCormack method) or for the Euler method (the explicit, upstream method without convection). When two or more effects are present, the situation is much more complicated. For example, with reaction and diffusion the finite difference method requires the following limit for a stable step-size:

$$\lambda = \max_i \left[\frac{4}{\Delta x^2} + \left|\frac{dG}{du}\right|_i \right], \quad \lambda \Delta t \le 2. \qquad (10.73)$$

It is prudent to calculate these various limits and use a smaller step-size than given by any of them.

It is possible to have moving fronts in a problem without having convection terms. Let us take the problem

$$\frac{\partial u}{\partial t} = \frac{\partial^2 u}{\partial x^2} + G(u). \qquad (10.74)$$

First we look at the limitations on Δx when only reaction and diffusion take place:

$$\frac{d^2 c}{dx^2} + G(c) = 0, \quad c(0) = 1, \quad c(1) = 0. \qquad (10.75)$$

We note that the exact solution with $G = 0$ is not steep; in fact, it is $c = 1 - x$. For the purposes of analysis, we use a linear reaction rate term,

$$G(c) = -kc. \qquad (10.76)$$

A centered finite difference method applied to Eq. (10.75)-(10.76) gives

$$\frac{c_{i+1} - 2c_i + c_{i-1}}{\Delta x^2} - k c_i = 0 \qquad (10.77)$$

or

$$c_{i+1} - 2c_i + c_{i-1} - k \Delta x^2 c_i = 0. \qquad (10.78)$$

The solution to this difference equation is in the form

$$c_i = A \phi_1^{i-1} + B \phi_2^{i-1}, \tag{10.79}$$

where

$$\phi_{1,2} = 1 + \frac{k \Delta x^2}{2} \pm \sqrt{k \Delta x^2 \left(1 + \frac{k \Delta x^2}{4}\right)}. \tag{10.80}$$

The complete solution is (see Problem 10.5)

$$c_i = \frac{\phi_2^{i-1}}{1 - (\phi_2/\phi_1)^N} \left[1 - \left(\frac{\phi_2}{\phi_1}\right)^{N-i+1} \right]. \tag{10.81}$$

If $i = 1$, we get $c_0 = 1$, while if $i = N+1$, we get $c_N = 0$. For intermediate values of i we get $c_i > 0$ and c_i is monotone from node to node. Thus the solution to the difference equation will not oscillate, no matter what value of $k\Delta x$ is chosen. It may be necessary to use a small Δx value to resolve the front if it is steep, but it is not necessary to use a small Δx value to avoid oscillations. Thus the case of reaction and diffusion is very different from the case of reaction and convection. They both have moving fronts, but one will oscillate while the other will not (at least by this simple analysis).

The stability of the unsteady-state problem can be studied using the von Neumann analysis when the reaction rate is linear. In Eq. (10.64)-(10.69) we take $F = 0$ and $G = -Da_I u$. The Fourier transform of Eq. (10.65) is

$$\hat{c}^{n+1} = \left(1 - Da_I \Delta t - 4 \frac{\Delta t}{\Delta x^2} \sin^2 \frac{\xi}{2}\right) \hat{c}^n. \tag{10.82}$$

The amplification factor is then

$$\rho = \left(1 - Da_I \Delta t - 4 \frac{\Delta t}{\Delta x^2} \sin^2 \frac{\xi}{2}\right). \tag{10.83}$$

The Fourier transforms of the other equations lead to the following amplification factors:

MacCormack method

$$\rho = \frac{1}{2}\left[1 + \left(1 - Da_I \Delta t - 4 \frac{\Delta t}{\Delta x^2} \sin^2 \frac{\xi}{2}\right)^2\right], \tag{10.84}$$

Taylor-finite difference method

$$\rho = \left(1 - Da_I \Delta t + \frac{(Da_I \Delta t)^2}{2} - 4 \frac{\Delta t}{\Delta x^2} \sin^2 \frac{\xi}{2}\right), \tag{10.85}$$

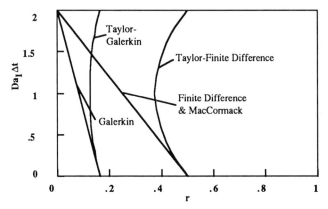

Figure 10.6. Stability Diagram for Finite Difference and Finite Element Methods

Galerkin method

$$\rho = (1 - Da_I \, \Delta t) - \frac{24 \, \Delta t}{\Delta x^2} \frac{\sin^2(\xi/2)}{2 \cos \xi + 4}, \tag{10.86}$$

Taylor-Galerkin method

$$\rho = \left(1 - Da_I \, \Delta t + \frac{(Da_I \, \Delta t)^2}{2}\right) - \frac{24 \, \Delta t}{\Delta x^2} \frac{\sin^2(\xi/2)}{2 \cos \xi + 4}. \tag{10.87}$$

The applification factors are plotted in Figures A.27-31. For a centered finite difference method when $r = \Delta t/\Delta x^2 = 0.5$, there is significant dissipation (particularly for $\xi = \pi/2$ and $3\pi/2$; the value of $k\Delta t$ must be reduced below that which would be acceptable when diffusion is absent. For the MacCormack method there is less dissipation and less variation with ξ. The Taylor-finite difference method is slightly improved over the centered finite difference method. The Galerkin method gives the amplification factor shown in Figure A.30; the values of r are smaller for finite element methods than for finite difference methods. There is also significant dissipation at the largest values of r. The Taylor-Galerkin method improves on the Galerkin method in that there is less dissipation, the growth factors are not as big, and there is less variation with ξ. A graph of the stability region with respect to the two parameters

$$Da_I \, \Delta t \quad \text{and} \quad r = \frac{\Delta t}{\Delta x^2} \tag{10.88}$$

is shown in Figure 10.6. The amplification factor is less than or equal to 1.0 under the following circumstances:

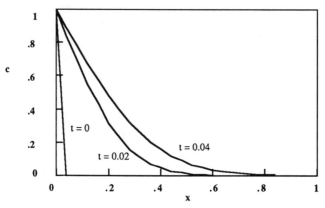

Figure 10.7. Reaction with Diffusion, MacCormack Method, $\Delta x = 0.04$, $\Delta t = 0.0008$ Reaction Rate in Eq. (10.92), $Da_I = -0.1$, $\alpha = 2$

$$Da_I \Delta t \leq 2 - 4r \quad \text{for finite difference methods,}$$
$$Da_I \Delta t \leq 2 - 12r \quad \text{for finite element methods.} \quad (10.89)$$

A good review of stability properties for nonlinear problems is given by Stuart [1989].

An example problem is taken as

$$\frac{\partial c}{\partial t} = \frac{\partial^2 c}{\partial x^2} + Da_I R(c), \quad (10.90)$$

$$c = 0 \text{ at } t = 0, \quad c = 1 \text{ at } x = 0, \quad \frac{\partial c}{\partial x} = 0 \text{ at } x = 1, \quad (10.91)$$

where the reaction rate term is

$$R = \frac{c}{1 + \alpha c}. \quad (10.92)$$

The derivative of the reaction rate is

$$\frac{dR}{dc} = \frac{1}{(1 + \alpha c)^2}. \quad (10.93)$$

Results are given for the reaction rate [Eq. (10.92)] with $\alpha = 2$. Figure 10.7 shows the results when $Da_I = -0.1$. This graph is the counterpart of Figure 10.4b; Figure 10.4b has convection and reaction, while Figure 10.7 has diffusion and reaction. Clearly the case with diffusion and reaction is much easier; no oscillations are evident despite the use of fewer nodes. Although the front is still steep in some regions, it is resolved to the extent possible by the grid, without oscillations. When the reaction rate is increased, as in Figure 10.8 with $Da_I = -10$, similar results are achieved. Figure 10.8 can be compared with Figure 10.5.

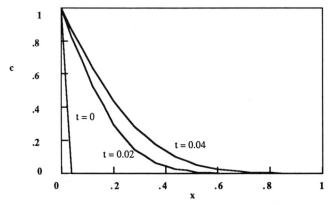

Figure 10.8. Reaction with Convection, MacCormack Method, $\Delta x = 0.01$, $\Delta t = 0.005$ Reaction Rate in Eq. (10.92), $Da_I = -10$, $\alpha = 2$

When reaction predominates over either diffusion (Figure 10.8) or convection (Figure 10.5), the results are similar. The Galerkin method (with the same Δt) gave similar results to those shown in Figures 10.7 and 10.8.

Next let us consider a case where the reaction, diffusion, and convection terms are all present:

$$\frac{\partial c}{\partial t} + Pe \frac{\partial c}{\partial x} = \frac{\partial^2 c}{\partial x^2} + Da_I R(c). \qquad (10.94)$$

In this problem there are three time constants:

$$t_{conv} = \frac{L}{u}, \quad t_{rxn} = \frac{1}{k}, \quad t_{diff} = \frac{L^2}{D}. \qquad (10.95)$$

The ratios of the time constants provide the dimensionless numbers in Eq. (10.94),

$$\frac{t_{diff}}{t_{conv}} = \frac{Lu}{D} = Pe, \quad \frac{t_{conv}}{t_{rxn}} = \frac{Lk}{u} = Da_I. \qquad (10.96)$$

Thus the problem is expected to be more difficult for large Peclet numbers or large Damköhler numbers. With Eq. (10.94) in this form, we can compare it with earlier chapters. If we want to compare with the results from Section 10.1 (which had no diffusion), we must put the equation in a different form. We divide Eq. (10.94) by Pe and use

$$\tau = \frac{t}{Pe}, \qquad (10.97)$$

$$\frac{\partial c}{\partial \tau} + \frac{\partial c}{\partial x} = \frac{1}{Pe} \frac{\partial^2 c}{\partial x^2} + Da'_I R(c), \quad Da'_I = \frac{Da_I}{Pe}. \qquad (10.98)$$

We can solve Eq. (10.94) for large Peclet numbers (convection > diffusion) and use $Da_I = Da'_I Pe$. The result is a problem with strong convection and a small amount of diffusion.

The same methods can be applied to Eq. (10.94) by comparing it with Eq. (10.64). In Eq. (10.64) we take $F(u) = Pe\, c$ and $u = c$. Then we have the following Taylor form [derived from Eq. (10.67)]:

$$\frac{c^{n+1} - c^n}{\Delta t} = -Pe\frac{\partial c}{\partial x} + \frac{\partial^2 c}{\partial x^2} + Da_I R(c) + \frac{Pe^2 \Delta t}{2}\frac{\partial^2 c}{\partial x^2} + \frac{Da_I \Delta t}{2} R(c)\frac{dR}{dc} - Pe\,\Delta t \frac{\partial R(c)}{\partial x}. \quad (10.99)$$

In this case, two of the terms of Eq. (10.67) combine. The Taylor-finite difference method then gives

$$\frac{c_i^{n+1} - c_i^n}{\Delta t} = -\frac{Pe}{2\Delta x}(c_{i+1}^n - c_{i-1}^n) + \frac{1}{\Delta x^2}(c_{i+1}^n - 2c_i^n + c_{i-1}^n) + Da_I R(c_i^n) +$$

$$+ \frac{Pe^2 \Delta t}{2}(c_{i+1}^n - 2c_i^n + c_{i-1}^n) + \frac{Da_I \Delta t}{2} R(c_i^n)\frac{dR}{dc}(c_i^n) - \frac{Pe\,\Delta t}{2\Delta x} Da_I [R(c_{i+1}^n) - R(c_{i-1}^n)], \quad (10.100)$$

while the Taylor-Galerkin method gives

$$\frac{1}{6}\frac{c_{i+1}^{n+1} - c_{i+1}^n}{\Delta t} + \frac{2}{3}\frac{c_i^{n+1} - c_i^n}{\Delta t} + \frac{1}{6}\frac{c_{i-1}^{n+1} - c_{i-1}^n}{\Delta t} = -\frac{Pe}{2\Delta x}(c_{i+1}^n - c_{i-1}^n) +$$

$$+ \frac{1}{\Delta x^2}(c_{i+1}^n - 2c_i^n + c_{i-1}^n) + Da_I\left[\frac{1}{6}R(c_{i+1}^n) + \frac{2}{3}R(c_i^n) + \frac{1}{6}R(c_{i-1}^n)\right] +$$

$$+ \frac{Da_I \Delta t}{2}\left[\frac{1}{6}R(c_{i+1}^n)\frac{dR}{dc}(c_{i+1}^n) + \frac{2}{3}R(c_i^n)\frac{dR}{dc}(c_i^n) + \frac{1}{6}R(c_{i-1}^n)\frac{dR}{dc}(c_{i-1}^n)\right] + \quad (10.101)$$

$$+ \frac{Pe^2 \Delta t}{2}(c_{i+1}^n - 2c_i^n + c_{i-1}^n) - \frac{Pe\,\Delta t}{2\Delta x} Da_I[R(c_{i+1}^n) - R(c_{i-1}^n)].$$

Figure 10.9 shows calculations using the MacCormack method for $Pe = 1000$ and $Da = -100$. The value of the reaction rate term is

$$\frac{|Da_I|}{Pe} R(c_0) = \frac{100}{1000}\frac{1}{3^2} = 0.011. \quad (10.102)$$

Since the reaction rate term is small, the case is convection-dominated. Convection is greater than diffusion (because of the large Peclet number) and is also greater than reaction. Figures 10.4b and 10.9 are comparable; Figure 10.9 shows a case with a small amount of diffusion added. The MacCormack method gives oscillations in both figures since both cases are dominated by convection. When the reaction rate is increased by changing Da to $-10{,}000$, we get the results shown in Figure 10.10. Figures 10.5 and 10.10 are comparable, with a small amount of

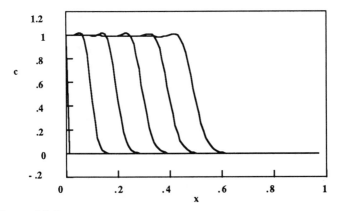

Figure 10.9. Reaction with Convection and Diffusion, MacCormack Method
$\Delta x = 0.01$, $\Delta t = 5E\text{-}6$, Pe = 1000, Reaction Rate in Eq. (10.92), $Da_t = -100$, $\alpha = 2$

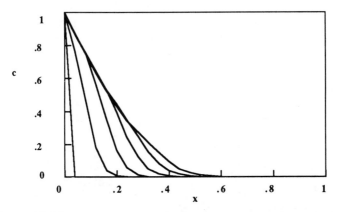

Figure 10.10. Reaction with Convection and Diffusion, MacCormack Method
$\Delta x = 0.04$, $\Delta t = 0.00002$, Pe = 1000, Reaction Rate in Eq. (10.92), $Da_t = -10{,}000$, $\alpha = 2$

diffusion added for the case in Figure 10.10. Figures 10.9 and 10.10 show the effect of drastically increasing the reaction rate term. In Figure 10.10 the reaction rate term is 100 times larger and the case is reaction-dominated rather than convection-dominated. The MacCormack method gives a good solution in this case.

Thus, the only factor that gives rise to oscillations is the large convection term. In Figures 10.4 and Figure 10.9 the convection term predominates. In Figures 10.8 and 10.10 the reaction rate term dominates. In Figure 10.7 the diffusion term dominates. Only when the convection term dominates did the solutions oscillate. The oscillations can be eliminated, of course, by using the MacCormack method with flux-correction.

10.3. Combustion in Flames

We next solve the problem for a burner-stabilized flame, as shown in Figure 10.11.

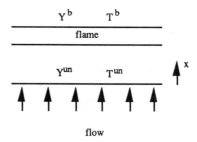

Figure 10.11. Burner-stabilized Flame Geometry

The flame propagates from one boundary ($x = -\infty$) to the other ($x = +\infty$). If the mixture is stationary, the fluid velocity ahead of the flame is zero. We transform the coordinates into a Lagrangian coordinate system in order to eliminate the convection term. This is possible because of the infinite extent of the physical domain. The coordinates go from

$$(t, x) \rightarrow (\tau, \psi), \tag{10.103}$$

where

$$\tau = t, \quad \frac{\partial \psi}{\partial x} = \rho, \quad \frac{\partial \psi}{\partial t} = \int_{-\infty}^{x} \frac{\partial \rho}{\partial t} dx = -\int_{-\infty}^{x} \frac{\partial (\rho u)}{\partial x} dx = -\rho u. \tag{10.104}$$

These equations are derived by Ramos [1987] and are contained in the electronic text "Combustion in Flames." The steady-state forms of the equations are

$$\dot{m} \frac{dY_i}{d\psi} = R_i - \frac{d}{d\psi}(\rho Y_i V_i), \tag{10.105}$$

$$\dot{m} \frac{dT}{d\psi} = \frac{1}{C_p} \left[-\sum_{i=1}^{N} h_i R_i - \frac{dq_1}{d\psi} - \sum_{i=1}^{N} \rho Y_i V_i C_{pi} \frac{dT}{d\psi} \right], \tag{10.106}$$

together with the boundary conditions

$$T(0) = T^{un}, \quad Y_i(0) = Y_i^{un},$$
$$\frac{dT}{dx}(0) = \frac{dT}{dx}(\infty) = 0, \quad \frac{dY_i}{dx}(0) = \frac{dY_i}{dx}(\infty) = 0. \tag{10.107}$$

We must solve these equations for the distribution of mole fractions, Y_i, and temperature, T; we must also solve for the mass flux, $\dot{m}$.

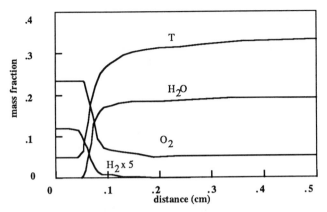

Figure 10.12. Temperature and Concentrations for Hydrogen-Air Flame
Data from Smooke [1986]

First let us consider solutions of the burner-stabilized flames, Eq. (10.105)-(10.107). Smooke [1986] discusses such solutions for a situation slightly more complicated than that derived above (he includes thermal diffusion effects). The problem is for a hydrogen-air flame and the kinetics are complicated expressions of several concentrations [Smooke, et al., 1983]. To find the flame velocity, we append the equation

$$\frac{d\dot{m}}{dx} = 0. \tag{10.108}$$

The boundary conditions are

$$T(0) = T^{un}, \quad \varepsilon_i(0) = Y_{i0}, \quad \frac{dT}{d\psi}(L) = 0, \quad \frac{dY_i}{d\psi}(L) = 0, \tag{10.109}$$

where

$$\varepsilon_i = Y_i + \frac{\rho Y_i V_i}{\dot{m}}. \tag{10.110}$$

The location of the exit region is placed far enough from the flame zone so that the results are not affected. We need, in addition, a boundary condition for the added equation, Eq. (10.108). Smooke [1986] recommends setting the temperature at some interior node,

$$T(x_f) = T_f. \tag{10.111}$$

The location x_f is chosen such that the flame is far enough from the flame holder that the derivatives in T and the mole fraction are zero at $x = 0$ and at the end of the domain ($x = \infty$). The value of T_f apparently falls between the inlet temperature and the adiabatic temperature. Typical finite difference results for the concentrations of various species are shown in Figure 10.12. There is a region of steep

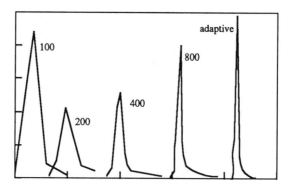

Figure 10.13. Concentration Profiles for Different Grid Sizes
Data from Smooke [1986]

change in the solution. The flame speed varies with the hydrogen concentration in the inlet, which has a maximum of approximately 5 mass per cent hydrogen. The flame speed for different grids is shown in Table 10.1. When only a few nodes are used, the predicted velocity is much higher than the actual velocity. Even with 800 nodes, the velocity is approximately 15 per cent too high. When an adaptive mesh is used with 40 nodes and the equidistribution principles discussed in Section 7.3, the flame speed is quite close to the one measured experimentally.

The concentration profile of one of the minor species, HO_2, is shown in Figure 10.13 for various grids. Again, a large number of nodes is required unless the nodes are located adaptively. Finally, the computation time is illustrated in Table 10.2. For solutions with somewhat comparable accuracies (800 fixed nodes and 40 adaptive nodes), there is a 20-fold reduction in the number of nodes and a 17-fold reduction in the computer time when the adaptive scheme is used. Thus, in this application most of the speed advantage of an adaptive mesh is realized through a reduction in the number of nodes.

Table 10.1. Flame Speed for Hydrogen-Air Flame
[Data from Smooke, 1986]

number of nodes	velocity (cm/sec)
100	600
200	400
400	360
800	320
40, adapt.	280
exp.	275

In burner-stabilized flames, the wave speed is an unknown and is found as a parameter in the problem. The set of equations, however, represent a set of nonlinear equations that are difficult to solve unless the initial guess is good. If

Table 10.2. Computation Time for Hydrogen-Air Flame [Smooke, 1986]

method	number of nodes	CPU time (sec)
adaptive	40	59
fixed	800	1050

an initial guess is not available, another option must be used, such as integrating the time-dependent equations until the profiles are unchanged. The velocity of movement of the fronts is then the flame speed. Presented here are the results for some simplified models. Spalding [1957] proposed considering a case with only one reacting species. The model is then

$$\frac{\partial Y}{\partial \tau} = \beta \frac{\partial^2 Y}{\partial \psi^2} + S(Y). \tag{10.112}$$

For a particular reaction rate expression,

$$S(Y) = a Y (1 - Y), \tag{10.113}$$

we get the Fisher equation [Fisher, 1937]. We use the boundary conditions

$$\frac{\partial Y}{\partial \psi} = 0 \text{ at } \psi = \pm \infty \tag{10.114}$$

and the initial conditions $Y(-\infty,0) = 1$ and $Y(\infty,0) = 0$. Thus the reaction rate term is zero at both ends of the domain.

The Fisher equation has exact solutions. When $\beta = a = 1$, the solution is [Reitz, 1981 and Lee and Ramos, 1983]

$$Y(\psi,\tau) = Y(\psi - S\tau) = \frac{1}{[1 + \exp(\psi - S\tau)/\sqrt{6}]^2}, \tag{10.115}$$

where the asymptotic wave speed is

$$S = \frac{5}{\sqrt{6}}. \tag{10.116}$$

When the reaction rate is

$$S(Y) = Y^2 (1 - Y), \tag{10.117}$$

the exact solution is [Reitz, 1981]

$$Y(\psi,\tau) = Y(\psi - S\tau) = \frac{1}{\{1 + \exp[S(\psi - S\tau)]\}}, \text{ where } S = \frac{1}{\sqrt{2}}. \tag{10.118}$$

Ramos [1987] describes several methods applied to these equations. An explicit, predictor-corrector method on a fixed grid gives

$$\bar{Y}_j^n = Y_j^n + \frac{\beta \Delta \tau}{\Delta \psi^2}(Y_{j-1}^n - 2Y_j^n + Y_{j+1}^n) + \Delta \tau\, S(Y_j^n)$$

$$Y_j^{n+1} = Y_j^n + \frac{\beta \Delta \tau}{2\Delta \psi^2}(Y_{j-1}^n - 2Y_j^n + Y_{j+1}^n) + \frac{\Delta \tau}{2} S(Y_j^n) + \frac{\beta \Delta \tau}{2\Delta \psi^2}(\bar{Y}_{j-1}^n - 2\bar{Y}_j^n + \bar{Y}_{j+1}^n) + \frac{\Delta \tau}{2} S(\bar{Y}_j^n).$$

(10.119)

In each case the nonlinear reaction rate term is handled explicitly. If the second and third terms of the second equation are replaced using the first equation, we get

$$Y_j^{n+1} = \frac{1}{2}(Y_j^n + \bar{Y}_j^n) + \frac{\beta \Delta \tau}{2\Delta \psi^2}(\bar{Y}_{j-1}^n - 2\bar{Y}_j^n + \bar{Y}_{j+1}^n) + \frac{\Delta \tau}{2} S(\bar{Y}_j^n). \qquad (10.120)$$

This is equivalent to the second step of the MacCormack method. Comparing the first step of Eq. (10.119) and Eq. (10.120) with Eq. (10.66) indicates that the explicit predictor-corrector method is a MacCormack method.

Another method applied to this problem is an operator-splitting method [Yanenko, 1971]. In this method we divide the calculation into two steps: one that involves reaction only,

$$\frac{dY_j}{d\tau} = S(Y_j) \text{ to give the solution } \bar{Y}_j, \qquad (10.121)$$

and one that involves diffusion only,

$$\frac{\partial Y}{\partial \tau} = \beta \frac{\partial^2 Y}{\partial \psi^2} \text{ applied as } \frac{Y_j^{n+1} - \bar{Y}_j}{\Delta \tau} = \beta \left.\frac{\partial^2 Y}{\partial \psi^2}\right|_j. \qquad (10.122)$$

The terms in Eq. (10.122) can be handled in a variety of ways: explicit or implicit and with second-order or higher-order spatial approximations. Ramos [1985a,b, 1986, 1987] gives details of these methods as well as others. Ramos [1987] discusses an adaptive mesh method that is different from the one described here; it has the effect of concentrating the nodes where the gradient is large.

Computations were perfomed by Ramos [1987] on a grid from $\psi = -50$ to $\psi = 400$ with $a = \beta = 1$ and the reaction rate in Eq. (10.113). The initial conditions were taken approximately as

$$\begin{aligned} Y &= 1 - 0.5 \exp(0.307\,\psi) \quad \text{for } \psi < 0 \\ Y &= 0.5 \exp(-0.307\,\psi) \quad \text{for } \psi \geq 0 \end{aligned} \qquad (10.123)$$

The solution obtained with $\Delta\psi = 0.25$ is shown in Figure 10.14. The profile is initially steep, but it becomes less steep as time proceeds. The wave speed at $\tau = 20$ is given in Table 10.3. The use of an adaptive grid reduces the number of nodal points by a factor of approximately five and reduces the computation time by a factor of seven or eight. (The fact that the ratio of computation times is larger than the ratio of number of nodal points is probably because of differences in handling the nonlinear terms in the time-dependent calculations.) These results

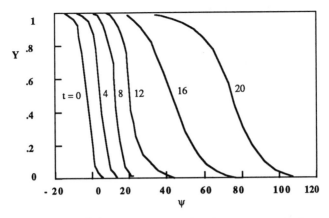

Figure 10.14. Composition in Burner-stabilized Flame, Data from Ramos [1987]

show again that the use of an adaptive mesh can reduce computation time and the number of nodal points necessary to achieve good results. Similar results are reported by Ramos [1987] for a more complicated case involving ozone decomposition. Sheintuch [1990] uses a method which transforms the domain to keep the front stationary and then applies the Galerkin method to the resulting equations.

Table 10.3. Computation Time for the Fisher Equation, [Ramos 1987]
Reprinted with permission of Hemisphere Publishing Corp.

method	number of nodes	no. nodes in front	CPU time (sec)	wave speed
adaptive	101	70	24	1.9675
adaptive	171	140	30	2.0330
fixed	451	70	171	1.9655
fixed	901	140	252	2.0331

Smooke [1986] solved a similar problem with one-step chemistry, but he included temperature as well. The equations are

$$\frac{\partial T}{\partial t} = \frac{\partial^2 T}{\partial x^2} + R \tag{10.124}$$

$$\frac{\partial Y}{\partial t} = \frac{1}{Le} \frac{\partial^2 Y}{\partial x^2} - R. \tag{10.125}$$

The reaction rate is taken as

$$R = \frac{\beta^2}{2Le} Y \exp\left(-\frac{\beta(1-T)}{1-\alpha(1-T)}\right). \tag{10.126}$$

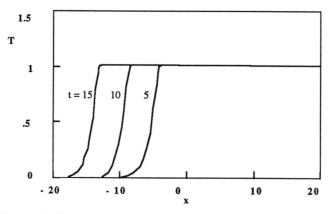

Figure 10.15a. Temperature Profiles for 75-node Adaptive Calculation

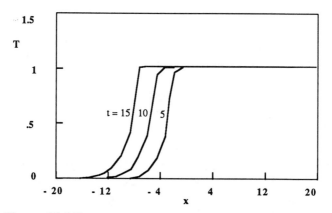

Figure 10.15b. Temperature for 75-node Equispaced Calculation, Data from Smooke [1986]

The initial conditions are

$$T = e^x, \quad Y = 1 - e^{Le\, x}, \quad x \le 0$$
$$T = 1, \quad Y = 0, \quad x > 0, \tag{10.127}$$

while the boundary conditions are

$$T = 0, \quad Y = 1, \quad x \to -\infty$$
$$\frac{\partial T}{\partial x} = 0, \quad \frac{\partial Y}{\partial x} = 0, \quad x \to \infty. \tag{10.128}$$

The problem is solved on the range $-50 \le x \le 50$ with $\alpha = 0.8$, $\beta = 10.0$, and Le = 1. When the Lewis number (Le) is 1.0, as is the case here, it is possible to add Eq. (10.124)-(10.125) and show that $T + Y = 1$ for all location and time values. Then the equations can be reduced to a single equation for either T or Y. An adaptive mesh is used with the equidistribution principle discussed in Section 7.3. The solution using 75 adaptive nodes is shown in Figure 10.15a. The solution on a fixed grid of 75 nodes is shown in Figure 10.15b; it is not as steep and the reaction front is in the wrong place. It is necessary to use 500 nodes on a fixed grid to obtain solutions as good as the ones obtained using 75 adaptive nodes. The computation times and flame velocities derived from these results are listed in Table 10.4. Again it is necessary to use many fixed nodes to achieve results comparable with a few adaptive nodes. Even with 1000 fixed nodes the wave speed is not as accurate as that achieved with 75 adaptive nodes. (Calculations with 100, 200 and 300 adaptive nodes confirm the accuracy of the wave speed to two digits.) Thus we reduce the number of nodes necessary by at least a factor of 14 when adaptive meshes are used. The computational time is reduced by a factor of 10, indicating that there is considerable overhead associated with the adaptive mesh calculation. With other parameters ($\alpha = 0.8$, $\beta = 20.0$, and Le = 2) the flame is unsteady and oscillates in time. Even though adaptive meshes allowed a reduction in the number of nodal points by a factor of 10-20 for these parameters, computation time was reduced by only a factor of 3-4. Again, the overhead associated with the adaptive mesh was considerable. Another application is given by Mack, *et al.* [1991], who solve for a non-stationary flame propagation using an equidistribution principle that is adjusted to prevent the nodes from coming too close to each other.

Table 10.4. Computation Time for a Freely Propagating Flame
Smooke [1986]
Reproduced by permission of the American Institute of Chemical Engineers © 1986 AIChE

method	number of nodes	CPU time (sec)	wave speed
fixed	75		-0.70
fixed	150		-0.84
fixed	250		-0.88
fixed	500	245	-0.89
fixed	1000	512	-0.91
adaptive	75	48	-0.96

The final example involves flame propagation with a different reaction rate [Dwyer and Sanders 1978], as solved by Galinas, *et al.* [1979]:

$$\frac{\partial T}{\partial t} = \frac{\partial^2 T}{\partial x^2} + R, \quad \frac{\partial Y}{\partial t} = \frac{\partial^2 Y}{\partial x^2} - R. \tag{10.129}$$

The reaction rate is taken as

$$R = Y\, 3.52 \cdot 10^6 \exp\left(-\frac{4}{T}\right). \tag{10.130}$$

The initial conditions are

$$Y(x,0) = 1,\ T(x,0) = 0.2, \quad 0 \le x \le 1 \tag{10.131}$$

and the boundary conditions are

$$\frac{\partial Y}{\partial x}(0,t) = \frac{\partial T}{\partial x}(0,t) = 0, \quad \frac{\partial Y}{\partial x}(1,t) = 0, \quad T(1,t) = f(t), \quad t > 0, \qquad (10.132)$$

$$f(t) = \begin{cases} 0.2 + \dfrac{t}{2 \cdot 10^{-4}}, & t \leq 2 \cdot 10^{-4} \\ 1.2 & t > 2 \cdot 10^{-4}. \end{cases} \qquad (10.133)$$

First we notice that we cannot collapse this problem into a single equation because the boundary and initial conditions do not maintain the constant ratio $T + Y = 1.2$. Next we notice that the term in the exponential is $-4/T$. To compare the difficulty of this problem with that expressed by Eq. (10.126), we need to simplify Eq. (10.126). We use $\alpha = 1$ and write Eq. (10.126) as

$$\exp\left[-\frac{\beta(1-T)}{T}\right] = e^{\beta} \exp\left[-\frac{\beta}{T}\right]. \qquad (10.134)$$

Smooke uses $\beta = 10$ (rather than 4), and the pre-exponential factor is 1.1×10^6, while in Eq. (10.130) it is 3.5×10^6. The pre-exponential factor can, of course, be absorbed into the distance and time measures for these problems, which have no natural length scale. The value of β is the primary indicator of difficulty in this problem. Thus Smooke's problem [1986] is more difficult than the one treated by Galinas, *et al.* [1979]. Galinas, *et al.* [1979] used the method of moving finite elements to solve this problem. They use 11, 21, and 31 nodes, with comparable results. By using 101 nodes on a fixed finite element mesh, they were able to obtain similar results. When the pre-exponential factor was increased to 4×10^7 (on the same grid length), they obtained the solutions shown in Figure 10.16a and 10.16b, which have quite steep fronts. These solutions were obtained using 11 moving nodes, while 101 fixed nodes gave a poor solution for the same case. Galinas, *et al.* [1979] estimated that 3000 fixed nodes would have been necessary to obtain a good solution.

The examples discussed above are all one-dimensional. Two-dimensional problems have been solved using adaptive meshes by Smooke, *et al.* [1984], who placed a rectangular grid on a square. The Δr increments in the r direction and the Δz increments in the z direction are variable, but if a small Δr is needed for some z value it is used for all z values. Thus the grid has a rectangular pattern (see Figure 10.17a). Computational considerations are very important in two-dimensional problems and Smooke *et al.* organize the calculations to take advantage of vector machines. Typical profiles of temperature and concentration are shown in Figure 10.17b for a methane-air flame. (See also

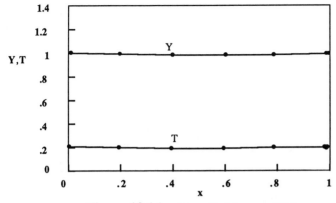

Figure 10.16a. Flame Problem, t = 2E-4

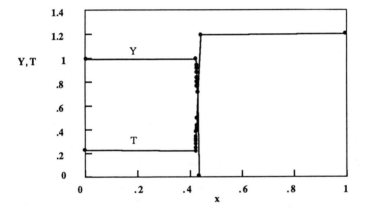

Figure 10.16b. Flame Problem Eq. [10.129-10.134], t = 1.2E-3
Data from Galinas, *et al.* [1979]

Degreve, *et al.* [1987] for a two-dimensional combustion example with transformation of the grid, as described in Sections 8.1 and 8.7.)

Next we apply the methods discussed here to Eq. (10.112) and Eq. (10.113) with $a = \beta = 1$. We expect that all of the methods we have discussed give good results, since there is no convection term. Figures 10.18a-c show the results obtained with the MacCormack method and 26, 51, and 100 nodes. No oscillations are evident in any of the curves, so the smallest number of nodes suffices except if a smooth profile is desired. The fronts move with a constant velocity (after an initial period), as is given in Table 10.5; the velocity is the same (to within a few percent) with 26, 51, or 100 nodes. The curves are, however, different from those in Figure 10.14, which show a transition from one traveling wave to another at approximately $\tau = 14$. This difference may be caused by a slightly different initial condition.

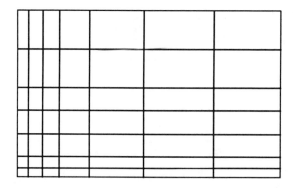

Figure 10.17a. Two-dimensional Grid Refinement

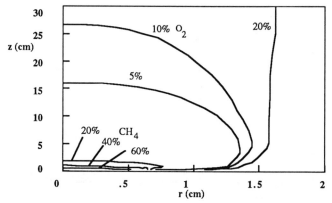

Figure 10.17b. Two-dimensional Flame Fronts, Data from Smooke, et al. [1984]

Table 10.5. Wave Speed and Gradient at c = 0.5

Method	Figure No.	No. nodes	Speed	Gradient
	Initial Conditions with C = 0.307			
MacCormack	10.18a	26	3.494	0.0624
MacCormack	10.18b	51	3.501	0.0683
MacCormack	10.18c	100	3.538	0.0686
Explicit	10.21a	51	2.877	0.0834
Galerkin	10.21b	51	2.826	0.0821
Taylor-Galerkin	10.21c	51	3.306	0.0738
	Comparisons with Different Initial Conditions			
Mac, C = 1	10.19	51	2.299	0.0956
Mac, C = 0.307	10.18b	51	3.501	0.0683
Mac, C = 0.2	10.20	51	5.050	0.0471

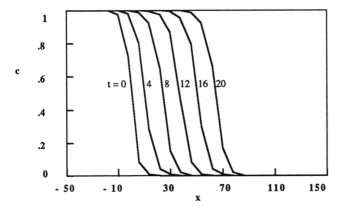

Figure 10.18a. Solution to Fisher's Equation, 26 nodes, $\Delta t = 1$

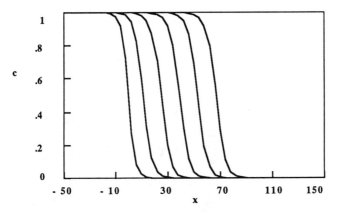

Figure 10.18b. Solution to Fisher's Equation, 51 nodes, $\Delta t = 0.5$

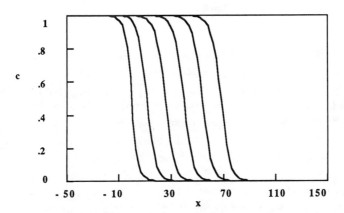

Figure 10.18c. Solution to Fisher's Equation, Eq. (10.112-10.113) MacCormack Method, 101 nodes, $\Delta t = 0.25$

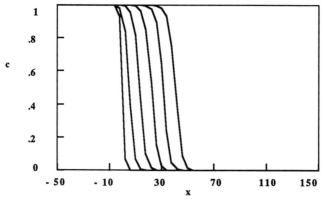

Figure 10.19. Solution to Fisher's Equation, Eq. (10.112-10.113) MacCormack Method, C = 1, 51 nodes, $\Delta t = 0.5$

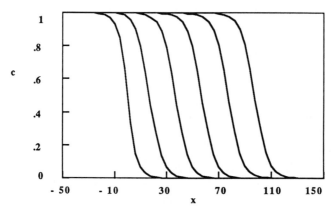

Figure 10.20. Solution to Fisher's Equation, Eq. (10.112-10.113) MacCormack Method, C = 0.2, 51 nodes, $\Delta t = 0.5$

The effect of the initial conditions is illustrated by comparing Figures 10.18b, 10.19, and 10.20. The parameter in the exponents of Eq. (10.123) is changed from 0.307 to 1.0 or 0.2, respectively. The impact on the front is dramatic: with a larger value (giving a sharper front initially), the front is much steeper and the traveling wave moves more slowly.

The explicit, Euler method gives the results shown in Figure 10.21a. Compared with the MacCormack method, the velocity differs slightly so that the profiles are displaced from each other a small amount. The Euler method is first-order in Δt while the MacCormack method is second-order in Δt. Thus we expect the MacCormack method to be the most accurate. In this problem, then, the spatial truncation error is not as important as the temporal truncation error. The Galerkin method and the Taylor-Galerkin method give the results shown in Figures 10.21b and 10.21c, respectively. The Galerkin method is also first-order

CHEMICAL REACTIONS 389

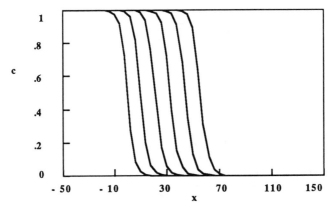

Figure 10.21a. Solution to Fisher's Equation, Euler Method
51 nodes, $\Delta t = 0.5$

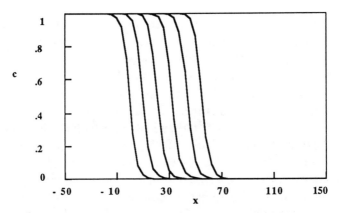

Figure 10.21b. Solution to Fisher's Equation, Galerkin Method
51 nodes, $\Delta t = 0.5$

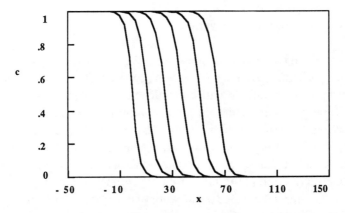

Figure 10.21c. Solution to Fisher's Equation, Eq. (10.112-10.113)
Taylor-Galerkin Method, 51 nodes, $\Delta t = 0.5$

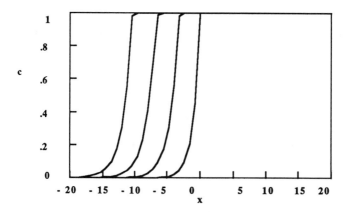

Figure 10.22. Solution to Fisher's Equation, Eq. (10.124-10.128) MacCormack Method, 51 nodes, $\Delta t = 0.02$, Le = 1, $\alpha = 0.8$, $\beta = 10$

in Δt; it gives results that agree with the results obtained with the Euler method. The Taylor-Galerkin method, which is second-order, gives results that are closer to those obtained with the MacCormack method.

The last calculation shown in Figure 10.22 is for the MacCormack method applied to Eq. (10.124)-(10.128) with Le = 1, $\alpha = 0.8$, and $\beta = 10.0$. This can be compared with Figures 10.15a (the best solution) and 10.15b. Figure 10.15b and 10.22 both use a fixed mesh, with Figure 10.15b using 75 nodes and Figure 10.22 using 51 nodes. Figure 10.22 has a steep front that moves with a velocity between that shown in Figures 10.15a and 10.15b (the front moves to the left and the velocity can be ascertained from the front location). Figure 10.22 is better than Figure 10.15b, presumably because of the use of the second-order MacCormack method rather than a first-order method. Based on the limited number of calculations presented here, when a higher-order method is used in time, the spatial discretization does not seem to be as crucial.

10.4. Modeling a Catalytic Converter

A chemical reaction sometimes occurs at a significant rate only when the temperature is high. To make the reaction occur rapidly at a lower temperature, a catalyst is used. Problems with catalytic reaction cause some of the same numerical difficulties as those that arise with combustion in flames and convection and reaction, but now additional phases are involved. Presented here are mathematical models of catalytic converters on automobiles.

The catalytic converter in an automobile is frequently made from a ceramic base structure with long tubes running through it (see Figure 10.23). Highly porous alumina is deposited on the walls of the tubes and platinum catalyst is deposited on the alumina. The reactants then enter one end of the tube and flow down the tube. Reactants near the wall come in contact with the catalyst and react,

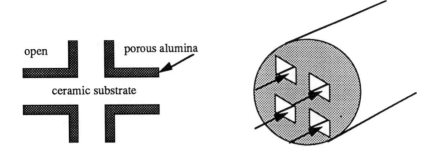

Figure 10.23. Geometry of a Catalytic Converter

giving off energy and products. The products are then swept downstream in the gas phase. The energy can be swept downstream in the gas phase or conducted longitudinally in the ceramic solid. The usual reaction is the oxidation of carbon monoxide to form carbon dioxide; this reaction is highly exothermic. Thus significant quantities of heat are released, causing the temperature to rise dramatically in a small spatial region. These steep gradients are the source of the difficulty in solving problems with catalytic converters.

Let us write equations for temperature and one species concentration. When oxygen is in great excess compared with the stoichiometric requirements of the carbon monoxide, the oxygen concentration hardly changes, even when all the carbon monoxide reacts. Thus we can solve for the concentration of carbon monoxide alone. Within the gas phase, the energy and mass balance give

$$\frac{\partial T}{\partial z} = 2\alpha_1 \, \text{Nu} \, (T^s - T), \tag{10.135}$$

$$\frac{\partial Y}{\partial z} = 2\alpha_2 \, \text{Sh} \, (Y^s - Y), \tag{10.136}$$

where T is the fluid temperature, T^s is the solid temperature of the wall, Y is the carbon monoxide mole fraction, and Y^s is the carbon monoxide mole fraction within the pores of the catalytic layer. Nu is the Nusselt number and Sh is the Sherwood number, while the α's are a collection of physical properties and physical dimensions. In the first equation, the convection of heat is balanced by the transfer of heat between the fluid and the solid. Here that transfer is governed by a Nusselt number, which is taken as a given quantity. In a series of papers, Young and Finlayson [1976a,b] and Finlayson and Young [1979] examined models in which the Nusselt number is predicted by the model rather than being assumed. The heat transfer is thus proportional to the Nusselt number (employing a heat transfer coefficient) multiplied by the temperature difference between the average temperature in the fluid and the solid wall temperature. Similar statements can be made about the mass transfer equation. The transient terms have been left out of Eq. (10.135)-(10.136) because the coefficients of the missing time derivatives are very small. Thus, this form of Eq. (10.135)-(10.136) represents a quasi-static approximation for the gas phase.

The equations for the temperature of the solid and the concentration in the porous solid are

$$\alpha_5 \left[\alpha_3 \frac{\partial T^s}{\partial t} - \left(\frac{r_{hs}}{L}\right)^2 \frac{\partial^2 T^s}{\partial z^2} \right] + \frac{1}{2} \text{Nu} \, (T^s - T) + \beta_5 <r> = 0, \quad (10.137)$$

$$-\alpha_6 \beta_4 <r> + \frac{1}{2} \text{Sh} \, (Y^s - Y) = 0, \quad (10.138)$$

where the β's are Damköhler numbers for heat and mass. The energy equation [Eq. (10.137)] contains the accumulation term (time derivative) since the solid heat capacity is very large and the solid can store significant quantities of energy. The second term represents heat conduction axially through the solid. The ratio r_{hs}/L accounts for the fact that heat conduction occurs only through a portion of the cross-sectional area, since most of it is space through which the gas flows. The next term represents the transfer of energy between the solid and the fluid; energy gained here is lost in Eq. (10.137). The final term represents the energy released by chemical reaction. The mass balance looks different because two terms, the time derivative term and the axial diffusion term, are neglected. The time derivative term is neglected because of the small capacity of the thin catalytic layer to contain mass while the axial diffusion term is neglected because it is small compared to the reaction and transfer between the solid and the fluid.

The reaction rate is for the combustion of carbon monoxide on a platinum catalyst. In the absence of diffusion effects, the rate of reaction is

$$\text{rate} = \frac{D \, y_{O_2} y_{CO} \, e^{-A/T}}{(1 + C' \, y_{CO} \, e^{B/T})^2} = \frac{k_0 \, (E + y) \, y}{(1 + \alpha y)^2}. \quad (10.139)$$

The term E comes from the oxygen concentration, which is in great excess. The reaction is so fast, however, that diffusion slows down the overall rate of reaction. This is usually accounted for by multiplying the reaction rate by an effectiveness factor. Here the effectiveness factor is obtained by solving the boundary value problem:

$$\frac{d^2 y}{ds^2} = \phi^2 f(y), \quad y = y_0 \text{ at } s = 1, \quad \frac{dy}{ds} = 0 \text{ at } s = 0, \quad (10.140)$$

$$\phi^2 = \frac{x_c^2 \, k_0 \, g(y_0)}{D_s \, C_0}, \quad f(y) = \frac{g(y)}{g(y_0)}, \quad g(y) = \frac{(E + y) \, y}{(1 + \alpha y)^2}. \quad (10.141)$$

The Thiele modulus squared, ϕ^2, is proportional to the absolute reaction rate multiplied by the layer thickness squared divided by the effective diffusion coefficient. The effectiveness factor is then simply the average rate of reaction divided by the rate of reaction at the surface conditions. The result given by

Finlayson and Young [1979] is

$$\eta = \frac{\sqrt{2}}{\phi}\left[\int_0^{y_0} f(y)\,dy\right]^{1/2} = \frac{\sqrt{2}}{\phi}\frac{1}{g(y_0)}\left\{\frac{E}{\alpha^2}\left[\frac{1}{1+\alpha y} + \log_e(1+\alpha y)\right]\Big|_0^{y_0} + \right.$$
$$\left. + \frac{1}{\alpha^3}\left[1 + \alpha y - 2\log_e(1+\alpha y) - \frac{1}{1+\alpha y}\right]\Big|_0^{y_0}\right\}. \quad (10.142)$$

The variable y_0 is the concentration of carbon monoxide in the fluid external to the catalyst, in this case $Y(z,t)$. The reaction rate in Eq. (10.137)-(10.138) is then

$$<r> = \eta \text{ rate.} \quad (10.143)$$

Let us first consider steady-state equations when axial conduction is neglected. The equations are then Eq. (10.135)-(10.136) together with Eq. (10.137)-(10.138) minus the axial conduction term.

$$\frac{dT}{dz} = 2\alpha_1 \text{Nu}(T^s - T), \quad \frac{dY}{dz} = 2\alpha_2 \text{Sh}(Y^s - Y) \quad (10.144)$$

$$\frac{1}{2}\text{Nu}(T^s - T) + \beta_5 <r> = 0, \quad -\alpha_6 \beta_4 <r> + \frac{1}{2}\text{Sh}(Y^s - Y) = 0 \quad (10.145)$$

We can solve these equations by using a robust method to integrate the ordinary differential equations and solve the algebraic constraints at each z-location, as needed. The step-size control can be automated to guarantee a solution, as in the packages LSODE and RKF45. Thus, numerical problems are easily handled using standard packages that have adjustable step-sizes. When axial conduction is included, however, the steady-state equations become

$$\frac{dT}{dz} = 2\alpha_1 \text{Nu}(T^s - T), \quad \frac{dY}{dz} = 2\alpha_2 \text{Sh}(Y^s - Y) \quad (10.146)$$

$$-\alpha_5 \left(\frac{r_{hs}}{L}\right)^2 \frac{d^2 T^s}{dz^2} + \frac{1}{2}\text{Nu}(T^s - T) + \beta_5 <r> = 0$$
$$-\alpha_6 \beta_4 <r> + \frac{1}{2}\text{Sh}(Y^s - Y) = 0. \quad (10.147)$$

Young and Finlayson [1976a,b] solved these by first solving the fluid equations in terms of the solid quantities T^s and Y^s.

$$T(z) = e^{-2\alpha_1 \text{Nu} z} T(0) + 2\alpha_1 \text{Nu} \int_0^z e^{-2\alpha_1 \text{Nu}(z-z')} T^s(z')\,dz' \quad (10.148)$$

$$Y(z) = e^{-2\alpha_2 \text{Sh} z} Y(0) + 2\alpha_2 \text{Sh} \int_0^z e^{-2\alpha_2 \text{Sh}(z-z')} Y^s(z')\,dz' \quad (10.149)$$

The equations for the solid temperature and concentration then become

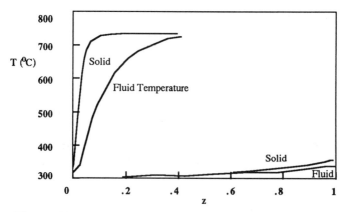

Figure 10.24. Multiple Steady-State Solutions to Eq. (10.148-10.151) Data from Young and Finlayson [1976a,b]

$$-\alpha_5 \left(\frac{r_{hs}}{L}\right)^2 \frac{d^2 T^s}{dz^2} + \beta_5 <r> +$$

$$+ \frac{1}{2} \text{Nu} \left(T^s - e^{-2\alpha_1 \text{Nu} z} T(0) - 2\alpha_1 \text{Nu} \int_0^z e^{-2\alpha_1 \text{Nu}(z-z')} T^s(z') \, dz' \right) = 0$$

(10.150)

$$-\alpha_6 \beta_4 <r> +$$ (10.151)

$$+ \frac{1}{2} \text{Sh} \left(Y^s - e^{-2\alpha_2 \text{Sh} z} Y(0) - 2\alpha_2 \text{Sh} \int_0^z e^{-2\alpha_2 \text{Sh}(z-z')} Y^s(z') \, dz' \right) = 0.$$

Eq. (10.150)-(10.151) are integro-differential equations needing two boundary conditions at each end of the device, $z = 0$ and $z = L$. Young and Finlayson [1976a,b] approximated the solid temperature and concentration with finite element polynomials using the method of orthogonal collocation on finite elements. This reduces Eq. (10.150)-(10.151) to a large set of algebraic equations that were solved iteratively.

A solution for a typical set of parameters is given in Figure 10.24. The curve on the left shows that the solid temperature rises rapidly near the inlet to an asymptotic value. The fluid temperature rises less steeply since the fluid must get its energy from the wall, which it does gradually. We notice, however, that there are two pairs of curves. Each of these pairs is a solution to Eq. (10.135)-(10.138) together with reasonable boundary conditions. The problem has more than one solution even when all parameters take the same values. Which solution we obtain numerically depends on how the solution is found iteratively. The multiple solutions arise because of the presence of the axial conduction term in Eq. (10.137)-(10.138) or Eq. (10.150)-(10.151); without those terms, only one solution is possible for a given set of parameters. The axial conduction term

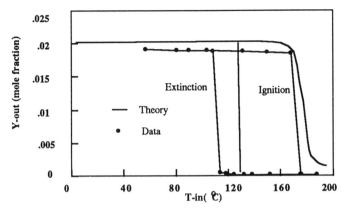

Figure 10.25. Hysteresis Curve, 2% Carbon Monoxide
Data from Young and Finlayson [1976b]

provides a method of feedback of energy upstream of the reaction zone.

The existence of multiple solutions depends on the parameters of the problem; the most important parameters are the feed temperature and concentration and the flow rate of gas. If the reactor is too short, only one of the solutions may be possible unless the temperature is raised. Let us suppose that the reactor is cold; we send hot gas down the tubes and wait until steady-state is approached. There may be no reaction zone at all. If the inlet temperature is raised and steady-state is again approached, eventually there will be a reaction zone in the device. The temperature sufficient to cause this reaction zone is called the ignition temperature. Then we start to introduce inlet gases with a lower temperature than the ignition temperature; again we wait for steady-state. The reaction zone remains in the device until the temperature is lowered still further. At some point, there is no reaction in the device; the lowest temperature sustaining a reaction zone is called the extinction temperature. The extinction temperature is lower than the ignition temperature. A typical diagram of ignition and extinction is shown in Figure 10.25. For inlet temperatures between 140 °C and 190 °C, there are two solutions to the equations (10.150)-(10.151): one solution with nearly everything reacting and one solution with almost nothing reacting. For temperatures higher than 190 °C, everything will react, while for temperatures lower than 140 °C, nothing will react. Also shown in Figure 10.25 is experimental data from a laboratory device presented by Hlavacek, V. and Votruba [1974]. In this case, one constant was fit to the experimental data when the carbon monoxide concentration was 4% (to essentially determine the strength of the reaction or the amount of platinum); the same constants were used without adjustment for the 2% case shown in Figure 10.25. The phenomena are clearly modeled correctly. These ignition-extinction profiles are moved to lower temperatures when the flow rate is decreased and to higher temperatures when the flow rate is increased [Finlayson and Young, 1979].

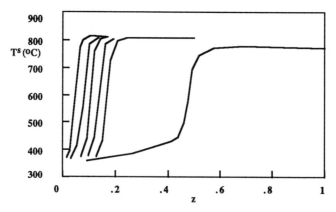

Figure 10.26. Thermal Cooling at t = 5, 10, 20, and 25 seconds and Steady-State Data from Young and Finlayson [1976b]

A transient case was modeled by Young and Finlayson [1976b] that gives the results shown in Figure 10.26. The reactor is initially hot, at 773 °C. At time zero the inlet gas, with a temperature of 344 °C, is introduced. An alternative solution method is provided in the electronic text entitled "Transient Catalytic Muffler." The case presented in Figure 10.26 is for a slightly different reaction rate expression and includes the calculation of Nu and Sh. The reaction zone is initially at the inlet of the device but as time proceeds, the reaction zone moves down the device. The temperature is very steep in the reaction zone, which moves with a constant velocity until steady-state is reached. This is an example of a steep profile that is caused by chemical reaction rather than convection. The movement is associated with combined conduction, diffusion, and reaction; the velocity of movement is then not the convection velocity. Other examples solved with moving fronts are by Ramachandrau and Dudukovíc [1984], who stretched the entire coordinate system using one function in time, Bhattacharya and Joseph [1988], who used collocation with an adaptive method, and Gawdzik and Rakowski [1989], who combined the method of characteristics with the orthogonal collocation method.

10.5. Solid-Gas Reactions

In the previous section, the reaction occurred on a solid, but the solid served as a catalyst to speed up the reaction and was not consumed during the reaction. Sometimes the solid itself reacts with the gas. Such situations lead to sharp, moving fronts since the solid disappears. The method of moving coordinate systems is applied here to this type of gas-solid reactions.

We begin with a solid particle (taken as spherical) and bring reactants to the surface of the sphere. Reaction takes place, consuming part of the solid and leaving behind a porous char. The reactants must now diffuse through the porous char in order to react at an inner reaction zone. The mathematical description of this problem is taken from Yoshida, et al. [1975].

$$\gamma \frac{\partial c}{\partial t} = \frac{\partial^2 c}{\partial r^2} + \frac{2}{r}\frac{\partial c}{\partial r} \qquad (10.152)$$

$$c(1,t) = 1, \; c(r_c, t) = 0 \qquad (10.153)$$

$$\frac{dr_c}{dt} = -\frac{\partial c}{\partial r}\bigg|_{r=r_c}, \; r_c(0) = 1 \qquad (10.154)$$

The variable c is the concentration of reactants and r_c is the radius of the unreacted solid. The fourth equation represents a balance expressing that the rate of diffusion to the reaction zone equals the rate of reaction at the zone, as represented by the movement of the reaction front. Yoshida, et al. [1975] solve this problem by first making the transformation

$$x = \frac{r - r_c(t)}{1 - r_c(t)}, \; \tau = \sqrt{t}, \qquad (10.155)$$

leading to the equations

$$\frac{\gamma}{2\tau}\frac{\partial c}{\partial \tau} = \frac{1}{(1-r_c)^2}\frac{\partial^2 c}{\partial x^2} + \left\{\frac{2}{(1-r_c)x + r_c}\frac{1}{1-r_c} - \gamma \frac{1-x}{(1-r_c)}\left[\frac{\partial c}{\partial x}\right]_{x=0}\right\}\frac{\partial c}{\partial x} \qquad (10.156)$$

$$\frac{1}{2\tau}\frac{dr_c}{d\tau} = -\frac{1}{1-r_c}\left[\frac{\partial c}{\partial x}\right]_{x=0} \qquad (10.157)$$

$$c(1,\tau) = 1, \; c(0,\tau) = 0, \; r_c(0) = 1. \qquad (10.158)$$

Eq. (10.156) can be discretized using a finite difference method or the orthogonal collocation method, giving a set of ordinary differential equations to solve in time. Sketches of the solution as a function of time are shown in Figure 10.27. Clearly the profile is steep when the reaction begins and the diffusion layer is contained in a small region near the boundary of the sphere. The moving coordinate takes this feature into account easily and moves the collocation points into the interior of the sphere as the solid is consumed. This represents a case in which the nodes move in time, but their movement is specified *a priori*. In some cases, this is a viable technique and leads to simple models. Similar models have also been obtained for combustion in flames [Kurylko and Essenhigh, 1973]. A finite difference solution of a similar problem is given in Section 11.2 and uses the technique of a moving grid governed by Eq. (10.157).

For planar geometry, the solution method given above leads to an exact solution. The equation is Eq. (10.152) without the last term, written here with x as the independent variable:

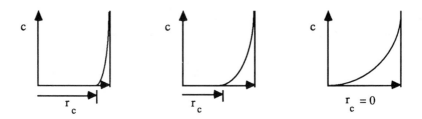

Figure 10.27. Concentration Profiles During a Gas-Solid Reaction
(All the collocation points are between $r = r_c$ and $r = 1$.)

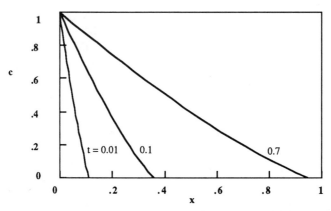

Figure 10.28. Exact Solution to Eq. (10.159-10.160), $\gamma = 2$

$$\gamma \frac{\partial c}{\partial t} = \frac{\partial^2 c}{\partial x^2}, \quad \frac{dx_c}{dt} = -\frac{\partial c}{\partial x}\bigg|_{x=x_c}, \tag{10.159}$$

$$c(x,0) = 0, \; c(x_c,t) = 0, \; c(0,t) = 1. \tag{10.160}$$

The solution, as derived in Section 11.2, is

$$c = 1 - \frac{\operatorname{erf} z}{\operatorname{erf} z_c}, \quad z = \sqrt{\frac{\gamma}{4t}}\, x, \; z_c \text{ is the root to } 1 = \frac{\sqrt{\pi}}{\gamma} z_c \exp(z_c^2) \operatorname{erf} z_c. \tag{10.161}$$

For $\gamma = 2.0$, the solution is plotted in Figure 10.28.

Problems

13. Derive Eq. (10.33). [Hint: Differentiate Eq. (10.30) with respect to time. To evaluate F_{xt}, interchange the order of differentiation and use

$$\frac{\partial F}{\partial t} = \frac{dF}{du} \frac{\partial u}{\partial t}. \tag{10.162}$$

2_2. Derive Eq. (10.45).

3_2. Derive Eq. (10.47).

4_2. Use the option REACT in the program APPLICATIONS to redo the calculations in Figure 10.4 using only 26 grid points.

5_3. Derive Eq. (10.81). [Hint: try

$$c_i = \phi^{i-1} \qquad (10.163)$$

and solve for ϕ. Then write the solution as in Eq. (10.79) and solve for A and B by setting $c_1 = 1$ and $c_{N+1} = 0$.]

6_2. Use the option REACT in the program APPLICATIONS for an initial condition corresponding to a concentration of zero everywhere except for between $x = 0.25$ and $x = 0.75$, where it is 1.0 (i.e., like those problems in Part I). Use $\alpha = 2$, $Da_I = -0.1$, as in Figure 10.4.

7_2. Use the option REACT in the program APPLICATIONS for the best methods of Problem 10.6 with $\alpha = 2$ and $Da_I = -0.1$ (like Figure 10.4) and $Da_I = -1$ and $Da_I = -10$ (like Figure 10.5). Use an initial condition corresponding to a concentration of zero everywhere except for between $x = 0.25$ and $x = 0.75$ (i.e., like those in Part I).

8_2. Verify Eq. (10.115) is a solution to Eq. (10.112-10.113) when $\beta = a = 1$.

9_2. Verify Eq. (10.118) is a solution to Eq. (10.112) and Eq. (10.117) when $\beta = 1$.

10_1. Write the orthogonal collocation form of Eq. (10.156)-(10.158).

References

Bhattacharya, A. and Joseph, B., "Simulation of Fixed-Bed Gas-Solid Reactors Using an Adaptive-Spline Collocation Method," Comp. Chem. Eng. 12 351-353 [1988].

Chung, T. J., Kim, Y. M., and Sohn, J. L., "Finite Element Analysis in Combustion Phenomena," Int. J. Num. Methods Fluids 7 989-1012 [1987].

Degreve, J., Dimitriou, P., Puszynski, J., and Hlavacek, V., "Use of 2-D Adaptive Mesh in Simulation of Combustion Front Phenomena," Comp. Chem. Eng. 11 749-755 [1987].

Dwyer, H. A. and Sanders, B. R., "Numerical Modeling of Unsteady Flame Propagation," Sandia Livermore Laboratories Report SAND77-8275 [1978].

Finlayson, B. A. and Young, L. C., "Mathematical Models of the Monolith Catalytic Converter. Part III. Hysteresis in Carbon Monoxide Reactor," A. I. Ch. E. J. 25 192-196 [1979].

Fisher, R. A., "The Wave of Advance of Advantageious Genes," Annals of Eugenics 7 355-369 [1937].

Galinas, R. J., Doss, S. K., Miller, K., "The Moving Finite Element Method: Applications to General Partial Differential Equations with Multiple Large Gradients," J. Comp. Phys. 40 202-249 [1981].

Gatica, J. E., Puszynski, J., Hlavacek, V., "Reaction Front Propagation in Nonadiabatic Exotermic Reaction Flow Systems," A. I. Ch. E. J. 33 819-833 [1987].

Gawdzik, A., Rakowski, L., "The Methods of Analysis of the Dynamic Properties of the Adiabatic Tubular Reactor with Switch Flow," Comp. Chem. Eng. 13 1165-1173 [1989].

Hlavacek, V. and J. Votruba, "Experimental Study of Multiple Steady States in Adiabatic Catalytic Systems," in Chemical Reaction Engineering - II, H. M. Hulburt, Ed. pp. 545-558, Am. Chem. Soc. Ser. 133 [1974].

Kurylko, L. and Essenhigh, R. H., "Steady and Unsteady Combustion of Carbon," 14th Symp. on Combustion, 1375-1386 [1973].

Lee, D. N. and Ramos, J. I., "Application of the Finite Element Method to One-Dimensional Flame Propagation Problems," AIAA J 21 262-269 [1983].

Mack, A., Weber, H. J., and Roth, P., "A Moving Grid Method Applied to One-Dimensional Non-Stationary Flame Progagation," Int. J. Num. Methods Fluids 13 869-882 [1991].

Ramachandrau, P. A. and Dudukovic, M. P., "A Moving Finite Element Collocation Method for Transient Problems with Steep Gradients," Chem. Eng. Sci. 39 1321-1324 [1984].

Ramos, J. I., "Numerical Solution of Reactive-Diffusive Systems. Part 1: Explicit Methods," Int. J. Comp. Math. 18 43-65 [1985].

Ramos, J. I., "Numerical Solution of Reactive-Diffusive Systems. Part 2: Methods of Lines and Implicit Algorithms," Int. J. Comp. Math. 18 141-161 [1985].

Ramos, J. I., "Numerical Solution of Reactive-Diffusive Systems. Part 3: Time Linearization and Operator-Splitting Techniques," Int. J. Comp. Math. 18 289-309 [1986].

Ramos, J. I., "Numerical Methods for One-Dimensional Reaction-Diffusion Equations Arising in Combustion Theory," pp. 150-261 in Ann. Rev. Num. Fluid Mech. Heat Transfer, Vol. 1 (T. C. Chawla, ed.), Hemisphere [1987].

Reitz, R. D., "A Study of Numerical Methods for Reaction-Diffusion Equations," SIAM J. Sci. Stat. Comp. 2 95-106 [1981].

Rhee, H. K., Aris, R., Amundson, N. R., *First-Order Partial Differential Equations. Volume I. Theory and Application of Single Equations*, Prentice-Hall, Englewood Cliffs, N. J., [1986].

Sheintuch, M., "Numerical Approaches for Computation of Fronts," Num. Methods Partial Diff. Eqn. 6 43-58 [1990].

Smooke, M. D., Miller, J. A., and Kee, R. J., "Determination of Adiabatic Flame Speeds by Boundary Value Methods," Combustion Sci. Tech. 34 79-90 [1983].

Smooke, M. D., "On the Use of Adaptive Methods in Premixed Combustion,", AIChE. J. 32 1233-1241 [1986].

Smooke, M. D., Mitchell, R. E., Grcar, J. F., "Numerical Solution of a Confined Laminar Diffusion Flame," pp. 557-568 in Elliptic Problem Solvers II, (ed. G. Birkhoff and A. Schoenstadt) Academic Press [1984].

Spalding, D. B., "Predicting the Laminar Flame Speed in Gases with Temperature-Explicit Reaction Rates," Combust. Flame 1 287-295 [1957].

Stuart, A., "Nonlinear Instability in Dissipative Finite Difference Schemes," SIAM Review 31 191-220 [1989].

Tezduyar, T. E., Park, Y. J., Deans, H. A., "Finite Element Procesures for Time-dependent Convection-Diffusion-Reaction Systems," Int. J. Num. Methods Fluids 7 1013-1033 [1987].

Yanenko, N. N., "The Method of Fractional Steps," Springer-Verlag, New York [1971].

Yoshida, K., Kunii, D. and Shimizu, F., "Application of Collocation Technique for Moving-Boundary Problems in Solid-Gas Reactions," J. Chem. Eng. Japan 8 417-419 [1975].

Young, L. C. and Finlayson, B. A., "Mathematical Models of the Monolith Catalytic Converter. Part I. Development of Model and Application of Orthogonal Collocation," A. I. Ch. E. J. 22 331-343 [1976a].

Young, L. C. and Finlayson, B. A., "Mathematical Models of the Monolith Catalytic Converter. Part II. Application to Automobile Exhaust," A. I. Ch. E. J. 22 343-353 [1976b].

CHAPTER
ELEVEN

PROBLEMS WITH PHASE CHANGE

When a liquid freezes or a solid melts, the phase change takes place at an interface that is at the freezing or melting point; this interface moves with time. Such freezing and melting problems are classic examples of moving boundary problems and they thus provide natural applications of numerical methods with moving nodes. Problems of this type arise in metal casting, welding, food processing, freeze-coating of fibers, laser glazing, and the use of latent heat devices for energy conservation [Gupta and Arora, 1988]. In this chapter we will review techniques applied to these problems in both one and two dimensions. The first four sections will limit consideration to heat conduction, while the fifth section will include the complicating effect of natural convection, which is important in crystal growth for making electronic chips. The methods to be discussed include the analytical movement of nodes and moving finite element methods, as well as methods on fixed grids. Some of the finite difference methods for these problems are given in a book by Crank [1984], while Pelcé [1988] gives several physical applications.

11.1. Equations

For the phase change problems discussed in most of this chapter, we will assume that convection does not occur in the liquid phase. In this case, the governing equations are the heat conduction equation in each phase. We take first the one-dimensional equations

$$\rho_i C_i \frac{\partial T_i}{\partial t} = k_i \frac{\partial^2 T_i}{\partial x^2}, \quad i = 1, 2. \tag{11.1}$$

The temperature is specified at two boundaries,

$$T_1(0,t) = T_0 \text{ and } T_2(L,t) = T_L. \tag{11.2}$$

(A subscript 1 refers to one phase and a subscript 2 to the other phase.) The initial condition is taken as

$$T_i(x,0) = T_{initial}(x). \tag{11.3}$$

If the domain is semi-infinite, the second boundary condition is applied at infinity and the initial condition must be compatible with the boundary condition at infinity. At the freezing or melting interfaces, the boundary conditions are that (a) the temperature is the melting temperature

$$T = T_m \text{ at } x = X(t), \tag{11.4}$$

and (b) energy is conserved. The net heat flux to the boundary equals the amount of energy used in the melting or freezing process.

$$\Delta H_f \frac{dX}{dt} = \left[k_1 \frac{\partial T_1}{\partial x} - k_2 \frac{\partial T_2}{\partial x} \right]_{x = X(t)} \tag{11.5}$$

ΔH_f is the latent heat. This problem [Eq. (11.1)-(11.5)] is called the two-phase problem.

Let us consider a special case where the initial temperature and the temperature at $x = L$ (or x approaching ∞) are the melting temperatures:

$$T_1(0,t) = T_0 \text{ and } T_2(L,t) = T_m \text{ and } T_i(x,0) = T_m. \tag{11.6}$$

Phase 2 is maintained at the melting or freezing point; we need only to solve the heat transfer problem in Phase 1, but the domain is always changing. Then the problem is

$$\rho_1 C_1 \frac{\partial T_1}{\partial t} = k_1 \frac{\partial^2 T_1}{\partial x^2} \tag{11.7}$$

$$T_1(0,t) = T_0, \quad T_1(X(t),t) = T_m, \quad T_1(x,0) = T_m$$

$$\Delta H_f \frac{dX}{dt} = k_1 \frac{\partial T_1}{\partial x} \quad \text{at } x = X(t). \tag{11.8}$$

This is called the one-phase problem. The time-dependent phenomena in the one-phase problem are governed by the following time constants:

$$\text{conduction: } t_c = \frac{\rho C_p L^2}{k_1}, \text{ and melt movement: } t_m = \frac{\Delta H_f L^2}{k_1 (T_0 - T_m)}. \tag{11.9}$$

The ratio of these time constants can indicate the difficulty of this problem. It is

$$\frac{t_c}{t_m} = \frac{\rho C_p}{\Delta H_f} (T_0 - T_m) = \text{Stefan number}. \tag{11.10}$$

For the problem treated below, the Stefan number is only 0.35; in this case, the difficulty is provided by the moving front. In the two-phase problem, the difficulty results from both the moving front and the discontinuity at the front.

We apply the Galerkin method to this problem by making the residual orthogonal to the weighting functions W_j:

$$\int_0^{X(t)} W_j(x) \left[\rho_1 C_1 \frac{\partial T_1}{\partial t} - k_1 \frac{\partial^2 T_1}{\partial x^2} \right] dx = 0. \tag{11.11}$$

We integrate the last term by parts to obtain

$$-\int_0^{X(t)} W_j(x) k_1 \frac{\partial^2 T_1}{\partial x^2} dx = -\int_0^{X(t)} \frac{\partial}{\partial x}\left[W_j(x) k_1 \frac{\partial T_1}{\partial x} \right] dx + \int_0^{X(t)} k_1 \frac{\partial W_j}{\partial x} \frac{\partial T_1}{\partial x} dx$$

$$= \int_0^{X(t)} k_1 \frac{\partial W_j}{\partial x} \frac{\partial T_1}{\partial x} dx - \left[W_j(x) k_1 \frac{\partial T_1}{\partial x} \right]_{x=X(t)} + \left[W_j(x) k_1 \frac{\partial T_1}{\partial x} \right]_{x=0}. \tag{11.12}$$

The last term drops out because $W_j(0) = 0$ for $j = 2, \ldots, NT$ and we do not use the equation for $j = 1$. Combining all terms gives the set of equations

$$\int_0^{X(t)} W_j(x) \rho_1 C_1 \frac{\partial T_1}{\partial t} dx + \int_0^{X(t)} k_1 \frac{\partial W_j}{\partial x} \frac{\partial T_1}{\partial x} dx - \left[W_j(x) k_1 \frac{\partial T_1}{\partial x} \right]_{x=X(t)} = 0 \tag{11.13}$$

$$\Delta H_f \frac{dX}{dt} = k_1 \frac{\partial T_1}{\partial x} \quad \text{at } x = X(t). \tag{11.14}$$

It is also useful to have the integral energy balance; this is obtained from Eq. (11.13) by setting $W_j = 1$:

$$\int_0^{X(t)} \rho_1 C_1 \frac{\partial T_1}{\partial t} dx = \left[k_1 \frac{\partial T_1}{\partial x} \right]_{x=X(t)}. \tag{11.15}$$

This can be combined with Eq. (11.14) to give

$$\Delta H_f \frac{dX}{dt} = \int_0^{X(t)} \rho_1 C_1 \frac{\partial T_1}{\partial t} dx. \tag{11.16}$$

If a finite element method is applied at NT nodes, the conditions at each node are as illustrated in Figure 11.1. The temperatures are set at $j = 1$ and $j = NT$ and the residual equations [Eq. (11.13)] are used for $j = 2, \ldots, NT-1$. This gives NT conditions. But we have NT+1 unknowns, including the position of the melting front, $X(t)$. The equation for this variable is given by Eq. (11.14). As discussed by Hrymak and Westerberg [1986], the calculation may be inaccurate if linear finite elements are used to calculate the derivative; only first-order accuracy would be obtained. This difficulty can be overcome by using a three-point one-sided difference expression in this equation or by using the integrated equation, Eq. (11.16).

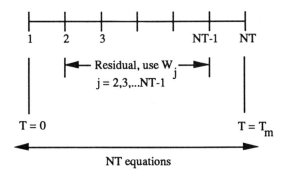

Figure 11.1. Equations for Phase Change Problems

11.2. One-Phase Problems in One Dimension

In this section we solve one-phase problems using several methods. The first method transforms the domain so that the coordinate system moves with the freezing front. This was apparently done first by Landau [1950]. A global approximation of the temperature is made on this transformed grid and the problem is then solved on a grid that is stretching in time. The second method moves the coordinate system in the same way but uses a finite element method for the temperature approximation. The third method uses a moving finite element method to move the nodes independently (rather than with a velocity characterized by the velocity of the moving front). A variety of methods for one-phase problems are reviewed by Furzeland [1980].

We first solve Eq. (11.13) by transforming the grid into one that extends from $x = 0$ to $x = X(t)$, with the position of $X(t)$ to be determined. This idea was developed using a heat balance idea by Goodman [1958, 1964] and Murray and Landis [1959]. We will assume a trial function that fits the boundary conditions at $x = 0$ and $x = X(t)$:

$$T_1 = T_0 + (T_m - T_0)\frac{x}{X(t)} + c(t)\frac{x}{X(t)}\left(\frac{x}{X(t)} - 1\right). \quad (11.17)$$

The first term of Eq. (11.17) satisfies the inhomogeneous equations, while the second term satisfies the homogeneous conditions; thus T satisfies the boundary conditions for any value of $c(t)$. The various derivatives are

$$\frac{\partial T_1}{\partial x} = \frac{T_m - T_0}{X(t)} + c(t)\left[\frac{2x}{X^2} - \frac{1}{X}\right] = \frac{T_m - T_0}{X(t)} + \frac{c(t)}{X(t)}\left[2\frac{x}{X(t)} - 1\right], \quad (11.18)$$

$$\frac{\partial T_1}{\partial t} = \frac{\dot{X}}{X}\left\{-(T_m - T_0)\frac{x}{X(t)} + c\left[-2\frac{x^2}{X^2} + \frac{x}{X(t)}\right]\right\} + \dot{c}\frac{x}{X(t)}\left[\frac{x}{X(t)} - 1\right]. \quad (11.19)$$

The first spatial derivative, evaluated at $x = X(t)$, is

$$\left.\frac{\partial T_1}{\partial x}\right|_{x=X(t)} = \frac{T_m - T_0}{X(t)} + \frac{c(t)}{X(t)}. \tag{11.20}$$

We take the following weighting function for the Galerkin method:

$$W = \frac{x^2}{X^2} - \frac{x}{X(t)}. \tag{11.21}$$

Its first derivative is

$$\frac{\partial W}{\partial x} = \frac{2x}{X^2} - \frac{1}{X}. \tag{11.22}$$

Combining all the terms in Eq. (11.13) gives

$$\rho_1 C_1 \frac{\dot{X}}{X}\left\{\int_0^{X(t)}\left[-(T_m - T_0)\frac{x}{X(t)}\left(\frac{x^2}{X^2} - \frac{x}{X(t)}\right)\right]dx + c\int_0^{X(t)}\left(\frac{x^2}{X^2} - \frac{x}{X(t)}\right)\left(-2\frac{x^2}{X^2} + \frac{x}{X(t)}\right)dx\right\}$$

$$+ \rho_1 C_1 \dot{c}\int_0^{X(t)}\left(\frac{x^2}{X^2} - \frac{x}{X(t)}\right)^2 dx + \frac{k_1}{X^2}\int_0^{X(t)}\left(\frac{2x}{X} - 1\right)\left[T_m - T_0 + c(t)\left(2\frac{x}{X(t)} - 1\right)\right]dx = 0.$$

$$\tag{11.23}$$

The integrals are changed using the following formula:

$$\int_0^{X(t)} f\left(\frac{x}{X}\right) dx = \int_0^1 f(u)\, X(t)\, du, \quad u = \frac{x}{X(t)}. \tag{11.24}$$

Then Eq. (11.23) becomes

$$\rho_1 C_1 \dot{X}\left\{\int_0^1 [-(T_m - T_0)\, u\,(u^2 - u)]\, du + c\int_0^1 (u^2 - u)(-2u^2 + u)\, du\right\}$$

$$+ \rho_1 C_1 \dot{c} X\int_0^1 (u^2 - u)^2 du + \frac{k_1}{X}\int_0^1 (2u - 1)[T_m - T_0 + c(t)(2u - 1)]\, du = 0. \tag{11.25}$$

The integrals are

$$\int_0^1 u(u^2 - u)\, du = -\frac{1}{12}, \qquad \int_0^1 (u^2 - u)(-2u^2 + u)\, du = \frac{1}{60},$$

$$\int_0^1 (u^2 - u)^2 du = \frac{1}{30}, \qquad \int_0^1 (2u-1)\, du = 0, \qquad \int_0^1 (2u-1)^2 du = \frac{1}{3}. \tag{11.26}$$

The combined result is

$$\rho_1 C_1 \dot{X} \left\{ \frac{c}{60} + \frac{T_m - T_0}{12} \right\} + \rho_1 C_1 \dot{c} X \frac{1}{30} + \frac{k_1}{X} \frac{c}{3} = 0. \tag{11.27}$$

In addition, we have the following boundary condition at the freezing or melting point:

$$\Delta H_f \frac{dX}{dt} = \frac{k_1 (T_m - T_0 + c(t))}{X(t)}. \tag{11.28}$$

Eq. (11.27)-(11.28) are two ordinary differential equations to be solved for c(t) and X(t). The parameters (Table 11.1) are appropriate to the freezing of dense sand saturated with water [Lynch and O'Neill, 1981]. It is assumed that the sand and water are in thermal equilibrium.

Table 11.1 Physical Parameters for Melting Problem One

$\rho\, C_p = 0.62$ cal/ °C cm^3

$k = 0.0096$ cal/cm sec °C

$\Delta H_f = 17.68$ cal/cm^3

$T_m - T_0 = -10$ °C

The solution is

$$X = a\, t^{0.5} \text{ and } c = \text{constant.} \tag{11.29}$$

The solution for T[x/X(t)] compares favorably with the exact solution, as shown in Figure 11.2. Typical profiles for T(x,t) are shown in Figure 11.3. The growth of the melt interface is proportional to $t^{0.5}$ in both cases; the exact solution has a = 0.0988213 whereas the approximate solution has a = 0.102. The function X(t) is shown in Figures 11.4 and 11.5.

Another way to approach this problem is to transform the entire problem onto a new coordinate system using

$$u = \frac{x}{X(t)}, \quad \tau = t. \tag{11.30}$$

Then the derivatives with respect to x and t are

$$\frac{\partial u}{\partial x} = \frac{1}{X}, \quad \frac{\partial u}{\partial t} = -\frac{x}{X^2} \dot{X} = -u \frac{\dot{X}}{X}, \quad \frac{\partial \tau}{\partial t} = 1, \quad \frac{\partial \tau}{\partial x} = 0. \tag{11.31}$$

The function T is now considered a function of u and τ. The derivatives with respect to x and t are

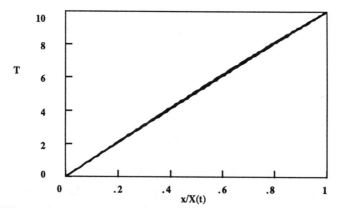

Figure 11.2. Exact and Analytical Approximation to Melting Problem One (in Moving Coordinates)

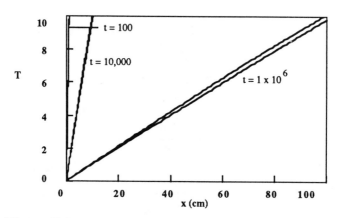

Figure 11.3. Exact and Analytical Approximation to Melting Problem One (in Spatial Coordinates)

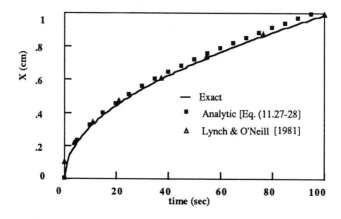

Figure 11.4. Comparison of Solutions for Melt Thickness (Short Times)

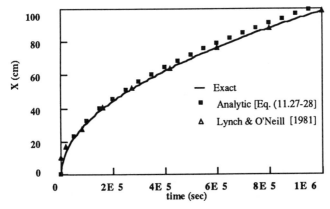

Figure 11.5. Comparison of Solutions for Melt Thickness (Long Times)

$$\frac{\partial T(\tau,u)}{\partial t} = \frac{\partial T}{\partial \tau}\frac{\partial \tau}{\partial t} + \frac{\partial T}{\partial u}\frac{\partial u}{\partial t} = \frac{\partial T}{\partial \tau} - u\frac{\dot{X}}{X}\frac{\partial T}{\partial u},$$

$$\frac{\partial T(\tau,u)}{\partial x} = \frac{\partial T}{\partial \tau}\frac{\partial \tau}{\partial x} + \frac{\partial T}{\partial u}\frac{\partial u}{\partial x} = \frac{1}{X}\frac{\partial T}{\partial u}, \quad (11.32)$$

$$\frac{\partial^2 T}{\partial x^2} = \frac{1}{X^2}\frac{\partial^2 T}{\partial u^2}.$$

The differential equation, Eq. (11.7), and boundary conditions, Eq. (11.8), then become

$$\rho_1 C_1 \left(\frac{\partial T}{\partial \tau} - u\frac{\dot{X}}{X}\frac{\partial T}{\partial u} \right) = \frac{k_1}{X^2}\frac{\partial^2 T}{\partial u^2}, \quad (11.33)$$

$$T(0,\tau) = T_0, \quad T(1,\tau) = T_m, \quad \Delta H_f \frac{dX}{dt} = \frac{k_1}{X}\frac{\partial T_1}{\partial u}\bigg|_{u=1}. \quad (11.34)$$

When the problem is in an infinite medium, it has an exact solution [Carslaw and Jaeger 1959]. This exact solution is obtained by trying the form

$$T = T_0 + \phi(\eta), \quad \eta = \frac{x}{\delta(t)}. \quad (11.35)$$

Thus, the solution depends only on the coordinate η and not on t by itself. By using the transformation rules illustrated in Eq. (11.32), we can transform Eq. (11.33) into

$$\rho_1 C_1 \left(-\eta \frac{\dot{\delta}}{\delta}\frac{d\phi}{d\eta} \right) = \frac{k_1}{\delta^2}\frac{d^2\phi}{d\eta^2}, \quad (11.36)$$

which is to be solved together with the boundary conditions

$$\Delta H_f \frac{dX}{dt} = \frac{k_1}{\delta} \frac{d\phi}{d\eta} \quad \text{and} \quad \phi = T_m - T_0 \quad \text{at } \eta = \frac{X(t)}{\delta(t)}. \tag{11.37}$$

Eq. (11.36) can be rearranged to give

$$\frac{\rho_1 C_1}{k_1} \delta \dot{\delta} = -\frac{\frac{d^2\phi}{d\eta^2}}{\eta \frac{d\phi}{d\eta}} = \text{constant} = \alpha. \tag{11.38}$$

The left-hand side of the equation is a function of time alone while the right-hand side is a function of η alone. Thus, both functions must be constant. We call the constant α; the numerical value of the constant appears in the formulas and affects the function $\phi(\eta)$, but not the solution in the form $T(x,t)$. Thus the value of α is arbitrary; we take it to equal 2.0. Then $\delta(t)$ satisfies

$$\frac{d\left(\frac{1}{2}\delta^2\right)}{dt} = \alpha \frac{k_1}{\rho_1 C_1}. \tag{11.39}$$

With $\alpha = 2$ and $\delta(0) = 0$, we get

$$\delta = \sqrt{4 \frac{k_1 t}{\rho_1 C_1}}. \tag{11.40}$$

The function $\phi(\eta)$ is given by the solution to

$$\frac{d^2\phi}{d\eta^2} + 2\eta \frac{d\phi}{d\eta} = 0. \tag{11.41}$$

The general solution to Eq. (11.41) is

$$\phi = A + B \,\text{erf}\, \eta, \tag{11.42}$$

where erf is the error function. This function is a tabulated function and takes the value 0 at the origin and 1.0 at infinity. At the freezing or melting front, the boundary condition is

$$\phi(0) = 0 \text{ or } A = 0. \tag{11.43}$$

The boundary conditions of Eq. (11.37) are satisfied if we take X proportional to δ,

$$X = \lambda \delta \text{ and } \frac{dX}{dt} = \lambda \frac{d\delta}{dt}, \tag{11.44}$$

because then

$$\delta \frac{dX}{dt} \text{ is constant} = \lambda \delta \dot{\delta} = \lambda \frac{2 k_1}{\rho_1 C_1}. \tag{11.45}$$

The derivative of the error function is

$$\frac{d}{d\eta} \operatorname{erf} \eta = \frac{2}{\sqrt{\pi}} e^{-\eta^2}. \tag{11.46}$$

When Eq. (11.46) is introduced into Eq. (11.37) we get

$$\lambda \frac{2 \Delta H_f k_1}{\rho_1 C_1} = k_1 \frac{2}{\sqrt{\pi}} e^{-\lambda^2} B \quad \text{and} \quad T_m - T_0 = B \operatorname{erf} \lambda. \tag{11.47}$$

Thus the solution is

$$T = (T_m - T_0) \frac{\operatorname{erf} \eta}{\operatorname{erf} \lambda}, \quad \lambda \text{ from} \quad \lambda \operatorname{erf} \lambda \, e^{\lambda^2} = \frac{\rho_1 C_1}{\Delta H_f \sqrt{\pi}} (T_m - T_0),$$

$$\eta = \frac{x}{\delta(t)}, \quad X = \lambda \delta, \quad \delta = \sqrt{4 \frac{k_1 t}{\rho_1 C_1}}. \tag{11.48}$$

For the set of parameters in Table 11.1, we get $\lambda = 0.39708$; the exact solution is plotted in Figure 11.2. The exact solution in an semi-infinite domain is also the exact solution in a finite domain until the interface reaches the boundary $x = L$. The approximate solution in a finite domain must be revised from Eq. (11.17) once the melt interface reaches the boundary.

Lynch and O'Neill [1981] have solved this same problem but replace the trial function, Eq. (11.17), with a finite element approximation:

$$T(x,t) = \sum_{i=1}^{N} T_i(t) N_i(x,t). \tag{11.49}$$

They use Hermite polynomials in their calculation. In cubic Hermite polynomials, the unknown values are the function and its first derivative at the ends of each element. The resulting trial function is continuous and has first derivatives that are continuous. The nodes are allowed to move with a velocity proportional to the velocity of the front. For an x-position equal to

$$x = \beta X, \tag{11.50}$$

the velocity is taken as

$$\frac{dx}{dt} = \beta \frac{dX}{dt}. \tag{11.51}$$

The boundary condition is written as

$$\Delta H_f \frac{dX}{dt} = k_1 \frac{\partial T_1}{\partial x}\bigg|_{x=X} = k_1 T_N, \tag{11.52}$$

where T_N is the coefficient in the trial function corresponding to the first derivative. Lynch and O'Neill [1981] use several different methods of integrating Eq. (11.52) including explicit methods, implicit methods with iterative solutions, and predictor-corrector methods. Typical solutions are shown in Figure 11.4 for the interface movement in time (and the same parameters as were used in Figures 11.2 and 11.3). In the numerical solution, Lynch and O'Neill took the initial thickness of the melted front as 0.1 cm and integrated until $X = 1$ cm. Shown in Figure 11.4 is the analytical solution for an initial thickness of zero. The solution shown in Figure 11.5 is for a melting front at a distance of 1 meter, or 100 cm. We notice that the spatial region of the problem changes from 0.1 cm to 100 cm, or expands 1000 times, although only 9 nodes are used. Since the trial functions are cubic polynomials, this corresponds to about 28 nodes. This finite element approximation is extremely efficient for this simple problem.

Another method involves moving finite elements in which each node is allowed to move with its own velocity. The method is that described in Section 7.4 as a Petrov-Galerkin method, Eq. (7.126)-(7.127); the results are taken from work by Hrymak and Westerberg [1986]. Table 11.2 shows the location of the freezing front at $t = 1$ second. The solutions obtained with only 5 nodes are the same (to within three significant digits) as those obtained with 10, 20, or 30 nodes. The approximate analytical solution is accurate to within 3% and is quite easy to derive.

Table 11.2. Location of the Freezing Front at $t = 1$ Seconds
[Hrymak and Westerberg, 1986] Reproduced by permission of the American Institute of Chemical Engineers © 1986 AIChE Pergamm in

Method	Location (cm)
Exact	0.0988213
Moving finite elements, N = 5	0.098822
Moving finite elements, N = 10	0.098830
Moving finite elements, N = 20	0.098832
Moving finite elements, N = 30	0.098833
Analytic [Eq. (11.27)-(11.28)]	0.102

The space-time finite element method has been applied to similar problems by Bonnerot and Jamet [1974]. The finite element method is used in both time and spatial directions and the nodes are allowed to move with a velocity proportional to that of the front.

We next apply a centered finite difference method to Eq. (11.33) and see what happens when the front reaches the other side. The grid points are moving; they are uniform on the grid $u = 0$ to $u = 1$, or $x = 0$ to $x = X(t)$. It is necessary to start from some non-zero initial condition (i.e., $X(0) = 0$ is not allowed, since the equation would then be indefinite). The finite difference representation of Eq. (11.33)-(11.34) is

$$\rho_1 C_1 \left(\frac{dT_i}{d\tau} - u_i \frac{\dot{X}}{X} \frac{T_{i+1} - T_{i-1}}{2\,\Delta u} \right) = \frac{k_1}{X^2} \frac{T_{i+1} - 2T_i + T_{i-1}}{\Delta u^2}, \qquad (11.53)$$

$$T_1(\tau) = T_0, \quad T_{NT}(\tau) = T_m, \quad \Delta H_f \frac{dX}{d\tau} = \frac{k_1}{X} \frac{T_{NT-1} - 4T_{NT} + 3T_{NT+1}}{2\,\Delta u}. \qquad (11.54)$$

A one-sided derivative has been used to evaluate the flux at the freezing front in order to retain the second-order accuracy of the method. When the freezing front reaches x = 1, we must use the simpler equations

$$\rho_1 C_1 \frac{\partial T_i}{\partial \tau} = k_1 \frac{T_{i+1} - 2T_i + T_{i-1}}{\Delta x^2}, \quad i = 2, \ldots, NT, \qquad (11.55)$$

$$T_1(t) = T_0, \quad \frac{T_{NT-1} - 4T_{NT} + 3T_{NT+1}}{2\,\Delta x} = 0. \qquad (11.56)$$

Equation (11.56) is written for a boundary condition of no heat flux across the boundary at x = 1 (or i = NT+1). The only difficulty here is to make the transition from Eq. (11.53)-(11.54) to Eq. (11.55)-(11.56). A Runge-Kutta code can be programmed to stop when the solution reaches a certain value rather than when the time reaches a certain value. Thus we use Eq. (11.53)-(11.54) and tell the code to stop when X(t) = 1; we then start over with Eq. (11.55)-(11.56), using the final solution of Eq. (11.53)-(11.54) as the initial conditon. The final solution is shown in Figure 11.6 for the parameters in Table 11.1. The exact solution is also plotted for the semi-infinite boundary. We can see that the finite domain has no effect until the freezing or melting front actually reaches the other boundary. When other complications are involved, such as the second boundary used here or variable physical properties, this finite difference method on a moving grid can still be applied, even if an exact solution is not known.

Similar approaches can be taken when there is combined heat transfer and mass transfer. Chawla, *et al.* [1984] applied the orthogonal collocation method on the transformed moving grid for problems involving the meltdown of a nuclear reactor. Tzanos [1987] solves a problem that involves two adjacent phases with heat transfer between them. He uses an adaptive mesh technique to include more nodes where the curvatuve of the solution is large; this is monitored by the rate of heat transfer from one phase to another. See also Fukusako and Seki [1987].

11.3. Two-Phase Problems in One Dimension

Next let us consider the case where heat conduction must be considered in both phases [i.e., Eq. (11.1)-(11.5)]. This occurs when the initial temperature is not at the freezing or melting point. The Galerkin finite element method is applied to give Eq. (11.13) for each phase. For the other phase we have

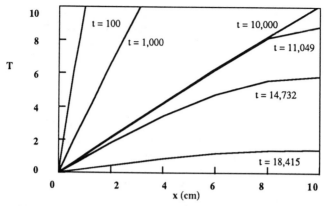

Figure 11.6. Melting Problem One in a Finite Domain, Finite Difference Method on Moving Nodes

$$\int_{X(t)}^{\infty} W_j \rho_2 C_2 \frac{\partial T_2}{\partial t} dx + \int_{X(t)}^{\infty} k_2 \frac{\partial W_j}{\partial x} \frac{\partial T_2}{\partial x} dx + \left[W_j k_2 \frac{\partial T_2}{\partial x} \right]_{x=X(t)} = 0. \quad (11.57)$$

The condition for energy conservation at the freezing or melting front is

$$\Delta H_f \frac{dX}{dt} = \left[k_1 \frac{\partial T_1}{\partial x} - k_2 \frac{\partial T_2}{\partial x} \right] \text{ at } x = X(t). \quad (11.58)$$

Lynch and O'Neill [1981] solved this problem under the boundary conditions

$$T = -10 \text{ at } x = 0, \ T = +4 \text{ at } x = \infty, \ T = +4 \text{ at } t = 0 \quad (11.59)$$

with the parameter values given in Table 11.3.

Table 11.3 Physical Parameters for Melting Problem Two

$\rho_1 C_1 = 0.49$ cal /°C cm^3
$\rho_2 C_2 = 0.62$ cal /°C cm^3
$k_1 = 9.6 \times 10^{-3}$ cal/cm sec °C
$k_2 = 6.9 \times 10^{-3}$ cal/cm sec °C
$\Delta H_f = 19.2$ cal/cm^3

They applied a Galerkin finite element method in the second phase from $X(t)$ to $X_2(t)$ and kept $X_2(t)$ ahead of $X(t)$ by using the same expression for velocity, Eq. (11.51). Cubic Hermite polynomials were used in both the solid and liquid phases and the boundary condition was taken as

$$\Delta H_f \frac{dX}{dt} = k_1 T_{N1} - k_2 T_{N2}. \quad (11.60)$$

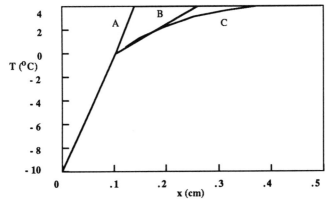

Figure 11.7. Initial Conditions for a Two-Phase Stefan Problem
(A - naive start, B - gentle start, C - analytical solution for frozen thickness = 0.1 cm)

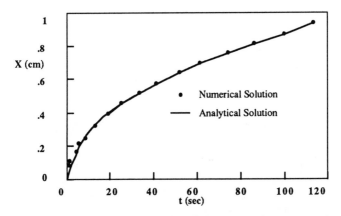

Figure 11.8. Two-Phase Stefan Problem, Frozen Thickness at Small Times, Naive Start
Data from Lynch and O'Neill [1981]

The values of T_{Ni} are the slopes at the nodes of the Hermite trial function. The initial profile was taken with a temperature distribution that is linear in Region 0 to the melting temperature and then is linear within the first element of the second phase, as shown in Figure 11.7. The movement of the freezing front is shown in Figure 11.8 for time values near zero and in Figure 11.9 for larger time values. The solutions are shown in Figure 11.10. Clearly the method has no difficulty obtaining smooth results.

If we solve this problem on a fixed grid, then obviously there are discontinuities in the slope of the temperature at the freezing or melting fronts. When using moving nodes, as in Eq. (11.51), the location of the discontinuity is tracked exactly. Another approach to this problem is to change the thermodynamics slightly so that the sharp discontinuity goes away. The change in thermodynamics is not as drastic as it might appear, since it corresponds to the actual thermodynamics of a dirty system or a system with multicomponents. The

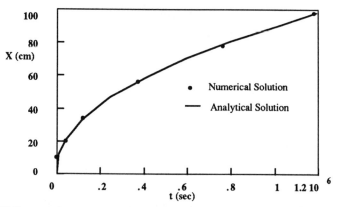

Figure 11.9. Two-Phase Stefan Problem, Frozen Thickness at Long Times, Naive Start Data from Lynch and O'Neill [1981]

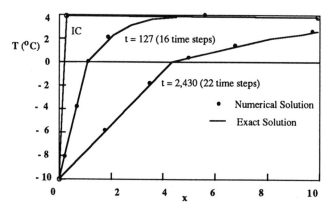

Figure 11.10. Two-Phase Stefan Problem, Temperature Profiles Data from Lynch and O'Neill [1981]

results of Comini, *et al.* [1974] will be presented first.

When a phase transition takes place, normally the heat capacity and thermal conductivity of each phase take on distinct values in separate phases. If we calculate a heat capacity during the phase change itself, it would be infinite because a change of enthalpy occurs even with no temperature change. Rather than have discontinuities and infinite values, let us consider a material in which the phase change takes place gradually over a small temperature interval, $2\Delta T$. Typical physical parameters are plotted in Figure 11.11. The infinite value of heat capacity is replaced by a very large value that is chosen such that the energy contained in the heat capacity pulse equals the latent heat of the phase change.

$$2\Delta T \, \rho C_p = \Delta H_f \qquad (11.61)$$

A finite element method applied to Eq. (11.13) is

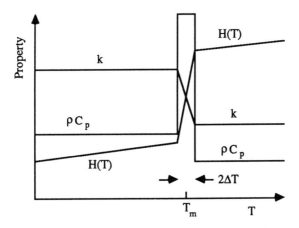

Figure 11.11. Physical Properties for Phase Change Problems

$$\sum_e \sum_I \int \rho C\, N_J\, N_I\, dx\, \frac{dT_I^e}{dt} + \sum_e \sum_I \int k\, \frac{dN_J}{dx}\, \frac{dN_I}{dx}\, dx\, T_I^e = 0. \quad (11.62)$$

Now, however, the terms in the integral depend on the solution T in a highly nonlinear way. We handle this difficulty by noting that the enthalpy,

$$H = \int_{T_0}^{T} \rho\, C\, dT, \quad (11.63)$$

is a smooth function of temperature. Thus, from a finite element solution for temperature at the nodes

$$T(x,t) = \sum_{i=1}^{NT} T_i(t)\, N_i(x) \quad (11.64)$$

we compute the enthalpy at the nodes

$$H(x,t) = \sum_{i=1}^{NT} H_i(t)\, N_i(x), \quad \text{where } H_i = H[T_i(t)]. \quad (11.65)$$

Then, since

$$\rho C = \frac{dH}{dT}, \quad (11.66)$$

we evaluate the heat capacity within an element with a linear trial function by

$$<\rho\, C_p> = \frac{\partial H}{\partial x}\bigg/\frac{\partial T}{\partial x}. \quad (11.67)$$

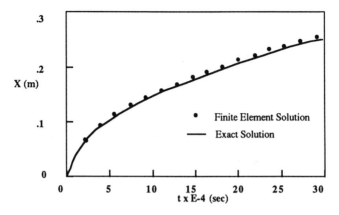

Figure 11.12. Solidification Front, Data from Comini, *et al.* [1974]

This averaging process ensures that the energy is balanced, since the integral under the enthalpy-temperature diagram is kept constant. A similar technique is used for thermal conductivity. We want the net heat flux across an element to be correct. Thus we need

$$k = \frac{\left[k_1 \frac{\partial T}{\partial x}\right]_1 - \left[k_2 \frac{\partial T}{\partial x}\right]_2}{\frac{T_2 - T_1}{x_2 - x_1}}, \tag{11.68}$$

where the subscript 1 represents the left side of an element (the temperature gradient is evaluated using the element to the left) and the subscript 2 represents the right side of the element (the temperature gradient is evaluated using the element to the right).

The following example is for a space initially filled with liquid above its freezing point at 10 °C. The left boundary is set equal to –20 °C. The physical properties of the liquid are given in Table 11.4.

Table 11.4 Physical Parameters for Melting Problem Three

$\rho_1 C_1 = 1.762 \times 10^6 \text{ J/m}^3 \text{ K}$

$\rho_2 C_2 = 4.226 \times 10^6 \text{ J/m}^3 \text{ K}$

$k_1 = 2.221 \text{ W/m K}$

$k_2 = 0.556 \text{ W/m K}$

$\Delta H_f = 1.5613 \text{ J/m}^3$

The temperature interval $2\Delta T$ shown in Figure 11.11 is taken as 0.5 °C. The movement of the freezing front is shown in Figure 11.12 and the temperature

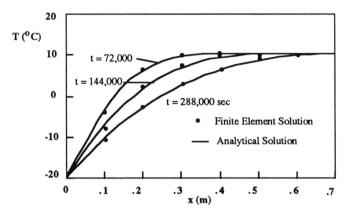

Figure 11.13. Temperature Distribution During Solidification
Data from Comini, *et al.* [1974]

profiles are shown in Figure 11.13. The solutions are reasonable approximations to the exact solution. Comparisons of such enthalpy methods with interface-following methods are given by Griffith and Nassersharif [1990].

Phase change problems with the heat capacity function taken as in Figure 11.11 have also been solved using the moving finite element method, as described in Section 7.4. Considerable adjustment was necessary to choose the viscosity and elastic parameters in order to prevent the nodes from bunching up near the front [Miller, 1981].

11.4. Two-dimensional Heat Conduction

The two-dimensional problem is given by

$$\nabla \cdot (k_1 \nabla T_1) = \rho_1 C_{p1} \frac{\partial T_1}{\partial t} \text{ in } \Omega_1$$

$$\nabla \cdot (k_2 \nabla T_2) = \rho_2 C_{p2} \frac{\partial T_1}{\partial t} \text{ in } \Omega_2. \tag{11.69}$$

$$\Delta H_f \mathbf{V} \cdot \mathbf{n} = k_1 \mathbf{n} \cdot \nabla T_1 - k_2 \mathbf{n} \cdot \nabla T_2 \tag{11.70}$$

In two-dimensional problems nodal movement is very important. If the nodes can move independently of one another in response to some equation, it is possible that they would coalesce or an element would become distorted. For this reason, good nodal movement is a very important goal. The first example of a two-phase freezing problem is freezing around a pipe [Lynch, 1982]. Lynch's scheme involves moving all of the nodes in the entire mesh. The initial situation is shown in Figure 11.14.

As time proceeds, the phase boundary moves out in time. Because of the problem specifications, the front should move in a way that maintains the

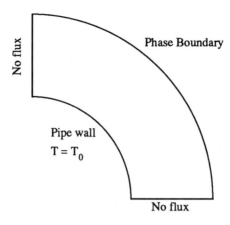

Figure 11.14. Geometry of a Freezing Pipe

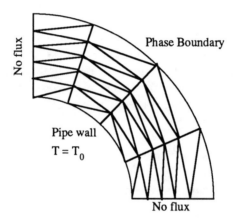

Figure 11.15. Initial Mesh for the Freezing Pipe Problem

phase boundary as a circle. The problem is symmetric in the azimuthal direction and is thus one-dimensional in cylindrical coordinates. The initial mesh is composed of triangles, as shown in Figure 11.15.

The nodal movement for this problem uses the equations derived in Chapter 7, namely Eq. (7.78) and the two-dimensional analog of Eq. (7.134). Here the nodal velocities are taken as the solution to the following equations:

$$\frac{\partial^2 \dot{U}}{\partial x^2} + \nu \frac{\partial^2 \dot{V}}{\partial x \partial y} + \frac{(1-\nu)}{2}\left[\frac{\partial^2 \dot{U}}{\partial y^2} + \frac{\partial^2 \dot{V}}{\partial x \partial y}\right] = 0, \quad \frac{\partial^2 \dot{V}}{\partial x^2} + \nu \frac{\partial^2 \dot{U}}{\partial x \partial y} + \frac{(1-\nu)}{2}\left[\frac{\partial^2 \dot{V}}{\partial y^2} + \frac{\partial^2 \dot{U}}{\partial x \partial y}\right] = 0.$$

(11.71)

$\dot{U}$ and $\dot{V}$ are the x- and y-velocity of a node. The Poisson ratio, ν, is arbitrarily set to zero. The nodal velocities move according to the equation

$$\frac{d\mathbf{X}}{dt} = \mathbf{V}, \quad \mathbf{V} = (\dot{U}, \dot{V})^T. \tag{11.72}$$

Lynch [1982] employs a predictor-corrector method. The predictor uses

$$\frac{\mathbf{X}(t+\Delta t) - \mathbf{X}(t)}{\Delta t} = \mathbf{V}(t) \tag{11.73}$$

to get the positions at the new time-step, $t + \Delta t$. Next Eq. (11.71) is solved to give an estimate of the velocity, $\mathbf{V}^{t+\Delta t}$. Then Eq. (11.72) is applied as

$$\frac{\mathbf{X}(t+\Delta t) - \mathbf{X}(t)}{\Delta t} = \theta \mathbf{V}(t+\Delta t) + (1-\theta) \mathbf{V}(t) \tag{11.74}$$

and Eq. (11.71) is solved again for improved values of $\mathbf{V}^{t+\Delta t}$. Additional iteration could be done, but it is not suggested. It is often more accurate to solve the problem for two steps with a time-step half as large rather than iterate once more with a larger time-step.

The Galerkin formulation is given by Eq. (7.78), which is written here as

$$\sum_i \int \rho C N_j N_i \, dV \frac{dT_i}{dt} - \sum_i \int \rho C \mathbf{V}^e \cdot \nabla N_j N_i \, dV \, T_i + \\ + \sum_i \int k \nabla N_j \cdot \nabla N_i \, dV \, T_i = \sum_i \int k \nabla N_i \cdot \mathbf{n} \, N_j \, dS \, T_i. \tag{11.75}$$

The Galerkin formulation of the phase boundary condition [Eq. (11.70)] is

$$\Delta H_f \sum_j \int \mathbf{V}_j N_j \cdot \mathbf{n} \, dS = \sum_i \int k \nabla N_i \cdot \mathbf{n} \sum_j N_j \, dS \, T_i. \tag{11.76}$$

If we require the equation to be satisfied for each j value (rather than for the sum over all j values), then

$$\Delta H_f \int \mathbf{V}_j N_j \cdot \mathbf{n} \, dS = \sum_i \int k \nabla N_i \cdot \mathbf{n} \, N_j \, dS \, T_i \tag{11.77}$$

or

$$\Delta H_f \mathbf{V}_i \cdot \mathbf{n}_i = f_i. \tag{11.78}$$

The "weighted" normal and "weighted" flux are given by

$$\mathbf{n}_i = \frac{\int N_i \mathbf{n} \, dS}{\int N_i \, dS}, \quad f_i = \frac{\int k \nabla N_i \cdot \mathbf{n} \, N_j \, dS \, T_i}{\int N_j \, dS}. \tag{11.79}$$

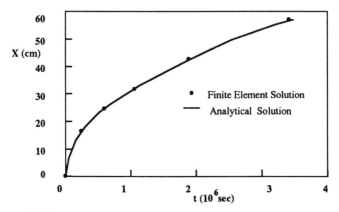

Figure 11.16. Phase Boundary for Freezing Pipe Problem, Data from Lynch [1982]

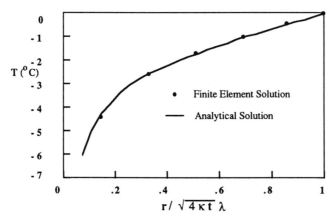

Figure 11.17. Temperature Distribution for Freezing Pipe Problem
Data from Lynch [1982]

Eq. (11.78) thus gives the velocities of the nodes on the phase boundary. Nodes on a solid boundary are either fixed or allowed to move tangential to the boundary.

A typical solution is given in Figure 11.16 for the location of the phase boundary and in Figure 11.17 for the temperature. Since Figure 11.17 is plotted using the similarity variable, $r/t^{1/2}$, the profile is the same for all time values; only the distance it covers changes. The mesh extends from a radius of approximately 2 cm to a radius of 60 cm.

The next problem is freezing of a liquid in a corner. The problem and initial grid is shown in Figure 11.18. Along the fixed boundaries ($y = 0$ and $y = x$) the nodes cannot move normal to the boundary but are allowed to move tangential to the boundary. Along the phase boundary the nodes can move with free tangential motion, but the normal motion is governed by Eq. (11.78). Along the semi-infinite boundary, free tangential motion is allowed and the normal motion is set by $x = \beta t^{1/2}$ so that the nodes remain in front of the phase boundary.

PHASE CHANGE 423

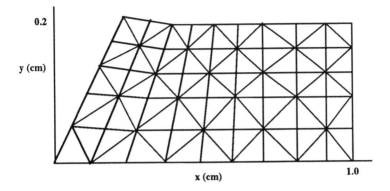

Figure 11.18. Mesh for Freezing in a Corner, Data from Lynch [1982]

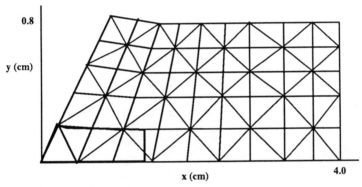

Figure 11.19. Mesh for Freezing in a Corner, Later Time, Data from Lynch [1982]

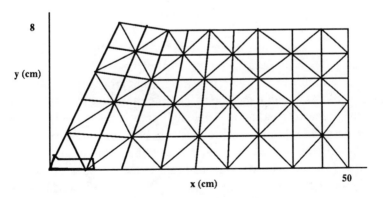

Figure 11.20. Mesh for Freezing in a Corner, Much Later Time
Data from Lynch [1982]

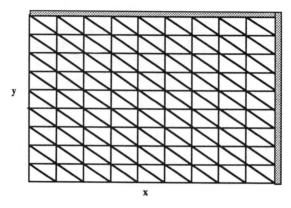

Figure 11.21. Grid for Solidification in a Corner

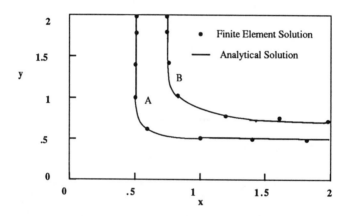

Figure 11.22. Location of Phase Boundary, Solidification in a Corner
A - Liquid Initially at Melting Temperature
B - Liquid Initially Hotter than Melting Temperature
Data from Comini, *et al.* [1974]

The growth of the solution is illustrated in Figures 11.19 and 11.20. In Figure 11.19 we can see how the domain has grown from the initial grid in the lower left-hand corner. In Figure 11.20 we see how the grid has grown even from that shown in Figure 11.19. An implicit method of integration was used; only 14 time-steps were needed to reach the solution shown in Figure 11.19 and only 40 total time-steps were needed to reach the solution shown in Figure 11.20.

Solidification at an internal corner has been solved on a fixed grid by Comini, *et al.* [1974]. The thermodynamic and transport properties they used are listed in Table 11.5. Now the phase change interval (in Figure 11.11) is $2\Delta T = 0.01$. The grid is shown in Figure 11.21 and a typical phase boundary is shown in Figure 11.22. One of the curves in Figure 11.22 is for the case when the liquid is initially at the freezing temperature while the other curve is for the case when the liquid is initially at a temperature above the freezing temperature. The

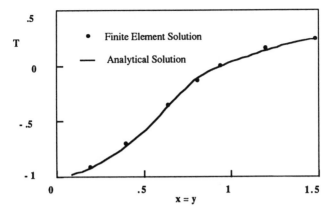

Figure 11.23. Temperature Along Diagonal During Freezing in a Corner

temperature along the diagonal is shown versus time in Figure 11.23. This solution method also leads to smooth solutions, even on a fixed domain.

Table 11.5 Physical Parameters for Melting Problem Four

$\rho_1 C_1 = 1 \text{ J/m}^3 \text{ K}$	$\rho_2 C_2 = 1 \text{ J/m}^3 \text{ K}$
$k_1 = 1 \text{ W/m K}$	$k_2 = 1 \text{ W/m K}$
$\Delta H_f = 1.5613 \text{ J/m}^3$ for Case I,	$\Delta H_f = 0.25 \text{ J/m}^3$ for Case II

Bonnerot and Jamet [1977] used a space-time finite element method for similar two-dimensional problems, while Alexander, *et al.* [1979] used the moving finite element method for two-dimensional problems. There are considerable difficulties, however, in applying the moving finite element method in two dimensions in the manner used by Alexander, *et al.* [1979], since many adjustable constants must be arbitrarily chosen for the numerical method to work.

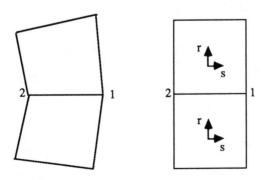

Figure 11.24. Geometry for Two Elements Adjoining an Interface

Yoo and Rubinsky [1983] treat nodal movement at the front as follows. Let us consider the surface S as the isotherm representing the melt-solid interface. The movement of the interface is based on a finite element mesh that is embedded in the interface. Since the temperature is constant within the interface, we can write

$$T(S) = T(x,y,t) = T(r,s,t), \qquad (11.80)$$

where r is normal and s is tangential to the interface. Thus, $\partial T(S)/\partial s = 0$. The velocity of a node on the interface is obtained by satisfying Eq. (11.70) in each direction:

$$\Delta H_f \frac{dX}{dt} = k_1 \frac{\partial T_1}{\partial x} - k_2 \frac{\partial T_2}{\partial x} \quad \text{on S}.$$

$$\Delta H_f \frac{dY}{dt} = k_1 \frac{\partial T_1}{\partial y} - k_2 \frac{\partial T_2}{\partial y} \quad \text{on S}. \qquad (11.81)$$

Since the temperature is constant within the surface, there is no nodal movement in the x direction if the surface lies along a line of constant y and vice versa. The incremental movement of the interface is represented by a finite element basis set on one boundary between the two phases, as illustrated in Figure 11.24. The differential movement of the interface is represented by

$$dX = \mathbf{H}^s\, dx_i^s, \quad dY = \mathbf{H}^s\, dy_i^s, \qquad (11.82)$$

where

$$\mathbf{H}^s = \left[\, \tfrac{1}{2}(1+s),\, \tfrac{1}{2}(1-s)\, \right]. \qquad (11.83)$$

Here x_i^S and y_i^S are the x and y coordinates of the nodal points of the element whose edge forms the interface. The variable s is a distance along the interface. (Here the r-s notation is reversed from that used by Yoo and Rubinsky [1983] in order to be consistent with the notation used in Chapter 8.) The velocity of the moving interface is then given by

$$\frac{dX}{dt} = \mathbf{H}^s \frac{dx_i^s}{dt}, \quad \frac{dY}{dt} = \mathbf{H}^s \frac{dy_i^s}{dt}. \qquad (11.84)$$

We write the velocity matrix as

$$\mathbf{U}^s = \begin{bmatrix} \tfrac{1}{2}(1+s) & 0 & \tfrac{1}{2}(1-s) & 0 \\ 0 & \tfrac{1}{2}(1+s) & 0 & \tfrac{1}{2}(1-s) \end{bmatrix} \qquad (11.85)$$

and the displacement matrix as

$$d\mathbf{S}^T = [\, dx_1 \quad dy_1 \quad dx_2 \quad dy_2\,] \qquad (11.86)$$

so that the mesh movement is given by

$$\frac{d\mathbf{X}}{dt} = \mathbf{U}^s \frac{d\mathbf{S}}{dt}. \tag{11.87}$$

In the finite element formulation we use

$$\int_S W_s \, \Delta H_f \, \mathbf{V} \cdot \mathbf{n} \, ds = \int_S W_s \, \{k_1 \, \mathbf{n} \cdot \nabla \, T_1 - k_2 \, \mathbf{n} \cdot \nabla \, T_2 \} \, ds. \tag{11.88}$$

The weighting function is $\mathbf{U}^s$ and the transformation matrix is

$$[\mathbf{J}^s]^{-1} = \begin{bmatrix} \frac{\partial s}{\partial x} & \frac{\partial r}{\partial x} \\ \frac{\partial s}{\partial y} & \frac{\partial r}{\partial y} \end{bmatrix}, \quad \begin{bmatrix} \frac{\partial}{\partial x} \\ \frac{\partial}{\partial y} \end{bmatrix} = [\mathbf{J}^s]^{-1} \begin{bmatrix} \frac{\partial}{\partial s} \\ \frac{\partial}{\partial r} \end{bmatrix}, \tag{11.89}$$

which gives

$$\Delta H_f \int_{-1}^{1} \mathbf{U}^{sT} \mathbf{U}^s \frac{d\mathbf{S}}{dt} \det \mathbf{J}^s \, ds = \Delta H_f \int_{-1}^{1} \mathbf{U}^{sT} \{k_1 \, \mathbf{n} \cdot \nabla \, T_1 - k_2 \, \mathbf{n} \cdot \nabla \, T_2 \} \det \mathbf{J}^s \, ds. \tag{11.90}$$

Yoo and Rubinsky [1983] move the melt-solid interface through a fixed mesh and re-mesh in the vicinity of the interface, but not throughout the domain. Thus, only some of the nodes are moving. This contrasts with Lynch [1982], who moves all the nodes. Both approaches seem to work for sample problems. Yoo and Rubinsky's approach is similar to that taken in the random choice method, where an elliptic grid and a hyperbolic grid are superimposed; here the elliptic grid is for the heat conduction problem and the hyperbolic grid is for the interface and the elements near it.

Gilmore and Guceri [1988] treat three-dimensional solidification problems by using a grid transformation from $\{x, y, z\}$ to $\{\xi, \eta, \nu\}$, such that the external boundaries lie along surfaces of constant coordinate values of ξ, η, or ν. The grid is generated by solving the equations

$$\nabla^2 \xi = 0, \quad \nabla^2 \eta = 0, \quad \nabla^2 \nu = 0. \tag{11.91}$$

Temperature gradients are neglected in the liquid, assuming that the temperature quickly equilibrates to the solidification temperature. Gilmore and Guceri employed a procedure to allow a redistribution of grid points on the melt-solid interface. The computations were extensive, but this procedure could be used for irregular base surfaces with the solidification front passing across the irregular surface. Lacroix and Voller [1990] compare phase change problems that are solved on transformed grids versus fixed grids. An application in which a crater is formed because of laser heating is given by Barillot [1990].

11.5. Two-dimensional Heat Conduction and Convection

The method of front tracking described in Section 11.4 can also be applied to more complicated problems. Tsai and Rubinsky [1984] have treated problems involving mass transfer and heat transfer during phase transformations. Yoo and Rubinsky [1986] and Lan and Kuo [1991] have analyzed a melting problem when natural convection occurs in the fluid phase. Presented here is the approach taken by Sackinger, *et al.* [1989], which uses a finite element method and locates the interfaces using a spine representation. An article by Sackinger, *et al.* [1989] is a model of the Czochralski crystal growth process for making silicon for electronic chips.

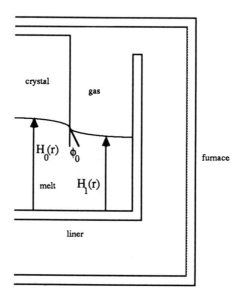

Figure 11.25. Crystal Growth Furnace

A schematic representation of the crystal growth process is shown in Figure 11.25. The crystal is being pulled upward and the entire assembly is being heated in the furnace. The solid phase is governed by the energy equation

$$\rho_s C_{ps} V_g \frac{\partial T}{\partial z} = \nabla \cdot (k_s \nabla T). \tag{11.92}$$

The velocity term appears in this equation because the solid is being pulled vertically as the crystal is forming. There are additional energy equations for the crucible and liner. The energy equation for the melt phase is more complicated than Eq. (11.92) because it also includes a convection term.

$$\rho_m C_{pm} \mathbf{v} \cdot \nabla T = \nabla \cdot (k_m \nabla T) \tag{11.93}$$

To find the convection velocity, we solve the Navier-Stokes equation:

$$\rho_m \mathbf{v} \cdot \nabla \mathbf{v} = -\nabla p + g \rho_m \beta_m (T - T_{mp}) \mathbf{e}_z + \nabla \cdot [\mu (\nabla \mathbf{v} + \nabla \mathbf{v}^T)],$$
$$\nabla \cdot \mathbf{v} = 0. \tag{11.94}$$

The details of solving this equation are given in Chapter 12. Here we will concentrate on the connection to the melt-solid problem. In the crystal growth process, there are two key interfaces whose locations must be determined: the melt-crystal interface and the melt-gas interface. These interfaces are represented by means of a spine representation, as discussed in Section 8.7. There are additional complications where the two interfaces join. In particular, the point where the melt, crystal, and gas join is forced to have a particular contact angle, ϕ_0, while the point where the melt-gas interface intersects the crucible wall is forced to have a contact angle of 90°. Because of the two interfaces, it is necessary to use a spine representation so that a complete Newton-Raphson method can be used on the entire interface. The location of the interfaces is then given by the functions $H_0(r)$ and $H_1(r)$. Along the melt-crystal interface the function $H_0(r)$ satisfies an energy balance, which is

$$\mathbf{n}_0 \cdot (k_m \nabla T_m - k_s \nabla T_s) = V_g \Delta H_f \mathbf{n}_0 \cdot \mathbf{e}_z, \tag{11.95}$$

where $\mathbf{n}_0$ is the normal to the interface and the temperature there is taken to be the melt temperature. The boundary conditions at the sides are given by the heat transfer boundary condition,

$$k_i \mathbf{n}_k \cdot \nabla T = h(T - T_a) + \varepsilon_k \sigma (T_{im}^4 - T_a^4). \tag{11.96}$$

At the melt-gas interface, surface tension is important; the boundary condition for the balance of normal forces is

$$2H\gamma = \rho_m g(H_1 + \lambda) - \mathbf{n}_1 \mathbf{n}_1 : [\mu_m (\nabla \mathbf{v} + \nabla \mathbf{v}^T)], \tag{11.97}$$

where 2H is the mean curvature, given in terms of $H_1(r)$ by

$$2H = \frac{1}{\left[1 + \left(\frac{dH_1}{dr}\right)^2\right]^{3/2}} \frac{d^2 H_1}{dr^2} + \frac{1}{r\left[1 + \left(\frac{dH_1}{dr}\right)^2\right]^{1/2}} \frac{dH_1}{dr}. \tag{11.98}$$

The boundary condition for the balance of tangential forces is

$$\mathbf{t}_1 \mathbf{n} : [\mu_m (\nabla \mathbf{v} + \nabla \mathbf{v}^T)] = -\frac{1}{\left[1 + \left(\frac{dH_1}{dr}\right)^2\right]^{1/2}} \frac{d\gamma}{dT} \frac{\partial T}{\partial r}. \tag{11.99}$$

The weighted energy equation for the melt is

$$\int_{\Omega_m} N_j \rho_m C_{pm} \mathbf{v} \cdot \nabla T \, d\Omega = \int_{\Omega_m} N_j \nabla \cdot (k_m \nabla T) \, d\Omega. \quad (11.100)$$

This is integrated by parts to give

$$\int_{\Omega_m} N_j \rho_m C_{pm} \mathbf{v} \cdot \nabla T \, d\Omega = -\int_{\Omega_m} \nabla N_j \cdot \nabla T \, d\Omega + \int_{S_m} N_j \mathbf{n} \cdot (k_m \nabla T) \, dS. \quad (11.101)$$

The same operations are performed on the solid equation, Eq. (11.92), giving

$$\int_{\Omega_s} N_j \rho_s C_{ps} V_g \frac{\partial T}{\partial z} \, d\Omega = -\int_{\Omega_s} \nabla N_j \cdot \nabla T \, d\Omega + \int_{S_s} N_j \mathbf{n} \cdot (k_s \nabla T) \, dS. \quad (11.102)$$

On one part of the boundary S_m the boundary condition is Eq. (11.96). This is substituted into Eq. (11.101) to give

$$\int_{\Omega_m} N_j \rho_m C_{pm} \mathbf{v} \cdot \nabla T \, d\Omega = -\int_{\Omega_m} \nabla N_j \cdot \nabla T \, d\Omega + \int_{S_m} N_j [h(T-T_a) + \varepsilon_k \sigma(T_{im}^4 - T_a^4)] \, dS. \quad (11.103)$$

On the boundary between the melt and the crystal the condition is Eq. (11.95). Equations (11.103) and (11.102) are added to each other to give a combined equation; Eq. (11.95) is inserted to give

$$-\int_{\Omega_m} \nabla N_j \cdot \nabla T \, d\Omega - \int_{\Omega_s} \nabla N_j \cdot \nabla T \, d\Omega - \int_{\Omega_m} N_j \rho_m C_{pm} \mathbf{v} \cdot \nabla T \, d\Omega -$$
$$-\int_{\Omega_s} N_j \rho_s C_{ps} V_g \frac{\partial T}{\partial z} \, d\Omega + \int_{S_m} N_j V_g \Delta H_f n_0 \cdot \mathbf{e}_z \, dS - \quad (11.104)$$
$$-\int_{S_m} N_j [h(T-T_a) + \varepsilon_k \sigma(T_{im}^4 - T_a^4)] \, dS = 0.$$

In addition, we must force the temperature to be the melt temperature along the interface. Thus we use

$$\int_0^R \Gamma^j [T(r, H_0(r)) - 1] r \, dr = 0. \quad (11.105)$$

This formulation of the problem is called the isotherm method [Ettouney and Brown, 1983]. The Navier-Stokes equation is solved by Ettouney and Brown in the direct fashion described in Chapter 12.

The solution to the crystal growth problem is a function of many dimensionless groups. Key parameters include the Grashof number, which represents the force of natural convection, the Prandtl number, which is generally small and leads to oscillatory time-dependent solutions, and various Reynolds numbers, which represent the velocity of rotation and pulling. Solutions to the crystal growth problem will not be given here, but the problem as outlined shows the importance of spine representation for modeling the location of the interface. Since the complete Newton-Raphson method can be used with spine representation, the number of iterations can be minimized. Because of the complexity of the interface and the double interface, it is doubtful that less powerful methods would converge.

Problems

1_1. Verify Eq. (11.28) [Steps of Eq. (11.17)-(11.28)].

2_2. Verify that Eq. (11.48) is the solution to Eq. (11.36)-(11.37).

3_2. In deriving Eq. (11.48), α was taken as 2. Write the solution T(x,t) in terms of the functions $\delta(t)$ and X(t), along with the equations defining those functions. Do the same thing when α is taken as 1. Are the solutions for $\delta(t)$, X(t), and T(x,t) the same?

References

Alexander, R., Manselli, P., and Miller, K., "Moving Finite Element Solution of the Stefan Problem in Two Dimensions," Rend. Accad. Naz. Lincei (Rome), serie VIII, vol. LXVII, fasc.1-2, pp. 57-61 [1979].

Barillot, Ph., "Numerical Simulation of Crater Formation Heating by Laser Beam," Num. Heat Trans. Part B 17 245-256 [1990].

Bonnerot, R. and Jamet, P., "Numerical Computation of the Free Boundary for the Two-Dimensional Stefan Problem by Space-Time Finite Elements," J. Comp. Phys. 25 163-181 [1977].

Bonnerot, R. and Jamet, P., "A Second Order Finite Element Method for the One-dimensional Stefan Problem," Int. J. Num. Methods Engn. 8 811-820 [1974].

Carslaw, H. S. and Jaeger, J. C., *Conduction of Heat in Solids,* Clarendon Press, Oxford [1959].

Chawla, T. C., Pedersen, D. R., Leaf, G., Minkowycz, W. J., "Adaptive Collocation Method for Simultaneous Heat and Mass Diffusion with Phase Change," J. Heat Trans. 106 491-497 [1984].

Comini, G., Del Guidice, S., Lewis, R. W., and Zienkiewicz, O. C., "Finite Element Solution of Non-linear Heat Conduction Problems with Special Reference to Phase Change," Int. J. Num. Methods Eng. 8 613-624 [1974].

Crank, J., *Free and Moving Boundary Problems,* Carendeon Press, Oxford [1984].

Ettouney, H. M. and Brown, R. A., "Finite-element methods for steady solidification problem," J. Comp. Phys. 49 118-150 [1983].

Fukusako, S. and Seki, N., "Fundamental Aspects of Analytical and Numerical Methods on

Freezing and Melting," *Annual Rev. Num. Fluid Mech. & Heat Transf.*, (Ed. T. C. Chawla) Ch. 7, pp. 351-402, Hemisphere [1987].

Furzeland, R. M., "A Comparative Study of Numerical Methods for Moving Boundary Problems," J. Inst. Maths. Applics. 26 411-429 [1980].

Gilmore, S. D., and Guceri, S. I., "Three-dimensional Solidification, A Numerical Approach," Num. Heat Transf. 14 165-186 [1988].

Goodman, T. R., "The Heat-Balance Integral and Its Application to Problems Involving a Change in Phase," Trans. ASME 80 335-342 [1958].

Goodman, T. R., "Application of Integral Methods to Transient Nonlinear Heat Transfer," Adv. Heat Transf. 1 51-122 [1964].

Griffith, R. and Nassersharif, B., "Comparison of One-dimensional Interface-following and Enthalpy Methods for the Numerical Solution of Phase Change," Num. Heat Trans., Part B, 18 169-187 [1990].

Gupta, S. C. and Arora, P. R., "Outward Spherical Solidification of a Superheated Melt with Time Dependent Boundary Flux," Appl. Sci. Res. 45 17-31 [1988].

Hrymak, A. N. and Westerberg, A. W., "Computational Considerations for Moving Finite Element Methods," Chem. Eng. Sci. 41 1673-1680 [1986].

Lacroix, M. and Voller, V. R., "Finite Difference Solutions of Solidification Phase Change Problems: Transformed versus Fixed Grids," Num. Heat Trans., Part B 17 25-41 [1990].

Lan, C. W. and Kuo, S., "Thermocapillary Flow and Natural Convection in a Melt Column with an Unknown Melt/Solid Interface," Int. J. Num. Methods Fluids 12 59-80 [1991].

Landau, H. G., "Heat Conduction in a Melting Solid," Q. Appl. Math. 8 81-94 [1950].

Lynch, D. R., and O'Neill, K., "Continuously Deforming Finite Elements for the Solution of Parabolic Problems, with and without Phase Change," Int. J. Num. Meth. Engn. 17 81-96 [1981].

Lynch, D. R., "Unified Approach to Simulation on Deforming Elements with Application to Phase Change Problems," J. Comp. Phys. 47 387-411 [1982].

Miller, K. and Miller, R., "Moving Finite Elements, Part II," SIAM J. Numer. Anal. 18 1033-1057 [1981].

Murray, W. D. and Landis, F., "Numerical and Machine Solutions of Transient Heat Conduction Problems Involving Melting or Freezing. Part I. Method of Analysis and Sample Solutions," Trans. ASME, J. Heat Transf. 81 106-112 [1959].

Pelcé, P. (ed.), *Dynamics of Curved Fronts*, Academic Press, New York [1988].

Sackinger, P. A., Brown, R. A. and Derby, J. J., "A Finite Element Method for Analysis of

Fluid Flow, Heat Transfer and Free Interfaces in Czochralski Crystal Growth," Int. J. Num. Methods Fluids 9 453-492 [1989].

Tsai, H. L., and Rubinsky, B., "A "Front Tracking" Finite Element Study on Change of Phase Intedrface Stability During Solidification Processes in Solutions," J. Crys. Growth 70 56-63 [1984].

Tzanos, C. P., "Liquid-metal Fast Breeder Reactor Intermediate Heat Exchange Transient Modeling for Faster Than Real-time Analysis," Nucl. Tech. 76 337-351 [1987].

Yoo, J. and Rubinsky, B., "A Finite Element Method for the Study of Solidification Processes in the Presence of Natural Convection," Int. J. Num. Methods Engn. 23 1785-1805 [1986].

Yoo, J. and Rubinsky, B., "Numerical Computation Using Finite Elements for the Moving Interface in Heat Transfer Problems with Phase Transformation," Num. Heat Trans. 6 209-222 [1983].

CHAPTER
TWELVE

THE NAVIER-STOKES EQUATION

Numerical methods have now advanced to the stage whereby a wide variety of phenomona can be simulated that in the past could only be studied experimentally. Fluid mechanics has traditionally been divided into groups such as laminar or turbulent flow, while laminar flow has been subdivided into slow flow situations (Stokes flow) and high speed situations (leading to boundary layers, which require special boundary layer methods). Stokes flow applies for zero Reynolds numbers; in this situation, analytic solutions can occasionally be obtained. As the Reynolds number increases, however, numerical solutions are required. Boundary layer methods are useful at high speeds (large Reynolds numbers) when a thin boundary layer develops; in a few situations analytic solutions are possible. Numerical methods can be used to bridge the gap between a Reynolds number of zero and a large Reynolds number. In addition, numerical methods allow the solution of cases with more complicated boundary conditions than can usually be handled analytically, even for a zero Reynolds number. As the Reynolds number increases to the point that boundary layer methods are applicable, the numerical method must use appropriate mesh refinement in the boundary layer or change to a boundary layer method. We can regard the tools available to solve the Navier-Stokes equation for laminar flow as essentially well developed. Turbulent flow is modeled using equations that have their parameters determined by experiment. Usually these equations are not fundamental ones, in that they are not derivable from first principles, but they are useful nonetheless. It is possible to directly simulate turbulence, but the mesh must be very small and the calculations are time-consuming and thus very expensive. In this book only laminar flows are discussed. In this chapter, we will study both finite element methods and finite difference methods for solving the Navier-Stokes equation.

12.1. Equations

The Navier-Stokes equation and the energy equation are listed in Table 12.1. A few assumptions have been made:

Table 12.1. Equations

Navier-Stokes Equations:

$$\rho \frac{\partial \mathbf{u}}{\partial t} + \rho \mathbf{u} \cdot \nabla \mathbf{u} = \rho \mathbf{f} - \nabla p + \nabla \cdot [\mu(\nabla \mathbf{u} + \nabla \mathbf{u}^T)]$$

Energy Equation:

$$\rho C_p \left(\frac{\partial T}{\partial t} + \mathbf{u} \cdot \nabla T \right) = \nabla \cdot (k \nabla T) + \Phi_v$$

Navier-Stokes Equations in Cartesian Coordinates (2D):

$$\rho \left(\frac{\partial u}{\partial t} + u \frac{\partial u}{\partial x} + v \frac{\partial u}{\partial y} \right) = \rho f_x - \frac{\partial p}{\partial x} + \frac{\partial}{\partial x}\left(2\mu \frac{\partial u}{\partial x}\right) + \frac{\partial}{\partial y}\left[\mu \left(\frac{\partial u}{\partial y} + \frac{\partial v}{\partial x}\right)\right]$$

$$\rho \left(\frac{\partial v}{\partial t} + u \frac{\partial v}{\partial x} + v \frac{\partial v}{\partial y} \right) = \rho f_y - \frac{\partial p}{\partial x} + \frac{\partial}{\partial y}\left(2\mu \frac{\partial v}{\partial y}\right) + \frac{\partial}{\partial x}\left[\mu \left(\frac{\partial u}{\partial y} + \frac{\partial v}{\partial x}\right)\right]$$

Energy Equation in Cartesian Coordinates (2D):

$$\rho C_p \left(\frac{\partial T}{\partial t} + u \frac{\partial T}{\partial x} + v \frac{\partial T}{\partial y} \right) = \frac{\partial}{\partial x}\left(k \frac{\partial T}{\partial x}\right) + \frac{\partial}{\partial y}\left(k \frac{\partial T}{\partial y}\right) + \Phi_v$$

---

1. Newton's law of motion
2. Law of conservation of energy
3. Incompressible fluid
4. Continuum
5. Newtonian fluid

No two-dimensional or three-dimensional experiment should be used to test these assumptions; these assumptions are either axioms or can be tested with simple experiments. Newton's law and the law of conservation of energy are axioms, which we accept unless relativistic effects are important. The incompressibility of a fluid can be tested by measurement or by looking up a compressibility values in a table. Indeed, if desired, simulations can be done for compressible fluids. Whether or not a fluid is a continuum can be decided based on kinetic theory and the perfect gas law. The fifth assumption, a Newtonian fluid, is an important one. We can measure a fluid's viscosity in a viscometer and determine if the shear stress is directly proportional to the shear rate and if elastic effects are absent. If so, the fluid is Newtonian and it will be discussed in this chapter; if not, it will be discussed in Chapter 13.

Let us further simplify the equations by also assuming the following:

6. Two-dimensional flow
7. Laminar flow

The Navier-Stokes equation is written in Cartesian coordinates in Table 12.1. It is important to do an experiment (or to read literature where experimental information is reported) to determine whether or not the flow is two-dimensional and laminar. If the flow is two-dimensional and laminar, then it is easier, faster, and cheaper to solve the Navier-Stokes equation numerically than it is to measure some aspect of the flow. This statement could not have been made twenty years ago, but it proves true today. The most important part of any model (or experiment) is how the system interacts with its surroundings, i.e., the boundary conditions. Thus an experiment may be necessary to show that the boundary conditions of the theory are appropriate or that an important phenomenon is not left out. On the other hand, calculations can be used to assess the validity of assumptions made in interpreting the experimental results.

12.2. Finite Element Methods for Steady-State Problems

Here we concentrate on two-dimensional, laminar flows. A finite element method can be applied to the Navier-Stokes equation in a variety of ways:

1. Primitive variables (velocity and pressure)
2. Primitive variables (velocity) with pressure from a penalty method
3. Stream function-vorticity

In the first and second cases, the velocity variables are expanded in a series and the decision between the two cases is essentially an economic one (for Newtonian fluids). In the third case, the primary variables are derivatives or integrals of velocity; sometimes the boundary conditions are difficult to specify because the variable of interest is not a physical quantity. Finite element methods applied to the Navier-Stokes equation are described in detail in a book by Cuvelier, *et al.* [1986].

Let us first consider the case when both velocity and pressure are expanded in a series:

$$u = \sum_{i=1}^{NU} u_i \, N_i(x,y), \quad v = \sum_{i=1}^{NU} v_i \, N_i(x,y), \quad p = \sum_{i=1}^{NP} p_i \, N_i'(x,y). \qquad (12.1)$$

Typically the velocity is expanded using an order that is one higher than the pressure. Thus the velocity may be quadratic and the pressure linear or the velocity may be linear and the pressure constant. The reason different degree polynomials are used for pressure and velocity is so that the discrete equations will have a solution for pressure when the velocity is completely specified (Jackson and Cliffe [1981], Sani, *et al.* [1981]). The trial function may be defined on triangles or quadrilaterals with constant, linear, or quadratic interpolation.

Next we put the trial function into the Navier-Stokes and continuity equations to define the residuals. To illustrate the procedure, let us use the equations in non-dimensional form:

$$\text{Re}\, \mathbf{u} \cdot \nabla \mathbf{u} = -\nabla p + \nabla \cdot [\mu (\nabla \mathbf{u} + \nabla \mathbf{u}^T)], \tag{12.2}$$

$$\nabla \cdot \mathbf{u} = 0 \tag{12.3}$$

with the boundary conditions

$$\mathbf{n} \cdot \mathbf{u} = u_n \text{ or } \mathbf{n} \cdot \tau \cdot \mathbf{n} = -p + \mu\, \mathbf{n} \cdot [\nabla \mathbf{u} + \nabla \mathbf{u}^T] \cdot \mathbf{n} = f_n,$$
$$\mathbf{t} \cdot \mathbf{u} = u_t \text{ or } \mathbf{n} \cdot \tau \cdot \mathbf{t} = \mu\, \mathbf{n} \cdot [\nabla \mathbf{u} + \nabla \mathbf{u}^T] \cdot \mathbf{t} = f_t, \tag{12.4}$$

where the Reynolds number is $\rho u_s x_s / \mu_s$, with u_s, x_s, and μ_s as reference quantities representing an average velocity, a certain characteristic length, and a viscosity. Here the symbol μ refers to the actual viscosity (which may depend on temperature) divided by the reference viscosity (a constant). The momentum residuals are multipled by $\delta \mathbf{u}$ (which is $N_j \mathbf{e}_x$ or $N_j \mathbf{e}_y$) and integrated over the domain, giving

$$\text{Re} \int_V \delta \mathbf{u} \cdot (\mathbf{u} \cdot \nabla \mathbf{u})\, dV = -\int_V \delta \mathbf{u} \cdot \nabla p\, dV + \int_V \delta \mathbf{u} \cdot \{\nabla \cdot [\mu (\nabla \mathbf{u} + \nabla \mathbf{u}^T)]\} dV. \tag{12.5}$$

The viscous terms are integrated by parts and the divergence theorem is applied, giving

$$\int_V \delta \mathbf{u} \cdot \{\nabla \cdot [\mu (\nabla \mathbf{u} + \nabla \mathbf{u}^T)]\} dV = \int_V \nabla \cdot \{\delta \mathbf{u} \cdot [\mu (\nabla \mathbf{u} + \nabla \mathbf{u}^T)]\} dV -$$
$$\int_V \nabla \delta \mathbf{u} : [\mu (\nabla \mathbf{u} + \nabla \mathbf{u}^T)] dV$$
$$= \int_S \{\delta \mathbf{u} \cdot [\mu (\nabla \mathbf{u} + \nabla \mathbf{u}^T)]\} \cdot \mathbf{n}\, dS - \int_V \nabla \delta \mathbf{u} : [\mu (\nabla \mathbf{u} + \nabla \mathbf{u}^T)] dV \tag{12.6}$$

The pressure term is also integrated by parts and the divergence theorem is used:

$$\int_V \delta \mathbf{u} \cdot \nabla p\, dV = \int_V \nabla \cdot (\delta \mathbf{u}\, p) dV - \int_V p \nabla \cdot \delta \mathbf{u}\, dV$$
$$= \int_S \mathbf{n} \cdot \delta \mathbf{u}\, p\, dS - \int_V p \nabla \cdot \delta \mathbf{u}\, dV \tag{12.7}$$

The complete result is then

$$\text{Re} \int_V \delta \mathbf{u} \cdot (\mathbf{u} \cdot \nabla \mathbf{u})\, dV = \int_V p \nabla \cdot \delta \mathbf{u}\, dV - \int_V \nabla \delta \mathbf{u} : [\mu (\nabla \mathbf{u} + \nabla \mathbf{u}^T)] dV -$$
$$\int_S \mathbf{n} \cdot \delta \mathbf{u}\, p\, dS + \int_S \{\delta \mathbf{u} \cdot [\mu (\nabla \mathbf{u} + \nabla \mathbf{u}^T)]\} \cdot \mathbf{n}\, dS \tag{12.8}$$

We would not use this equation for a node on the boundary for which the velocity must take a specified value. In the last two terms we separate the velocity into its normal and tangential components.

$$\delta u = n \cdot \delta u + t \cdot \delta u \, t \qquad (12.9)$$

Then we substitute the following into the last two terms of Eq. (12.8):

$$-\int_S n \cdot \delta u \, p \, dS + \int_S \{ \delta u \cdot [\mu (\nabla u + \nabla u^T)]\} \cdot n \, dS =$$

$$\int_S n \cdot \delta u \, [f_n - n \cdot \mu (\nabla u + \nabla u^T) \cdot n \,] dS +$$

$$+ \int_S n \cdot \delta u \, n \cdot \mu (\nabla u + \nabla u^T) \cdot n \, dS + \int_S t \cdot \delta u \, f_t \, dS \qquad (12.10)$$

to give

$$\text{Re} \int_V \delta u \cdot (u \cdot \nabla u) \, dV = \int_V p \nabla \cdot \delta u \, dV - \int_V \nabla \delta u : [\mu (\nabla u + \nabla u^T)]\} dV +$$

$$+ \int_{S_1} n \cdot \delta u \, f_n \, dS + \int_{S_2} t \cdot \delta u \, f_t \, dS \qquad (12.11)$$

The essential boundary conditions are velocities specified on the boundary. If the normal velocity is specified, then we do not use Eq. (12.11) for the normal component of velocity or use the integral over S_1. Similar considerations apply to the tangential component of velocity and the integral over S_2. Since the essential boundary conditions must be applied to the normal and tangential components of velocity, it is clearly useful to have the flow boundaries parallel to the coordinate axes if an essential boundary condition is used for the normal direction and a natural boundary condition is used in the tangential direction (and vice versa). Finally, the continuity equation is multiplied by N_j', the trial function for pressure. The continuity equation is essentially an equation for pressure; furthermore, this treatment gives the same result as is given by the variational principle for slow flow.

$$\int_V \delta p \, \nabla \cdot u \, dV = 0 \qquad (12.12)$$

We then solve Eq. (12.11) and Eq. (12.12). These represent a set of nonlinear algebraic equations that must be solved iteratively. The most common way that these equations are solved is by using the Newton-Raphson method: the equations are linearized about the current iterate and only the first-order terms are kept. Denoting the iteration number with a superscript then gives

$$\text{Re} \int_V \delta u \cdot (u^{n+1} \cdot \nabla u^n) \, dV + \text{Re} \int_V \delta u \cdot (u^n \cdot \nabla u^{n+1}) \, dV - \int_V p^{n+1} \nabla \cdot \delta u \, dV +$$

$$+ \int_V \nabla \delta u : [\mu (\nabla u^{n+1} + \nabla u^{n+1,T})]\} dV +$$

$$= -\text{Re} \int_V \delta u \cdot (u^n \cdot \nabla u^n) \, dV + \int_{S_1} n \cdot \delta u \, f_n \, dS + \int_{S_2} t \cdot \delta u \, f_t \, dS \qquad (12.13)$$

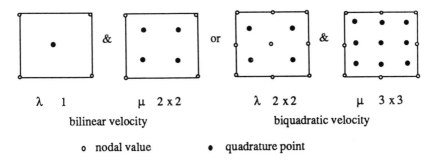

Figure 12.1. Quadrature Points to Use with the Penalty Method

An initial guess is supplied for u^0, v^0, and p^0. Then Eq. (12.13) is applied once to find u^1, v^1, and p^1; this process is repeated until the successive iterates are within some error tolerance (such as 10^{-6} when the calculations are done in double precision). If the Reynolds number is small (e.g., 0.00001), only one iteration is necessary. If the Reynolds number is 100, usually only four or five iterations are necessary if a reasonable initial guess is provided. For this reason, a good strategy is to always solve the case for Re = 0 first and use that as the initial guess for larger Reynolds numbers. If the Reynolds number is 1000 or more, then the method may not converge at all, depending on the initial guess and the mesh. One important point is that the pressure must be specified for at least one node, since the differential (and algebraic) equations are indeterminant to within a constant pressure.

In the penalty method, the incompressibility condition is handled as a Lagrange multiplier. We write the pressure equation as

$$p = -\lambda \nabla \cdot \mathbf{u}, \qquad (12.14)$$

where λ is a large constant, chosen by the analyst. Then Eq. (12.13) becomes

$$\text{Re} \int_V \delta \mathbf{u} \cdot (\mathbf{u}^{n+1} \cdot \nabla \mathbf{u}^n) dV + \text{Re} \int_V \delta \mathbf{u} \cdot (\mathbf{u}^n \cdot \nabla \mathbf{u}^{n+1}) dV + \lambda \int_V (\nabla \cdot \mathbf{u}^{n+1}) \nabla \cdot \delta \mathbf{u} \, dV +$$
$$+ \int_V \nabla \delta \mathbf{u} : [\mu (\nabla \mathbf{u}^{n+1} + \nabla \mathbf{u}^{n+1,T})]\} dV +$$
$$= -\text{Re} \int_V \delta \mathbf{u} \cdot (\mathbf{u}^n \cdot \nabla \mathbf{u}^n) dV \int_{S_1} \mathbf{n} \cdot \delta \mathbf{u} \, f_n \, dS + \int_{S_2} \mathbf{t} \cdot \delta \mathbf{u} \, f_t \, dS$$

(12.15)

We thus have to solve for fewer unknowns since the only unknown values are the velocities u_i and v_i. If we want the pressure, we can calculate it using Eq. (12.14), although smoothing is necessary. However, there is a problem. For a Reynolds number of zero if we solve the problem as stated, the solution converges to the zero solution (i.e., zero velocity everywhere) as λ approaches infinity (which is the appropriate value for the incompressible problem). One practical way around this difficulty is to use inexact numerical quadrature when evaluating the integrals. It is recommended by Hughes, et al. [1979] that we use the quadrature as shown in Figure 12.1. This figure shows the types of trial functions and the number of Gauss points that should be used for terms multipled by μ and λ. The Newton-Raphson method is also applied to Eq. (12.15) to obtain the equations to be solved iteratively. This completes the formulation of the penalty method for the Navier-Stokes equation. Comparisons of this method with the artificial compressibility method are available [Shih, *et al.*, 1989].

The primitive (u-p-v) method for the Navier-Stokes equation for an incompressible fluid is unusual in that the equation for pressure (the continuity equation) does not have the pressure variable in it. The continuity equation acts like a constraint on the velocity field, leading to the restriction that the pressure and velocities must be expanded in different trial functions. The stream function-vorticity method avoids this difficulty because the incompressibility constraint is satisfied automatically. This method introduces another problem, however, because on a solid surface there are two boundary conditions for stream function and none for vorticity. This presents little difficulty for a finite element method. When solving these equations, it is highly recommended that we solve them together as a coupled system since the boundary conditions are not provided for vorticity on all surfaces. It is possible to separate the problems, as is done with finite difference methods, but experience suggests it is better not to do so (Campion-Renson and Crochet [1978] and Stevens [1982]).

12.3. Finite Difference Methods

There are a variety of finite difference methods for the Navier-Stokes equation. Usually the various methods are designed to operate in an iterative fashion, even for a zero Reynolds number. This is advantageous because large computer memories are not necessary since no matrices are "inverted." However, there are also disadvantages, as indicated below for each finite difference method. In contrast, in finite element methods we obtain a large matrix problem to be solved at each iteration; for Re = 0 only one iteration is needed.

The first finite difference method, which originated with Chorin [1967], allows the fluid to have a small value of compressibility. The chief problem with the Navier-Stokes equation and the continuity equation is that the Navier-Stokes equation is an equation for velocity and it includes the pressure gradient, while the continuity equation is a constraint on the velocity but is in essence the equation for pressure. In the artificial compressibility method a time derivative of pressure is added to the continuity equation, as shown in Table 12.2. The constant parameter c^2 is chosen to aid in convergence; the transient itself does not have any physical significance. This method is called the artificial compressibility method because the pressure equation is the continuity equation for a compressible fluid with an equation of state

$$p = c^2 \rho. \qquad (12.16)$$

(See the electronic text entitled "Compressibility Method.")

Table 12.2. Finite Difference Methods for the Navier-Stokes Equations

Artificial Compressibility

$$\frac{\partial u}{\partial t} = -u \cdot \nabla u - \nabla p + \frac{1}{Re} \nabla^2 u$$

$$\frac{\partial p}{\partial t} = -c^2 \nabla \cdot u$$

Unsteady-State

$$\frac{u^* - u^n}{\Delta t} = -u^n \cdot \nabla u^n + \frac{1}{Re} \nabla^2 u^n$$

$$\nabla^2 p^{n+1} = \frac{1}{\Delta t} \nabla \cdot u^*$$

$$\frac{u^{n+1} - u^*}{\Delta t} = -\nabla p^{n+1}$$

Stream Function-Vorticity

$$u \cdot \nabla \xi - \frac{1}{Re} \nabla^2 \xi = 0, \quad \nabla^2 \psi = -\xi$$

$$\tau \cdot \nabla \psi = u_s \cdot n, \quad n \cdot \nabla \psi = -u_s \cdot \tau$$

The second finite difference method is a pseudo-steady-state method that splits the Navier-Stokes equation into two parts, one involving the velocity terms and one involving the pressure terms (see Table 12.2). After computing u^* using the first equation, we solve a Poisson equation for pressure. This equation is derived by taking the divergence of the third equation and setting the divergence of u^{n+1} equal to zero. After solving the Poisson equation for pressure, we solve for u^{n+1} using the third equation. The calculations are straightforward for the velocity terms. For the pressure terms we must solve the Poisson equation. This can be done using iterative techniques, such as Gauss-Seidel or overrelaxation methods. One boundary condition that can be used with this Poisson equation is a zero flux condition (Peyret and Taylor [p. 162, 1983]), provided that the velocity u^* satisfies the same boundary conditions as u^{n+1}. The actual boundary condition is obtained by taking the normal component of the third equation but the value of u^* drops out of the Poisson equation. What is being solved for in the pseudo-steady-state method is not pressure but velocity, and this velocity is the correct only at steady-state. The correct pressure can be obtained by taking the divergence of the Navier-Stokes equation to give

$$\nabla^2 p = -\nabla \cdot (u \cdot \nabla u), \qquad (12.17)$$

$$n \cdot \nabla p = -n \cdot (u \cdot \nabla u) + \frac{1}{Re} n \cdot \nabla^2 u. \qquad (12.18)$$

The divergence of the Laplacian of the velocity is the same as the Laplacian of the divergence of the velocity, which is zero according to the continuity equation. The finite difference equations are therefore

$$\frac{u^*_{i+1/2,j} - u^n_{i+1/2,j}}{\Delta t} = -u^n_{i+1/2,j}\frac{u^n_{i+3/2,j} - u^n_{i-1/2,j}}{2\Delta x} +$$

$$- \frac{1}{4}(v^n_{i+1,j+1/2} + v^n_{i,j+1/2} + v^n_{i,j-1/2} + v^n_{i+1,j-1/2})\frac{u^n_{i+1/2,j+1} - u^n_{i+1/2,j-1}}{2\Delta y} +$$

$$+ \frac{1}{Re}\left(\frac{u^n_{i+3/2,j} - 2u^n_{i+1/2,j} + u^n_{i-1/2,j}}{\Delta x^2} + \frac{u^n_{i+1/2,j+1} - 2u^n_{i+1/2,j} + u^n_{i+1/2,j-1}}{\Delta y^2}\right) \quad (12.19)$$

$$\frac{v^*_{i,j+1/2} - v^n_{i,j+1/2}}{\Delta t} = -v^n_{i,j+1/2}\left(\frac{v^n_{i,j+3/2} - v^n_{i,j-1/2}}{2\Delta y}\right)$$

$$- \frac{1}{4}(u^n_{i+1/2,j} + u^n_{i+1/2,j+1} + u^n_{i-1/2,j} + u^n_{i-1/2,j+1})\frac{v^n_{i+1,j+1/2} - v^n_{i-1,j+1/2}}{2\Delta x} +$$

$$+ \frac{1}{Re}\left(\frac{v^n_{i+1,j+1/2} - 2v^n_{i,j+1/2} + v^n_{i-1,j+1/2}}{\Delta x^2} + \frac{v^n_{i,j+3/2} - v^n_{i,j+1/2} + v^n_{i,j-1/2}}{\Delta y^2}\right), \quad (12.20)$$

$$\frac{p^{n+1}_{i+1,j} - 2p^{n+1}_{i,j} + p^{n+1}_{i-1,j}}{\Delta x^2} + \frac{p^{n+1}_{i,j+1} - 2p^{n+1}_{i,j} + p^{n+1}_{i,j-1}}{\Delta y^2} =$$

$$= -\frac{1}{\Delta t}\left[\frac{u^*_{i+1/2,j} - u^*_{i-1/2,j}}{\Delta x} + \frac{v^*_{i,j+1/2} - v^*_{i,j-1/2}}{\Delta y}\right] \quad (12.21)$$

$$\frac{u^{n+1}_{i+1/2,j} - u^*_{i+1/2,j}}{\Delta t} = -\frac{p^{n+1}_{i+1,j} - p^{n+1}_{i,j}}{\Delta x} \quad (12.22)$$

$$\frac{v^{n+1}_{i,j+1/2} - v^*_{i,j+1/2}}{\Delta t} = -\frac{p^{n+1}_{i,j+1} - p^{n+1}_{i,j}}{\Delta y}. \quad (12.23)$$

Finite difference methods can also be applied to the stream function-vorticity equations, which are usually solved iteratively.

12.4. Finite Element Methods for Unsteady-State Problems

Finite element methods are usually applied to steady-state problems by solving the set of algebraic equations [Eq. (12.11)-(12.12)] using the Newton-Raphson method. Finite difference equations are usually solved using relaxation or other iterative methods, or a pseudo-unsteady-state method is used. New finite element methods must be developed for the transient case. The first method discussed below has been developed by Gresho and co-workers [1984]. It is a classical finite element method, although several approximations are made to improve the efficiency. The second method employs the method of characteristics to move material points and is in fact similar to some of the finite difference methods. Three other finite element methods are briefly mentioned at the end of this section. The next section incorporates the ideas of Taylor-Galerkin methods.

Modified finite element method. For transient problems, we expand the velocity and pressures as in Eq. (12.1), except that now the coefficients are functions of time rather than constants. When these expansions are inserted into the Navier-Stokes equation and the continuity equation, we obtain the following set of differential-algebraic equations:

$$M \dot{u} + N(u) u + K u + C P = f, \qquad (12.24)$$

$$C^T u = 0. \qquad (12.25)$$

The first term of equation (12.24) represents the mass matrix, which appears in the finite element treatment of time derivatives. The second term represents the inertial terms, the third term is the viscous term, and the remaining terms come from the pressure gradient and body forces. The second equation is a discrete representation of the continuity equation. These equations are inconvenient to solve because they have time derivatives for the velocities in the two velocity equations, but no time derivative for pressure in the continuity equation. Thus we derive a discrete Poisson equation for pressure in the manner used to derive Eq. (12.17). We write the discrete Navier-Stokes equation as

$$C P + M \dot{u} = f - N(u) u - K u. \qquad (12.26)$$

We multiply this by M^{-1}, giving

$$M^{-1} C P + \dot{u} = M^{-1} [f - N(u) u - K u]. \qquad (12.27)$$

Next we multiply Eq. (12.27) by C^T, giving

$$C^T M^{-1} C P + C^T \dot{u} = C^T M^{-1} [f - N(u) u - K u] \qquad (12.28)$$

and differentiate the continuity equation with respect to time. Thus

$$C^T \dot{u} = 0 \qquad (12.29)$$

and the equation for pressure becomes

$$C^T M^{-1} C P = C^T M^{-1} [f - N(u) u - K u]. \qquad (12.30)$$

This is a discrete approximation of Eq. (12.17). To solve this equation we must apply boundary conditions on the pressure, which should be the Neumann conditions given in Eq. (12.18) [Gresho and Sani, 1987].

In order to solve Eq. (12.24) using Eq. (12.30), we need an efficient way to calculate M^{-1} as well as to solve Eq. (12.30). The first method is an explicit method. Gresho, *et al.* [1984] use a lumped mass matrix so that M is diagonal and its inverse is easy to find. Then, for further efficiency, they evaluate the coefficient matrices using one-point quadrature; an hourglass correction is needed for the diffusion terms to improve the results. Finally, they use the Euler method to integrate Eq. (12.24) in time, with the pressure equation [Eq. (12.30)] solved only occasionally. We note that the matrix problem in Eq. (12.30) need be solved only once per problem since it does not change from time-step to time-step. Even so, the matrix problem is a time-consuming and expensive calculation; the pressure does not change rapidly, so it is updated on a longer time-step than the velocity. In order to do this, it is necessary for the initial approximation of velocity to be solenoidal,

$$C^T u_0 = 0, \tag{12.31}$$

so that the initial transients arising from Eq. (12.29) are not severe.

The algorithm is sufficiently important to be written in detail. First let us look at the continuity equation, Eq. (12.25). We suppose that we have N+M number of velocity unknowns, with M of them specified on boundaries. We separate the vector of velocity nodal values into two vectors, one of length N and the other of length M. The continuity equation can then be written in the form of Figure 12.2. The first N equations can be written as

$$C_{11}^T u = -C_{12}^T u_B, \tag{12.32}$$

where the right-hand side is known. We call it g(t). Then Eq. (12.32) becomes

$$C_{11}^T u = g(t) \tag{12.33}$$

and Eq. (12.29) becomes

$$C_{11}^T \dot{u} = \frac{dg}{dt}. \tag{12.34}$$

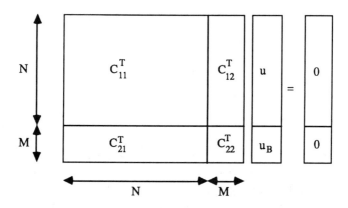

Figure 12.2. Matrix Form for the Continuity Equation

We start with a velocity that satisfies the essential boundary conditions (specified velocities) and Eq. (12.33) at time zero. Then we solve for pressure using the equation

$$C^T M^{-1} CP^n = C^T M^{-1} [\, f^n - N(u^n)\, u^n - K\, u^n \,] - \frac{g^{n+1} - g^n}{\Delta t}. \tag{12.35}$$

All terms on the right-hand side are known since g^{n+1} is known from the boundary conditions. Eq. (12.35) is derived from Eq. (12.28) using Eq. (12.34). It is the equivalent of Eq. (12.30) but with the boundary conditions taken into account. There are two types of boundary conditions. On the part of the boundary where the velocity is specified, the boundary conditions are the Neumann conditions, derived as the normal component of Eq. (12.24),

$$\frac{\partial p}{\partial n} = n \cdot \left(f + \mu \nabla^2 u - \rho \frac{\partial u}{\partial t} - \rho\, u \cdot \nabla\, u \right). \tag{12.36}$$

On the rest of the boundary, the pressure is calculated from Eq. (12.4).

$$p = \mu\, n \cdot [\nabla\, u + \nabla\, u^T] - f_n \tag{12.37}$$

Finally, the velocity is updated to give

$$M \frac{u^{n+1} - u^n}{\Delta t} = -CP^n + f^n - N(u^n)\, u^n - K\, u^n. \tag{12.38}$$

For efficiency in this version, the mass matrix is lumped in all occurrences. The final algorithm is then (where M_L represents a lumped mass matrix)

$$C^T M_L^{-1} CP^n = C^T M_L^{-1} [\, f^n - N(u^n)\, u^n - K\, u^n \,] - \frac{g^{n+1} - g^n}{\Delta t}, \tag{12.39}$$

$$M_L \frac{u^{n+1} - u^n}{\Delta t} = -CP^n + f^n - N(u^n)\, u^n - K\, u^n. \tag{12.40}$$

This algorithm is a finite element version of the finite difference method discussed in Section 12.3, based on the unsteady-state version listed in Table 12.2. If a finite element method were applied directly to the unsteady-state equations as listed in Table 12.2, we would get the algorithm

$$M \frac{\tilde{u}^{n+1} - u^n}{\Delta t} = f^n - N(u^n)\, u^n - K\, u^n, \tag{12.41}$$

$$C^T M^{-1} CP^{n+1} = \frac{1}{\Delta t} C^T \tilde{u}^{n+1}, \tag{12.42}$$

$$M \frac{u^{n+1} - \tilde{u}^{n+1}}{\Delta t} = -C P^{n+1}. \tag{12.43}$$

If we add Eq. (12.41) to Eq. (12.43), we get the momentum equation:

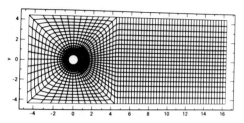

Figure 12.3a. Finite Element Mesh for Flow Past a Cylinder

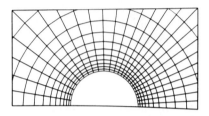

Figure 12.3b. Enlargement of the Finite Element Mesh for Flow Past a Cylinder
From Gresho, *et al.*, [1984]; copyright 1984; reprinted by permission of
John Wiley & Sons, Ltd.

$$M \frac{u^{n+1} - u^n}{\Delta t} = - C P^{n+1} + f^n - N(u^n) u^n - K u^n. \tag{12.44}$$

How are the finite element equations, Eq. (12.41)-(12.44), related to the modified finite element method as summarized in Eq. (12.35)-(12.38)? To see the relation, we take a case where the velocity boundary conditions are constant in time; then $g^{n+1} = g^n$ in Eq. (12.35). We define a new variable, $\tilde{u}^{n+1}$, such that

$$f^n - N(u^n) u^n - K u^n = M \frac{\tilde{u}^{n+1} - u^n}{\Delta t}. \tag{12.45}$$

Then Eq. (12.35) is written as

$$C^T M^{-1} C P^n = C^T M^{-1} M \frac{\tilde{u}^{n+1} - u^n}{\Delta t} = C^T \frac{\tilde{u}^{n+1} - u^n}{\Delta t}. \tag{12.46}$$

At the initial time, the equation $C^T u^n = 0$ holds; this equation is then forced to be true for subsequent times. Eq. (12.46) is equivalent to Eq. (12.42), except that the result is called P^{n+1} instead of P^n. Thus the final equations, Eq. (12.38) and Eq. (12.44), are in reality the same; the method used by Gresho and co-workers [Gresho and Chan, 1985] is the finite element analog to the finite difference method listed as "Unsteady-State" in Table 12.2.

Gresho and co-workers [Gresho and Chan, 1985] have developed their method further to allow larger time-steps. The algorithm represented by Eq.

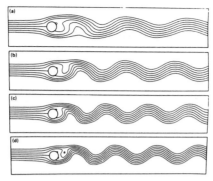

Figure 12.4. Streamlines During Vortex Shedding; Re = (a) 50, (b) 100, (c) 200, (d) 400. From Gresho, *et al.*, [1984]; copyright 1984; reprinted by permission of John Wiley & Sons, Ltd.

(12.38) has a step-size that is limited by diffusion, i.e., $\Delta t \leq c \, \mu \, \Delta x^2$ for some value of c. When the viscosity is small but non-zero, this limitation requires a very small step-size. To improve on this, the semi-implicit method was adopted. We merely replace the diffusion terms in Eq. (12.40) with an implicit version:

$$C^T M_L^{-1} CP^n = C^T M_L^{-1} [f^n - N(u^n) u^n - K u^n] - \frac{g^{n+1} - g^n}{\Delta t}, \quad (12.47)$$

$$M_L \frac{u^{n+1} - u^n}{\Delta t} + K u^{n+1} = -CP^n + f^n - N(u^n) u^n. \quad (12.48)$$

The method can be improved even more if the mass matrix is not lumped in the momentum equation. The pressure equation is easily and quickly solved when the mass matrix is lumped, so we keep that approximation in the pressure equation. Then Eq. (12.47)-(12.48) are replaced by

$$C^T M_L^{-1} CP^n = C^T M_L^{-1} [f^n - N(u^n) u^n - K u^n] - \frac{g^{n+1} - g^n}{\Delta t}, \quad (12.49)$$

$$M \frac{u^{n+1} - u^n}{\Delta t} + K u^{n+1} = -CP^n + f^n - N(u^n) u^n. \quad (12.50)$$

An additional viscosity term is added that is proportional to $\Delta t u_i u_j/2$; it is called a balancing tensor diffusivity. The reason for using this term is closely connected to the Taylor-Galerkin method, which is outlined in Section 12.5. Eq. (12.49)-(12.50) gives good results, too [Gresho and Chan, 1988].

The explicit method [Eq. (12.47)-(12.48)] has been applied to the problem of flow past a cylinder at large Reynolds numbers. The mesh is as shown in Figure 12.3 [Gresho, *et al.*, 1984]. For Reynolds numbers above 50, the steady-state solution to this problem is unstable and an unsteady flow develops. Vortices are shed behind the cylinder one after another with a frequency f. A typical view

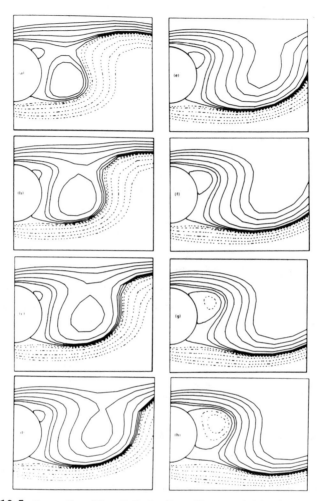

Figure 12.5. Streamlines Near Cylinder, Re = 50, $\Delta t = \tau/16$ for Successive Pictures From Gresho, *et al.*, [1984]; copyright 1984; reprinted by permission of John Wiley & Sons, Ltd.

of the streamlines behind the cylinder is shown in Figure 12.4. For one Reynolds number, the successive streamlines are shown as a function of time in Figure 12.5. The frequency at which the vortices are shed is a convenient measure for comparing experiments; this dimensionless frequency is called the Strouhal number,

$$St = \frac{u_0 f}{D}, \qquad (12.51)$$

where u_0 is the upstream velocity of the fluid past a cylinder with diameter D. This agrees reasonably well with experimental correlations (0.14 versus 0.12 for a Reynolds number = 50). The calculations using the first algorithm discussed,

Eq. (12.24)-(12.40), took about two hours on a CRAY computer. See also Kovacs and Kawahara [1991] for additional applications.

Donea, *et al.* [1982] presented a fractional step finite element method in which the momentum equation was split into two steps. The convective and diffusive terms were used to calculate the velocity at the n+1/2-th time-step and a Poisson equation for pressure was used to obtain the pressure at the n+1-th step. Then the velocity at the n+1-th step was obtained, including the pressure term. Similar ideas for the Taylor-Galerkin method are described in detail in Section 12.5.

Method of characteristics with operator splitting. The second method is one developed by Huffenus and Khaletzky [1984] that involves operator splitting. They employ a method of characteristics to follow the movement of fluid elements. If we solve the hyperbolic equation

$$\frac{\partial F}{\partial t} + u \frac{\partial F}{\partial x} = 0 \qquad (12.52)$$

to move from time level n to time level n+1, then the solution at node i and time level n+1 is the value of the function at F*, as illustrated in Figure 12.6.

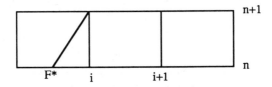

Figure 12.6. Method of Characteristics to Go from n-th Time level to n+1-th Time Level

Thus, when solving

$$\frac{dM}{dt} = V(M,t), \qquad (12.53)$$

we use

$$M^* = M_i^{n+1} - V\Delta t, \qquad (12.54)$$

where the velocity is constant during the time interval from time level n to time level n+1. The method works as follows. We use the method of characteristics to obtain u* and v* at each node (i.e., the velocity of the fluid at node i and time level n+1). Then we solve

$$\nabla^2 p = \frac{\nabla \cdot v^*}{\Delta t} \qquad (12.55)$$

using the Neumann boundary conditions from the momentum equation. Finally, we solve

$$\frac{v^{n+1} - v^*}{\Delta t} = -\nabla p^{n+1} + \frac{1}{Re} \nabla^2 v^{n+1} \qquad (12.56)$$

to determine the velocity at the new time level. The steps of the method are similar to those used in the unsteady-state finite difference method listed in Table 12.2, except that the convection terms are handled using the method of characteristics, Eq. (12.54), rather than in the standard way. Huffenus and Khaletzky [1984] use triangular elements with quadratic velocities and linear pressures to solve the driven cavity problem. Yang, et al. [1991] use a similar method for the convective diffusion equation.

Compressible-fluid method. The third method, based on work by Kawahara [1983], allows for density to depend on pressure so that the fluid is effectively compressible (although the compressibility may be small). In this way, the continuity equation is changed from a differential equation constraint to a time-dependent equation for pressure. After the compressible fluid finite element method is applied in space, the unsteady-state problem has three ordinary differential equations for u, v, and p, which can be solved using a variety of methods. Kawahara [1983] uses an explicit method; he also solves the problem of vortex shedding behind a cylinder.

Stream function-vorticity method. The fourth method is a time-dependent application of the stream function-vorticity method by Miyauchi, *et al.* [1983]. In this case, the time-dependent vorticity equation is integrated using either the Euler method or a Lax-Wendroff method in time.

Upstream method. The final method comes from work by Hughes, *et al.* [1979]: a penalty method is used along with upstream evaluation of the convection matrices. Lobatto quadrature is used, which in the lowest order is the trapezoid rule and in the next highest order is Simpson's rule. This has the effect of lumping the mass. When the steady-state equations are written as

$$\mathbf{C}\,\mathbf{v} + \mathbf{N}(\mathbf{v}) = \mathbf{F}, \qquad (12.57)$$

then the transient method is as follows. From the n-th time level we know the value of v_n and a_n. The algorithm is then

$$\begin{aligned}
\mathbf{v}_{n+1} &= \mathbf{v}_n + (1-\gamma)\,\mathbf{a}_n, \\
\mathbf{v}^0_{n+1} &= \tilde{\mathbf{v}}_{n+1}, \\
(\mathbf{M} + \gamma\,\Delta t\,\mathbf{C})\,\mathbf{v}^{k+1}_{n+1} &= \mathbf{M}\,\tilde{\mathbf{v}}_{n+1} + \gamma\,\Delta t\,[\,\mathbf{F} - \mathbf{N}\,(\mathbf{v}^k_{n+1})], \qquad (12.58)\\
\mathbf{a}_{n+1} &= (\,\mathbf{v}_{n+1} - \tilde{\mathbf{v}}_{n+1}\,)/\gamma\,\Delta t\,.
\end{aligned}$$

Apparently the equation is iterated until it converges (at each time-step). The parameter γ is set as $\geq 1/2$ for stability reasons. The penalty parameter takes values similar to the parameter for a steady-state simulation. Since upstream effects are included, the Reynolds number can be very large. One solution is given for flow past a step in the boundary; starting from a steady-state solution for a Reynolds number of 30, the flow rate is increased to an effective Reynolds number of 10^7. No attempt is made to determine the influence of upstream weighting on the type of solution, but the solution is "reasonable-looking," even if there is so much upstream diffusion that it is incorrect. Since real flows usually become turbulent at such large Reynolds numbers, determining whether or not it is an appropriate solution of the equations must await more detailed studies with finer meshes.

For unsteady flows it is also possible to move the mesh in some way. Ramaswamy and Kawahara [1987] use a combined Lagrangian-Eulerian finite element method for treating time-dependent flows with free surfaces. Gopalakrishnan [1988] moved the mesh using local kinematics.

The **flux-correction method** has also been applied to compressible flow equations and the Navier-Stokes equation [Löhner, et al., 1987 and Thareja, et al., 1989].

12.5. The Taylor-Galerkin Method

The Taylor-Galerkin method can be applied to the unsteady-state Navier-Stokes equation,

$$\rho \left(\frac{\partial \mathbf{u}}{\partial t} + \mathbf{u} \cdot \nabla \mathbf{u} \right) = \rho \mathbf{f} - \nabla p + \nabla \cdot [\mu (\nabla \mathbf{u} + \nabla \mathbf{u}^T)], \qquad (12.59)$$

which must be solved together with the continuity equation,

$$\nabla \cdot \mathbf{u} = 0. \qquad (12.60)$$

As before, we expand the velocity in a Taylor series in time giving

$$\mathbf{u}^{n+1} = \mathbf{u}^n + \frac{\partial \mathbf{u}}{\partial t}\bigg|^n \Delta t + \frac{\partial^2 \mathbf{u}}{\partial t^2}\bigg|^n \frac{\Delta t^2}{2}. \qquad (12.61)$$

This equation is rearranged to give

$$\frac{\partial \mathbf{u}}{\partial t}\bigg|^n = \frac{\mathbf{u}^{n+1} - \mathbf{u}^n}{\Delta t} - \frac{\Delta t}{2} \frac{\partial^2 \mathbf{u}}{\partial t^2}\bigg|^n. \qquad (12.62)$$

We evaluate the second time derivative from the unsteady-state Navier-Stokes equation by using Cartesian tensor notation and transforming the final result into vector notation to make it general. Let us consider only the convective term in the Navier-Stokes equation:

$$\frac{\partial u_i}{\partial t} + u_j \frac{\partial u_i}{\partial x_j} = 0 \qquad (12.63)$$

or

$$\frac{\partial u_i}{\partial t} = - u_j \frac{\partial u_i}{\partial x_j}. \qquad (12.64)$$

We can also write this in conservative form as

$$\frac{\partial u_i}{\partial t} + \frac{\partial}{\partial x_j}(u_j u_i) = 0 \tag{12.65}$$

or

$$\frac{\partial u_i}{\partial t} = -\frac{\partial}{\partial x_j}(u_j u_i). \tag{12.66}$$

We differentiate Eq. (12.66) once with respect to time, giving

$$\frac{\partial^2 u_i}{\partial t^2} = -\frac{\partial}{\partial x_j}\left(u_j \frac{\partial u_i}{\partial t}\right) - \frac{\partial}{\partial x_j}\left(\frac{\partial u_j}{\partial t} u_i\right). \tag{12.67}$$

Then we insert the non-conservative form, Eq. (12.64), which gives

$$\frac{\partial^2 u_i}{\partial t^2} = \frac{\partial}{\partial x_j}\left(u_j u_k \frac{\partial u_i}{\partial x_k}\right) + \frac{\partial}{\partial x_j}\left(u_k \frac{\partial u_j}{\partial x_k} u_i\right). \tag{12.68}$$

Finally, we write this in vector notation:

$$\frac{\partial^2 \mathbf{u}}{\partial t^2} = \nabla \cdot [\mathbf{u}(\mathbf{u} \cdot \nabla \mathbf{u})] + \nabla \cdot [(\mathbf{u} \cdot \nabla \mathbf{u})\mathbf{u}]. \tag{12.69}$$

The unsteady-state Navier-Stokes equation is then

$$\rho\left(\frac{\mathbf{u}^{n+1} - \mathbf{u}^n}{\Delta t} + \mathbf{u} \cdot \nabla \mathbf{u}\right) = \rho \mathbf{f} - \nabla p + \nabla \cdot [\mu(\nabla \mathbf{u} + \nabla \mathbf{u}^T)] + \frac{\rho \Delta t}{2}\frac{\partial^2 \mathbf{u}}{\partial t^2}\bigg|^n, \tag{12.70}$$

which becomes

$$\rho\left(\frac{\mathbf{u}^{n+1} - \mathbf{u}^n}{\Delta t} + \mathbf{u} \cdot \nabla \mathbf{u}\right) = \rho \mathbf{f} - \nabla p + \nabla \cdot [\mu(\nabla \mathbf{u} + \nabla \mathbf{u}^T)] +$$
$$+ \frac{\rho \Delta t}{2}\{\nabla \cdot [\mathbf{u}(\mathbf{u} \cdot \nabla \mathbf{u})] + \nabla \cdot [(\mathbf{u} \cdot \nabla \mathbf{u})\mathbf{u}]\}. \tag{12.71}$$

For us to see the effect of the additional terms, we should write the equations in component notation. We use cylindrical geometry and take a case with azimuthal symmetry. We use z for the axial coordinate and r for the radial coordinate and also use the condition

$$u_\theta = 0 \tag{12.72}$$

and allow no θ-variation in the solution. The velocities are

$$\mathbf{e}_z \cdot \mathbf{u} = u \quad \text{and} \quad \mathbf{e}_r \cdot \mathbf{u} = v. \tag{12.73}$$

The convective terms are given on the next page.

The $\mathbf{e}_z$ component of $\mathbf{u} \cdot \nabla\ \mathbf{u}$ is $v \frac{\partial u}{\partial r} + u \frac{\partial u}{\partial z}$.

The $\mathbf{e}_r$ component of $\mathbf{u} \cdot \nabla\ \mathbf{u}$ is $v \frac{\partial v}{\partial r} + u \frac{\partial v}{\partial z}$. (12.74)

The Taylor terms involve the dyadic $\mathbf{u}\,(\mathbf{u} \cdot \nabla\ \mathbf{u}) + (\mathbf{u} \cdot \nabla\ \mathbf{u})\,\mathbf{u}$. Its components are

$\mathbf{e}_z\mathbf{e}_z$ $\qquad 2\left(uv\frac{\partial u}{\partial r} + u^2 \frac{\partial u}{\partial z}\right)$,

$\mathbf{e}_z\mathbf{e}_r$ and $\mathbf{e}_r\mathbf{e}_z$ $\qquad u\left(v\frac{\partial v}{\partial r} + u\frac{\partial v}{\partial z}\right) + \left(v\frac{\partial u}{\partial r} + u\frac{\partial u}{\partial z}\right)v$, (12.75)

$\mathbf{e}_r\mathbf{e}_r$ $\qquad 2\left(v^2 \frac{\partial v}{\partial r} + uv \frac{\partial v}{\partial z}\right)$,

$\mathbf{e}_\theta\mathbf{e}_\theta$ $\qquad 0$. (12.76)

We note that the dyadic is symmetric. Now a second-order dyadic can be written in terms of its components (in this geometry) as

$$\mathbf{A} = \mathbf{e}_z\mathbf{e}_z A_{zz} + \mathbf{e}_z\mathbf{e}_r A_{zr} + \mathbf{e}_r\mathbf{e}_z A_{rz} + \mathbf{e}_r\mathbf{e}_r A_{rr} + \mathbf{e}_\theta\mathbf{e}_\theta A_{\theta\theta}.$$ (12.77)

For the divergence, the components of $\nabla \cdot \mathbf{A}$ are

$\mathbf{e}_z \qquad \frac{1}{r}\frac{\partial}{\partial r}(r A_{rz}) + \frac{\partial}{\partial z}(A_{zz})$,

$\mathbf{e}_r \qquad \frac{1}{r}\frac{\partial}{\partial r}(r A_{rr}) + \frac{\partial}{\partial z}(A_{zr}) - \frac{A_{\theta\theta}}{r}$. (12.78)

For the Taylor terms, the components of $\nabla \cdot [\mathbf{u}\,(\mathbf{u} \cdot \nabla \mathbf{u}) + (\mathbf{u} \cdot \nabla\ \mathbf{u})\,\mathbf{u}]$ are

$\mathbf{e}_z \quad \frac{1}{r}\frac{\partial}{\partial r}\left\{r\left[v^2\frac{\partial u}{\partial r} + uv\frac{\partial u}{\partial z} + vu\frac{\partial v}{\partial r} + u^2\frac{\partial v}{\partial z}\right]\right\} + \frac{\partial}{\partial z}\left\{2uv\frac{\partial u}{\partial r} + 2u^2\frac{\partial u}{\partial z}\right\}$,

$\mathbf{e}_r \quad \frac{1}{r}\frac{\partial}{\partial r}\left\{r\left[2v^2\frac{\partial v}{\partial r} + 2uv\frac{\partial v}{\partial z}\right]\right\} + \frac{\partial}{\partial z}\left\{uv\frac{\partial v}{\partial r} + u^2\frac{\partial v}{\partial z} + v^2\frac{\partial u}{\partial r} + uv\frac{\partial u}{\partial z}\right\}$. (12.79)

Let us next consider the case where the dyadic $\mathbf{A}$ represents a shear stress for a Newtonian fluid. The components are

$$\tau_{zz} = -2\mu\frac{\partial u}{\partial z}, \quad \tau_{rz} = \tau_{zr} = -\mu\left(\frac{\partial u}{\partial r} + \frac{\partial v}{\partial z}\right), \quad \tau_{rr} = -2\mu\frac{\partial v}{\partial r}, \quad \tau_{\theta\theta} = -2\mu\frac{v}{r}. \quad (12.80)$$

By comparing Eq. (12.75)-(12.76) with Eq. (12.80) we can see that the Taylor terms add viscous-like terms to the equations. In fact, we can write Eq. (12.75)-(12.76) as

$$\mathbf{A} = \mathbf{T} : \nabla\ \mathbf{u} \qquad (12.81)$$

and deduce the fourth-order dyadic $\mathbf{T}$. The Taylor term can then be regarded as an anisotropic viscous term.

An alternative formulation of the Taylor-Galerkin method for the Navier-Stokes equation was given by Laval [1988]. She expresses the method in terms of operator splitting. First there is a convection step,

$$\rho \left(\frac{u^* - u^n}{\Delta t} + u \cdot \nabla u \right) = \frac{\rho \Delta t}{2} \{ \nabla \cdot [u^n(u^n \cdot \nabla u^n)] + \nabla \cdot [(u^n \cdot \nabla u^n)u^n] \},$$
(12..82)

$$u^* = b^{n+1} \text{ on inflow boundaries,} \quad (12..83)$$

where **b** represents the inlet velocity. Then there is a viscous step,

$$\rho \frac{u^{**} - u^*}{\Delta t} = \rho f + \nabla \cdot [\mu (\nabla u^* + \nabla u^{*T})],$$
(12.84)

$$u^{**} = b^{n+1} \text{ on boundaries with velocity specified.} \quad (12.85)$$

The pressure is obtained by solving

$$\rho \frac{\nabla \cdot u^{**}}{\Delta t} = -\nabla^2 p^{n+1}.$$
(12.86)

Finally, u^{n+1} can be calculated:

$$\rho \frac{u^{n+1} - u^{**}}{\Delta t} = -\nabla p^{n+1}, \text{ and } \nabla \cdot u^{n+1} = 0,$$
(12.87)

$$u^{n+1} \cdot n = b^{n+1} \cdot n \text{ on boundaries where velocity is specified.} \quad (12.88)$$

Comparison of Eq. (12.82)-(12.88) with Eq. (12.41)-(12.43) shows that they are the same, except that the convection and viscous step are separated into two steps and the diffusion is evaluated at the predicted velocity u^*. This means that the method is similar to Gresho's formulation when the same finite element decisions are made (i.e., the same trial functions, decisions on lumping, etc.). In the two-step version, Eq. (12.82)-(12.88), methods optimized for each step could be used. Other examples of similar methods are by Ramaswamy [1990], Hawken, et al. [1990], and Pepper and Singer [1990].

For the compressible high speed flow equations, Löhner, et al. [1985] used a two-step formulation of the Taylor-Galerkin method that is similar to the MacCormack and Lax-Wendroff methods. The two-step formulation proved faster than the one-step formulation because the Jacobian was less complicated. The compressible flow equations were without viscosity but artificial viscosity was added proportional to the velocity gradients. Löhner, et al. also did adaptive remeshing by using a criterion that the local error should be constant. The local error was given by

$$\| u - u^h \|_0^e < C h_e^2 |u|_2^e.$$
(12.89)

Thus they chose the mesh size so that

$$h_e^2 \, |u|_2^e = \text{constant.} \tag{12.90}$$

The norm was defined as

$$|u|_2 = \sqrt{\int_\Omega \left[\left(\frac{\partial^2 u}{\partial x^2}\right)^2 + 2\left(\frac{\partial^2 u}{\partial x \partial y}\right)^2 + \left(\frac{\partial^2 u}{\partial y^2}\right)^2 \right] d\Omega} \tag{12.91}$$

and Löhner, et al. [1985] used a finite element approximation to this norm. (See also Löhner [1987]). Oden, et al. [1986] treated the same equations, using artificial viscosity and adaptive mesh refinement. For a group of four elements, the center node could be moved or the elements could be subdivided to achieve an equidistribution criterion based on the estimated error.

References

Campion-Renson, A. and Crochet, M. J., "On the Stream Function-Vorticity Finite Element Solutions of Navier-Stokes Equations, Int. J. Num. Methods Engn. 12 1809-1818 [1978].

Chorin, A. J., "A Numerical Method for Solving Incompressible Viscous Flow Problems," J.Comp.Phys.2 12-26 [1967].

Cuvelier, C., Segal, A. and van Steenhoven, A. A., *Finite Element Methods and Navier-Stokes Equations*, Reidel Publishing, Dordrecht [1986].

Donea, J., Giuliani, S., Laval, H., Quartapelle, L., "Finite Element Solution of the Unsteady Navier-Stokes Equations by a Fractional Step Method," Comp. Methods Appl. Mech. Eng. 30 53-73 [1982].

Gopalakrishnan, T. C., "Comparison of Fixed and Moving Boundary Models for Coastal Circulation," Proc. First Int. Conf. Comput. Methods Flow Analysis, Vol. 1, pp. 587-593, Okayama Univ., Japan [1988].

Gresho, P. M., Chan, S. T., Upson, C., and Lee, R. L., "A Modified Finite Element Method for Solving the Time-dependent, Incompressible Navier-Stokes Equations," Int. J. Num. Meth. Fluids 4, "Part 1: Theory", pp. 557-597; "Part 2: Applications", pp. 619-640 [1984].

Gresho, P. M. and Chan, S., "Semi-Consistent Mass Matrix Techniques for Solving the Incompressible Navier-Stokes Equations," LLNL Report UCRL-99503 [1988].

Gresho, P. M. and Chan, S., "A New Semi-Implicit Method for Solving the Time-Dependent Conservation Equations for Incompressible Flow," Proc. Num. Methods in Laminar and Turbulent Flow, Pineridge Press Lt., Swansea, U. K., pp. 3-21 [1985].

Gresho, P. M. and Sani, R. L., "On Pressure Boundary Conditions for the Incompressible Navier-Stokes Equations," Int. J. Num. Methods Fluids 7 1111-1145 [1987].

Hawken, D. M., Tamaddon-Jahromi, H. R., Townsend, P. and Webster, M. F., "A Taylor-Galerkin-based Algorithm for Viscous Incompressible Flow," Int. J. Num. Methods Fluids 10 327-351 [1990].

Huffenus, J. P. and Khaletzky, D., "A Finite Element Method to Solve the Navier-Stokes Equations Using the Method of Characteristics," Int. J. Num. Methods Fluids 4 247-269 [1984].

Hughes, T. J. R., Liu, W. K., Brooks, A., "Finite Element Analysis of Incompressible Viscous Flows by the Penalty Function Formulation," J. Comp. Phys. 30 1-60 [1979].

Jackson, C. P., and Cliffe, K. A., "Mixed Interpolation in Primitive Variable Finite Element Formulation for Incompressible Flow," Int. J. Num. Methods Engn 17 1659-1688 [1981]

Kawahara, M. and Hirano, H., "A Finite Element Method for High Reynolds Number Viscous Fluid Flow Using Two Step Explicit Scheme," Int. J. Num. Methods Fluids 3 137-163 [1983].

Kovacs, A. and Kawahara, M., "A Finite Element Scheme Based on the Velocity Correction Method for the Solution of the Time-dependent Incompressible Navier-Stokes Equations," Int. J. Num. Methods Fluids 13 403-423 [1991].

Laval, H. "Taylor-Galerkin Solution of the Time-Dependent Navier-Stokes Equations," Proc. First Int. Conf. Comput. Methods Flow Analysis, pp. 414-421 Okayama Univ., Japan [1988].

Löhner, R., Morgan, K., and Zienkiewicz, O. C., "An Adaptive Finite Element Procddure for Compressible High Speed Flows," Comp. Methods Appl. Mech. Eng. 51 441-465 [1985].

Löhner, R., "An Adaptive Finite Element Scheme for Transient Problems in CFD," Comp. Methods Appl. Mech. Eng. 61 323-338 [1987].

Löhner, R., Morgan, K., Peraire, J. and Vahdati, M., "Finite Element Flux-corrected Transport (FEM-FCT) for the Euler and Navier-Stokes Equations," Int. J. Num. Methods Fluids 7 1093-1109 [1987].

Miyauchi, Y., Masuda, M., and Shimizu, M., "A Transient Finite Element Analysis of Natural Convection Around a Horizontal Hot Cylinder," Int. J. Num. Methods Fluids 3 429-443 [1983].

Oden, J. T., Strouboulis, T., and Devloo, P., "Adaptive Finite Element Methods for the Analysis of Inviscid Compressible Flow: Part 1. Fast Refinement/Unrefinement and Moving Mesh Methods for Unstructured Meshes," Comp. Meth. Appl. Mech. Eng. 59 327-362 [1986].

Pepper, D. W. and Singer, A. P, "Calculation of Convective Flow on the Person al Computer Using a Modified Finite-Element Method," Num. Heat Trans., Part A 17 379-400 [1990].

Peyret, R. and T. D. Taylor, *Computational Methods for Fluid Flow*, Springer-Verlag, [1983].

Ramaswamy, B., "Efficient Finite Element Method for Two-Dimensional Fluid Flow and Heat Transfer Problems," Num. Heat Trans., Part B 17 123-154 [1990].

Ramaswamy, B. and Kawahara, M., "Arbitrary Lagrangian-Eulerian Finite Element Method for Unsteady, Convective, Incompressible Viscous Free Surface Fluid Flow," Int. J. Num. Methods Fluids 7 1053-1075 [1987].

Sani, R. L., Gresho, P. M., Lee, R. L., Griffiths, D. F. and Engleman, M., "The Cause and Cure(!) of the Spurious Pressures Generated by Certain FEM Solutions of the Incompressible Navier-Stokes Equations: Part 2", Int. J. Num. Methods Fluids 1 171-204 [1981].

Shih, T. M., Tan, C. H. and Hwang, B. C., "Equivalence of Artificial Compressibility Method and Penalty-Function Method," Num. Heat Trans., Part B 15 127-130 [1989].

Stevens, W. N. R., "Finite Element, Stream Function-Vorticity Solution of Steady Laminar Natural Convection," Int. J. Num. Methods Fluids 2 349-366 [1982].

Thareja, R. R., Stewart, J. R., Hassan, O., Morgan, K., and Peraire, J., "A Point Implicit Unstructured Grid Solver for the Euler and Navier-Stokes Equations," Int. J. Num. Methods Fluids 9 405-425 [1989].

Yang, J. C., Chen, K. N., and Lee, H. Y., "Investigation of Use of Reach-back Characteristics Method for 2D Dispersion Equation," Int. J. Num. Methods Fluids 13 841-855 [1991].

CHAPTER
THIRTEEN

POLYMER FLOW

The mathematical modeling of polymer flows presents difficult numerical challenges. Model problems that have an exact solution can be used to demonstrate those difficulties. For simplicity we will consider three such model problems in one dimension, one of which has a moving front. These problems include integrating stress under a constant gradient, integrating stress under a variable gradient, and unsteady Couette flow. Since two-dimensional steady-state problems are difficult to solve and two-dimensional problems with moving fronts are the subject of current research, they will not be discussed here.

13.1. Introduction

Polymers are molten plastics or fluids with additives that are able to flow but also appear to behave like a solid. If a force is applied, the fluid moves. When the force is removed, however, the fluid may recoil towards its original position before the force was applied. A solid would return completely to its original state (like a rubber band), but the elastic fluid or polymer only returns part of the way. Thus the polymer has both a viscous component and an elastic component.

The field of rheology studies the relation between stress and strain or rate-of-strain in polymers and other non-Newtonian fluids. Several classical experiments can be performed to illustrate the behavior of polymeric liquids. Let us first characterize the polymer by examining the Couette flow illustrated in Figure 13.1. The top plate is moving at a fixed velocity, V, while the bottom plate is held fixed. We assume that

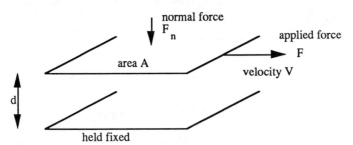

Figure 13.1. Measurement Geometry for Viscosity and Normal Stress

the plates are infinite in extent (or at least very large) in the x- and z-directions. First we will measure the force needed to move the top plate. The shear stress, τ, and shear rate, $\dot{\gamma}$, are

$$\tau = \frac{F}{A} \quad \text{and} \quad \dot{\gamma} = \frac{V}{d}. \tag{13.1}$$

Experimentally, we find that these are related:

$$\frac{F}{A} = \text{function}\left(\frac{V}{d}\right) \quad \text{or} \quad \tau = f(\dot{\gamma}). \tag{13.2}$$

The viscosity is defined by

$$\eta = \frac{\tau}{\dot{\gamma}} = \frac{f(\dot{\gamma})}{\dot{\gamma}}. \tag{13.3}$$

We may also find experimentally that it is necessary to apply a normal force to keep the plates together and that this normal force is also a function of shear rate.

$$\frac{F_n}{A} = g(\dot{\gamma}) \tag{13.4}$$

The normal force divided by the area is called the normal stress. The first normal stress coefficient is defined as the first normal stress difference divided by the shear rate squared:

$$\Psi = \frac{N_1}{\dot{\gamma}^2} = \frac{g(\dot{\gamma})}{\dot{\gamma}^2}. \tag{13.5}$$

The character of the fluid is determined by how the stresses depend on the shear rate. If the relationship is linear for shear stress and there is no normal stress, the fluid is called a Newtonian fluid. If the relation between shear stress and shear rate is nonlinear but the normal stress is zero, we call the fluid a purely viscous, non-Newtonian fluid or a generalized Newtonian fluid. If the normal stress is not equal to zero, we call it an elastic non-Newtonian fluid.

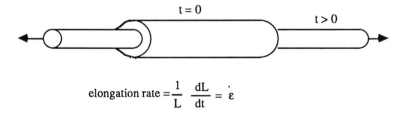

Figure 13.2. Elongation Experiment

Another experiment to characterize a polymer is to take a fiber and pull on it (see Figure 13.2). Since polymers are generally very viscous, it is possible to do this. The force per unit area exerted on the ends of the fiber is the normal stress [a force acting in the direction of a normal (or perpendicular) to the ends of the fiber]. The elongation rate is measured and the elongational viscosity is defined as

$$\eta_e = \frac{\tau_{11} - \tau_{22}}{\dot{\varepsilon}}. \tag{13.6}$$

Since the fiber lengthens as it is pulled, it eventually would break. Thus, transient experiments can be done where the force and length of the fiber are measured versus time.

Transient experiments using the flow field shown in Figure 13.1 are also possible. The fluid is initially at rest and at t = 0 the top plate begins to move with a constant velocity. The force needed to do this depends on the fluid; two examples are shown in Figure 13.3.

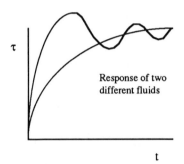

Figure 13.3. Shear Stress Response to a Plate Moving Suddenly with Velocity V

All flow fields are locally composed of both shear components (Figure 13.1) and elongation components (Figure 13.2). Typical behavior of polymers is shown in Figure 13.4. The model problems solved below are carefully constructed to be one-dimensional problems that either are entirely shear or entirely elongational in character.

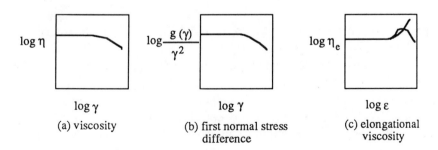

Figure 13.4. Steady Rheological Behavior of Real Fluids

How do we model a substance that is part fluid and part solid? A brief outline of one way to build a model will be given here. In the experiment illustrated by Figure 13.1, a Newtonian fluid satisfies

$$\tau = \mu \dot{\gamma}. \tag{13.7}$$

Here we write the shear rate as

$$\dot{\gamma} = \frac{d\gamma}{dt}. \tag{13.8}$$

Next let us suppose that the material shown in Figure 13.1 is a solid. If we start with the solid at rest and apply a force, we find that the force needed is proportional to the strain (how far the solid has moved) rather than to its time derivative, the rate of strain.

$$\tau = G \gamma \tag{13.9}$$

Eq. (13.9) is the equation for a Hookean solid where G is the shear modulus. We differentiate Eq. (13.9) to obtain

$$\frac{d\tau}{dt} = G \frac{d\gamma}{dt} = G \dot{\gamma}. \tag{13.10}$$

Now we combine Eq. (13.7) and Eq. (13.10) as a linear combination to get

$$\frac{1}{\mu} \tau + \frac{1}{G} \frac{d\tau}{dt} = \dot{\gamma}. \tag{13.11}$$

We rearrange this to give

$$\tau + \frac{\mu}{G} \frac{d\tau}{dt} = \mu \dot{\gamma} \quad \text{or} \quad \tau + \lambda \frac{d\tau}{dt} = \mu \dot{\gamma}. \tag{13.12}$$

This is a linear Maxwell model. The variable λ is the time constant and the variable μ is the viscosity. If this equation is integrated for constant λ and μ and no stress at time approaching $-\infty$, we get

$$\tau = \frac{\mu}{\lambda} \int_{-\infty}^{t} e^{-(t-t')/\lambda} \dot{\gamma}(t') \, dt'. \tag{13.13}$$

The role of λ is to determine how fast the material "forgets"; the rate of strain far in the past (relative to λ) is less important.

The linear Maxwell model, Eq. (13.12), is a simplistic model; it is the only one that will be treated here. It is apparently the hardest model to solve (in two dimensions) and it contains all the important mathematical features. (See Bird, *et al*. [1987] for further discussion of constitutive equations.) The equations governing flow are the momentum equation,

$$\rho \mathbf{u} \cdot \nabla \mathbf{u} = -\nabla p - \nabla \cdot \tau, \tag{13.14}$$

coupled with the continuity equation,

$$\nabla \cdot \mathbf{u} = 0, \tag{13.15}$$

and the constitutive equation for a co-deformational Maxwell model,

$$\tau + \lambda \left[\frac{\partial \tau}{\partial t} + \mathbf{v} \cdot \nabla \tau - \nabla \mathbf{v}^T \cdot \tau - \tau \cdot \nabla \mathbf{v} \right] = -\dot{\gamma}. \tag{13.16}$$

The sign convention corresponds to that usually used within the chemical engineering community.

The time constants used in non-dimensionalization are

$$\text{diffusion of vorticity:} \quad t_d = \frac{x_s^2}{\nu} = \frac{\rho x_s^2}{\mu},$$

$$\text{convection:} \quad t_c = \frac{x_s}{u_s}, \tag{13.17}$$

$$\text{elasticity:} \quad t_e = \lambda_s.$$

The non-dimensional variables are defined by dividing the real variable by a standard variable:

$$t = \frac{\text{time}}{t_s}, \quad \tau = \frac{\text{stress}}{\tau_s}, \quad u = \frac{\text{velocity}}{u_s}, \quad \text{and} \quad x = \frac{\text{position}}{x_s}. \tag{13.18}$$

We choose the velocity standard, u_s, to be an average velocity or inlet velocity and the length standard, x_s, to be some dimension, such as the radius. The time and stress standards are taken as

$$t_s = \frac{x_s}{u_s} \quad \text{and} \quad \tau_s = \frac{\mu u_s}{x_s}. \tag{13.19}$$

The ratios of time constants then give the dimensionless numbers, the Reynolds number and the Weissenberg number:

$$\frac{t_d}{t_c} = \frac{\rho u_s x_s}{\mu} = Re, \quad \frac{t_e}{t_c} = \frac{\lambda_s u_s}{x_s} = We, \quad \frac{t_e}{t_d} = \frac{\lambda \mu}{\rho x_s^2} = \frac{We}{Re} \equiv E. \tag{13.20}$$

Naturally, we expect the problem to be difficult when the ratios of time constants are very large or very small.

13.2. Model Problem One: Stress with a Constant Gradient

Let us consider the 4:1 contraction problem illustrated in Figure 13.5 where the radius of the tube is suddenly decreased to one-fourth of its former value. The boundary conditions used are shown in Figure 13.6; symmetry allows us to solve the problem on only half the domain. We want to determine the flow

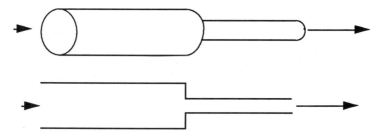

Figure 13.5. Flow Geometry in a 4:1 Contraction in Pipe Radius

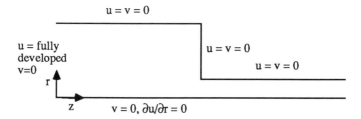

Figure 13.6. Boundary Conditions for a 4:1 Contraction in Pipe Radius

field and stress distribution.

Let us consider the stress equation along the centerline. The variable τ_{zz} is governed by

$$\tau_{zz} + \lambda \left(u \frac{d\tau_{zz}}{dz} - 2 \frac{du}{dz} \tau_{zz} \right) = -2 \frac{du}{dz}$$

$$\text{or} \quad \lambda u \frac{d\tau_{zz}}{dz} + \left(1 - 2\lambda \frac{du}{dz} \right) \tau_{zz} = -2 \frac{du}{dz}. \tag{13.21}$$

If the velocity $u(z,r = 0)$ is known, then this is an ordinary differential equation with variable coefficients for τ_{zz} on $r = 0$. The $u(z)$ is taken as an approximation to $u(z,r = 0)$ from the solution to the Newtonian problem in the 4:1 contraction. In non-dimensional form the problem is

$$\left[1 - 2 \text{ We} \frac{du}{dz} \right] \tau + \text{We } u \frac{d\tau}{dz} = -2 \frac{du}{dz}. \tag{13.22}$$

The first model problem is a simplification of Eq. (13.22).

$$\text{We} \frac{d\tau}{dz} + \tau = 1 \tag{13.23}$$

Eq. (13.23) is also the problem for a transient elongation (Figure 13.2) of a Maxwell fluid (with z becoming t and the shear rate becoming the elongation rate).

Let us first consider the initial value problem of Eq. (13.23) with the initial condition

$$\tau(0) = \tau_0. \qquad (13.24)$$

Eq. (13.23) is similar to the equation for stress along the centerline, but it allows analytical manipulation. We will first solve this initial value problem exactly. Then we will apply several methods to the solution of the ordinary differential equation: the Euler method, the backward Euler method, the centered finite difference method, the Galerkin method, and the Petrov-Galerkin method. The Galerkin method can be applied either as an initial value method or as a boundary value method; the difference is shown below. In each case we are interested in the conditions such that the solution does not oscillate from node to node. Some of this material follows work by Josse and Finlayson [1984].

The exact solution is

$$\tau = 1 + (\tau_0 - 1) e^{-z/We}. \qquad (13.25)$$

If the exact solution is evaluated at a set of equally spaced points, it is

$$\tau_i = 1 + (\tau_0 - 1) e^{-i\Delta z/We}. \qquad (13.26)$$

This gives

$$\tau_i = 1 + (\tau_0 - 1)(e^{-\Delta z/We})^i. \qquad (13.27)$$

Expanding this for small Δz values gives

$$\tau_i = 1 + (\tau_0 - 1) \left(1 - \frac{\Delta z}{We} + \frac{\Delta z^2}{2\,We^2} \right)^i + \ldots \qquad (13.28)$$

Next we apply the Euler method to Eq. (13.23) by using a grid that is equally spaced:

$$We\,\frac{\tau_{i+1} - \tau_i}{\Delta z} + \tau_i = 1. \qquad (13.29)$$

This is a difference equation for τ_i. A particular solution to the nonhomogeneous equation is

$$\tau_i = 1. \qquad (13.30)$$

The homogeneous equation is Eq. (13.29) with 0 on the right-hand side. It is solved by assuming a form of the solution ϕ raised to the power i, or ϕ^i. This gives

$$\frac{We}{\Delta z}(\phi^{i+1} - \phi^i) + \phi^i = 0, \qquad (13.31)$$

which has the solution

$$\phi^{i+1} = \frac{(We/\Delta z) - 1}{We/\Delta z} \phi^i = \left(1 - \frac{\Delta z}{We}\right)\phi^i, \quad \phi = 1 - \frac{\Delta z}{We}. \tag{13.32}$$

The complete solution is then

$$\tau_i = 1 + A\phi^i. \tag{13.33}$$

The constant is found by applying the initial condition at $i = 0$, which gives

$$\tau_i = 1 + (\tau_0 - 1)\left(1 - \frac{\Delta z}{We}\right)^i. \tag{13.34}$$

Applying the backward Euler method gives

$$We \frac{\tau_{i+1} - \tau_i}{\Delta z} + \tau_{i+1} = 1 \tag{13.35}$$

and hence the solution

$$\tau_i = 1 + (\tau_0 - 1)\left(\frac{1}{1 + \Delta z/We}\right)^i. \tag{13.36}$$

A centered difference expression gives

$$We \frac{\tau_{i+1} - \tau_{i-1}}{2\Delta z} + \tau_i = 1 \tag{13.37}$$

and hence the solution

$$\phi_{1,2} = \frac{-1 \pm [1 + (We/\Delta z)^2]^{1/2}}{We/\Delta z},$$
$$\tau_i = 1 + A\phi_1^i + B\phi_2^i. \tag{13.38}$$

This equation requires two constants, A and B, only one of which is determined from the initial conditions. The exact form of the solution is deferred until the Galerkin method is applied as a boundary value method.

We are interested in whether or not these solutions oscillate from node to node as i is increased and whether or not the solutions grow without bound as i is increased. The solution using Euler's method, Eq. (13.34), will oscillate if Δz > We. The solution has a negative number raised to the power i. Thus the term oscillates in sign as i is an even or odd number. The backward Euler method does not lead to an oscillating solution because the term taken to the power i in Eq. (13.36) is always positive. The centered difference expressions gives a solution in which one of the roots ϕ is positive and one is negative. Thus, the complete solution oscillates as i changes. These solutions represent the type of solutions that we would obtain in more complicated cases if the convective term of the stress equation is evaluated in an explicit, implicit, or centered difference expression, respectively.

Next we solve Eq. (13.23) using the Galerkin method for an initial value problem. The weighted residual over one interval, Δz, is

$$\int_0^{\Delta z} N_j \left(We \frac{d\tau}{dz} + \tau - 1 \right) dz = 0. \tag{13.39}$$

The Galerkin expansion is

$$\tau = \sum \tau_i N_i(\xi) \quad \xi = \frac{2(z - z_i)}{\Delta z} - 1 \tag{13.40}$$

and the element-based equation is

$$\sum_e \sum_I \int_{-1}^{1} N_J \left(We \frac{dN_I}{d\xi} + \frac{\Delta z}{2} N_I \right) d\xi \, \tau_I^e = \sum_e \frac{\Delta z}{2} \int_{-1}^{1} N_J \, d\xi,$$

$$\sum_e \sum_I (We \, A_{JI}^e + C_{JI}^e) \, \tau_I^e = \frac{\Delta z}{2} \int_{-1}^{1} N_J \, d\xi. \tag{13.41}$$

If only one element is used with linear trial functions, we have

$$\sum_{I=1}^{2} \int_{-1}^{1} N_2 \left(We \frac{dN_I}{d\xi} + \frac{\Delta z}{2} N_I \right) d\xi \, \tau_I = \frac{\Delta z}{2} \int_{-1}^{1} N_2 \, d\xi,$$

$$\left(-\frac{1}{2} We + \frac{\Delta z}{2} \frac{1}{3} \right) \tau_1 + \left(\frac{1}{2} We + \frac{\Delta z}{2} \frac{2}{3} \right) \tau_2 = \frac{\Delta z}{2}. \tag{13.42}$$

Only the equation with N_2 as the weighting function is used because the value of the solution at the first node is known from the initial conditions. We then start with $\tau = 1$ at node 0, apply Eq. (13.42) to get τ at node 1, use that value as the initial condition in the next element and apply Eq. (13.42) again. We then proceed for increasing z values in an element-by-element fashion. The solution at the i-th node is then

$$\tau_i = 1 + (\tau_0 - 1) \left[\frac{1 - \frac{\Delta z}{3 \, We}}{1 + \frac{2 \Delta z}{3 \, We}} \right]^i. \tag{13.43}$$

This solution oscillates if $\Delta z > 3 \, We$.

When solving the polymer flow equations (for the full two-dimensional problem) the Galerkin method is not applied as an initial value method but as a boundary value method. If we solve Eq. (13.23) as a boundary value method, too, we get the following equations (derived from the matrices defined in Chapter 2):

$$\text{We} \begin{bmatrix} -\frac{1}{2} & \frac{1}{2} & \cdots & \cdots & \\ -\frac{1}{2} & 0 & \frac{1}{2} & \cdots & \\ 0 & -\frac{1}{2} & 0 & \frac{1}{2} & \cdots \\ \cdots & \cdots & \cdots & \cdots & \cdots \\ \cdots & \cdots & \cdots & \cdots & \cdot \end{bmatrix} \begin{bmatrix} \tau_0 \\ \tau_1 \\ \tau_2 \\ \cdot \\ \cdot \end{bmatrix} + \Delta z \begin{bmatrix} \frac{1}{3} & \frac{1}{6} & \cdots & \cdots & \\ \frac{1}{6} & \frac{2}{3} & \frac{1}{6} & \cdots & \\ \cdot & \frac{1}{6} & \frac{2}{3} & \frac{1}{6} & \cdots \\ \cdot & \cdot & \frac{1}{6} & \frac{2}{3} & \frac{1}{6} \\ \cdots & \cdots & \cdots & \cdots & \cdot \end{bmatrix} \begin{bmatrix} \tau_0 \\ \tau_1 \\ \tau_2 \\ \cdot \\ \cdot \end{bmatrix} = \Delta z \begin{bmatrix} \frac{1}{2} \\ 1 \\ 1 \\ \cdot \\ \cdot \end{bmatrix}. \quad (13.44)$$

The final equation in Eq. (13.44) is a partial equation because only one element is summed.

$$\left(\frac{1}{6} - \frac{We}{2\Delta z} \right) \tau_{N-1} + \left(\frac{1}{3} + \frac{We}{2\Delta z} \right) \tau_N = \frac{1}{2} \quad (13.45)$$

We replace the first row of the matrix with the boundary condition. Thus we solve

$$\tau_0 = \tau_0, \quad (13.46)$$

$$\tau_{i-1}\left(\frac{1}{6} - \frac{We}{2\Delta z} \right) + \frac{2}{3}\tau_i + \tau_{i+1}\left(\frac{1}{6} + \frac{We}{2\Delta z} \right) = 1, \quad i = 2, \ldots, N-1, \quad (13.47)$$

$$\left(\frac{1}{6} - \frac{We}{2\Delta z} \right) \tau_{N-1} + \left(\frac{1}{3} + \frac{We}{2\Delta z} \right) \tau_N = \frac{1}{2}. \quad (13.48)$$

In contrast with Eq. (13.42), we must solve these equations simultaneously since they are all coupled (see Problem 13.5). The solution is

$$\tau_i = 1 + A \phi_1^i + B \phi_2^i, \quad (13.49)$$

where

$$\phi_{1,2} = \frac{-2 \pm [\, 3 + 9\,(We/\Delta z)^2\,]^{1/2}}{1 + 3\,We/\Delta z}. \quad (13.50)$$

A and B are obtained from the initial condition and Eq. (13.48). The initial condition is

$$\tau_0 = 1 + A + B. \quad (13.51)$$

Eq. (13.49) substituted into Eq. (13.48) gives

$$A\left[\frac{1}{6}\phi_1^{N-1} + \frac{1}{3}\phi_1^N + \frac{We}{2\Delta z}(\phi_1^N - \phi_1^{N-1}) \right] + B\left[\frac{1}{6}\phi_2^{N-1} + \frac{1}{3}\phi_2^N + \frac{We}{2\Delta z}(\phi_2^N - \phi_2^{N-1}) \right] = 0. \quad (13.52)$$

As Δz approaches zero, we get

$$A\,\phi_1^{N-1}\left(-\frac{\Delta z}{We} \right) + B\,\phi_2^{N-1}\left(-2 + \frac{\Delta z}{3\,We} \right) = 0. \quad (13.53)$$

For $\Delta z = 0$ this gives

$$B \phi_2^{N-1}(-2) = 0 \text{ or } B = 0. \tag{13.54}$$

The complete solution is then

$$\tau_i = 1 + (\tau_0 - 1)\phi_1^i. \tag{13.55}$$

The unstable root, ϕ_2, drops out when Δz is small enough. As a result, the solution does not oscillate as long as $\Delta z < \text{We}$. Thus, the chief result of applying the Galerkin method over the entire domain, $0 \le z \le 4$, is that a smaller Δz value is required for no oscillations ($\Delta z < \text{We}$) than that for the Galerkin method applied as an initial value method ($\Delta z < 3\text{We}$).

If the mass is lumped in Eq. (13.47)-(13.48), we get

$$\tau_0 = \tau_0, \tag{13.56}$$

$$\tau_{i-1}\left(-\frac{\text{We}}{2\Delta z}\right) + \tau_i + \tau_{i+1}\left(+\frac{\text{We}}{2\Delta z}\right) = 1, \quad i = 2,\ldots,N-1, \tag{13.57}$$

$$\left(-\frac{\text{We}}{2\Delta z}\right)\tau_{N-1} + \left(\frac{1}{2} + \frac{\text{We}}{2\Delta z}\right)\tau_N = \frac{1}{2}. \tag{13.58}$$

Eq. (13.57) is the same as would be derived with the centered difference expression, so the roots ϕ_1 and ϕ_2 are the same as in Eq. (13.38). The general solution is also the same; again, the negative root drops out when Δz is small enough.

The Petrov-Galerkin method is sometimes used for viscoelastic equations; for this model problem, the Petrov-Galerkin method gives the following equations in place of Eq. (13.46)-(13.48):

$$\tau_0 = \tau_0,$$

$$\tau_{i-1}\left[\frac{1}{6} + \frac{\alpha}{4} - \frac{\text{We}}{2\Delta z}(1+\alpha)\right] + \left[\frac{2}{3} + \frac{\alpha\,\text{We}}{\Delta z}\right]\tau_i + \tau_{i+1}\left[\frac{1}{6} - \frac{\alpha}{4} + \frac{\text{We}}{2\Delta z}(1-\alpha)\right] = 1, \quad i=2,\ldots,N-1,$$

$$\left[\frac{1}{6} + \frac{\alpha}{4} - \frac{\text{We}}{2\Delta z}(1+\alpha)\right]\tau_{N-1} + \left[\frac{1}{3} + \frac{\alpha}{4} + \frac{\text{We}}{2\Delta z}(1+\alpha)\right]\tau_N = \frac{1}{2}(1+\alpha). \tag{13.59}$$

This gives the indicial equation

$$\left[\frac{1}{6} + \frac{\alpha}{4} - \frac{\text{We}}{2\Delta z}(1+\alpha)\right] + \left[\frac{2}{3} + \frac{\alpha\,\text{We}}{\Delta z}\right]\phi + \left[\frac{1}{6} - \frac{\alpha}{4} + \frac{\text{We}}{2\Delta z}(1-\alpha)\right]\phi^2 = 0. \tag{13.60}$$

As $\Delta z \to 0$ the solution to the indicial equation is

$$\phi_1 = 1, \quad \phi_2 = \frac{\alpha+1}{\alpha-1}. \tag{13.61}$$

Unfortunately, $\phi_1 = 1$ is the particular solution. Thus the particular solution must be generalized. The complete solution is

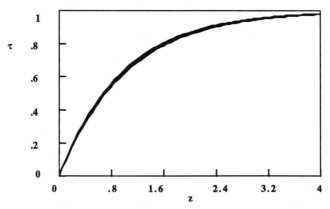

Figure 13.7a. Initial Value Problem Solved with Several Methods, We = 1, $\Delta z = 0.05$
Exact Solution, Euler, Backward Euler, Centered Difference, Galerkin,
Petrov-Galerkin Methods

$$\tau = \frac{\alpha - 1}{\alpha(\alpha+1)} i \phi_2^i + A + B \phi_2^i . \tag{13.62}$$

Substitution of Eq. (13.62) into the two boundary conditions gives

$$\tau_0 = A + B,$$
$$B = -\frac{N(\alpha-1)}{\alpha(\alpha+1)}. \tag{13.63}$$

When Δz is not small, the solution is still equal to Eq. (13.49), with ϕ_1 and ϕ_2 given as the roots of Eq. (13.60). The constants A and B satisfy

$$\tau_0 = 1 + A + B \tag{13.64}$$

and

$$A \phi_1^{N-1} \left\{ \frac{1}{6} + \frac{\alpha}{4} - \frac{We}{2\Delta z}(1+\alpha) + \left[\frac{1}{3} + \frac{\alpha}{4} + \frac{We}{2\Delta z}(1+\alpha)\phi_1\right] \right\} +$$
$$B \phi_2^{N-1} \left\{ \frac{1}{6} + \frac{\alpha}{4} - \frac{We}{2\Delta z}(1+\alpha) + \left[\frac{1}{3} + \frac{\alpha}{4} + \frac{We}{2\Delta z}(1+\alpha)\phi_2\right] \right\} = 0. \tag{13.65}$$

The solutions derived with the various methods discussed in this section are illustrated in Figure 13.7a. In this figure the Δz value is relatively small, 0.05, and 80 nodes are used between $z = 0$ and $z = 4$. All the solutions are good, with no oscillations visible. Naturally, the first-order solutions (Euler and backward Euler) are less accurate. When only 20 nodes are used and $\Delta z = 0.2$, the disparity between the solutions grows, as shown in Figure 13.7b, but the basic conclusions remain: all the second-order methods are good. When only 10 nodes are used for $\Delta z = 0.4$, the solutions are distinct on the scale of Figure 13.7c, but none of them oscillate. If Δz is increased to 2, the Euler method oscillates (not shown). The impact of solving with the Galerkin method as an initial value problem (IVP) rather than a boundary value problem (BVP) is illustrated in Figure 13.8a. The boundary value problem is much more accurate, which

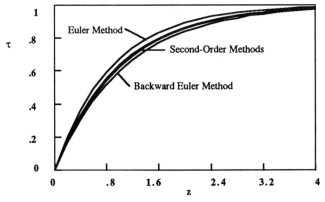

Figure 13.7b. Initial Value Problem Solved with Several Methods, We = 1, $\Delta z = 0.2$

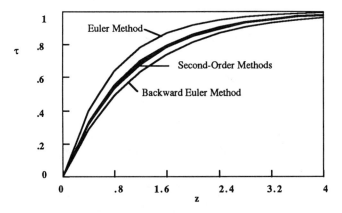

Figure 13.7c. Initial Value Problem Solved with Several Methods, We = 1, $\Delta z = 0.4$

is shown clearly in Table 13.1. The impact of the Petrov-Galerkin parameter is illustrated in Figure 13.8b and Table 13.1; as α increases, the solution becomes less accurate but it is still extremely accurate. These examples suggest that the additions of the Petrov-Galerkin method are not necessary since the Galerkin method is already so accurate.

Table 13.1. Solution to Eq. (13.24) at z = 1, $\Delta z = 0.2$, and We = 1

Method	Value	Error
Exact	0.6321205	0
Galerkin, IVP	0.6212125	0.011
Galerkin, BVP	0.6321521	0.000032
Petrov-Galerkin, $\alpha = 0.0$	0.6321521	0.000032
Petrov-Galerkin, $\alpha = 0.25$	0.6321535	0.000033
Petrov-Galerkin, $\alpha = 0.5$	0.6321816	0.000061
Petrov-Galerkin, $\alpha = 1.0$	0.632237	0.00012

470 NUMERICAL METHODS - MOVING FRONTS

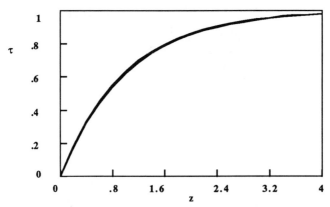

Figure 13.8a. Initial Value Problem Solved With Boundary Value Methods Exact Solution, Galerkin IVP (Eq. 13.43), Galerkin BVP (Eq. 13.44), We = 1, $\Delta z = 0.2$

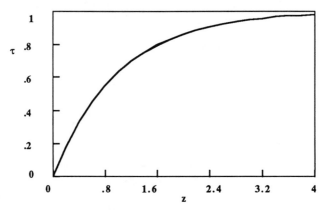

Figure 13.8b. Initial Value Problem Solved With Boundary Value Methods Exact Solution, Petrov-Galerkin Method, $\alpha = 0, 0.25, 0.5, 1$, We = 1, $\Delta z = 0.2$

13.3. Model Problem Two: Stress Along a Centerline

The second model problem solves for the normal stress along the centerline when the velocity is known. The equation is [Eq. (13.22)]

$$\left[1 - 2\,\text{We}\,\frac{du}{dz} \right] \tau + \text{We}\,u\,\frac{d\tau}{dz} = -2\,\frac{du}{dz}. \tag{13.66}$$

The velocity profile is an approximate one; it is defined in Eq. (13.67) and plotted in Figure 13.9.

$$u = \begin{cases} 2 & z \leq -1 \\ 2 + 15(z+1)^2 & -1 < z \leq 0 \\ 17 + 30z - 15z^2 & 0 < z < 1 \\ 32 & z \geq 1 \end{cases} \tag{13.67}$$

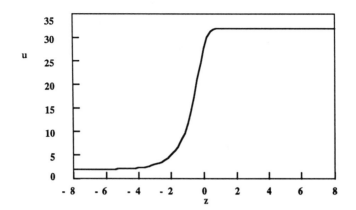

Figure 13.9. Velocity Profile, Eq. (13.67)

The exact solution for stress (obtained with a Runge-Kutta method for initial value problems and variable step-sizes) is shown in Figure 13.10 for Weissenberg numbers of 0.01, 0.02, 0.04, 0.1, 0.2, and 0.4. (For those readers familiar with the polymer flow literature, these Weissenberg numbers correspond to 0.64, 1.28, 2.56, 6.4, 12.8, and 25.6 in terms of the shear rate at the exit of the 4:1 axisymmetric contraction shown in Figure 13.6.) The peak values of stress are ordered according to We = 0.01, 0.02, 0.04, 0.4, 0.2, and 0.1. The peak values do not occur for the largest Weissenberg number; in these cases, the fluid does not have time to respond to the increased velocity profile before it leaves the region of large gradient.

The Galerkin finite element method (as a boundary value problem) gives the equations

$$\sum_{i=1}^{N} \int_0^z N_j \left[1 - 2We \frac{du}{dz} \right] N_i \, dz \, \tau_i + We \sum_{i=1}^{N} \int_0^z N_j u \frac{dN_i}{dz} \, dz \, \tau_i = -2 \sum_{i=1}^{N} \int_0^z N_j \frac{du}{dz} \, dz. \quad (13.68)$$

The element form of this equation is

$$\sum_e \sum_I \frac{\Delta z}{2} \int_{-1}^{1} N_J \left[1 - 2 \frac{2We}{\Delta z} \frac{du}{d\xi} \right] N_I \, d\xi \, \tau_I^e + We \sum_e \sum_I \int_{-1}^{1} N_J u \frac{dN_I}{d\xi} \, d\xi \, \tau_I^e = -2 \sum_e \sum_I \int_{-1}^{1} N_J \frac{du}{d\xi} \, d\xi. \quad (13.69)$$

Within an element, the velocity gradient is a linear function (since the velocity is interpolated with quadratic functions), so the element integrals are

$$\int_{-1}^{1} N_J \left[1 - 2We \frac{2}{\Delta z} \frac{du}{d\xi} \right] N_I \, d\xi = \frac{1}{3} \begin{bmatrix} 2a - b & a \\ a & 2a - b \end{bmatrix},$$

$$\text{with} \left[1 - 2We \frac{2}{\Delta z} \frac{du}{d\xi} \right] = a + b\xi \text{ within an element} \quad (13.70)$$

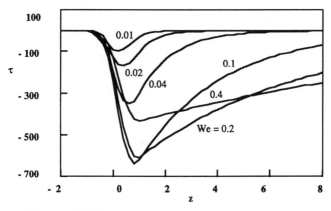

Figure 13.10. Stress Solution for a Simple Velocity Profile

and

$$\int_{-1}^{1} N_J \, u \, \frac{dN_I}{d\xi} \, d\xi = \frac{1}{2} \begin{bmatrix} -a + \frac{b}{3} - \frac{c}{3} & a - \frac{b}{3} + \frac{c}{3} \\ -a - \frac{b}{3} - \frac{c}{3} & a + \frac{b}{3} + \frac{c}{3} \end{bmatrix} \quad (13.71)$$

with $u = a + b\xi + c\xi^2$ within an element

and

$$\int_{-1}^{1} N_J \, \frac{du}{d\xi} \, d\xi = \begin{bmatrix} b - \frac{2c}{3} \\ b + \frac{2c}{3} \end{bmatrix} \quad (13.72)$$

with $\frac{du}{d\xi} = b + 2c\xi$ within an element.

For the first node we have the boundary condition

$$\tau(z=0) = \tau_0. \quad (13.73)$$

These element equations [Eq. (13.69)-(13.72)] must be assembled into matrix form and the resulting problem solved. The solutions for We = 0.02 and Δz = 25We, 12.5We, and 6.25We are shown in Figure 13.11a. The only oscillations present are those near the start of the steep drop, despite the large value of Δz/We. The peak value of stress also reproduces reasonably well, taking values from -164 to -169.

When the Petrov-Galerkin method is used, the element equations become

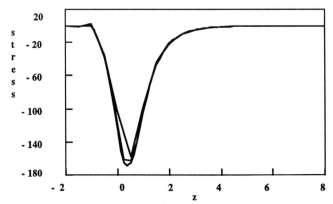

Figure 13.11a. Solution of Eq. (13.66), Galerkin Method, Eq. (13.69), $\Delta z = 0.1, 0.2, 0.4$

$$\sum_e \sum_I \frac{\Delta z}{2} \int_{-1}^{1} W_J \left[1 - 2 \frac{2We}{\Delta z} \frac{du}{d\xi} \right] N_I \, d\xi \, \tau_I^e + We \sum_e \sum_I \int_{-1}^{1} W_J u \frac{dN_I}{d\xi} d\xi \, \tau_I^e =$$

$$= -2 \sum_e \sum_I \int_{-1}^{1} W_J \frac{du}{d\xi} d\xi. \tag{13.74}$$

The element integrals are

$$\int_{-1}^{1} W_J \left[1 - 2 We \frac{2}{\Delta z} \frac{du}{d\xi} \right] N_I \, d\xi = \frac{1}{3} \begin{bmatrix} 2a - b & a \\ a & 2a - b \end{bmatrix} + \alpha \begin{bmatrix} -\frac{a}{2} + \frac{b}{10} & -\frac{a}{2} - \frac{b}{10} \\ \frac{a}{2} - \frac{b}{10} & \frac{a}{2} + \frac{b}{10} \end{bmatrix} \tag{13.75}$$

with $\left[1 - 2 We \frac{2}{\Delta z} \frac{du}{d\xi} \right] = a + b \xi$ within an element

$$\int_{-1}^{1} W_J u \frac{dN_I}{d\xi} d\xi = \frac{1}{2} \begin{bmatrix} -a + \frac{b}{3} - \frac{c}{3} & a - \frac{b}{3} + \frac{c}{3} \\ -a - \frac{b}{3} - \frac{c}{3} & a + \frac{b}{3} + \frac{c}{3} \end{bmatrix} + \frac{\alpha}{2} \begin{bmatrix} a + \frac{c}{5} & -a - \frac{c}{5} \\ -a - \frac{c}{5} & a + \frac{c}{5} \end{bmatrix} \tag{13.76}$$

with $u = a + b \xi + c \xi^2$ within an element

and

$$\int_{-1}^{1} W_J \frac{du}{d\xi} d\xi = \begin{bmatrix} b - \frac{2c}{3} \\ b + \frac{2c}{3} \end{bmatrix} + \alpha \begin{bmatrix} -b \\ b \end{bmatrix} \tag{13.77}$$

with $\frac{du}{d\xi} = b + 2c \xi$ within an element.

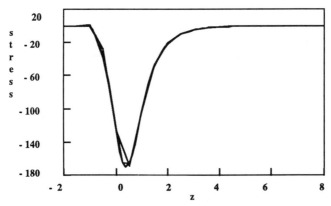

Figure 13.11b. Solution of Eq. (13.66), Petrov-Galerkin Method, $\alpha = 1$, Eq. (13.74) $\Delta z = 0.1, 0.2, 0.4$

Solutions for the Petrov-Galerkin method are shown in Figure 13.11b for We = 0.02; they oscillate approximately as much as the results from the Galerkin method; the Petrov-Galerkin method also reproduces the peak value of stress. Thus, the Petrov-Galerkin method and the Galerkin method give similar results; the extra work necessary for the Petrov-Galerkin method is not worthwhile.

13.4. Model Problem Three: Unsteady Couette Flow

The third model problem considers the case illustrated in Figure 13.1, but in a transient situation. The fluid is initially at rest; in this case we apply a force on the bottom plate while holding the top plate fixed in position, as illustrated in Figure 13.12.

Figure 13.12. Sudden Movement of the Bottom Plate

The goal is to predict the velocity and stress distribution as a function of time throughout the domain. If we assume that the Couette flow occurs only in the x-direction and that all variables only depend on the y-variable, then the momentum equation becomes

$$\rho \frac{\partial u}{\partial t} = -\frac{\partial \tau_{xy}}{\partial y}, \qquad (13.78)$$

while the constitutive equations are

$$\tau_{xy} + \lambda \left\{ \frac{\partial \tau_{xy}}{\partial t} \right\} = -\mu \frac{\partial u}{\partial y}, \tag{13.79}$$

$$\tau_{xx} + \lambda \left\{ \frac{\partial \tau_{xx}}{\partial t} \right\} = 2\lambda \tau_{xy} \frac{\partial u}{\partial y}, \tag{13.80}$$

$$\tau_{yy} + \lambda \frac{\partial \tau_{yy}}{\partial t} = 0. \tag{13.81}$$

The τ_{yy} stress is zero. The equations for the velocity u and the τ_{xy} stress must be solved together; the τ_{xx} equation can then be solved, if desired.

Each variable is made non-dimensional by using some standard measurement of that variable:

$$u' = \frac{u}{u_s}, \quad \tau'_{xy} = \frac{\tau_{xy}}{\tau_s}, \quad \tau'_{xx} = \frac{\tau_{xx}}{\tau_s}, \quad y' = \frac{y}{x_s}, \quad t' = \frac{t}{t_s}. \tag{13.82}$$

We insert these variables into Eq. (13.79)-(13.81) and define the following standards:

$$\tau_s = \frac{\mu u_s}{x_s}, \quad t_s = \frac{\rho u_s x_s}{\tau_s} = \frac{\rho x_s^2}{\mu}. \tag{13.83}$$

After dropping the primes and subscripts on the stress variable, we obtain (see Problem 13.9)

$$\frac{\partial u}{\partial t} = -\frac{\partial \tau}{\partial y}, \tag{13.84}$$

$$\tau + E \frac{\partial \tau}{\partial t} = -\frac{\partial u}{\partial y}, \tag{13.85}$$

$$\tau_{xx} + E \frac{\partial \tau_{xx}}{\partial t} = 2 \operatorname{We} \tau \frac{\partial u}{\partial y}. \tag{13.86}$$

The exact solution to this problem was derived by Weinberger [1965], while its implications for polymer flow were described by Denn and Porteus [1971]. The parameter E that appears is related to the Weissenberg number and Reynolds number.

$$\frac{\lambda \tau_s}{\mu} = \frac{\lambda u_s}{x_s} = \text{We}, \quad \text{Re} = \frac{\rho u_s x_s}{\mu}, \quad \frac{\lambda \tau_s x_s}{t_s \mu u_s} = \frac{\lambda \mu}{\rho x_s^2} \equiv E = \frac{\text{We}}{\text{Re}}. \tag{13.87}$$

The equation for τ_{xx} can be solved once the solution for u and τ_{xy} are known.

One way to solve Eq. (13.85) involves combining the velocity and stress equations. We differentiate Eq. (13.84) with respect to time and substitute in Eq. (13.85), giving

$$\frac{\partial^2 u}{\partial t^2} = -\frac{\partial^2 \tau}{\partial t \partial y} = -\frac{\partial}{\partial y}\left(\frac{\partial \tau}{\partial t}\right) = \frac{1}{E}\frac{\partial}{\partial y}\left(\frac{\partial u}{\partial y}+\tau\right). \tag{13.88}$$

But this is

$$\frac{\partial^2 u}{\partial t^2} = \frac{1}{E}\frac{\partial^2 u}{\partial y^2} + \frac{1}{E}\frac{\partial \tau}{\partial y} \tag{13.89}$$

and with the first equation substituted in, it becomes

$$\frac{\partial^2 u}{\partial t^2} = \frac{1}{E}\frac{\partial^2 u}{\partial y^2} - \frac{1}{E}\frac{\partial u}{\partial t}. \tag{13.90}$$

Thus we can write the problem as

$$E\frac{\partial^2 u}{\partial t^2} + \frac{\partial u}{\partial t} = \frac{\partial^2 u}{\partial y^2},$$
$$u(0,t) = 1, \ u(1,t) = 0, \ u(y,0) = 0. \tag{13.91}$$

This is known as the telegrapher's equation [Weinberger, 1965]. For this application, discontinuities propagate with the velocity $1/E^{0.5}$. It thus takes the dimensionless time $E^{0.5}$ to traverse the distance between the solid plates shown in Figure 13.12. In dimensional terms, this is

$$\frac{\text{time}}{t_s} = E^{0.5},$$

$$\text{time} = \frac{\rho x_s^2}{\mu}\left(\frac{\lambda \mu}{\rho x_s^2}\right)^{0.5} = \left(\frac{\rho x_s^2 \lambda}{\mu}\right)^{0.5}. \tag{13.92}$$

Typical values for a polymer melt are $x_s = 1$ mm $= 10^{-3}$ m, $\lambda = 1$ s, $\rho = 1000$ kg/m^3, and $\mu = 10^4$ Pa s; the characteristic time is then 3.2×10^{-4} seconds. If the characteristic velocity is 1 cm/sec $= 0.01$ m/s, we get

$$\text{We} = 10, \ \text{Re} = 10^{-6}, \ E = 10^7. \tag{13.93}$$

For a polymer solution, typical values are $x_s = 1$ cm $= 10^{-2}$ m, $\lambda = 10$ s, $\rho = 1000$ kg/m^3, and $\mu = 1$ Pa s; the characteristic time is then 1 second. If the characteristic velocity is 10 cm/sec $= 0.1$ m/s, we get

$$\text{We} = 100, \ \text{Re} = 1, \ E = 100. \tag{13.94}$$

The problem is well posed by Eq. (13.91). It is instructive to try to solve it without combining equations to form one equation. Let us consider the problem

$$\frac{\partial u}{\partial t} = -\frac{\partial \tau}{\partial y}, \tag{13.95}$$

$$E\frac{\partial \tau}{\partial t} = -\frac{\partial u}{\partial y} - \tau. \tag{13.96}$$

This is similar to two advection equations with a source term and fits the pattern of other problems in this book. For us to apply the problem as two advection equations, however, we need a boundary condition on both u and τ at y = 0. The problem is posed with boundary conditions on u at y = 0 and y = 1; in fact, the velocity is maintained at y = 0 by applying whatever force is needed to maintain the velocity; this force is proportional to τ. We cannot specify both u and τ at y = 0. Consequently, we cannot solve the equations in this way. However, we can still apply the basic ideas, as described below.

To analyze Eq. (13.95)-(13.96), we follow the treatment of Vichnevetsky and Bowles [§1.2, 1982]. We write the coupled set of equations in the form

$$\frac{\partial U}{\partial t} + F\frac{\partial U}{\partial x} = G. \tag{13.97}$$

In this case we have

$$F = \begin{bmatrix} 0 & 1 \\ \frac{1}{E} & 0 \end{bmatrix}, \quad G = \begin{bmatrix} 0 \\ -\frac{\tau}{E} \end{bmatrix}. \tag{13.98}$$

For the system to be hyperbolic, the matrix F must have real eigenvalues. Here they are

$$\det\begin{bmatrix} -\lambda & 1 \\ \frac{1}{E} & -\lambda \end{bmatrix} = 0, \quad \lambda^2 - \frac{1}{E} = 0, \quad \lambda = \pm\sqrt{\frac{1}{E}}. \tag{13.99}$$

The eigenvectors, W_k, satisfy

$$W_k^T F = W_k^T \lambda_k, \quad \lambda_1 = +\sqrt{\frac{1}{E}}, \quad \lambda_2 = -\sqrt{\frac{1}{E}}. \tag{13.100}$$

For k = 1, we have

$$[\, u_1 \;\; \tau_1 \,] \begin{bmatrix} 0 & 1 \\ \frac{1}{E} & 0 \end{bmatrix} = \frac{1}{\sqrt{E}} [\, u_1 \;\; \tau_1 \,]. \tag{13.101}$$

This gives

$$\frac{\tau_1}{E} = \frac{1}{\sqrt{E}} u_1,$$

$$u_1 = \frac{1}{\sqrt{E}} \tau_1. \tag{13.102}$$

If we normalize the eigenfunctions so that

$$u_1^2 + \tau_1^2 = 1, \qquad (13.103)$$

we get

$$u_1 = \sqrt{\frac{1}{1+E}}, \quad \tau_1 = \sqrt{\frac{E}{1+E}}, \quad W_1^T = (u_1, \tau_1). \qquad (13.104)$$

Similarly, for $k = 2$ we have

$$u_2 = -\sqrt{\frac{1}{1+E}}, \quad \tau_2 = \sqrt{\frac{E}{1+E}}, \quad W_2^T = (u_2, \tau_2). \qquad (13.105)$$

When the eigenvectors satisfy Eq. (13.100), we can calculate

$$\begin{aligned} W_k^T \left(\frac{\partial U}{\partial t} + F \frac{\partial U}{\partial x} \right) &= W_k^T \frac{\partial U}{\partial t} + W_k^T F \frac{\partial U}{\partial x} \\ &= W_k^T \frac{\partial U}{\partial t} + W_k^T \lambda_k \frac{\partial U}{\partial x} \\ &= W_k^T \left(\frac{\partial}{\partial t} + \lambda_k \frac{\partial}{\partial x} \right) U \\ &= W_k^T D_k U. \end{aligned} \qquad (13.106)$$

Thus

$$\frac{dt}{dx} = \pm \sqrt{E} \qquad (13.107)$$

are the characteristic directions and

$$\frac{dx}{dt} = \pm \sqrt{\frac{1}{E}} \qquad (13.108)$$

is the characteristic velocity. Now we multiply Eq. (13.97) by W^T_k and use Eq. (13.106) to get a diagonal system:

$$W_k^T D_k U = W_k^T G. \qquad (13.109)$$

We define

$$V_k = W_k^T U \qquad (13.110)$$

and the equation is

$$W_k^T D_k U = D_k W_k^T U = \frac{\partial V_k}{\partial t} + \lambda_k \frac{\partial V_k}{\partial x} = W_k^T G = H_k. \qquad (13.111)$$

Now V_k is a scalar. The right-hand side is evaluated to give

$$\mathbf{W}_1^T \mathbf{G} = \left(\sqrt{\frac{1}{1+E}}, \sqrt{\frac{E}{1+E}}\right)\begin{pmatrix} 0 \\ -\frac{\tau}{E} \end{pmatrix} = -\tau\sqrt{\frac{1}{(1+E)E}},$$

$$\mathbf{W}_2^T \mathbf{G} = \left(-\sqrt{\frac{1}{1+E}}, \sqrt{\frac{E}{1+E}}\right)\begin{pmatrix} 0 \\ -\frac{\tau}{E} \end{pmatrix} = -\tau\sqrt{\frac{1}{(1+E)E}}. \quad (13.112)$$

Thus $H_1 = H_2$. Also

$$V_1 = \left(\sqrt{\frac{1}{1+E}}, \sqrt{\frac{E}{1+E}}\right)\begin{pmatrix} u \\ \tau \end{pmatrix} = \sqrt{\frac{1}{1+E}}\, u + \sqrt{\frac{E}{1+E}}\, \tau,$$

$$V_2 = \left(-\sqrt{\frac{1}{1+E}}, \sqrt{\frac{E}{1+E}}\right)\begin{pmatrix} u \\ \tau \end{pmatrix} = -\sqrt{\frac{1}{1+E}}\, u + \sqrt{\frac{E}{1+E}}\, \tau. \quad (13.113)$$

This gives

$$V_1 + V_2 = 2\sqrt{\frac{1}{1+E}}\, \tau, \quad (13.114)$$

so the eigenvector equations are

$$\frac{\partial V_1}{\partial t} + \frac{1}{\sqrt{E}}\frac{\partial V_1}{\partial x} = -\frac{1}{2\sqrt{E}}(V_1 + V_2),$$

$$\frac{\partial V_2}{\partial t} - \frac{1}{\sqrt{E}}\frac{\partial V_2}{\partial x} = -\frac{1}{2\sqrt{E}}(V_1 + V_2). \quad (13.115)$$

These equations are semi-linear but clearly display the convective nature of the problem. One set of characteristics has the velocity $1/E^{0.5}$ while the other has the velocity $-1/E^{0.5}$. We could solve Eq. (13.115) in place of Eq. (13.95)-(13.96), if desired. The stability results for both sets will be the same, since they are related by a linear transformation.

We prefer to solve the original equations, Eq. (13.95)-(13.96), but in a more symmetric form. We replace τ by $\tau'/E^{0.5}$. After rearrangement we get

$$\frac{\partial u}{\partial t} = -\frac{1}{\sqrt{E}}\frac{\partial \tau'}{\partial y},$$

$$\frac{\partial \tau'}{\partial t} = -\frac{1}{\sqrt{E}}\frac{\partial u}{\partial y} - \frac{\tau'}{E}. \quad (13.116)$$

In this form we can easily remember that the characteristic velocity is $\pm 1/E^{0.5}$. We drop the primes but use the equations in the form of Eq. (13.116). We study the stability of these equations by employing a Fourier analysis and looking at the amplification matrix.

Let us first consider using centered finite difference expressions for the derivatives. Then the difference equations are

$$\frac{u_i^{n+1} - u_i^n}{\Delta t} = -\frac{1}{\sqrt{E}} \frac{\tau_{i+1}^n - \tau_{i-1}^n}{2\Delta y},$$

$$\frac{\tau_i^{n+1} - \tau_i^n}{\Delta t} = -\frac{1}{\sqrt{E}} \frac{u_{i+1}^n - u_{i-1}^n}{\Delta y} - \frac{\tau_i^n}{E}. \tag{13.117}$$

We define the Courant number as

$$\text{Co} = \frac{\Delta t}{\sqrt{E}\,\Delta y}. \tag{13.118}$$

We then take the Fourier transform (see Table 4.1):

$$\hat{u}^{n+1} = \hat{u}^n + \text{Co}\, i \sin \xi\, \hat{\tau}^n,$$

$$\hat{\tau}^{n+1} = \hat{\tau}^n + \text{Co}\, i \sin \xi\, \hat{u}^n - \frac{\Delta t}{E} \hat{\tau}^n. \tag{13.119}$$

When rearranged, this is

$$\begin{bmatrix} \hat{u}^{n+1} \\ \hat{\tau}^{n+1} \end{bmatrix} = \begin{bmatrix} 1 & \text{Co}\, i \sin \xi \\ \text{Co}\, i \sin \xi & 1 - \frac{\Delta t}{E} \end{bmatrix} \begin{bmatrix} \hat{u}^n \\ \hat{\tau}^n \end{bmatrix}. \tag{13.120}$$

Previously, the Fourier analysis was applied to a single equation. Now we have multiple equations. The proper way to apply the analysis is given by Richtmyer and Morton [1967]. Eq. (13.120) is written with the amplification matrix, G, as

$$\begin{bmatrix} \hat{u}^{n+1} \\ \hat{\tau}^{n+1} \end{bmatrix} = [G] \begin{bmatrix} \hat{u}^n \\ \hat{\tau}^n \end{bmatrix}. \tag{13.121}$$

The stability of Eq. (13.121) depends on the eigenvalues of the amplification matrix. The result is called the von Neumann necessary condition for stability. It is [Richtmyer and Morton, p. 70, 1967]

$$|\lambda_i| \leq 1$$

or $\qquad$ for all $0 \leq \xi \leq 2\pi$, $i = 1, 2, \ldots$ $\qquad$ (13.122)

$$|\lambda_i| \leq 1 + O(\Delta t)$$

for all values of ξ between 0 and 2π. The term $O(\Delta t)$ allows for a component of the exact solution that grows exponentially with time. If the exact solution does not do this, then the upper bound is simply 1.0.

Here the eigenvalues are given as the solution to

$$\det \begin{bmatrix} 1-\lambda & \text{Co}\, i \sin \xi \\ \text{Co}\, i \sin \xi & 1 - \frac{\Delta t}{E} - \lambda \end{bmatrix} = 0. \qquad (13.123)$$

The eigenvalues depend on the sign of the discriminant.

$$\text{Discriminant} = \left(\frac{\Delta t}{E}\right)^2 - 4\,\text{Co}^2 \sin^2 \xi \qquad (13.124)$$

If the discriminant is positive, the eigenvalues are

$$|\lambda| = 1 - \frac{\Delta t}{2E} \pm \frac{1}{2}\sqrt{\left(\frac{\Delta t}{E}\right)^2 - 4\,\text{Co}^2 \sin^2 \xi}. \qquad (13.125)$$

If the discriminant is negative, we define

$$c^2 = -\frac{1}{4}\left[\left(\frac{\Delta t}{E}\right)^2 - 4\,\text{Co}^2 \sin^2 \xi\right] \qquad (13.126)$$

and the eigenvalues are

$$|\lambda| = \sqrt{\left(1 - \frac{\Delta t}{2E}\right)^2 + c^2}. \qquad (13.127)$$

The two parameters appearing in the analysis are

$$\frac{\Delta t}{E} \text{ and } \text{Co} = \frac{\Delta t}{\sqrt{E}\,\Delta y}. \qquad (13.128)$$

We note that Eq. (13.116) has both convection terms and source terms. The first parameter comes from the source term and the second one comes from the convection term. The role of the Fourier analysis is to provide stability information when both parameters are relevant. Figure 13.13 shows the magnitude of the eigenvalue as a function of ξ when $\Delta t/E = 0.01$. One of the eigenvalues is close to 0 and one is close to 1.0. For Co > 0.1, one of the eigenvalues is greater than 1.0, so the method is unstable. Figure 13.14 is for $\Delta t/E = 0.1$ and shows the case where Co = 0.4 is unstable; other calculations show that Co = 0.316 is stable while Co = 0.32 is unstable. The unusual patterns come about because for some values of ξ, the eigenvalues are real, while for other values of ξ, the eigenvalues are complex and there is only one magnitude. Figure 13.15 is for $\Delta t/E = 1$; all three Courant numbers are stable. With $\Delta t/E = 2$, Figure 13.16 shows that Courant numbers up to 1.0 are stable; other calculations show that Co = 1.4 is stable but Co = 1.5 is unstable. In fact, the dividing point is 1.414. In all cases, the stability limit is

$$\text{Co} \leq \sqrt{\frac{\Delta t}{E}}. \qquad (13.129)$$

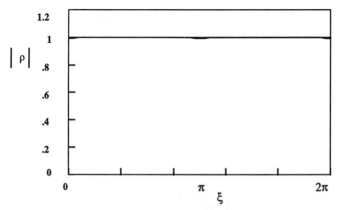

Figure 13.13. Magnitude of Eigenvalue, $\Delta t / E = 0.01$, Co = 0.05, 0.1, 0.11 (unstable)

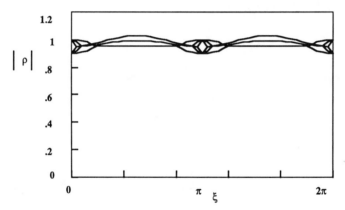

Figure 13.14. Magnitude of Eigenvalue, $\Delta t / E = 0.1$, Co = 0.1, 0.3, 0.4 (unstable)

When substituting in the definition for Co, Eq. (13.129) becomes

$$\Delta t \leq \Delta y^2. \tag{13.130}$$

Eq. (13.130) looks very much like a diffusion time-step limitation. We also have the condition

$$\frac{\Delta t}{E} \leq 2 \tag{13.131}$$

since if this condition is violated the eigenvalues become too negative (less than −1.0). Interestingly, we have a diffusion-type limitation, Eq. (13.130), even though we have a convection-type problem. This presumably is because the convection goes in both directions, so there is no preferred upstream direction.

To show that the form of the equations makes no difference to the stability analysis, let us consider the centered finite difference method applied to Eq.

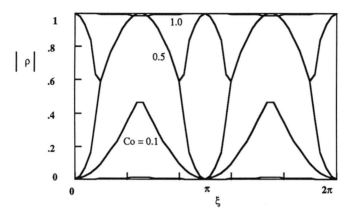

Figure 13.15. Magnitude of Eigenvalue, $\Delta t / E = 1$, Co =0.1, 0.5, 1.0

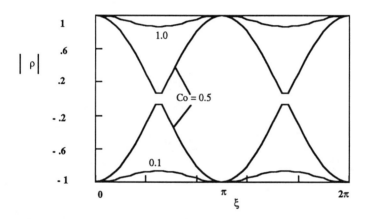

Figure 13.16. Magnitude of Eigenvalue, $\Delta t / E = 2$, Co = 0.1, 0.5, 1.0

(13.95)-(13.96):

$$\frac{u_i^{n+1} - u_i^n}{\Delta t} = -\frac{\tau_{i+1}^n - \tau_{i-1}^n}{2\Delta y},$$

$$\frac{\tau_i^{n+1} - \tau_i^n}{\Delta t} = -\frac{1}{E}\frac{u_{i+1}^n - u_{i-1}^n}{\Delta y} - \frac{\tau_i^n}{E}. \tag{13.132}$$

The Fourier analysis gives

$$\det \begin{bmatrix} 1 - \lambda & \frac{\Delta t}{\Delta y} i \sin \xi \\ \frac{\Delta t}{E \Delta y} i \sin \xi & 1 - \frac{\Delta t}{E} - \lambda \end{bmatrix} = 0, \tag{13.133}$$

which gives the same characteristic equation as before, Eq. (13.123).

Another method of solving this problem is to define a new variable, $v = \partial u/\partial t$, and solve for u and v. This is outlined in the electronic text "u-v formulation."

Upstream method. Next we develop the methods for the standard equations in the form of Eq. (13.95)-(13.96). When the derivatives are replaced with upstream derivatives we get

$$\frac{u_i^{n+1} - u_i^n}{\Delta t} = -\frac{\tau_i^n - \tau_{i-1}^n}{\Delta y},$$

$$E \frac{\tau_i^{n+1} - \tau_i^n}{\Delta t} = -\frac{u_i^n - u_{i-1}^n}{\Delta y} - \tau_i^n. \qquad i = 2,\ldots, N-1 \qquad (13.134)$$

The equations for u_1 and u_N come from the boundary conditions

$$\begin{aligned} u(0,t) = 1 \quad &\text{or} \quad u_1(t) = 1, \\ u(1,t) = 0 \quad &\text{or} \quad u_N(t) = 0. \end{aligned} \qquad (13.135)$$

We also need equations for τ_1 and τ_N. These are obtained from the original equations [Eq. (13.95)-(13.96)] using one-sided derivatives, giving

$$E \frac{\tau_1^{n+1} - \tau_1^n}{\Delta t} = -\frac{-3u_1^n + 4u_2^n - u_3^n}{2\Delta y} - \tau_1^n,$$

$$E \frac{\tau_N^{n+1} - \tau_N^n}{\Delta t} = -\frac{u_{N-2}^n - 4u_{N-1}^n + 3u_N^n}{2\Delta y} - \tau_N^n. \qquad (13.136)$$

Unfortunately, the upstream method is unstable (see Problem 13.12), because there is actually convection in both directions; the upstream method is unstable if the flow is in the wrong direction.

The **MacCormack method** uses the predictor equations,

$$\frac{u*_i^{n+1} - u_i^n}{\Delta t} = -\frac{\tau_{i+1}^n - \tau_i^n}{\Delta y},$$

$$E \frac{\tau*_i^{n+1} - \tau_i^n}{\Delta t} = -\frac{u_{i+1}^n - u_i^n}{\Delta y} - \tau_i^n, \qquad (13.137)$$

followed by the corrector equations

$$\frac{u_i^{n+1} - \frac{1}{2}(u*_i^{n+1} + u_i^n)}{\Delta t/2} = -\frac{\tau*_i^{n+1} - \tau*_{i-1}^{n+1}}{\Delta y},$$

$$E \frac{\tau_i^{n+1} - \frac{1}{2}(\tau*_i^{n+1} + \tau_i^n)}{\Delta t/2} = -\frac{u*_i^{n+1} - u*_{i-1}^{n+1}}{\Delta y} - \tau*_i^{n+1}. \qquad (13.138)$$

The source term is handled as source terms were handled in Ch. 10 for a predictor-corrector method. We also need equations for τ_1 and τ_N, which we take as

$$E \frac{\tau_1^{*n+1} - \tau_1^n}{\Delta t} = -\frac{-3u_1^n + 4u_2^n - u_3^n}{2\Delta y} - \tau_1^n$$

$$E \frac{\tau_N^{*n+1} - \tau_N^n}{\Delta t} = -\frac{u_{N-2}^n - 4u_{N-1}^n + 3u_N^n}{2\Delta y} - \tau_N^n \qquad (13.139)$$

for the predictor and

$$E \frac{\tau_1^{n+1} - \frac{1}{2}(\tau_1^{*n+1} + \tau_1^n)}{\Delta t / 2} = -\frac{3u_1^{*n+1} + 4u_2^{*n+1} - u_3^{*n+1}}{2\Delta y} - \tau_1^{*n+1}$$

$$E \frac{\tau_N^{n+1} - \frac{1}{2}(\tau_N^{*n+1} + \tau_N^n)}{\Delta t / 2} = -\frac{u_{N-2}^{*n+1} - 4u_{N-1}^{*n+1} + u_N^{*n+1}}{2\Delta y} - \tau_N^{*n+1} \qquad (13.140)$$

for the corrector. Unfortunately, the MacCormack method is also unstable (see Problem 13.13).

Derivation of the Taylor form of the equations. We first derive the following equation using a Taylor expansion:

$$\frac{\partial u}{\partial t} = \frac{u^{n+1} - u^n}{\Delta t} - \frac{\Delta t}{2} \frac{\partial^2 u}{\partial t^2}. \qquad (13.141)$$

This equation is substituted into the differential equation, Eq. (13.95), to obtain

$$\frac{u^{n+1} - u^n}{\Delta t} = -\frac{\partial \tau}{\partial y} + \frac{\Delta t}{2} \frac{\partial^2 u}{\partial t^2}. \qquad (13.142)$$

The Taylor form of the stress equation is derived in a similar manner, giving

$$\frac{\tau^{n+1} - \tau^n}{\Delta t} = -\frac{1}{E} \frac{\partial u}{\partial y} - \frac{\tau^n}{E} + \frac{\Delta t}{2} \frac{\partial^2 \tau}{\partial t^2}. \qquad (13.143)$$

Next we evaluate the second time derivatives. We differentiate the first equation [Eq. (13.95)] once, exchange the order of differentiation, and substitute in the second equation [Eq. (13.96)], which gives

$$\frac{\partial^2 u}{\partial t^2} = -\frac{\partial}{\partial t}\left(\frac{\partial \tau}{\partial y}\right) = -\frac{\partial}{\partial y}\left(\frac{\partial \tau}{\partial t}\right) =$$

$$= \frac{\partial}{\partial y}\left[\frac{1}{E}\left(\frac{\partial u}{\partial y} + \tau\right)\right] = \frac{1}{E}\frac{\partial^2 u}{\partial y^2} + \frac{1}{E}\frac{\partial \tau}{\partial y}. \qquad (13.144)$$

Combining Eq. (13.144) with Eq. (13.142) gives

$$\frac{u^{n+1} - u^n}{\Delta t} = -\frac{\partial \tau}{\partial y} + \frac{\Delta t}{2}\left[\frac{1}{E}\frac{\partial^2 u}{\partial y^2} + \frac{1}{E}\frac{\partial \tau}{\partial y}\right]$$

or

$$\frac{u^{n+1} - u^n}{\Delta t} = -\left(1 - \frac{\Delta t}{2E}\right)\frac{\partial \tau}{\partial y} + \frac{\Delta t}{2E}\frac{\partial^2 u}{\partial y^2}. \quad (13.145)$$

To obtain the Taylor version of the stress equation, we differentiate Eq. (13.96) once with respect to time:

$$\frac{\partial^2 \tau}{\partial t^2} = -\frac{1}{E}\frac{\partial^2 u}{\partial t \partial y} - \frac{1}{E}\frac{\partial \tau}{\partial t}. \quad (13.146)$$

We then exchange the order of differentiation and substitute in Eq. (13.96), giving

$$\frac{\partial^2 \tau}{\partial t^2} = -\frac{1}{E}\frac{\partial}{\partial y}\left(\frac{\partial u}{\partial t}\right) - \frac{1}{E}\left(-\frac{1}{E}\frac{\partial u}{\partial y} - \frac{\tau}{E}\right). \quad (13.147)$$

Including Eq. (13.95) then gives us

$$\frac{\partial^2 \tau}{\partial t^2} = -\frac{1}{E}\frac{\partial}{\partial y}\left(-\frac{\partial \tau}{\partial y}\right) + \frac{1}{E^2}\frac{\partial u}{\partial y} + \frac{1}{E^2}\tau =$$

$$= \frac{1}{E}\frac{\partial^2 \tau}{\partial y^2} + \frac{1}{E^2}\frac{\partial u}{\partial y} + \frac{1}{E^2}\tau. \quad (13.148)$$

Using Eq. (13.148) in Eq. (13.143) gives

$$\frac{\tau^{n+1} - \tau^n}{\Delta t} = -\frac{1}{E}\frac{\partial u}{\partial y} - \frac{\tau}{E} + \frac{\Delta t}{2}\left(\frac{1}{E}\frac{\partial^2 \tau}{\partial y^2} + \frac{1}{E^2}\frac{\partial u}{\partial y} + \frac{1}{E^2}\tau\right) \quad (13.149)$$

or

$$E\frac{\tau^{n+1} - \tau^n}{\Delta t} = -\left(1 - \frac{\Delta t}{2E}\right)\frac{\partial u}{\partial y} - \left(1 - \frac{\Delta t}{2E}\right)\tau + \frac{\Delta t}{2}\frac{\partial^2 \tau}{\partial y^2}. \quad (13.150)$$

The **Taylor-finite difference method** is obtained by applying centered finite difference expressions to Eq. (13.145) and Eq. (13.150):

$$\frac{u_i^{n+1} - u_i^n}{\Delta t} = -\left(1 - \frac{\Delta t}{2E}\right)\frac{\tau_{i+1}^n - \tau_{i-1}^n}{2\Delta y} + \frac{\Delta t}{2E}\frac{u_{i+1}^n - 2u_i^n + u_{i-1}^n}{\Delta y^2}, \quad (13.151)$$

$$\frac{\tau_i^{n+1} - \tau_i^n}{\Delta t} = -\frac{1}{E}\left(1 - \frac{\Delta t}{2E}\right)\frac{u_{i+1}^n - u_{i-1}^n}{2\Delta y} - \frac{1}{E}\left(1 - \frac{\Delta t}{2E}\right)\tau_i^n + \frac{\Delta t}{2E}\frac{\tau_{i+1}^n - 2\tau_i^n + \tau_{i-1}^n}{\Delta y^2}. \quad (13.152)$$

The equations for τ_1 and τ_N are obtained by lumping the corresponding one-sided Galerkin equations [Eq. (13.157) and (13.158)]. They are

$$\frac{1}{2}\frac{\tau_1^{n+1} - \tau_1^n}{\Delta t} = -\frac{1}{E}\left(1 - \frac{\Delta t}{2E}\right)\frac{u_2^n - u_1^n}{2\Delta y} - \frac{1}{2E}\left(1 - \frac{\Delta t}{2E}\right)\tau_1^n + \frac{\Delta t}{2E}\frac{\tau_2^n - \tau_1^n}{\Delta y^2}, \quad (13.153)$$

$$\frac{1}{2}\frac{\tau_N^{n+1} - \tau_N^n}{\Delta t} = -\frac{1}{E}\left(1 - \frac{\Delta t}{2E}\right)\frac{u_N^n - u_{N-1}^n}{2\Delta y} - \frac{1}{2E}\left(1 - \frac{\Delta t}{2E}\right)\tau_N^n + \frac{\Delta t}{2E}\frac{\tau_{N-1}^n - \tau_N^n}{\Delta y^2}. \quad (13.154)$$

The centered finite difference method uses Eq. (13.151)-(13.154) without the terms multiplied by $\Delta t/2E$ on the right-hand side.

The **Taylor-Galerkin method** is very similar but with a different treatment of the derivatives on the left-hand side and the source term, which gives

$$\frac{1}{6}\frac{u_{i+1}^{n+1} - u_{i+1}^n}{\Delta t} + \frac{2}{3}\frac{u_i^{n+1} - u_i^n}{\Delta t} + \frac{1}{6}\frac{u_{i-1}^{n+1} - u_{i-1}^n}{\Delta t} =$$

$$= -\left(1 - \frac{\Delta t}{2E}\right)\frac{\tau_{i+1}^n - \tau_{i-1}^n}{2\Delta y} + \frac{\Delta t}{2E}\frac{u_{i+1}^n - 2u_i^n + u_{i-1}^n}{\Delta y^2}, \quad (13.155)$$

$$\frac{1}{6}\frac{\tau_{i+1}^{n+1} - \tau_{i+1}^n}{\Delta t} + \frac{2}{3}\frac{\tau_i^{n+1} - \tau_i^n}{\Delta t} + \frac{1}{6}\frac{\tau_{i-1}^{n+1} - \tau_{i-1}^n}{\Delta t} = -\frac{1}{E}\left(1 - \frac{\Delta t}{2E}\right)\frac{u_{i+1}^n - u_{i-1}^n}{2\Delta y}$$

$$-\frac{1}{E}\left(1 - \frac{\Delta t}{2E}\right)\left(\frac{1}{6}\tau_{i+1}^n + \frac{2}{3}\tau_i^n + \frac{1}{6}\tau_{i-1}^n\right) + \frac{\Delta t}{2E}\frac{\tau_{i+1}^n - 2\tau_i^n + \tau_{i-1}^n}{\Delta y^2}. \quad (13.156)$$

The one-sided derivatives for τ_1 and τ_N are

$$\frac{1}{3}\frac{\tau_1^{n+1} - \tau_1^n}{\Delta t} + \frac{1}{6}\frac{\tau_2^{n+1} - \tau_2^n}{\Delta t} =$$

$$= -\frac{1}{E}\left(1 - \frac{\Delta t}{2E}\right)\frac{u_2^n - u_1^n}{2\Delta y} - \frac{1}{E}\left(1 - \frac{\Delta t}{2E}\right)\left(\frac{1}{3}\tau_1^n + \frac{1}{6}\tau_2^n\right) + \frac{\Delta t}{2E}\frac{\tau_2^n - \tau_1^n}{\Delta y^2}. \quad (13.157)$$

$$\frac{1}{3}\frac{\tau_N^{n+1} - \tau_N^n}{\Delta t} + \frac{1}{6}\frac{\tau_{N-1}^{n+1} - \tau_{N-1}^n}{\Delta t} = -\frac{1}{E}\left(1 - \frac{\Delta t}{2E}\right)\frac{u_N^n - u_{N-1}^n}{2\Delta y}$$

$$-\frac{1}{E}\left(1 - \frac{\Delta t}{2E}\right)\left(\frac{1}{3}\tau_N^n + \frac{1}{6}\tau_{N-1}^n\right) + \frac{\Delta t}{2E}\frac{\tau_{N-1}^n - \tau_N^n}{\Delta y^2}. \quad (13.158)$$

The **Galerkin method** is equivalent to Eq. (13.155)-(13.158) without the terms multiplied by $\Delta t/2E$ on the right-hand side.

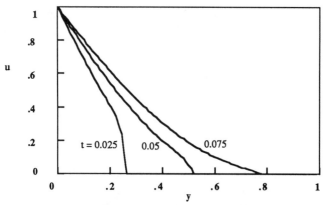

Figure 13.17a. Velocity Solution to the Polymer Problem
Taylor-Finite Difference Method
E = 0.01, 100 nodes, Δt = 0.001

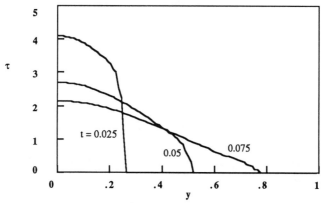

Figure 13.17b. Stress Solution to the Polymer Problem
Taylor-Finite Difference Method, E = 0.01, 100 nodes, Δt = 0.001

Results. Typical results are shown in Figures 13.17 for the case of E = 0.01, as derived from the Taylor-finite difference method. The convective nature of the problem is clear. Dispersion results for these methods are presented in Figures A.32-A.36. All the methods show little dispersion as long as the Courant number is small. Generally, as the value of Δt/E increases, the allowable Courant number increases. The stability regions are displayed in Figure 13.18. The centered finite difference method has the largest stability region when Δt/E is large, but the Taylor-finite difference method has the largest region when Δt/E is small.

Sample calculations are presented below for these methods. The exact solution is shown in Figure 13.19. The convective nature of the solution is evident even though the problem may be more like a diffusion-reaction problem than a convection-reaction problem because the characteristics go in both directions and the stability limits are similar to those of diffusion-reaction problems. Indeed, we

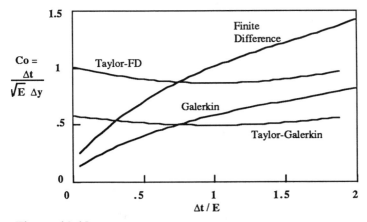

Figure 13.18. Stability Limits for the Polymer Problem, Eq. (13.95)-(13.96)

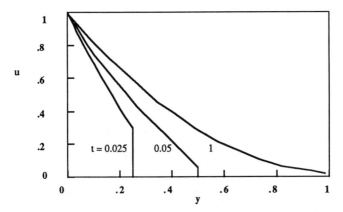

Figure 13.19. Exact Solution to Eq.(13.95)-(13.96) Data from Denn and Porteous [1971]

are unable to use the upstream method, the MacCormack method, or the MacCormack method with flux-correction.

The Galerkin method has severe stability limitations (see Figure 13.18). The Taylor-Galerkin method gives the results shown in Figure 13.20; these results are quite good and are similar to those obtained with the Taylor finite difference method. The difference in time steps, should be noted, however. If we want to solve for larger values of E, the stability limitations make the time step very small. Implicit methods are then needed. Results are presented in Figures 13.21-13.22 for implicit Galerkin methods applied with $E = 0.01$. When the trapezoid rule is used ($\theta = 0.5$), the solutions show small oscillations (Figure 13.21). The Galerkin results with a fully implicit integration ($\theta = 1$) are shown in Figure 13.22; these are excellent results with a large time-step. The implicit calculation, though, does introduce extra dispersion.

The chief advantage of the implicit method is that larger E values can be

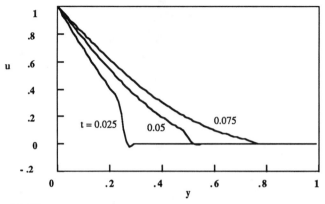

Figure 13.20a. Velocity Solution to the Polymer Problem, Taylor-Galerkin Method
E = 0.01, 100 nodes, Δt = 0.00025

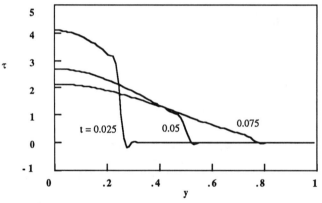

Figure 13.20b. Stress Solution to the Polymer Problem, Taylor-Galerkin Method
E = 0.01, 100 nodes, Δt = 0.00025

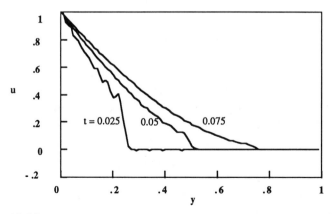

Figure 13.21a. Velocity Solution to the Polymer Problem, Implicit Galerkin Method
θ = 0.5, E = 0.01, 100 nodes, Δt = 0.001

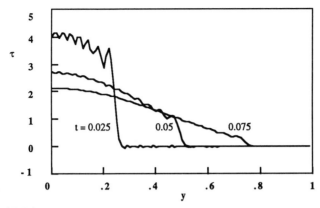

Figure 13.21b. Stress Solution to the Polymer Problem, Implicit Galerkin Method
$\theta = 0.5$, $E = 0.01$, 100 nodes, $\Delta t = 0.001$

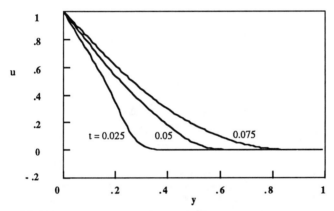

Figure 13.22a. Velocity Solution to the Polymer Problem, Implicit Galerkin Method
$\theta = 1$, $E = 0.01$, 100 nodes, $\Delta t = 0.001$

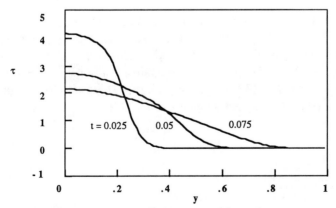

Figure 13.22b. Stress Solution to the Polymer Problem, Implicit Galerkin Method
$\theta = 1$, $E = 0.01$, 100 nodes, $\Delta t = 0.001$

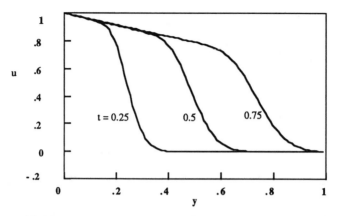

Figure 13.23a. Velocity Solution to the Polymer Problem, Implicit Galerkin Method
$\theta = 1$, $E = 1.0$, 100 nodes, $\Delta t = 0.01$

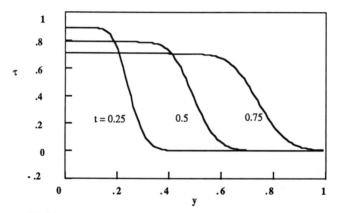

Figure 13.23b. Stress Solution to the Polymer Problem, Implicit Galerkin Method
$\theta = 1$, $E = 1.0$, 100 nodes, $\Delta t = 0.01$

considered. Consider the fully implicit Galerkin method ($\theta = 1$). Solutions for $E = 1$ are shown in Figure 13.23 and the convective nature of the solution is more evident. When the fully implicit finite difference method is used, the results are similar (see Figure 13.24). Solutions can be obtained for larger values of E ($E = 100$ in Figure 13.25 and $E = 10,000$ in Figure 13.26). These solutions do exhibit excess dispersion, but the convective nature of the solution is still evident.

Problems

1_1. Derive Eq. (13.36).

2_1. Derive Eq. (13.38).

3_1. Derive Eq. (13.43).

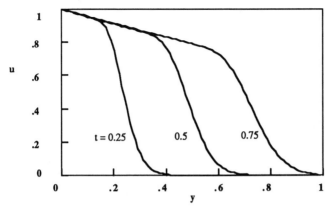

Figure 13.24a. Velocity Solution to the Polymer Problem, Implicit Finite Difference Method, $\theta = 1$, $E = 1.0$, 100 nodes, $\Delta t = 0.01$

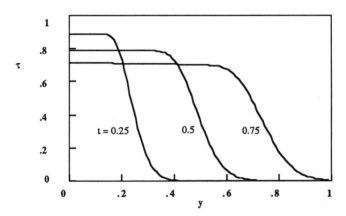

Figure 13.24b. Stress Solution to the Polymer Problem, Implicit Finite Difference Method, $\theta = 1$, $E = 1.0$, 100 nodes, $\Delta t = 0.01$

4$_2$. Derive Eq. (13.44).

5$_3$. Derive Eq. (13.55).

6$_3$. Derive Eq. (13.59).

7$_3$. Verify Eq. (13.70)-(13.72).

8$_3$. Verify Eq. (13.75)-(13.77).

9$_1$. Perform the non-dimensionalization to obtain Eq. (13.84)-(13.86).

10$_2$. Derive Eq. (13.105).

11$_2$. Use the POLYMER option in the program APPLICATIONS. For a centered finite difference method and the Galerkin finite element method, solve the problem with $E = 0.01$ and $\theta = 0, 0.5$, and 1.0. Discuss your results.

494 NUMERICAL METHODS - MOVING FRONTS

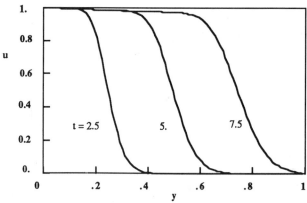

Figure 13.25a. Velocity Solution to the Polymer Problem, Implicit Galerkin Method
$\theta = 1$, $E = 100$, 100 nodes, $\Delta t = 0.1$

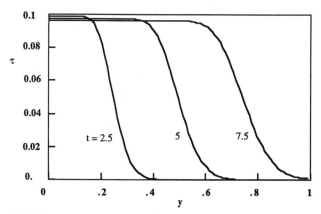

Figure 13.25b. Stress Solution to the Polymer Problem, Implicit Galerkin Method
$\theta = 1$, $E = 100$, 100 nodes, $\Delta t = 0.1$

12$_3$. Use the Fourier method to prove that the upstream method, Eq. (13.134), is unstable. [Hint: if the method is unstable for any value of ξ it is unstable. Choose a ξ value that simplifies the analysis.]

13$_3$. Use the Fourier method to prove that the MacCormack method, Eq. (13.137)-(13.138), is unstable. [Hint: Keep only terms of order Δt.]

14$_2$. Learn to use Figure 13.18 to deduce stable step-sizes. For a specified E and Δy draw a straight line from the origin to a point determined for an arbitrary choice of Δt. Anywhere on this line below the stability curve of the chosen method will give a stable step-size. Verify your results by using the POLYMER option in the program APPLICATIONS.

References

Bird, R. B., Armstrong, R. C., and Hassager, O., *Dynamics of Polymeric Liquids*, Vol. 1, Second Edition, Wiley-Interscience [1987].

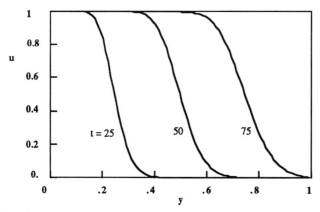

Figure 13.26a. Velocity Solution to the Polymer Problem, Implicit Galerkin Method
θ = 1, E = 10,000, 100 nodes, Δt = 1

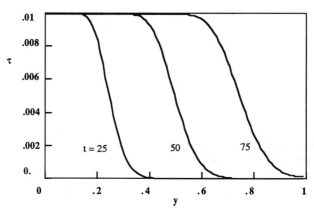

Figure 13.26b. Stress Solution to the Polymer Problem, Implicit Galerkin Method
θ = 1, E = 10,000, 100 nodes, Δt = 1

Denn, M. M. and Porteous, K. C., "Elastic Effects in Flow of Viscoelastic Liquids," Chem. Eng. J. 2 280-286 [1971].

Josse, S. and Finlayson, B. A., Reflections on the Numerical Viscoelastic Flow Problem, J. Non-Newtonian Fluid Mech. 16 13-36 [1984].

Richtmyer, R. D. and Morton, K. W., *Difference Methods for Initial-Value Problems*, Second Edition, Interscience/Wiley [1967].

Vichnevetsky, R. and Bowles, J. B., *Fourier Analysis of Numerical Approximations of Hyperbolic Equations*, SIAM [1982].

Weinberger, H. F., *A First Course in Partial Differential Equations.*, Blaisell Publishing [1965].

CHAPTER
FOURTEEN

FLOW THROUGH POROUS MEDIA

Porous media form highly irregular geometries through which fluids can flow. Rather than trying to describe the flow through each and every irregular space, we usually solve for a macroscopic flow, averaged over many pores. With that approximation, the velocities are average velocities and the pressures are average pressures, averaged over a space that is large with respect to the pores in the porous media. The relation between average velocity and average pressure is called Darcy's law, with the velocity proportional to the pressure gradient:

$$\mathbf{u} \propto \nabla p. \tag{14.1}$$

In this chapter we will consider a variety of situations. Flow through porous media may have only one phase (miscible displacement) or there may be two phases (immiscible displacement). In the case of two phases, we usually try to displace one fluid (oil) with another (water). More complicated cases than these will not be described here. An excellent discussion of this topic is in a book by Peaceman [1977]; his discussion will be followed in this chapter. The equations will first be derived and then various one-dimensional and two-dimensional examples will be treated.

14.1. Equations

The equations for two-phase flow are the mass balance for each phase, written here as o (for oil) and w (for water):

$$\phi \frac{\partial (\rho_o S_o)}{\partial t} = -\nabla \cdot (\rho_o \mathbf{q}_o),$$
$$\phi \frac{\partial (\rho_w S_w)}{\partial t} = -\nabla \cdot (\rho_w \mathbf{q}_w). \tag{14.2}$$

The variable ϕ represents the porosity, or the fraction of the total volume that is open for flow. S_i is the saturation of the i-th phase, either oil or water; the flow rate per unit area of each phase is q_i; and the density of the i-th phase is ρ_i. Usually the phase velocity is expressed as a fraction of the total velocity. We next use Darcy's law to relate the phase flow rate to the pressure gradient:

$$q_o = -\lambda_o (\nabla p_o - \rho_o g),$$
$$q_w = -\lambda_w (\nabla p_w - \rho_w g). \tag{14.3}$$

The mass balances are

$$\phi \frac{\partial (\rho_o S_o)}{\partial t} = \nabla \cdot [\rho_o \lambda_o (\nabla p_o - \rho_o g)],$$
$$\phi \frac{\partial (\rho_w S_w)}{\partial t} = \nabla \cdot [\rho_w \lambda_w (\nabla p_w - \rho_w g)]. \tag{14.4}$$

If the densities of the two phases are constant, we get

$$\phi \frac{\partial S_o}{\partial t} = \nabla \cdot [\lambda_o (\nabla p_o - \rho_o g)], \tag{14.5}$$

$$\phi \frac{\partial S_w}{\partial t} = \nabla \cdot [\lambda_w (\nabla p_w - \rho_w g)]. \tag{14.6}$$

Eq. (14.5)-(14.6) provides two equations for the saturations, S_o and S_w. Usually these equations are solved after they are put into a different form. The first rearrangement is based on the fact that the sum of the saturations equals one, or that all the non-solid space is filled with one fluid or another. Thus, we add the two equations [Eq. (14.5)-(14.6)] to obtain

$$\phi \left(\frac{\partial S_o}{\partial t} + \frac{\partial S_w}{\partial t} \right) = \nabla \cdot [\lambda_o (\nabla p_o - \rho_o g) + \lambda_w (\nabla p_w - \rho_w g)]$$

or $\nabla \cdot [\lambda_o (\nabla p_o - \rho_o g) + \lambda_w (\nabla p_w - \rho_w g)] = 0 \tag{14.7}$

or $\nabla \cdot q = 0.$

Next we define the capillary pressure as

$$p_c \equiv p_o - p_w. \tag{14.8}$$

The capillary pressure is a function of saturation. We eliminate ∇p_w and ∇p_o from the following equations:

$$q_o = -\lambda_o (\nabla p_o - \rho_o g)$$
$$q_w = -\lambda_w (\nabla p_w - \rho_w g) \tag{14.9}$$

using

$$\nabla p_c = \nabla p_o - \nabla p_w. \tag{14.10}$$

This gives

NUMERICAL METHODS - MOVING FRONTS

$$\lambda_w \lambda_o \nabla p_c = \lambda_o q_w - \lambda_w q_o + \lambda_w \lambda_o (\rho_o - \rho_w) g. \tag{14.11}$$

Substituting for q_o gives

$$q_o = q - q_w,$$
$$\lambda_w \lambda_o \nabla p_c = (\lambda_o + \lambda_w) q_w - \lambda_w q + \lambda_w \lambda_o (\rho_o - \rho_w) g. \tag{14.12}$$

Thus

$$q_w = f_w q + f_w \lambda_o [\nabla p_c - (\rho_o - \rho_w) g], \tag{14.13}$$

where

$$f_w = \frac{\lambda_w}{\lambda_o + \lambda_w}, \quad f_o = \frac{\lambda_o}{\lambda_o + \lambda_w}, \quad f_w + f_o = 1 \tag{14.14}$$

are functions of saturation. Eq. (14.12), (14.8), and (14.6) are combined to give

$$\phi \frac{\partial S_w}{\partial t} = -\nabla \cdot [q_w] = -\nabla \cdot [f_w q] - \nabla \cdot \{f_w \lambda_o [\nabla p_c - (\rho_o - \rho_w) g]\}. \tag{14.15}$$

We write

$$h_w = -\frac{\lambda_w \lambda_o}{\lambda_w + \lambda_o} \frac{dp_c}{dS_w},$$
$$\frac{\lambda_w \lambda_o}{\lambda_w + \lambda_o} \nabla p_c = -h_w \nabla S_w. \tag{14.16}$$

Then Eq. (14.13) becomes

$$q_w = f_w q - h_w \nabla S_w + f_w \lambda_o (\rho_w - \rho_o) g \tag{14.17}$$

and Eq. (14.6) can be written as

$$\phi \frac{\partial S_w}{\partial t} = -\nabla \cdot [q_w]$$
$$= \nabla \cdot [h_w \nabla S_w] - \nabla \cdot [f_w q + f_w \lambda_o (\rho_w - \rho_o) g]. \tag{14.18}$$

One term is expanded to give

$$\nabla \cdot [f_w q] = f_w \nabla \cdot q + q \cdot \nabla f_w$$
$$= \frac{df_w}{dS_w} q \cdot \nabla S_w, \text{ since } \nabla \cdot q = 0 \text{ and } f_w = f_w(S_w). \tag{14.19}$$

Eq. (14.6) can then be rewritten as

$$\phi \frac{\partial S_w}{\partial t} + \frac{df_w}{dS_w} q \cdot \nabla S_w = \nabla \cdot [h_w \nabla S_w] - \nabla \cdot [f_w \lambda_o (\rho_w - \rho_o) g], \tag{14.20}$$

$$\nabla \cdot q = 0. \tag{14.21}$$

In this way of solving the problem we see that the saturation equation is hyperbolic when the velocity is known but is parabolic with the h_w term included. Thus, capillarity adds dispersion to the problem. The continuity equation (14.21) is elliptic for the pressure when the saturation is known.

Sometimes it is more convenient to solve for the average pressure and the capillary pressure than to solve for the water and oil pressures. We define

$$p_{avg} = \frac{1}{2}(p_w + p_o),$$
$$p_c = p_o - p_w. \tag{14.22}$$

We solve for p_o and p_w in terms of p_{avg} and p_c, giving

$$p_{avg} + \frac{1}{2} p_c = p_o, \quad p_{avg} - \frac{1}{2} p_c = p_w. \tag{14.23}$$

Then the term

$$\nabla \cdot [\lambda_w \nabla p_w] \tag{14.24}$$

becomes

$$\nabla \cdot [\lambda_w \nabla p_{avg}] - \frac{1}{2} \nabla \cdot [\lambda_w \nabla p_c]. \tag{14.25}$$

The last term is neglected since p_c is small compared to p_{avg}. Then Eq. (14.6) becomes

$$\phi \frac{\partial S_w}{\partial t} = \nabla \cdot [\lambda_w \nabla p_{avg}] - \nabla \cdot [\lambda_w \rho_w g]. \tag{14.26}$$

The terms

$$\nabla \cdot [\lambda_o \nabla p_o] + \nabla \cdot [\lambda_w \nabla p_w] \tag{14.27}$$

become

$$\nabla \cdot [\lambda_o \nabla p_{avg}] + \frac{1}{2} \nabla \cdot [\lambda_o \nabla p_c] + \nabla \cdot [\lambda_w \nabla p_{avg}] - \frac{1}{2} \nabla \cdot [\lambda_w \nabla p_c] =$$

$$\nabla \cdot [(\lambda_o + \lambda_w) \nabla p_{avg}] + \frac{1}{2} \nabla \cdot [(\lambda_o - \lambda_w) \nabla p_c]$$

or $\nabla \cdot [\lambda \nabla p_{avg}] - \frac{1}{2} \nabla \cdot [(\lambda_w - \lambda_o) \nabla p_c]. \tag{14.28}$

Eq. (14.7) then becomes

$$\nabla \cdot [\lambda \nabla p_{avg}] - \frac{1}{2} \nabla \cdot [(\lambda_w - \lambda_o) \nabla p_c] + \nabla \cdot [(\lambda_w \rho_w - \lambda_o \rho_o) g] = 0. \tag{14.29}$$

Eq. (14.26) and Eq. (14.29) are then solved for S_w and p_{avg}. The variables p_c, λ_w, and λ_o are all functions of S_w.

When only one phase is flowing, Eq. (14.7) becomes

$$\nabla \cdot \lambda \nabla p = -\nabla \cdot \mathbf{q} = 0. \quad (14.30)$$

For the miscible case, the equation for the concentration of a solvent is easily derived by setting $S = 1$ and replacing density by concentration. Thus we get the following equation from Eq. (14.2):

$$\phi \frac{\partial c}{\partial t} = -\nabla \cdot (c\mathbf{q}). \quad (14.31)$$

If we want to add dispersion to the model, Eq. (14.31) can be rewritten as

$$\phi \frac{\partial c}{\partial t} = -\nabla \cdot (c\mathbf{q}) + \nabla \cdot [\alpha \,|\mathbf{q}|\nabla c]. \quad (14.32)$$

The dispersion coefficient is proportional to the velocity, with a proportionality constant of α. Other, more general dispersion relations are also possible, but we will not consider them here. Equations (14.30)-(14.32) are referred to as the miscible model. The initial concentration and pressure profile must be specified. The boundary conditions usually are no flux (or flow) across an external boundary, except where there is a well, in which case there is a point-source or point-sink. At production boundaries, no dispersion occurs.

In section 14.2, below, we will consider the miscible flow equations for one-dimensional cases. In these cases, the problem is similar to those treated in other chapters. In section 14.3 we will consider immiscible flow in one-dimensional cases; these cases give rise to phase boundaries and shocks. Section 14.4 considers partially saturated flow in one dimension. Sections 14.5 and 14.6 will discuss the expansion of miscible and immiscible cases to two-dimensional problems. These applications will be illustrated using the methods on fixed meshes as well as methods that move the nodes.

14.2. Miscible Displacement in One Dimension

In this section we will consider the dispersion errors that are introduced by different numerical methods, since these errors are critical when solving the more complicated immiscible displacement problems. The discussion follows work by Young [1984]. For a one-dimensional problem, Eq. (14.21) reduces to

$$\frac{dq}{dx} = 0, \quad (14.33)$$

which means that the velocity is constant. Then Eq. (14.32) becomes

$$\phi \frac{\partial c}{\partial t} + u \frac{\partial c}{\partial x} = \alpha u \frac{\partial^2 c}{\partial x^2}. \quad (14.34)$$

The non-dimensional form of this equation is

$$\frac{\partial c}{\partial \tau} + \frac{\partial c}{\partial \xi} = \frac{1}{Pe} \frac{\partial^2 c}{\partial \xi^2}, \quad (14.35)$$

where

$$\tau = \frac{t\,u}{\phi L}, \quad \xi = \frac{x}{L}, \quad Pe = \frac{L}{\alpha}. \quad (14.36)$$

This is equivalent to the convective diffusion equation, which we previously have solved in the form

$$\frac{\partial c}{\partial \eta} + Pe \frac{\partial c}{\partial \xi} = \frac{\partial^2 c}{\partial \xi^2}. \quad (14.37)$$

The only difference between Eq. (14.37) and Eq. (14.35) is in the definition of the dimensionless time. We can divide Eq. (14.37) by Pe and define $\tau = \eta Pe$ to obtain Eq. (14.35) This transformation will be useful when we adapt the results already derived for Eq. (14.37) to the solution of Eq. (14.35) (see Problem 2.18).

Let us consider two methods of solving Eq. (14.37). Finite difference methods can be applied with the convective derivative centered [Eq. (2.96)] or with upstream weighting [Eq. (2.99)]. For Eq. (14.37) we found that the upstream evaluation of derivatives was equivalent to replacing the dimensionless diffusion coefficient 1 with $1 + Pe\,\Delta x/2$. For equations in the form of Eq. (14.35), we start with the centered difference approximation and replace the coefficient on the right with

$$\frac{1}{Pe} \approx \frac{1 + \dfrac{Pe\,\Delta x}{2}}{Pe} \quad (14.38)$$

to obtain the upstream formulation of the equations. This is the same as using a new Peclet number, Pe*.

$$\frac{1}{Pe^*} = \frac{1}{Pe} + \frac{\Delta x}{2} \quad (14.39)$$

We wish to assess the error of using the upstream formulation. To do this, Young [1984] employs an approximation to the exact solution:

$$c = \frac{1}{2}\mathrm{erfc}\left[(\xi - \tau) \sqrt{\frac{Pe}{4\tau}} \right]. \quad (14.40)$$

This function is expanded in a Taylor series in terms of an inverse Peclet number, which gives

$$c(\xi,\tau,Pe^*) = c(\xi,\tau,Pe) + \left.\frac{\partial c}{\partial(1/Pe)}\right|_{Pe} \left(\frac{1}{Pe^*} - \frac{1}{Pe} \right). \quad (14.41)$$

The error is then the last term in Eq. (14.41); the maximum value for any x is

$$\varepsilon = \frac{1}{\sqrt{2\pi}} e^{-1/2} \frac{Pe - Pe^*}{Pe + Pe^*} \approx 0.242 \frac{Pe - Pe^*}{Pe + Pe^*}. \tag{14.42}$$

Young [1984] showed that the actual numerical error is within a factor of 2 of this estimate. By substituting Eq. (14.39) into Eq. (14.42), we get

$$\varepsilon \approx \frac{0.242}{1 + \frac{4N}{Pe}} \approx 0.06 \frac{Pe}{N} \approx 0.12 \frac{Pe \, \Delta x}{2}. \tag{14.43}$$

As expected, this states that the error when using upstream derivatives is proportional to the grid spacing (to the first power) and is also proportional to the Peclet number.

Another alternative is to use the optimal upstream weighting function as given by Eq. (6.140):

$$\bar{\xi} = \coth(\mu) - \frac{1}{\mu}, \quad \mu = \frac{Pe\Delta x}{2}. \tag{14.44}$$

Eq. (6.139) can be turned into a finite difference form by using mass lumping. It is clear that the effective Peclet number is

$$\frac{1}{Pe^*} = \frac{1 + \frac{Pe\Delta x}{2} \bar{\xi}}{Pe}. \tag{14.45}$$

This then gives

$$Pe^* = \frac{2}{\Delta x} \tanh \mu. \tag{14.46}$$

Young [1984] showed that again Eq. (14.42) estimates the numerical error within a factor of 2. For small Δx values, the error becomes

$$\varepsilon \approx 0.01 \left(\frac{Pe}{N}\right)^2 \approx 0.04 \left(\frac{Pe \, \Delta x}{2}\right)^2. \tag{14.47}$$

This suggests that the upstream method with optimal weighting has a truncation error that is second-order in Δx for small Δx values. Errors for both methods, Eq. (14.43) and Eq. (14.47), are constant in time. If a problem is to be solved keeping ε fixed, no matter what the Peclet number is, then the number of points must be proportional to the Peclet number, even with the optimal weighting function.

Young [1984] also looked at the numerical error for the centered finite difference method and evaluated this error using Eq. (14.40) to approximate the exact solution. He found that the error is approximately

$$\varepsilon \approx 0.023 \left(\frac{Pe}{N}\right)^2 (Pe\,\tau)^{-0.5} = 0.092 \left(\frac{Pe \, \Delta x}{2}\right)^2 (Pe\,\tau)^{-0.5}. \tag{14.48}$$

Here the numerical error decreases with time; if the required error at a certain time is specified, then the number of grid points need not be inversely proportional to the Peclet number. Young [1984] also gave results for higher-order methods that have even smaller errors. Using these errors, Young presented a table showing how many grid points are needed to achieve an accuracy of a maximum of 2% error at $\tau = 0.5$. His results are shown in Table 14.1. As we can see, if accuracy is the determining factor, then the smallest number of points are required by the higher-order methods. In the upstream methods accuracy is sacrificed to avoid oscillations. We have seen that even better solutions are obtained by the random choice method, the MacCormack method with flux-correction, and moving grid methods.

Table 14.1. Grid Requirements for the Convective Diffusion Equation

(The value shown is the number of grid points needed to have a maximum error of 2% at $\tau = 0.5$.)
Data from Young [1984]. Reprinted with permission of Elsevier Science Publishers B. V.

	Peclet Number			
Method	100	250	500	1000
Upstream Weighting	300	750	1500	1000
Optimal Upstream	71	177	354	707
Centered Difference	40	80	135	227
Quadratic Finite Element	18	32	48	70

14.3. Immiscible Displacement in One Dimension

Next we consider the one-dimensional problem when one fluid is injected into a porous medium to displace another fluid; the two fluids are immiscible. Neither fluid completely fills the space either before or after the injection front, so we must solve for the saturation, or the fraction of the space filled with one phase. The material balance is Eq. (14.15) without gravity and written in one dimension:

$$\phi \frac{\partial S_w}{\partial t} = -\frac{\partial}{\partial x}[\, f_w q\,] - \frac{\partial}{\partial x}\left[f_w \lambda_o \frac{\partial p_c}{\partial x} \right], \tag{14.49}$$

$$\frac{\partial q}{\partial x} = 0. \tag{14.50}$$

The continuity equation [Eq. (14.50)] states that the total velocity is constant. Usually the initial conditions are a constant water saturation, $S_w = S_{wc}$, while the boundary conditions are no flow at the external boundaries, injection at wells giving point-sources, and production at wells giving point-sinks. At the production wells, the capillary pressure has zero gradients (so that capillary flow is zero there). The dimensionless form of Eq. (14.49) is

504 NUMERICAL METHODS - MOVING FRONTS

$$\frac{\partial S}{\partial \tau} + \frac{\partial f(S)}{\partial \xi} = \frac{\partial}{\partial \xi}\left[g(S) \frac{\partial S}{\partial \xi} \right], \quad (14.51)$$

where

$$S = \frac{S_w - S_{wc}}{1 - S_{or} - S_{wc}}, \quad \tau = \frac{t\,q}{\phi L(1 - S_{or} - S_{wc})}, \quad \xi = \frac{x}{L},$$

$$g = -(1 - S_{or} - S_{wc})\frac{f_w \lambda_o}{q L}\frac{dp_c}{dS_w}. \quad (14.52)$$

S_{or} is the residual oil saturation (the oil saturation cannot be taken lower than this) and S_{wc} is the connate water saturation. (Connate water is the irreducible water left in the pores). The function f is

$$f(S) \equiv f_w = \frac{\lambda_w}{\lambda_w + \lambda_o} = \frac{\lambda_w}{\lambda}, \quad \lambda_i = \frac{k_i}{\mu_i}, \quad i = o, w, \quad (14.53)$$

where the mobilities are defined as the ratio of the relative permeability to the viscosity of the phase. When there are no capillarity effects, the function f_w is the fraction of the total flow that is water. If we take $f = S$ and $g = 1/Pe$ in Eq. (14.51), we get the convective diffusion equation. We solve the equation for the f and g values defined in Table 14.2, which are more realistic.

Table 14.2. Physical Properties for Immiscible Displacement, Case 1

Data from Young [1984]

$$k_{ro} = (1-S)^3, \quad k_{rw} = \frac{1}{2}S^3, \quad \frac{\mu_o}{\mu_w} = 3, \quad -\frac{dp_c}{dS} = \frac{2}{S^3}, \quad \frac{(1-S_{or}-S_{wc})}{\mu_w q L} = 0.03 \quad (14.54)$$

$$f = \frac{\frac{1}{2}S^3}{\frac{1}{2}S^3 + (1-S)^3/3} = \frac{S^3}{S^3 + \frac{2}{3}(1-S)^3}; \quad g = 0.03\,\frac{(1-S)^3}{1.5\,S^3 + (1-S)^3} \quad (14.55)$$

$$\frac{df}{dS} = \frac{2\left(\frac{1}{S}-1\right)^2 \frac{1}{S^2}}{\left[1 + \frac{2}{3}\left(\frac{1}{S}-1\right)^3\right]^2} = \frac{2 S^2 (1-S)^2}{\left[S^3 + \frac{2}{3}(1-S)^3\right]^2} \quad (14.56)$$

The properties of the solution are strongly dependent on the functions f and g. The fractional flow curve [f(S)] for this example is plotted in Figure 14.1(a); Figure 14.1(b) gives the derivative and Figure 14.1(c) shows the function g(S).

The problem posed above is called the Buckley-Leverett problem; both Buckley and Leverett [1942] and Welge [1952] presented exact solutions to this

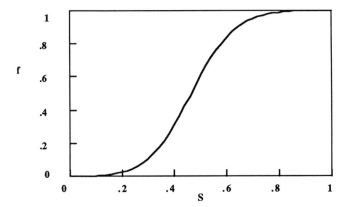

Figure 14.1a. Fractional Flow Curve, Table 14.2

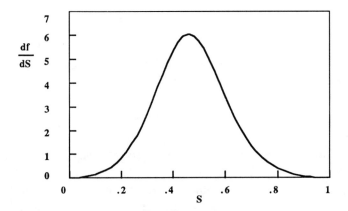

Figure 14.1b. Derivative of Fractional Flow Curve, Table 14.2

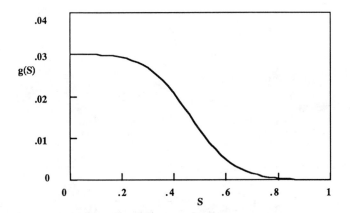

Figure 14.1c. Dispersion Function, Table 14.2

problem with $g = 0$. The equation is obtained from Eq. (14.51) by setting $g = 0$ and recognizing that f is a function of S:

$$\frac{\partial S}{\partial \tau} + \frac{df}{dS}\frac{\partial S}{\partial \xi} = 0. \tag{14.57}$$

Eq. (14.57) is a reducible homogeneous semi-linear equation [Rhee, et al. 1986]. We apply the method of characteristics (see Ch. 9) to give

$$\frac{d\tau}{ds} = 1, \quad \frac{d\xi}{ds} = \frac{df}{dS}, \quad \frac{dS}{ds} = 0,$$
$$\tau(0) = 0, \quad \xi(0) = \xi_0, \quad S(0) = S_0. \tag{14.58}$$

The characteristic directions are then

$$\frac{d\xi}{d\tau} = \frac{df}{dS} \quad \text{or} \quad \frac{d\tau}{d\xi} = \frac{dS}{df} = \sigma(S). \tag{14.59}$$

Since f is a function of S alone and S is constant along a characteristic, the characteristics are all straight. The entire solution is given by a group of straight characteristics. The solution to Eq. (14.58) is

$$\tau = s, \quad \xi = \frac{df}{dS}s + \xi_0, \quad S = S_0. \tag{14.60}$$

To further explore the solution, we consider a case in which the initial saturation everywhere is S_{init} and fluid is injected at the left-hand boundary with a higher saturation, S_{in}. The saturation in question is the water saturation. We assume that a shock forms at ξ_f; we wish to determine the conditions for such a shock and determine its velocity. First we make a mass balance around the front from $\xi_f - \varepsilon$ to $\xi_f + \varepsilon$:

$$\int_{\xi_f - \varepsilon}^{\xi_f + \varepsilon} \frac{\partial S}{\partial \tau} d\xi + \int_{\xi_f - \varepsilon}^{\xi_f + \varepsilon} \frac{\partial f}{\partial \xi} d\xi = 0. \tag{14.61}$$

The second term is evaluated as

$$\int_{\xi_f - \varepsilon}^{\xi_f + \varepsilon} \frac{\partial f}{\partial \xi} d\xi = f(\xi_f + \varepsilon) - f(\xi_f - \varepsilon) = f(S_{init}) - f(S_f). \tag{14.62}$$

The first term is evaluated using the following result from Leibnitz's rule:

$$\frac{d}{d\tau}\int_{\xi_f - \varepsilon}^{\xi_f + \varepsilon} S d\xi = \frac{d\xi_f}{d\tau}S(\xi_f + \varepsilon) - \frac{d\xi_f}{d\tau}S(\xi_f - \varepsilon) + \int_{\xi_f - \varepsilon}^{\xi_f + \varepsilon} \frac{\partial S}{\partial \tau}d\xi. \tag{14.63}$$

Combining all the results gives

$$\frac{d}{d\tau}\int_{\xi_f-\varepsilon}^{\xi_f+\varepsilon} S\,d\xi - \frac{d\xi_f}{d\tau}[\,S(\xi_f+\varepsilon) - S(\xi_f-\varepsilon)\,] + f(S_{init}) - f(S_f) = 0. \qquad (14.64)$$

If ε approaches zero, the first term is zero and we are left with

$$\frac{d\xi_f}{d\tau}[\,S(\xi_f) - S(\xi_{init})\,] = f(S_f) - f(S_{init}). \qquad (14.65)$$

Thus we get the velocity of the front:

$$\frac{d\xi_f}{d\tau} = \frac{[\,f\,]}{[\,S\,]}. \qquad (14.66)$$

Alternatively,

$$\frac{d\tau}{d\xi_f} = \frac{[\,S\,]}{[\,F\,]} \qquad (14.67)$$

gives the propagation direction of the discontinuity in the ξ–τ plane. In the notation of Section 9.2 this is

$$\sigma(S) \equiv \frac{dS}{df}, \quad \tilde{\sigma}(S^l, S^r) = \frac{[\,S\,]}{[\,F\,]}, \; S^l = S_f, \; S^r = S_{init}. \qquad (14.68)$$

The conditions for a shock are Eq. (9.70) and Eq. (9.71). Eq. (9.71) applies when the function σ has only one inflection point. In this case, Eq. (9.71) becomes

$$\sigma(S^l) \leq \sigma(S^r) \quad \text{or} \quad \frac{df}{dS}(S^l) \geq \frac{df}{dS}(S^r). \qquad (14.69)$$

If Eq. (14.69) holds a shock is formed. Eq. (14.69) applies even if the saturation before the front is not constant at S_{init}. Usually $S^l > S^r$ (for displacement of oil with water), and thus Eq. (14.69) is satisfied. When the shock condition is satisfied, the velocity of the front is [using Eq. (14.66) and Eq. (14.60)]

$$\frac{d\xi_f}{d\tau} = \frac{[\,f\,]}{[\,S\,]} = \frac{df}{dS}(S_f). \qquad (14.70)$$

When $S_{init} = 0$, this is

$$\frac{f(S_f)}{S_f} = \frac{df}{dS}(S_f). \qquad (14.71)$$

On Figure 14.1 we draw a line tangent to the f(S) curve from the point (0,0); the S value where the line intersects the curve is S_f. Eq. (14.71) serves to determine the velocity of the front when the shock condition is satisfied. The solution behind the front is given by Eq. (14.60). Buckley and Leverett determined Eq. (14.70) using the continuity equation, while Welge determined Eq. (14.70) using a graphical method. A typical solution is shown in Figure 14.2. Initially the water

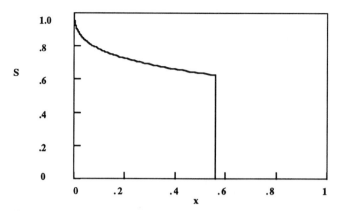

Figure 14.2. Exact Solution to the Buckley-Leverett Problem, Case I, t = 0.4

saturation is low. The value of the saturation at the front and the velocity at the front remain constant with time. The solution behind the front changes with time, however, as shown.

The effect of capillarity is to round the corners on the steep front shown in Figure 14.2. Clearly, capillarity has the same effect as dispersion; the problem without capillarity is more difficult since it leads to shock fronts.

From this point on, x-t notation will be used, with x and t both dimensionless:

$$\frac{\partial S}{\partial t} + \frac{\partial f(S)}{\partial x} = \frac{\partial}{\partial x}\left[g(S) \frac{\partial S}{\partial x} \right]. \qquad (14.72)$$

Equations are developed for the explicit, upstream finite difference method, the MacCormack method, the Taylor-finite difference method, the Galerkin method, and the Taylor-Galerkin method. Capillarity can be eliminated by taking g = 0. By comparing Eq. (14.51) with Eq. (4.41), we see that the correspondence is

$$u \to S, \quad F(u) \to f(S), \quad \frac{\Delta t}{2}\left(\frac{dF}{du}\right)^2 \to g(u), \text{ if used.} \qquad (14.73)$$

The TVD method is obtained by using Eq. (6.97) and Eq. (6.101) with the analogy of Eq. (14.73). When capillarity is included Eq. (6.103) is used with the viscosity replaced by $0.5(g_i + g_{i+1})$. The ENO method is obtained by using Eq. (6.124) together with Eq. (6.127); the viscosity replaced by $0.5(g_i + g_{i+1})$. The ENO method was applied to these problems first by Chen, *et al.* [1991]. Higher-order methods using QUICK schemes are applied by Saad, *et al.* [1990].

The **explicit, upstream finite difference method** is applied to Eq. (14.51) using an upstream evaluation of the first derivative:

$$\frac{S_i^{n+1} - S_i^n}{\Delta t} + \frac{f_i - f_{i-1}}{\Delta x} = \frac{g_{i+1/2}^n (S_{i+1}^n - S_i^n) - g_{i-1/2}^n (S_i^n - S_{i-1}^n)}{\Delta x^2},$$

$$g_{i+1/2}^n = \frac{1}{2}(g_{i+1}^n + g_i^n), \quad g_{i-1/2}^n = \frac{1}{2}(g_{i-1}^n + g_i^n). \qquad (14.74)$$

The **MacCormack method**, from Eq. (4.53)-(4.54), is applied to Eq. (14.72), giving

$$S*_i^{n+1} = S_i^n - \frac{\Delta t}{\Delta x}(f_{i+1}^n - f_i^n) + \frac{\Delta t}{\Delta x^2}[g_{i+1/2}^n(S_{i+1}^n - S_i^n) - g_{i-1/2}^n(S_i^n - S_{i-1}^n)], \quad (14.75\text{-a})$$

$$S_i^{n+1} = \frac{1}{2}(S_i^n + S*_i^{n+1}) - \frac{\Delta t}{2\Delta x}(f*_i^{n+1} - f*_{i-1}^{n+1}) +$$
$$+ \frac{\Delta t}{2\Delta x^2}[g*_{i+1/2}^{n+1}(S*_{i+1}^{n+1} - S*_i^{n+1}) - g*_{i+1/2}^{n+1}(S*_i^{n+1} - S*_{i-1}^{n+1})]. \quad (14.75\text{b})$$

As shown in Chapter 4, Eq. (14.75a)-(14.75b) include essentially the $\Delta t/2(dF/du)^2$ terms.

The **derivation of the Taylor formulation** of Eq. (14.72) is obtained by writing Eq. (14.72) in the form

$$\frac{\partial S}{\partial t} = -\frac{\partial f}{\partial x} = -\frac{df}{dS}\frac{\partial S}{\partial x}. \quad (14.76)$$

We differentiate it with respect to t; the result is (see Problem 14.1)

$$\frac{S^{n+1} - S^n}{\Delta t} = -\frac{\partial f}{\partial x} + \frac{\partial}{\partial x}\left\{\left[g(S) + \frac{\Delta t}{2}\left(\frac{df}{dS}\right)^2\right]\frac{\partial S}{\partial x}\right\}. \quad (14.77)$$

The **Taylor-finite difference method** uses

$$\frac{S_i^{n+1} - S_i^n}{\Delta t} = -\frac{1}{2\Delta x}[f_{i+1}^n - f_{i-1}^n]$$
$$+ \frac{\Delta t}{4\Delta x^2}\left\{\left[\left(\frac{df}{dS}\right)_{i+1}^2 + \left(\frac{df}{dS}\right)_i^2\right][S_{i+1}^n - S_i^n] - \left[\left(\frac{df}{dS}\right)_i^2 + \left(\frac{df}{dS}\right)_{i-1}^2\right][S_i^n - S_{i-1}^n]\right\}$$
$$+ \frac{1}{2\Delta x^2}\{[g_{i+1}^n + g_i^n][S_{i+1}^n - S_i^n] - [g_i^n + g_{i-1}^n][S_i^n - S_{i-1}^n]\}. \quad (14.78)$$

To get a **centered, finite difference method**, we leave off the terms multiplied by Δt on the right-hand side.

The **Taylor-Galerkin method** uses

$$\frac{1}{6}\frac{S_{i+1}^{n+1} - S_{i+1}^n}{\Delta t} + \frac{2}{3}\frac{S_i^{n+1} - S_i^n}{\Delta t} + \frac{1}{6}\frac{S_{i-1}^{n+1} - S_{i-1}^n}{\Delta t} = -\frac{1}{2\Delta x}[f_{i+1}^n - f_{i-1}^n]$$
$$+ \frac{\Delta t}{4\Delta x^2}\left\{\left[\left(\frac{df}{dS}\right)_{i+1}^2 + \left(\frac{df}{dS}\right)_i^2\right][S_{i+1}^n - S_i^n] - \left[\left(\frac{df}{dS}\right)_i^2 + \left(\frac{df}{dS}\right)_{i-1}^2\right][S_i^n - S_{i-1}^n]\right\} +$$
$$+ \frac{1}{2\Delta x^2}\{[g_{i+1}^n + g_i^n][S_{i+1}^n - S_i^n] - [g_i^n + g_{i-1}^n][S_i^n - S_{i-1}^n]\}. \quad (14.79)$$

To get the **Galerkin method**, we leave off the terms multiplied by Δt on the right-hand side.

These methods are first applied to the Buckley-Leverett problem

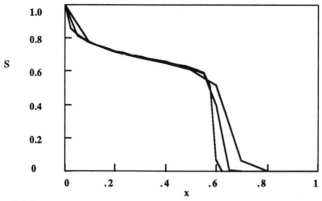

Figure 14.3a. Buckley-Leverett Equation, No Capillarity, Upstream Method at t = 0.4
11, 21, and 41 nodes with Δt = 0.025, 0.0125, and 0.00625, respectively

without capillarity by taking g = 0 and the parameters in Table 14.2. The explicit, upstream method gives the results shown in Figure 14.3a for several mesh sizes. The fronts are smeared when Δx is large but the front becomes steeper for simulations with a small Δx value. No oscillations are evident, however. The MacCormack method gives the results shown in Figure 14.3b. The plateau is in the wrong place and is much too flat compared with the exact solution. When flux-correction is used with the MacCormack method, the result is better but still incorrect. The TVD and ENO methods give excellent results as shown in Figure 14.3c. The Galerkin method without the Taylor term is unstable because of the centered difference expression for $\partial f/\partial x$. The Taylor-Galerkin method gives the results shown in Figure 14.3d; there are oscillations in the solution. If the time derivative is lumped but the Taylor terms are still retained, the results show a plateau that is unrealistic (see Figure 14.3d). Therefore, the only good solutions without capillaity were with the TVD and ENO methods.

When capillarity is included, the explicit, upstream method gives the results shown in Figure 14.4a. The effect of capillarity is to smooth the front in the same way as when using a large Δx value and no capillarity. The MacCormack method with and without flux-correction gives the results shown in Figure 14.4b. These results are reasonable; however, the flux-correction step leads to a slight irregularity near the grid point at $x = \Delta x$ and the saturation has been incorrectly moved up to 1.0. The TVD and ENO methods give the excellent results shown in Figure 14.4c. The Galerkin method gives poor results (see Figure 14.4d), whereas the Taylor-Galerkin method and lumped Taylor-Galerkin method give reasonable results (see Figure 14.4d). Thus the best methods for problems with capillarity are the TVD and ENO methods.

Next let us consider solving this problem using moving grids. The first technique moves the grid in accordance with the known velocity of the front, Eq. (14.70). Thus we define a new variable,

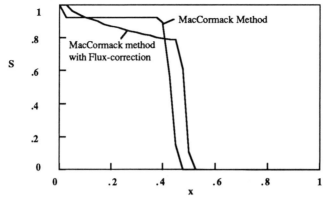

Figure 14.3b. Buckley-Leverett Equation, No Capillarity, MacCormack Method With and Without Flux-correction, 41 nodes, $\Delta t = 0.003125$ at $t = 0.4$

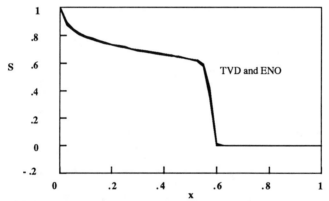

Figure 14.3c. Buckley-Leverett Equation, No Capillarity, TVD and ENO Methods 41 nodes, $\Delta t = 0.003125$

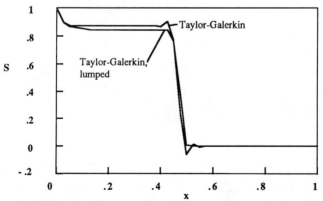

Figure 14.3d. Buckley-Leverett Equation, No Capillarity, Taylor-Galerkin Methods 41 nodes, $\Delta t = 0.00625$

512 NUMERICAL METHODS - MOVING FRONTS

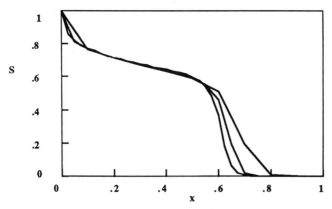

Figure 14.4a. Buckley-Leverett Equation, With Capillarity, Case I, Upstream Method 11, 21, and 41 nodes with $\Delta t = 0.025, 0.0125,$ and 0.003125

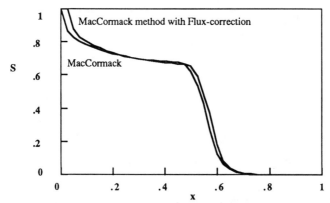

Figure 14.4b. Buckley-Leverett Equation, With Capillarity, Case I, MacCormack Method With and Without Flux-correction, 41 nodes, $\Delta t = 0.006215$

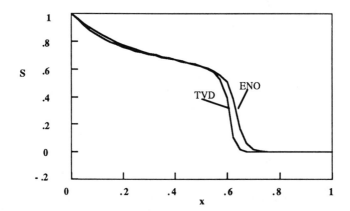

Figure 14.4c. Buckley-Leverett Equation, With Capillarity, Case I, TVD and ENO Methods, 41 nodes, $\Delta t = 0.003125$

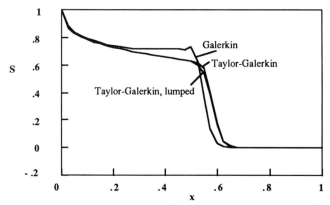

Figure 14.4d. Buckley-Leverett Equation, With Capillarity, Case I, Galerkin Methods 41 nodes, $\Delta t = 0.003125$

$$\eta = \frac{x}{ct}, \tag{14.80}$$

which is the location relative to the distance from $x = 0$ to the location of the front. We assume that the saturation is a function of this new variable, $S(\eta)$. Then we transform Eq. (14.72) using

$$\frac{\partial \eta}{\partial x} = \frac{1}{ct}, \quad \frac{\partial \eta}{\partial t} = -\frac{1}{ct^2} = -\frac{\eta}{t}. \tag{14.81}$$

The various derivatives are

$$\begin{aligned}
\frac{\partial S}{\partial t} &= \frac{dS}{d\eta} \frac{\partial \eta}{\partial t} = -\frac{\eta}{t} \frac{dS}{d\eta}, \\
\frac{\partial f}{\partial x} &= \frac{df}{dS} \frac{\partial S}{\partial x} = \frac{df}{dS} \frac{dS}{d\eta} \frac{\partial \eta}{\partial x} = \frac{df}{dS} \frac{dS}{d\eta} \frac{1}{ct} = \frac{1}{ct} \frac{df}{d\eta}, \\
\frac{\partial S}{\partial x} &= \frac{dS}{d\eta} \frac{\partial \eta}{\partial x} = \frac{dS}{d\eta} \frac{1}{ct}, \\
\frac{\partial^2 S}{\partial x^2} &= \frac{1}{ct} \frac{d}{d\eta} \left(\frac{1}{ct} \frac{dS}{d\eta} \right) = \frac{1}{(ct)^2} \frac{d^2 S}{d\eta^2}.
\end{aligned} \tag{14.82}$$

The capillary term is given by

$$\begin{aligned}
\frac{\partial}{\partial x} \left(g \frac{\partial S}{\partial x} \right) &= \frac{\partial g}{\partial x} \frac{\partial S}{\partial x} + g \frac{\partial^2 S}{\partial x^2} = \frac{dg}{dS} \left(\frac{\partial S}{\partial x} \right)^2 + g \frac{\partial^2 S}{\partial x^2} \\
\frac{\partial}{\partial x} \left(g \frac{\partial S}{\partial x} \right) &= \frac{1}{(ct)^2} \left[\frac{dg}{dS} \left(\frac{dS}{d\eta} \right)^2 + g \frac{d^2 S}{d\eta^2} \right] = \frac{1}{(ct)^2} \frac{d}{d\eta} \left(g \frac{dS}{d\eta} \right).
\end{aligned} \tag{14.83}$$

The final equation is then

$$-\frac{\eta}{t}\frac{dS}{d\eta} + \frac{1}{ct}\frac{df}{d\eta} = = \frac{1}{(ct)^2}\frac{d}{d\eta}\left(g\frac{dS}{d\eta}\right). \tag{14.84}$$

This equation does not simplify further if there is capillarity since the terms on the right-hand side have a different power of t than that which occurs on the left-hand side. Essentially, this means that the spreading caused by the capillarity effects does not follow a linear relation with respect to time. If there is no capillarity, however, this reduces to

$$\frac{1}{t}\frac{dS}{d\eta}\left[-\eta + \frac{1}{c}\frac{df}{dS}\right] = 0. \tag{14.85}$$

We thus can solve the ordinary differential equation

$$\frac{df}{dS} = c\eta, \tag{14.86}$$

where the constant c is the velocity of the shock.

$$c = \frac{[f]}{[S]} = \frac{f(S_f) - f(S_{init})}{S_f - S_{init}} \tag{14.87}$$

S_f is the saturation at the tangent to the fractional flow curve and S_{init} is the initial water concentration. If we choose a value of η and solve Eq. (14.86) for S, we get the exact solution. Here we apply the orthogonal collocation method by first expanding the saturation in a series:

$$S = S_0 + (S_1 - S_0)\eta + \eta(1-\eta)\sum_{i=1}^{NCOL} a_i P_i(\eta). \tag{14.88}$$

Then we apply collocation to Eq. (14.85) to give

$$\left.\frac{df}{dS}\right|_j = \eta_j c. \tag{14.89}$$

The collocation points, η_j are set by the collocation method. Thus we choose NCOL, the collocation method gives us η_j, and we solve Eq. (14.89) for S_j. This procedure can be thought of as the collocation solution to the original problem or as an interpolation of the exact solution. This method is applied to the Buckley-Leverett problem; the results are as shown in Figure 14.5. Nine and sixteen interior collocation points were used; the solution shows slight oscillations about the exact result.

Okuyiga and Ray [1985] used a similar method to model and estimate parameters for a oil-water displacement problem. The non-dimensional problem, expressed in terms of pressures of oil and water, is

$$\frac{\partial P_w}{\partial t} = \frac{\partial^2 P_w}{\partial x^2} \quad 0 \le x \le S(t), \quad \frac{\partial P_o}{\partial t} = \frac{1}{\alpha}\frac{\partial^2 P_o}{\partial x^2} \quad S(t) \le x \le 1, \quad \frac{dS}{dt} = -\beta_w \frac{\partial P_w}{\partial x}(S,t) \tag{14.90}$$

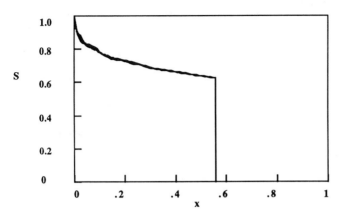

Figure 14.5. Buckley-Leverett Problem, No Capillarity, Exact Solution and Orthogonal Collocation Method

with the initial conditions

$$P_w(x,0) = P_{w0}(x), \quad P_o(x,0) = P_{o0}(x) \tag{14.91}$$

and the boundary conditions

$$P_w(0,t) = P_{wb}(t), \quad P_o(0,t) = P_{ob}(t). \tag{14.92}$$

The interface conditions at the front are

$$P_w(S,t) = P_o(S,t),$$

$$M \frac{\partial P_w}{\partial x}(S,t) = \frac{\partial P_o}{\partial x}(S,t). \tag{14.93}$$

If the front is at $S(t)$, then the coordinates are transformed according to

$$y = \frac{x}{S(t)}, \quad z = \frac{1-x}{1-S(t)}, \tag{14.94}$$

using y on one side of the front and z on the other side. The equations then become

$$\begin{aligned}
\frac{\partial P_w}{\partial t} &= \frac{1}{S^2} \frac{\partial^2 P_w}{\partial y^2} + y \frac{d \ln S}{dt} \frac{\partial P_w}{\partial y} & 0 \leq y \leq 1 \\
\frac{\partial P_o}{\partial t} &= \frac{1}{\alpha (1-S)^2} \frac{\partial^2 P_o}{\partial z^2} + z \frac{d \ln (1-S)}{dt} \frac{\partial P_w}{\partial z} & 0 \leq z \leq 1 \\
\frac{dS}{dt} &= -\frac{\beta_w}{S} \frac{\partial P_w}{\partial y}(1,t)
\end{aligned} \tag{14.95}$$

with the initial conditions of Eq. (14.91), the boundary conditions of Eq. (14.92), and the following interface conditions:

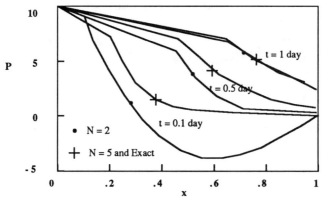

Figure 14.6. Solution to Eq. (14.95), Orthogonal Collocation Method
Data from Okuyiga and Ray [1985]

$$P_w(1,t) = P_o(1,t),$$

$$M \frac{1-S}{S} \frac{\partial P_w}{\partial y}(1,t) = \frac{\partial P_o}{\partial z}(1,t). \qquad (14.96)$$

Now the domain is fixed (in the y-z-t coordinates) but the equations are slightly more complicated. Okuyiga and Ray [1985] solved these equations using the orthogonal collocation method. Since the solutions were relatively smooth on each side of the front, only a few terms were needed in the orthogonal collocation method. Using five terms gave results that were extremely close to the exact solution, as shown in Figure 14.6. Since only a few terms were needed in the orthogonal collocation solution, it was possible to do parameter estimation calculations and optimization studies with relative ease.

The last two examples used grids that were transformed to keep the fronts fixed in the new coordinate systems. It would also be possible to solve these problems using grids that move with the velocity of the fronts but retain the original t-x coordinates. This is the approach that was used in Section 7.2.

Equidistribution principles are also applicable to this problem. Yortsos and Fokas [1983] have published an exact solution to the problem with capillarity that is useful for numerical comparisons. They solved the Buckley-Leverett problem with

$$\frac{df}{dS} = -\frac{\alpha}{(\beta S + \gamma)^2}, \quad g = \frac{1}{(\beta S + \gamma)^2}. \qquad (14.97)$$

Novy [1989] has solved this same problem using an adaptive method that concentrates the nodes in regions of high curvature in the solution [Benner et al., 1987]. (This is similar to using the second derivative, as illustrated in Section 7.3.) Typical results, which are quite good, are shown in Figure 14.7. Winkler, et al. [1985] discussed time-filtering techniques applied to the Buckley-Leverett

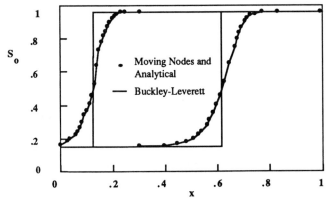

Figure 14.7. Buckley-Leverett Problem, Eq. (14.97); Data from Novy [1989]

problem with added dispersion; their method is useful for problems with multiple shocks.

14.4. Partially Saturated Flow in One Dimension

When water flows in a porous media, the air it displaces can often be ignored. Then two-phase equations reduce to equations for the saturation of only one phase, the water phase. Eq. (14.3)-(14.4), simplified for water saturation and a one-dimensional problem, becomes

$$q_w = -\frac{k_w}{\mu_w}\left(\frac{\partial p_w}{\partial x} - \rho_w g\right), \qquad (14.98)$$

$$\phi\frac{\partial(\rho_w S_w)}{\partial t} = \frac{\partial}{\partial x}\left[\frac{\rho_w k_w}{\mu_w}\left(\frac{\partial p_w}{\partial x} - \rho_w g\right)\right]. \qquad (14.99)$$

The capillary pressure is the difference between the pressure of the gas phase (air) and the water pressure:

$$p_c = p_{air} - p_w. \qquad (14.100)$$

The air pressure is essentially constant; we can take that constant to be zero, giving

$$p_c = -p_w. \qquad (14.101)$$

The water saturation depends on capillary pressure, so we have

$$\frac{\partial S_w}{\partial t} = -\frac{dS_w}{dp_c}\frac{\partial p_w}{\partial t}. \qquad (14.102)$$

The equation is made non-dimensional by using the following definitions:

$$p' = \frac{p_w}{\rho_w g L}, \quad k_r = \frac{k_w}{k_0}, \quad x' = \frac{x}{L}, \quad t' = \frac{t}{t_c}, \quad t_c = \frac{\phi L \mu_w}{k_0 \rho_w g}. \qquad (14.103)$$

The result is

$$-\frac{dS}{dp'_c}\frac{\partial p'}{\partial t'} = \frac{\partial}{\partial x'}\left(k_r \frac{\partial p'}{\partial x'}\right) - \frac{\partial k_r}{\partial x'}, \quad (14.104)$$

where we drop the w subscript on S. The relationship between relative permeability and capillary pressure and between saturation and capillary pressure is unique to a soil sample. The following functions fit experimental data for specific soils:

$$k_r = \frac{1}{1+\left(\frac{p'_c L}{B}\right)^\lambda}, \quad \frac{S-S_r}{1-S_r} = \frac{1}{1+\left(\frac{p'_c L}{A}\right)^\eta}. \quad (14.105)$$

The parameters are listed in Table 14.3. The functions $k_r(p_c)$ and $p_c(S)$ are plotted in Figure 14.8 and 14.9, respectively. The steepness of the pressure profile is related to the steepness of the capillary pressure versus the saturation curve [Finlayson, 1980, p. 247].

Table 14.3. Parameters for Typical Soil [Finlayson, 1980]

k_o (microns)	0.5
ϕ	0.485
S_r	0.32
A (cm)	231.0
B (cm)	146.0
η	3.65
λ	6.65

As the soil becomes very dry, the slope of the curve shown in Figure 14.8 becomes steeper. The drier the soil, the more difficult it is to simulate flow. We take the dimensionless boundary conditions

$$p' = BP1 = \text{constant at } x' = 0, \quad \frac{\partial p'}{\partial x'} = 0 \text{ at } x' = 1 \quad (14.106)$$

and the dimensionless initial conditions corresponding to a length of soil 100 cm long and a pressure of $p_c = 1000$ cm of water:

$$p = -1000 \text{ cm. water}, L = 100 \text{ cm}, p' = -10 \text{ at } t' = 0. \quad (14.107)$$

For simplicity, we solve the problem without gravity; we drop the primes and restate the problem as

$$-\frac{dS}{dp_c}\frac{\partial p}{\partial t} = \frac{\partial}{\partial x}\left(k_r \frac{\partial p}{\partial x}\right),$$

$$p(0,t) = 0, \frac{\partial p}{\partial x}(1,t) = 0, p(x,0) = -10. \quad (14.108)$$

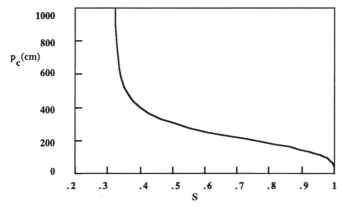

Figure 14.8. Capillarity Function, Eq. (14.105), Table 14.3

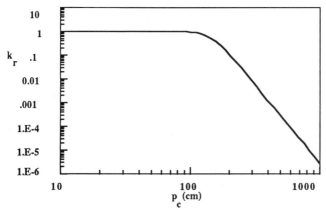

Figure 14.9. Relative Permeability, Eq. (14.105), Table 14.3

Finite difference methods. First let us consider several finite difference methods to solve this equation. We write the differential equation as

$$-\frac{dS}{dp_c}\bigg|_i \frac{dp_i}{dt} = \frac{1}{\Delta x^2} [\, k_{i+1/2} (p_{i+1} - p_i) - k_{i-1/2} (p_i - p_{i-1}) \,]. \quad (14.109)$$

Eventually we will use an ODE integrator to solve the equations. We focus on how the permeability between nodes is evaluated; four possibilities will be considered. The first method evaluates the permeability at the pressure between nodes:

$$k_{i-1/2} = k(p_{i-1/2}) \quad \text{exact.} \quad (14.110)$$

The second method uses an average of the permeabilities at the two nodes on either side of the chosen node:

$$k_{i-1/2} = \frac{1}{2}(k_i + k_{i-1}) \quad \text{average.} \quad (14.111)$$

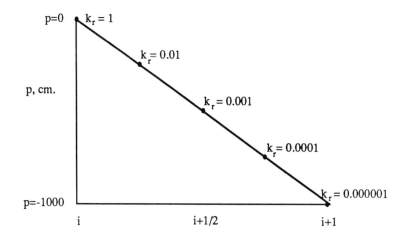

Figure 14.10. Permeability Values for a Sharp Front Within One Element

The third method uses an upstream evaluation, here written as

$$k_{i-1/2} = k_{i-1} \quad \text{when } k_{i-1} > k_i \quad \text{one–point upstream.} \quad (14.112)$$

The fourth method also uses an upstream evaluation, but it is a two-point evaluation. We then take

$$k_{i-1/2} = \tfrac{3}{2} k_{i-1} - \tfrac{1}{2} k_{i-2} \quad \text{two–point upstream,} \quad (14.113)$$

which corresponds to interpolating a curve through the permeabilities at the nodes i–1 and i–2 and extrapolating to the node i–1/2. The truncation error of all choices except the one-point upstream evaluation is second-order in Δx (see Problems 14.2-14.5).

Several ways of evaluating the permeability have been given. To illustrate the effect of the different choices, let us take the pressure profile shown in Figure 14.10, which gives typical values of permeability at the nodes. Then the finite difference equations become

$$-\frac{dS}{dp_c}\bigg|_i \frac{dp_i}{dt} \approx -\frac{10^{-3}}{\Delta x^2}(p_i - p_{i-1}) \quad \text{exact} \quad (14.114)$$

when the permeability is evaluated at the average pressure. When the permeability is evaluated at the upstream pressure, they are

$$-\frac{dS}{dp_c}\bigg|_i \frac{dp_i}{dt} \approx -\frac{1}{\Delta x^2}(p_i - p_{i-1}) \quad \text{one–point upstream.} \quad (14.115)$$

These provide the two extremes. Since the upstream permeability has a large value on the right-hand side, while the exact permeabilitiy has a much smaller value, we use a large value on the right-hand side to indicate a method that has appreciable dispersion.

The use of average permeabilities then leads to

$$-\left.\frac{dS}{dp_c}\right|_i \frac{dp_i}{dt} \approx -\frac{1}{2\Delta x^2}(p_i - p_{i-1}) \quad \text{average,} \tag{14.116}$$

while the use of two-point upstream permeabilities leads to

$$-\left.\frac{dS}{dp_c}\right|_i \frac{dp_i}{dt} \approx -\frac{1}{\Delta x^2}(p_{i+1} - p_i - p_i + p_{i-1}) = -\frac{1}{\Delta x^2}(p_{i+1} - 2p_i + p_{i-1}). \tag{14.117}$$

Since both of these last two methods are second-order but still have some dispersion, they are preferred over one-point upstream permeabilities.

These finite difference methods are applied to Eq. (14.108). The general features of the solution are shown in Figure 14.11. At successive times, the front moves (its position moves as $x = ct^{1/2}$) and retains its steep character. In order to study the numerical methods in more detail, we focus on the solution at $t = 0.015$. The solution obtained with averaged permeabilities and an implicit method with variable time-step-size is shown in Figure 14.12a for several different numbers of nodes. With fewer than 40 nodes the front is not very steep; however, as the mesh is refined by using 80-160 nodes, the profile becomes quite steep. When upstream permeabilities are used, the front is smoothed even more, as shown in Figure 14.12b. Figures 14.12a and 14.12b also show the effect of mesh refinement. When integration in time is explicit with a fixed time-step and average permeabilities are used, then the results do not oscillate, as shown in Figure 14.13.

MacCormack method or second-order Runge-Kutta method. The problem treated here looks like a diffusion problem [Eq. (14.108)] but acts like a convection problem. However, there is no conservative form of the flux with which to apply the MacCormack method. In addition, we have a nonlinear term multiplying the time derivative. The approach to handle the nonlinear term taken in Section 9.2 will not work here since we have only the diffusion operator, not the convection operator appearing in Eq. (9.78). Instead, we show the relationship between the MacCormack method and a second-order Runge-Kutta method, and then apply the Runge-Kutta method to Eq. (14.108). For the initial value problem

$$\frac{dy}{dt} = f(y), \tag{14.118}$$

a second-order Runge-Kutta method is [Finlayson, 1980, p. 30]

$$y^{n+1} = y^n + \frac{\Delta t}{2} f(y^n) + \frac{\Delta t}{2} f[y^n + \Delta t\, f(y^n)]. \tag{14.119}$$

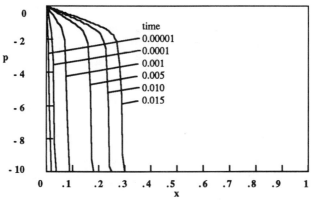

Figure 14.11. Exact Solution to the Soil Problem

The function on the right-hand side is taken as the average of the functions evaluated at both an old time and the projected solution at a new time. This can also be written as

$$\frac{y^{*n+1} - y^n}{\Delta t} = f(y^n) \qquad (14.120)$$

to find the projected solution at a new time and as

$$y^{n+1} = y^n + \frac{\Delta t}{2}[\,f(y^n) + f(y^{*n+1})\,] \qquad (14.121)$$

to correct it. Thus, this is a predictor-corrector scheme. The final step can be written as

$$\frac{y^{n+1} - \frac{1}{2}(y^n + y^{*n+1})}{\Delta t/2} = f(y^{*n+1}). \qquad (14.122)$$

If Eq. (14.122) is expanded and Eq. (14.120) is used for y^{*n+1} on the left-hand side, the result is Eq. (14.121). Thus we can regard the second-order Runge-Kutta method [Eq. (14.119)] as the same as Eq. (14.120) and Eq. (14.122), which are written in the format of the MacCormack method. Here we want to apply the method to the problem

$$P(y)\frac{dy}{dt} = f(y). \qquad (14.123)$$

If we regard f in Eq. (14.120) and Eq. (14.122) as f/P and rewrite the equations, we get

$$P(y^n)\frac{y^{*n+1} - y^n}{\Delta t} = f(y^n) \qquad (14.124)$$

and

POROUS MEDIA 523

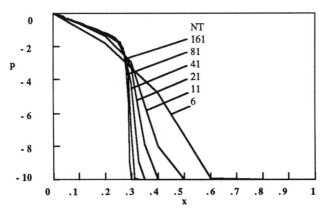

Figure 14.12a. Soil Problem, Finite Difference Method with Averaged Permeabilities, LSODE in time

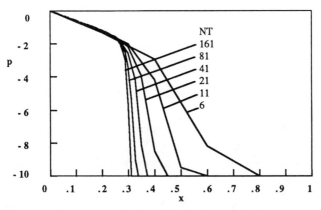

Figure 14.12b. Soil Problem, Finite Difference Method with Upstream Permeabilities, LSODE in time

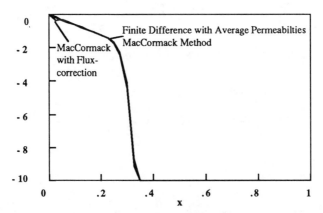

Figure 14.13. Soil Problem, Various Explicit Methods, Solution at t = 0.015
$\Delta x = 0.025$, $\Delta t = 1.25\text{E-}6$

$$P(y*^{n+1}) \frac{y^{n+1} - \frac{1}{2}(y^n + y*^{n+1})}{\Delta t/2} = f(y*^{n+1}). \qquad (14.125)$$

To apply the second-order Runge-Kutta method to the original problem, we incorporate the ideas in Eq. (14.124)-(14.125) to give

$$-\frac{dS}{dp_c}\bigg|_i^n \frac{p*_i^{n+1} - p_i^n}{\Delta t} = \frac{1}{\Delta x^2}[\, k_{i+1/2}^n(p_{i+1}^n - p_i^n) - k_{i-1/2}^n(p_i^n - p_{i-1}^n)\,]$$

$$k_{i+1/2}^n = \frac{1}{2}(k_{i+1}^n + k_i^n),\ k_{i-1/2}^n = \frac{1}{2}(k_i^n + k_{i-1}^n) \qquad (14.126)$$

for the predictor and

$$-\frac{dS}{dp_c}\bigg|_i^{*n+1} \frac{p_i^n - \frac{1}{2}(p_i^n + p*_i^{n+1})}{\Delta t/2} = \frac{1}{\Delta x^2}[\, k*_{i+1/2}^n(p*_{i+1}^{n+1} - p*_i^{n+1}) - k*_{i-1/2}^n(p*_i^{n+1} - p*_{i-1}^{n+1})\,]$$

$$k*_{i+1/2}^n = \frac{1}{2}(k*_{i+1}^n + k*_i^n),\ k*_{i-1/2}^n = \frac{1}{2}(k*_i^n + k*_{i-1}^n)$$

$$(14.127)$$

for the corrector. This also looks like a MacCormack method when the coefficient of the time derivative is constant.

Results obtained with this method are shown in Figure 14.13 both with and without flux-correction. All three methods displayed in Figure 14.13 give good solutions although the flux-correction step adds a small irregularity at the inlet. The flux-correction step is not needed here.

Galerkin finite element method. Next we apply the Galerkin finite element method. Since Eq. (14.104) is basically a diffusion-type equation, we will not include the Taylor terms here. The Galerkin equation is

$$-\int_0^1 \frac{dS}{dp_c} N_j \frac{\partial p}{\partial t} dx = \int_0^1 N_j \frac{\partial}{\partial x}\left(k_r \frac{\partial p}{\partial x}\right) dx = \left[N_j k_r \frac{\partial p}{\partial x}\right]_0^1 - \int_0^1 k_r \frac{\partial N_j}{\partial x} \frac{\partial p}{\partial x} dx. \qquad (14.128)$$

The flow terms at the two boundaries are evaluated as follows:

$$\text{flow in at left} \quad q_0 = -k_r \frac{\partial p}{\partial x}\bigg|_{x=0}$$
$$\text{flow in at right} \quad -q_1 = -k_r \frac{\partial p}{\partial x}\bigg|_{x=1}. \qquad (14.129)$$

Eq. (14.128) is rewritten on an element basis as

$$-\sum_e \sum_I \frac{\Delta x^e}{2} \int_{-1}^1 \frac{dS}{dp_c} N_J N_I d\xi \frac{dp_I}{dt} = \delta_{1j} q_0 + \delta_{N+1,j} q_1 - \sum_e \sum_I \frac{2}{\Delta x^e} \int_{-1}^1 k_r \frac{dN_J}{d\xi} \frac{dN_I}{d\xi} d\xi\, p_I$$

$$(14.130)$$

Eq. (14.130) is not used when j is a node on the boundary if the pressure is set at $x = 0$ and $x = 1$ (as it is here).

Next we evaluate the nonlinear integrals using several different methods. As with the Buckley-Leverett equation, we can evaluate nonlinear integrals using Gaussian quadrature or the trapezoid rule or we can interpolate the nonlinear term onto the basis functions or use upstream evaulations. The use of Gaussian quadrature gives

$$\int_{-1}^{1} \frac{dS}{dp_c} N_J N_I \, d\xi = \sum_{K=1}^{NG} \frac{dS}{dp_c}[p(\xi_K)] N_J(\xi_K) N_I(\xi_K) W_K,$$

$$\int_{-1}^{1} k_r \frac{dN_J}{d\xi} \frac{dN_I}{d\xi} \, d\xi = \sum_{K=1}^{NG} k_r[p(\xi_K)] \frac{dN_J}{d\xi}(\xi_K) \frac{dN_I}{d\xi}(\xi_K) W_K. \quad (14.131)$$

while the use of the trapezoid rule gives

$$\int_{-1}^{1} \frac{dS}{dp_c} N_J N_I \, d\xi = \begin{cases} -\frac{1}{2} \frac{dS}{dp_c}(-1) & \text{for } J = I = 1 \\ -\frac{1}{2} \frac{dS}{dp_c}(+1) & \text{for } J = I = 2 \\ 0 & \text{for } J \neq I \end{cases} \quad (14.132)$$

and

$$-\frac{dS}{dp_c}\bigg|_j \frac{dp_j}{dt} = \frac{1}{2\Delta x^2}[(k_{j+1} + k_j)(p_{j+1} - p_j) - (k_j + k_{j-1})(p_j - p_{j-1})]. \quad (14.133)$$

Eq. (14.133) is the same as is obtained using a finite difference method with averaged permeabilities.

If we interpolate permeability onto the basis set

$$k_r = \sum_K k_K N_K \quad \frac{dS}{dp_c} = \sum_K S'_K N_K, \quad (14.134)$$

then we get

$$\int_{-1}^{1} k_r \frac{dN_J}{d\xi} \frac{dN_I}{d\xi} \, d\xi = \sum_K k_K \int_{-1}^{1} N_K \frac{dN_J}{d\xi} \frac{dN_I}{d\xi} \, d\xi,$$

$$\int_{-1}^{1} \frac{dS}{dp_c} N_J N_I \, d\xi = \sum_K S'_K \int_{-1}^{1} N_K N_J N_I \, d\xi. \quad (14.135)$$

For linear trial functions, the integrals are

$$\int_{-1}^{1} N_K \frac{dN_J}{d\xi} \frac{dN_I}{d\xi} d\xi = \begin{cases} \frac{1}{2} & -\frac{1}{2} \\ -\frac{1}{2} & \frac{1}{2} \end{cases}, \quad K = 1,2$$

$$\int_{-1}^{1} N_1 N_J N_I d\xi = \begin{cases} \frac{1}{2} & \frac{1}{6} \\ \frac{1}{6} & \frac{1}{6} \end{cases}, \quad \int_{-1}^{1} N_2 N_J N_I d\xi = \begin{cases} \frac{1}{6} & \frac{1}{6} \\ \frac{1}{6} & \frac{1}{2} \end{cases}. \tag{14.136}$$

When these are assembled we get

$$\frac{\Delta x}{2}\left\{-\left[\frac{dS}{dp_c}\bigg|_{j-1} + \frac{dS}{dp_c}\bigg|_j\right]\frac{1}{6}\frac{dp_{j-1}}{dt} - \left[\frac{1}{6}\frac{dS}{dp_c}\bigg|_{j-1} + \frac{1}{2}\frac{dS}{dp_c}\bigg|_j + \frac{1}{6}\frac{dS}{dp_c}\bigg|_{j+1}\right]\frac{dp_j}{dt}\right.$$
$$\left. - \left[\frac{dS}{dp_c}\bigg|_j + \frac{dS}{dp_c}\bigg|_{j+1}\right]\frac{1}{6}\frac{dp_{j+1}}{dt}\right\} = \frac{1}{2\Delta x}[(k_{j+1}+k_j)(p_{j+1}-p_j) - (k_j+k_{j-1})(p_j-p_{j-1})].$$
$$\tag{14.137}$$

After rearrangement this is

$$-\left[\frac{dS}{dp_c}\bigg|_{j-1} + \frac{dS}{dp_c}\bigg|_j\right]\frac{1}{12}\frac{dp_{j-1}}{dt} - \left[\frac{1}{12}\frac{dS}{dp_c}\bigg|_{j-1} + \frac{1}{2}\frac{dS}{dp_c}\bigg|_j + \frac{1}{12}\frac{dS}{dp_c}\bigg|_{j+1}\right]\frac{dp_j}{dt}$$
$$-\left[\frac{dS}{dp_c}\bigg|_j + \frac{dS}{dp_c}\bigg|_{j+1}\right]\frac{1}{12}\frac{dp_{j+1}}{dt} = \frac{1}{2\Delta x^2}[(k_{j+1}+k_j)(p_{j+1}-p_j) - (k_j+k_{j-1})(p_j-p_{j-1})].$$
$$\tag{14.138}$$

When upstream permeabilities are used and the dS/dp_c is interpolated onto the basis functions, we get

$$-\left[\frac{dS}{dp_c}\bigg|_{j-1} + \frac{dS}{dp_c}\bigg|_j\right]\frac{1}{12}\frac{dp_{j-1}}{dt} - \left[\frac{1}{12}\frac{dS}{dp_c}\bigg|_{j-1} + \frac{1}{2}\frac{dS}{dp_c}\bigg|_j + \frac{1}{12}\frac{dS}{dp_c}\bigg|_{j+1}\right]\frac{dp_j}{dt}$$
$$-\left[\frac{dS}{dp_c}\bigg|_j + \frac{dS}{dp_c}\bigg|_{j+1}\right]\frac{1}{12}\frac{dp_{j+1}}{dt} = \frac{1}{\Delta x^2}[k_j(p_{j+1}-p_j) - k_{j-1}(p_j-p_{j-1})]. \tag{14.139}$$

These options give the following values for the case described in Figure 14.10:

$$-\frac{dS}{dp_c}\bigg|_i \frac{dp_i}{dt} \approx -\frac{10^{-2}}{2\Delta x^2}(p_i - p_{i-1}) \quad \text{Gaussian quadrature,}$$

$$\approx -\frac{1}{2\Delta x^2}(p_i - p_{i-1}) \quad \text{trapezoid rule,}$$

$$\approx -\frac{1}{2\Delta x^2}(p_i - p_{i-1}) \quad \text{interpolate onto basis set,} \tag{14.140}$$

$$\approx -\frac{1}{\Delta x^2}(p_i - p_{i-1}) \quad \text{upstream permeability.}$$

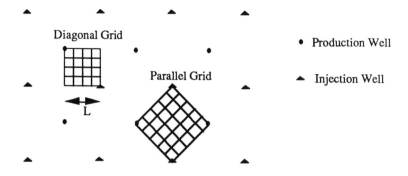

Figure 14.14. Geometry for a Repeated Five-spot Pattern

Again, we expect the use of the trapezoid rule or interpolation of the permeabilities to give results that have some dispersion, even though they are second-order. No Galerkin solutions are presented here.

14.5. Miscible Displacement in Two Dimensions

When one component (a solvent) displaces another (an oil) in a miscible displacement, it is necessary to solve for the concentration of the solvent in addition to the velocity of flow. The equations are Eq. (14.30) and Eq. (14.32):

$$\nabla \cdot \lambda \nabla p = -\nabla \cdot \mathbf{q} = 0, \tag{14.141}$$

$$\phi \frac{\partial c}{\partial t} = -\nabla \cdot (c\mathbf{q}) + \nabla \cdot [\alpha |\mathbf{q}| \nabla c]. \tag{14.142}$$

The situation considered here is injection into a five-spot pattern, as shown in Figure 14.14. The dispersivity is taken such that

$$Pe = \frac{L}{\alpha} = 40, \tag{14.143}$$

where L is the distance shown in Figure 14.14. The viscosity is taken as

$$\mu^{1/4} = \mu_o^{1/4} + \mu_w^{1/4}, \quad \frac{\mu_o}{\mu_w} = 10. \tag{14.144}$$

If the viscosity ratio is ten, then the mobility behind the front (where there is only solvent) is higher than the mobility of the oil. This is called an adverse mobility ratio since it leads to fingering. When fingering occurs, the injection fluid breaks through with a high velocity in a small region. In this case, physical instabilities are hard to differentiate from mathematical ones.

The solvent is injected at an injection well while material is removed at

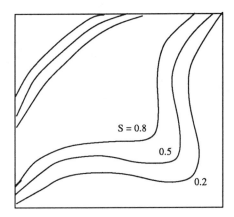

Figure 14.15. Miscible Displacement on Diagonal Grid, 20 x 20
Data from Young [1984]

a production well. Because of the repeated pattern evident in Figure 14.14, we are able to solve the problem on only a portion of the domain. There are two ways the domain can be divided, however; these are shown as a diagonal grid and a parallel grid in Figure 14.14. For the diagonal grid, the straight line from the injection well to the production well goes along a diagonal. For the parallel grid, the straight line from the injection well to the production well goes along a grid line. We show results with both grids; the results should be the same if the problem is solved correctly. Upstream weighting is used; thus, the convective term in Eq. (14.32) is evaluated with an upstream derivative and the mobility in Eq. (14.30) is evaluated with an upstream mobility. All solutions are from work by Young [1984].

The solution obtained on the diagonal grid is shown in Figure 14.15. It shows the front moving broadly across the whole domain. The solution obtained on the parallel grid has a front with sharp peaks, as shown in Figure 14.16. The behavior of these two solutions is completely different. Therefore, at least one solution is incorrect.

This problem has also been solved using a nine-point finite difference method. In the case of constant Δx values and constant (but different) Δy values, the difference expressions for Eq. (14.30) are [Yanosik and McCracken, 1970]

$$\lambda_{i+1/2,j} f_x (p_{i+1,j} - p_{i,j}) + \lambda_{i-1/2,j} f_x (p_{i-1,j} - p_{i,j}) +$$
$$+ \lambda_{i,j+1/2} f_y (p_{i,j+1} - p_{i,j}) + \lambda_{i,j-1/2} f_y (p_{i,j-1} - p_{i,j}) +$$
$$+ \lambda_{i+1/2,j-1/2} f_{xy} (p_{i+1,j-1} - p_{i,j}) + \lambda_{i-1/2,j-1/2} f_{xy} (p_{i-1,j-1} - p_{i,j}) +$$
$$+ \lambda_{i+1/2,j+1/2} f_{xy} (p_{i+1,j+1} - p_{i,j}) + \lambda_{i-1/2,j+1/2} f_{xy} (p_{i-1,j+1} - p_{i,j}) = 0 \quad (14.145)$$

where

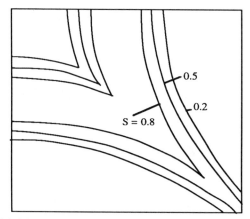

Figure 14.16. Miscible Displacement on Parallel Grid, 28 x 28
Data from Young [1984]

$$f_x = \frac{\Delta y}{\Delta x} - \frac{2}{3} \frac{\Delta x \, \Delta y}{(\Delta x^2 + \Delta y^2)}, \quad f_y = \frac{\Delta x}{\Delta y} - \frac{2}{3} \frac{\Delta x \, \Delta y}{(\Delta x^2 + \Delta y^2)}, \quad f_{xy} = \frac{1}{3} \frac{\Delta x \, \Delta y}{(\Delta x^2 + \Delta y^2)}. \quad (14.146)$$

Yanosik and McCracken also give formulas for variable grids. Calculations with this form of the equations eliminate the effect of grid orientation. If $\Delta x = \Delta y$, then we get

$$f_x = \frac{2}{3}, \quad f_y = \frac{2}{3}, \quad f_{xy} = \frac{1}{6}. \quad (14.147)$$

If $\lambda = 1$, then the resulting form for Laplace's equation is

$$\nabla^2 p_{ij} = \frac{1}{6 \, \Delta x^2} [4 \, (p_{i+1,j} + p_{i-1,j} + p_{i,j+1} + p_{i,j-1}) + (p_{i+1,j-1} + p_{i-1,j-1} + p_{i-1,j+1} + p_{i+1,j+1}) - 20 \, p_{i,j}] \quad (14.148)$$

This is a nine-point finite difference formula.

Similar results are obtained by using a Galerkin finite element method, linear trial functions, or the trapezoid rule for numerical integration. (Young [1984] refers to this method as a Lobatto-Galerkin method since the first Lobatto quadrature is the trapezoid rule). Presented here are the results for a variable grid spacing. The pressure equation is (Young [1981])

$$B_{i-1/2,j} \, (p_{i,j} - p_{i-1,j}) + B_{i+1/2,j} \, (p_{i,j} - p_{i+1,j}) + B_{i,j-1/2} \, (p_{i,j} - p_{i,j-1}) + B_{i,j+1/2} \, (p_{i,j} - p_{i,j+1}) = 0 \quad (14.149)$$

where

$$B_{i+1/2,j} = \frac{\Delta y_j + \Delta y_{j+1}}{4 \, \Delta x_{i+1}} \, (\lambda_{i,j} + \lambda_{i+1,j}), \quad B_{i,j+1/2} = \frac{\Delta x_i + \Delta x_{i+1}}{4 \, \Delta y_{j+1}} \, (\lambda_{i,j} + \lambda_{i,j+1}). \quad (14.150)$$

The concentration equation is

$$D_{i,j} \frac{dc_{i,j}}{dt} - \lambda_{i,j} \left\{ \frac{1}{2} (\Delta y_j + \Delta y_{j+1}) \left[\left(\frac{P_{i,j} - P_{i-1,j}}{\Delta x_i} \right) (c_{i,j} - c_{i-1,j}) + \left(\frac{P_{i+1,j} - P_{i,j}}{\Delta x_{i+1}} \right) (c_{i+1,j} - c_{i,j}) \right] \right.$$

$$\left. + \frac{1}{2} (\Delta x_i + \Delta x_{i+1}) \left[\left(\frac{P_{i,j} - P_{i,j-1}}{\Delta y_j} \right) (c_{i,j} - c_{i,j-1}) + \left(\frac{P_{i,j+1} - P_{i,j}}{\Delta y_{j+1}} \right) (c_{i,j+1} - c_{i,j}) \right] \right\}$$

$$+ C_{i-1/2,j} (c_{i,j} - c_{i-1,j}) + C_{i+1/2,j} (c_{i,j} - c_{i+1,j}) + C_{i,j-1/2} (c_{i,j} - c_{i,j-1})$$

$$+ C_{i,j+1/2} (c_{i,j} - c_{i,j+1}) = G_{i,j}, \qquad (14.151)$$

where the $G_{i,j}$ accounts for source and sink terms at the wells. The coefficient $D_{i,j}$ is given by

$$D_{i,j} = \frac{1}{4} \phi_{i,j} (\Delta x_i + \Delta x_i)(\Delta y_j + \Delta y_j). \qquad (14.152)$$

The coefficient $C_{i+1/2,j}$ is given by

$$4 \Delta x_{i+1} C_{i+1/2,j} = \alpha \left\{ \lambda_{i,j} \Delta y_j \left[\left(\frac{P_{i+1,j} - P_{i,j}}{\Delta x_{i+1}} \right)^2 + \left(\frac{P_{i,j} - P_{i,j-1}}{\Delta y_j} \right)^2 \right]^{1/2} \right.$$

$$+ \lambda_{i+1,j} \Delta y_j \left[\left(\frac{P_{i+1,j} - P_{i,j}}{\Delta x_{i+1}} \right)^2 + \left(\frac{P_{i+1,j} - P_{i+1,j-1}}{\Delta y_j} \right)^2 \right]^{1/2}$$

$$+ \lambda_{i,j} \Delta y_{j+1} \left[\left(\frac{P_{i+1,j} - P_{i,j}}{\Delta x_{i+1}} \right)^2 + \left(\frac{P_{i,j+1} - P_{i,j}}{\Delta y_{j+1}} \right)^2 \right]^{1/2} \qquad (14.153)$$

$$\left. + \lambda_{i+1,j} \Delta y_{j+1} \left[\left(\frac{P_{i+1,j} - P_{i,j}}{\Delta x_{i+1}} \right)^2 + \left(\frac{P_{i+1,j+1} - P_{i+1,j}}{\Delta y_{j+1}} \right)^2 \right]^{1/2} \right\}.$$

The other C's are determined with a similar formula.

The results obtained with these equations on both diagonal and parallel grids are nearly identical, as shown in Figure 14.17. Thus the results shown in Figure 14.17 are the correct ones. Young [1984] also does calculations on different grids and shows that the results become insensitive to the grid size after 25 or more grid blocks are used along the length L. Thus with an adverse mobility ratio of 10 and a Peclet number of 40, it is necessary to use $\Delta x = 0.04$ (in dimensionless terms). This corresponds to PeΔx = 1.6. We do not seem to be able to get away from the requirement of Pe$\Delta x/2$ = 1, even in these more complicated problems. Two-dimensional problems require a number of grid points that is the square of the number needed for a one-dimensional problem. Thus, two-dimensional problems are much more time-consuming and expensive to solve.

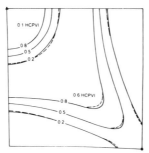

Figure 14.17. Contours of Solvent Fraction, Galerkin Finite Element Method, Linear Trial Functions, From Young [1984] with permission of Elsevier Science Publishers B. V.

14.6. Immiscible Displacement in Two Dimensions

The first example of immiscible displacement is one presented by Yanosik and McCracken [1979] and discussed by Young [1984]. The fractional flow curve is given in Figure 14.18. This gives a mobility ratio of approximately four; the numerator is calculated using the average saturation behind the front and the denominator is calculated using the saturation ahead of the front. The equations are Eq.(14.26) and Eq. (14.29), with capillarity and gravity effects eliminated from both equations:

$$\phi \frac{\partial S_w}{\partial t} = \nabla \cdot \lambda_w \nabla p_{avg}, \qquad (14.154)$$

$$\nabla \cdot \lambda \nabla p_{avg} = 0. \qquad (14.155)$$

As shown in Section 14.1, Eq. (14.154) can be rewritten in the form

$$\phi \frac{\partial S_w}{\partial t} + \frac{df_w}{dS_w} q \cdot \nabla S_w = 0. \qquad (14.156)$$

If the saturation is known, Eq. (14.155) is clearly elliptic in pressure. If the velocity is known (or the pressure is known), then Eq. (14.156) is clearly hyperbolic. When capillarity is included, Eq. (14.156) is parabolic.

Let us first consider the upstream finite difference method. The mobility of each phase is evaluated at the upstream location (parallel to a grid line). This is equivalent to solving the equations

$$\phi \frac{\partial S_w}{\partial t} = \nabla \cdot \lambda_w \nabla p_{avg} + \frac{1}{2} \Delta x \frac{\partial}{\partial x} \left(\left| \frac{\partial p}{\partial x} \right| \frac{\partial \lambda_w}{\partial x} \right) + \frac{1}{2} \Delta y \frac{\partial}{\partial y} \left(\left| \frac{\partial p}{\partial y} \right| \frac{\partial \lambda_w}{\partial y} \right) \qquad (14.157)$$

and

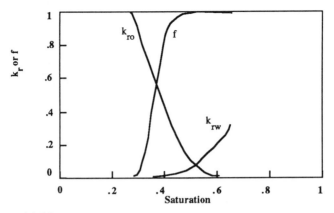

Figure 14.18. Fractional Flow and Total Mobility for Immiscible Displacement Data from Yanosik and McCracken [1979]

$$\nabla \cdot \lambda \nabla \, p_{avg} + \frac{1}{2} \Delta x \frac{\partial}{\partial x} \left(\left| \frac{\partial p}{\partial x} \right| \frac{\partial \lambda}{\partial x} \right) + \frac{1}{2} \Delta y \frac{\partial}{\partial y} \left(\left| \frac{\partial p}{\partial y} \right| \frac{\partial \lambda}{\partial y} \right) = 0. \quad (14.158)$$

The added terms in Eq. (14.157)-(14.158) [compared with Eq. (14.154)-(14.155)] are clearly dispersion terms. We can compare Eq. (14.157) with Eq. (14.49). Both have diffusion-type terms, one from upstream permeability and the other from capillary pressure. We notice, however, that the dispersion term depends on the coordinate direction in that the x-gradient of pressure is used in the x-derivatives and the y-gradient of pressure is used in the y-derivatives. With this evaluation, the dispersion terms vary when the coordinate system is rotated. This causes grid-orientation effects for the miscible displacement problem and the immiscible displacement problem. Yanosik and McCracken [1979] showed that the nine-point finite difference method eliminated the grid-orientation problems even when upstream permeabilities were used.

Presented here is the Galerkin method using linear trial functions and the trapezoid rule for integration. However, we need to add artificial dispersion. In analogy with Eq. (14.157)-(14.158), Young [1984] takes

$$\phi \frac{\partial S_w}{\partial t} = \nabla \cdot \lambda_w \nabla \, p_{avg} + \nabla \cdot [g_0 \, |\nabla \, p_{avg}| \, \nabla \, \lambda_w] \quad (14.159)$$

$$\phi \frac{\partial S_o}{\partial t} = \nabla \cdot \lambda_o \nabla \, p_{avg} + \nabla \cdot [g_0 \, |\nabla \, p_{avg}| \, \nabla \, \lambda_o] \quad (14.160)$$

The dispersion term g_0 does not vary when the grid is rotated, so grid-orientation effects are not expected (and, in fact, are not seen). Young [1984] considers several different ways of evaluating the dispersion term and derives the value of g_0 that guarantees a maximum principle for the equations for a uniform grid. The results obtained using the divergence formulation of the convection terms are presented here; the minimum value of g_0 is 0.5 Δx.

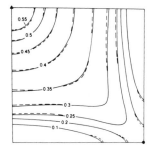

Figure 14.19. Saturation Contours for 0.3 Pore Volumes of Injection
-- 20 x 20 Diagonal Grid, - - - 28 x 28 Parallel Grid
From Young [1984] with permission of Elsevier Science Publishers B. V.

The equations are

$$V_{ij} \frac{dS_{ij}}{dt} + T_{i-1/2,j} (p_{i,j} - p_{i-1,j}) + T_{i+1/2,j} (p_{i,j} - p_{i+1,j}) + T_{i,j-1/2} (p_{i,j} - p_{i,j-1})$$
$$+ T_{i,j+1/2} (p_{i,j} - p_{i,j+1}) + D_{i-1/2,j} (f_{i,j} - f_{i-1,j}) + D_{i+1/2,j} (f_{i,j} - f_{i+1,j}) + \quad (14.161)$$
$$+ D_{i,j-1/2} (f_{i,j} - f_{i,j-1}) + D_{i,j+1/2} (f_{i,j} - f_{i,j+1}) = 0,$$

where the pore volume is given by

$$V_{ij} = \frac{1}{4} \phi_{ij} (\Delta x_{i-1} + \Delta x_i)(\Delta y_{j-1} + \Delta y_j). \quad (14.162)$$

The mesh sizes are

$$\Delta x_i = x_{i+1} - x_i \quad \text{and} \quad \Delta y_j = y_{j+1} - y_j. \quad (14.163)$$

The coefficient of the convective term is given by terms such as

$$T_{i+1/2,j} = \frac{\Delta y_{j-1} + \Delta y_j}{2 \Delta x_{i+1}} (\lambda f)_{i+1/2,j}, \quad \lambda_w = \lambda f. \quad (14.164)$$

In the divergence formulation, the coefficient is evaulated using

$$(\lambda f)_{i+1/2,j} = \frac{1}{2} (\lambda_{ij} f_{ij} + \lambda_{i+1,j} f_{i+1,j}). \quad (14.165)$$

The dispersion terms are

$$D_{i+1/2,j} = \frac{g_0}{4 \Delta x_i} [\Delta y_{j-1} (|q_{i,j-}| + |q_{i+1,j-}|) + \Delta y_j (|q_{i,j+}| + |q_{i+1,j+}|)], \quad (14.166)$$

where the following definitions are used:

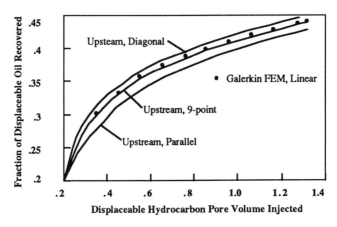

Figure 14.20. Immiscible Displacement, Upstream Finite Difference Methods, 5 and 9 Nodes, Galerkin Method with Linear Finite Elements, Data from Young [1984]

$$|q_{i+1,j-}| = \lambda_{i+1,j}\sqrt{\left(\frac{p_{i+1,j} - p_{ij}}{\Delta x_i}\right)^2 + \left(\frac{p_{i+1,j} - p_{i+1,j-1}}{\Delta y_{j-1}}\right)^2},$$

$$|q_{i,j-}| = \lambda_{i,j}\sqrt{\left(\frac{p_{i+1,j} - p_{ij}}{\Delta x_i}\right)^2 + \left(\frac{p_{i,j} - p_{i,j-1}}{\Delta y_{j-1}}\right)^2},$$

$$|q_{i,j+}| = \lambda_{i,j}\sqrt{\left(\frac{p_{i+1,j} - p_{ij}}{\Delta x_i}\right)^2 + \left(\frac{p_{i,j+1} - p_{i,j}}{\Delta y_j}\right)^2}, \quad (14.167)$$

$$|q_{i+1,j+}| = \lambda_{i+1,j}\sqrt{\left(\frac{p_{i+1,j} - p_{ij}}{\Delta x_i}\right)^2 + \left(\frac{p_{i+1,j+1} - p_{i+1,j}}{\Delta y_j}\right)^2}.$$

Typical results obtained with this formulation are shown in Figure 14.19. As we can see, the diagonal grid and parallel grid give almost identical results. The fractional recovery (coming out of the production well) is shown in Figure 14.20; the results are very close to those derived using the nine-point finite difference upstream method. Frauenthal, *et al.* [1985] showed that similar insensitivity to grid orientation can be obtained if the upstream mobility is evaluated in the direction of the flow rather than the direction of the grid lines. In all these cases, however, additional dispersion was used to dampen oscillations in order to obtain a physically realistic solution.

The random choice method does not require the introduction of numerical dispersion because it is designed to give good results when there is little dispersion or a shock. Presented here is the method as it is outlined by Glimm,

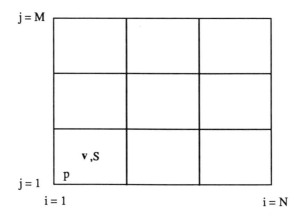

Figure 14.21. Grid for Flow Through Porous Media

et al. [1981a,b], with the recent extensions included. The grid is defined as shown in Figure 14.21; the velocities and saturations are assigned to the center of the grid block and the pressures are assigned to the corners of the grid block. Thus, we have the following variables:

$$q_{i+1/2,j+1/2} \quad S_{i+1/2,j+1/2} \quad p_{i,j}. \tag{14.168}$$

The solution method is to start with the saturation known everywhere and solve Eq. (14.155) for the pressure. Then the velocities are calculated using known pressures. Finally, the saturation equation, Eq. (14.156), is solved using the random choice method, which is explained below.

The mobilities are needed at the centers of the grid blocks; they are evaluated in terms of the saturations at the centers of the grid blocks.

$$\lambda_{i+1/2,j+1/2} = \lambda(S_{i+1/2,j+1/2}) \tag{14.169}$$

Let us consider an x-derivative and evaluate it with the notation shown in Figure 14.22.

We evaluate the first derivative using a centered derivative:

$$\lambda \frac{\partial p}{\partial x}\bigg|_{i+1/2,j} = \lambda_{i+1/2,j} \frac{p_{i+1,j} - p_{i,j}}{\Delta x}. \tag{14.170}$$

Then the mobility needed at the (i+1/2, j) point is evaluated by averaging the mobility at the (i+1/2, j+1/2) and (i+1/2, j-1/2) points. Thus we obtain

$$\lambda \frac{\partial p}{\partial x}\bigg|_{i+1/2,j} = \frac{\lambda_{i+1/2,j+1/2} + \lambda_{i+1/2,j-1/2}}{2} \frac{p_{i+1,j} - p_{i,j}}{\Delta x}. \tag{14.171}$$

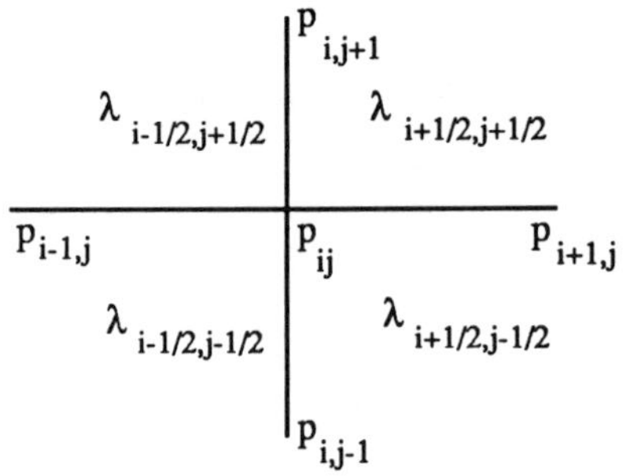

Figure 14.22. Notation for Relative Permeabilities

Continuing in the same fashion gives the following second derivative for pressure:

$$\frac{\partial}{\partial x}\left(\lambda \frac{\partial p}{\partial x}\right)\bigg|_{i,j} = \frac{\lambda \frac{\partial p}{\partial x}\big|_{i+1/2,j} - \lambda \frac{\partial p}{\partial x}\big|_{i-1/2,j}}{\Delta x}. \quad (14.172)$$

Then the pressure equation is

$$A_{ij} p_{ij} + B_{ij} p_{i+1,j} + C_{ij} p_{i-1,j} + D_{ij} p_{i,j+1} + E_{ij} p_{i,j-1} = F_{ij}, \quad (14.173)$$

where

$$A_{ij} = \frac{1}{2}\left(\frac{1}{\Delta x^2} + \frac{1}{\Delta y^2}\right) + \frac{1}{\Delta y^2}(\lambda_{i+1/2,j+1/2} + \lambda_{i+1/2,j-1/2} + \lambda_{i-1/2,j-1/2} + \lambda_{i-1/2,j+1/2}),$$

$$B_{ij} = -\frac{1}{2}\frac{1}{\Delta x^2}(\lambda_{i+1/2,j+1/2} + \lambda_{i+1/2,j-1/2}),\ C_{ij} = -\frac{1}{2}\frac{1}{\Delta x^2}(\lambda_{i-1/2,j-1/2} + \lambda_{i-1/2,j+1/2}),$$

$$D_{ij} = -\frac{1}{2}\frac{1}{\Delta y^2}(\lambda_{i+1/2,j+1/2} + \lambda_{i-1/2,j+1/2}),\ E_{ij} = -\frac{1}{2}\frac{1}{\Delta y^2}(\lambda_{i+1/2,j-1/2} + \lambda_{i-1/2,j-1/2}).$$

$$(14.174)$$

If the mobility falls outside of the domain, it is replaced by zero. Glimm, *et al.* [1981b] solved this equation using a preconditioned conjugate gradient method.

Next we calculate the velocities from the known pressures. The saturation equation, Eq. (14.156), requires the gradient at the center of the grid block. Glimm, *et al.* [1981b] evaluated this by using a finite element approximation on a single grid block. The x-component of Darcy's law is

$$u = -\lambda \frac{\partial p}{\partial x}. \quad (14.175)$$

We expand the pressure in a set of bilinear trial functions:

Figure 14.23. Pressure Interpolation

$$p = \sum_{i=1}^{4} p_i N_i(x,y). \tag{14.176}$$

The trial functions N_i are functions of x and y that take the value 1.0 at node i and 0.0 at the other nodes shown in Figure 14.23. The pressures p_i come from the pressure solution derived from Eq. (14.173). We assign only one velocity on the grid block, so we have only one function (a constant) for velocity. We substitute Eq. (14.176) into Eq. (14.175) and integrate over x and y to obtain

$$\int u \, dx dy = -\int \lambda \sum_{i=1}^{4} p_i \frac{\partial N_i}{\partial x} \, dx \, dy. \tag{14.177}$$

We evaluate the mobility as a constant at the centroid saturation. Evaluation of the terms gives

$$u = -\lambda \frac{1}{\Delta x} \frac{(p_2 - p_1) + (p_3 - p_4)}{2}. \tag{14.178}$$

In a general notation this is

$$u_{i+1/2, j+1/2}^n = -\lambda \, (S_{i+1/2, j+1/2}^{n-1}) \frac{p_{i+1,j+1}^n - p_{i,j+1}^n + p_{i+1,j}^n - p_{i,j}^n}{2 \Delta x}. \tag{14.179}$$

Another alternative would be to use the Galerkin method to solve Eq. (14.175) throughout the domain with p as a known function. Then the velocity components could be expressed by using continuous functions.

The final step in the solution is to solve the saturation equation, Eq. (14.156), using the random choice method. In early work, Glimm, *et al.* [1981b] solved the equation using operator splitting. We solve two subproblems in succession,

$$\frac{\partial S_w}{\partial t} + \frac{\partial}{\partial y} [v f(S_w)] = 0, \tag{14.180}$$

$$\frac{\partial S_w}{\partial t} + \frac{\partial}{\partial x} [u f(S_w)] = 0. \tag{14.181}$$

The solution to these equations is calculated using the algorithms outlined in Section 6.1. In particular, the velocity is taken as

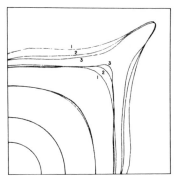

Figure 14.24. Solution to the Five-Spot Pattern
1, 2, 3 are for 16x16, 32x32, 48x48 nodal points, respectively; Mobility ratio = 5
From Glimm, *et al.* [1983]. Reprinted with permission from The Mathematics of Reservoir Simulation, pp. 107-136. Copyright 1983 by the Society for Industrial and Applied Mathematics. All rights reserved.

$$\tilde{u}_i = \frac{u^n_{i-1/2} + u^n_{i+1/2}}{2}, \tag{14.182}$$

$$\tilde{v}_j = \frac{v^n_{j-1/2} + v^n_{j+1/2}}{2}. \tag{14.183}$$

Glimm, *et al.* [1981b] solved problems in which the flow is essentially along the direction of the grid lines. In later work, Daripa, *et al.*, [1986] solved the saturation equation using the random choice method only in the direction of the velocity. The random choice method is essentially advancing a point whose saturation is known. If the advancement is in the direction of the velocity, the point to which the saturation is advanced may not be a grid point, so front tracking with grid refinement is neccessary. The procedure for this tracking was outlined in Section 8.6. Results for a five-spot pattern are shown in Figure 14.24 [Glimm, *et al.*, 1983]; a case with multiple wells has also been solved [Glimm, *et al.* 1987].

When polymer flooding is considered, there is more than one front to track. Then the Riemann problems used in the random choice method must be found. Isaacson [1980] dis this, and Daripa *et al.* [1986] then solved the equations for polymer flooding on a five-spot pattern. Typical results, where quite sharp saturation profiles are maintained, are shown in Figure 14.25.

The problem of viscous fingering can also be solved using the random choice method because the fronts are tracked extremely accurately. Flow in heterogeneous porous media was discussed by King, *et al.* [1985]. For a case with less heterogeneity, the results in Figure 14.26 are obtained [Glimm *et al.*, 1987].

An example from work by Parrott and Christie [1986], obtained with flux-corrected transport is shown in Figure 14.27. They applied the algorithm outlined in Section 8.4, with the high-order method lumped and achieved excellent results.

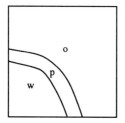 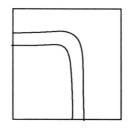 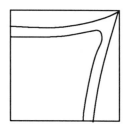

Figure 14.25. Sketch of Front Locations for Polymer Floods
Data from Daripa, *et al.* [1986]

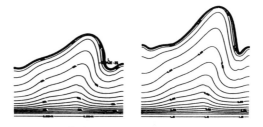

Figure 14.26. Viscous Fingering in Heterogeneous Porous Media, From Glimm, *et al.* [1987]. Reprinted with permission from <u>Mathematical and Computational Methods in Seismic Exploration and Reservoir Modeling</u>, pp.54-67. Copyright 1986 by the Society for Industrial and Applied Mathematics. All rights reserved.

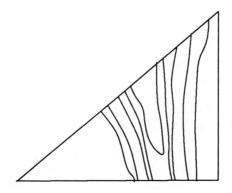

Figure 14.27. Chemical Flooding of an Oil Field
Concentration Contours, Flux-corrected Finite Element Method
Data from Parrott and Christie [1986]

Problems

1₂. Derive the Taylor-Galerkin form of the Buckley-Leverett equation, Eq. (14.77).

2₃. Derive the truncation error of the right-hand side of Eq. (14.109) when using Eq. (14.110) to evaluate the permeabilities and $p_{i+1/2} = 0.5\,(p_i + p_{i+1})$ to evaluate the pressure at the point i+1/2. [Hint: Write a Taylor series for $p_{i\pm1}$, $k_{i\pm1}$ about the point i.]

3₃. Derive the truncation error of the right-hand side of Eq. (14.109) when using Eq. (14.111).

4₃. Derive the truncation error of the right-hand side of Eq. (14.109) when using Eq. (14.112).

5₃. Derive the truncation error of the right-hand side of Eq. (14.109) when using Eq. (14.113). [Hint: Use a Taylor series for $k_{i\pm1/2}$ about the point i.]

References

Benner, R. E., Jr., Davis, H. T., and Scriven, L. E., "An Adaptive Finite Element Method for Steady and Transient Problems", SIAM J. Sci. Stat. Comput. 8 529-549 [1987].

Buckley, S. E. and Leverett, M. C., "Mechanism of fluid displacement in sands," Trans. AIME 146 107-116 [1942].

Chen, W. H., Durlofsky, L. J., Engquist, B., and Osher, S., "Minimization of Grid Orientation Effects", SPE paper 22887 [1991].

Daripa, P., Glimm, J., Lindquist, B., and McBryan, O., "Polymer Floods: A Case Study of Nonlinear Wave Analysis and of Instability Control in Tertiary Oil Recovery," SIAM J. Appl. Math. 48 353-373 [1988].

Finlayson, B. A., *Nonlinear Analysis in Chemical Engineering*, McGraw-Hill [1980].

Frauenthal, J. C., di Franco, R. B., and Towler, B. T., "Reduction of Grid-Orientation Effects in Reservoir Simulation," Soc. Pet. Eng. J. 25 902-908 [1985].

Glimm, J., Isaacson, E., Marchesin, D. and McBryan, O., "Front Tracking for Hyperbolic Systems," Adv. in Appl. Math 2 91-119 [1981a].

Glimm, J., Marchesin, D. and McBryan, O., "A Numerical Method for Two Phase Flow with an Unstable Interface" J. Comp. Phys. 39 179-200 [1981b].

Glimm, J., Lindquist, B., McBryan, O., Tryggvason, G., "Sharp and Diffuse Fronts in Oil Reservoirs: Front Tracking and Capillarity," pp. 54-67 in *Mathematical and Computational Methods in Seismic Exploration and Reservoir Modeling*, Fitzgibbon, W. E. (ed.), SIAM, Philadelphia [1987].

Glimm, J., Lindquist, B., McBryan, O., Padmanabhan, L., "A Front Tracking Reservoir Simulator, Five-Spot Validation Studies and the Water Coning Problem," pp 107-135 in *Frontiers in Applied Mathematics*, Glimm, Vol. 1, Ewing, R. (ed), SIAM, Philadelphia [1983].

Isaacson, E.,"Global Solution of a Riemann Problem for a Nonstrictly Hyperbolic System of Conservation Laws Arising in Enhanced Oil Recovery," unpublished manuscript [1980].

King, M. J., Lindquist, W. B., and Reyna, L., "Stability of Two Dimensional Immiscible Flow to Viscous Fingering," DOE/ER/03077-244 [1985].

Novy, R. A., private communication [1989].

Okuyiga, M. O. and Ray, W. H., "Modelling and Estimation for a Moving Boundary Problem," Int. J. Num. Methods Engn 21, 601-616 [1985].

Parrott, A. K., and Christie, M. A., "FCT Applied to the 2-D Finite Element Solution of Tracer Transport by Single Phase Flow in a Porous Medium," pp. 609-619 in *Numerical methods for Fluid Dynamics,* , (K. W. Morton and M. J. Baines, Eds.) Oxford University Press [1986].

Peaceman, D. W., *Fundamentals of Numerical Reservoir Simulation*, Elsevier [1977].

Rhee, H. K., Aris, R., Amundson, N. R., *First-Order Partial Differential Equations: Volume I. Theory and Application of Single Equations*, Prentice-Hall [1986].

Saad, N., Pope, G. A., Sepehrnoorl, K., "Application of Higher-Order Methods in Compositional Simulation," SPE Reservoir Engn. 5 623-630 [1990].

Welge, H. J., "A simplified method for computing oil recovery by gas or water drive," Trans. AIME 195 91-98 [1952].

Winkler, K. H. A., Mihalas, D., Norman, M. L., "Adaptive Grid Methods with Asymmetric Time-Filtering," Comp. Phys. Commun. 36 121-140 [1985].

Yanosik, J. L. and McCracken, T. A., "A Nine-point, finite-difference reservoir simulator for realistic prediction of adverse mobility ratio displacements," Soc. Pet. Eng. J. 19 253-262 [1979].

Yortsos, Y. C. and Fokas, A. S., "An Analytical Solution for Linear Waterflood Including the Effects of Capillary-Pressure," Soc. Pet. Eng. J. 23 115-124 [1983]; corrections 23 574 [1983].

Young, L. C., "A Finite-element Method for Reservoir Simulation," Soc. Pet. Eng. J. 21 115-127 [1981].

Young, L. C., "A Study of Spatial Approximations for Simulating Fluid Displacements in Petroleum Reservoirs," Comp. Methods Appl. Mech. Eng. 47 3-46 [1984].

APPENDIX

DISPERSION DIAGRAMS

The diagrams in this Appendix are all dispersion diagrams showing a 3D view of the amplification factor, |ρ|, as a function of the Fourier variable, ξ. The other parameters depend on the case in question. Figures A.1-A.26 are plotted versus Courant number, $Co = \Delta t/\Delta x$, and there are four plots, one for each value of $r = \Delta t/\Delta x^2$. Stability occurs for values of Co and r for which the amplification factor is less than or equal to one for <u>all</u> ξ. Thus if you can hold an imaginary pencil along the line |ρ| = 1 for ξ from 0 to 2π and it is above the curve for all ξ, then the case represented by that Co and r is stable.

Figures A.1-A.26 show |ρ| as a function of the Fourier variable ξ (to the right) and Courant number = $\Delta t/\Delta x$ (back). Different curves are for different values of $r = \Delta t/\Delta x^2$.

Figures A.27-A.31 show |ρ| as a function of the Fourier variable ξ (to the right) and the reaction number = $Da_I \Delta t$ (back). Different curves are for different values of $r = \Delta t/\Delta x^2$.

Figures A.32-A.36 show |ρ| as a function of the Fourier variable ξ (to the right) and the elastic number = $\Delta t /\Delta y \sqrt{E}$ (back). Different curves are for different values of $\Delta t / E$.

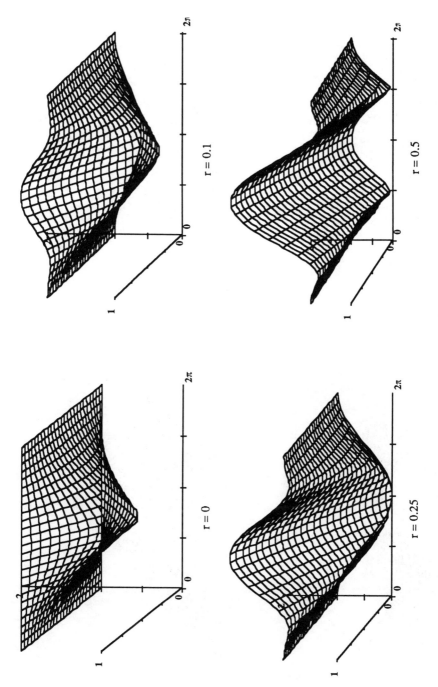

Figure A.1. Dispersion Diagram for the Upstream Finite Difference Method

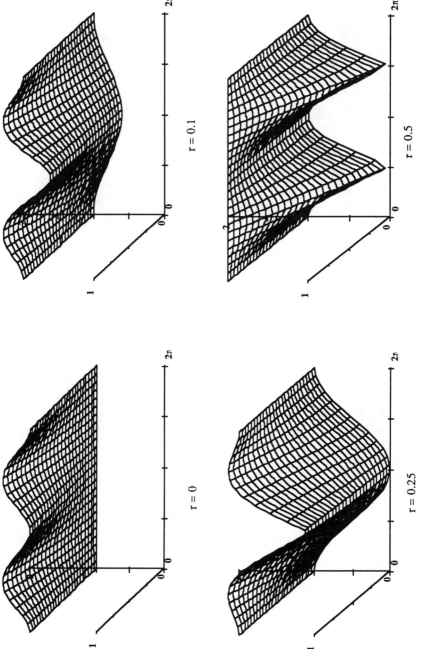

Figure A.2. Dispersion Diagram for the Centered Finite Difference Method

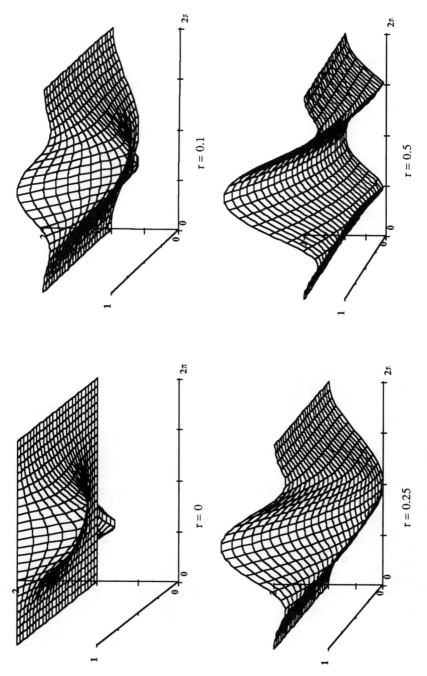

Figure A.3. Dispersion Diagram for the Lax-Wendroff Method

546 NUMERICAL METHODS - MOVING FRONTS

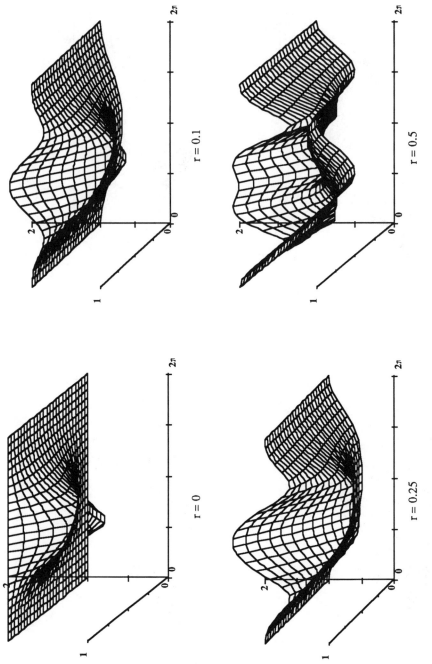

Figure A.4. Dispersion Diagram for the MacCormack Method

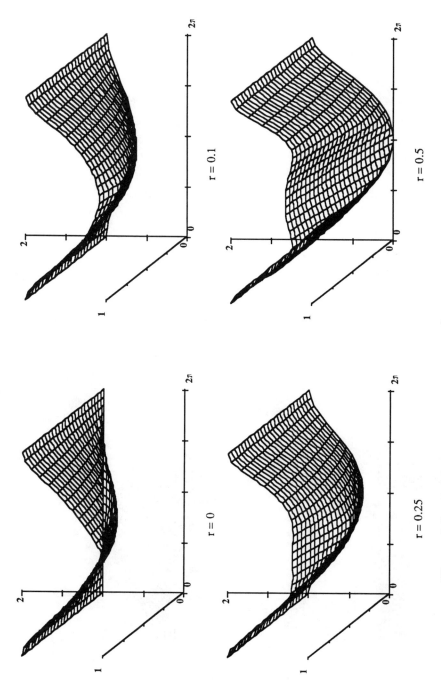

Figure A.5. Dispersion Diagram for the Upstream FD Method, Implicit, $\theta = 0.5$

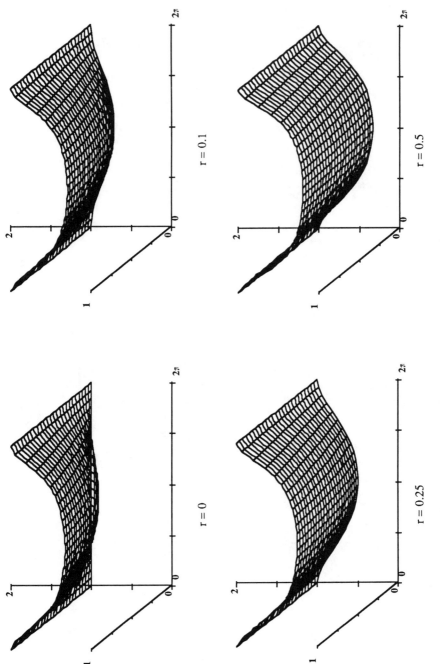

Figure A.6. Dispersion Diagram for the Upstream FD Method, Implicit, θ = 1.0

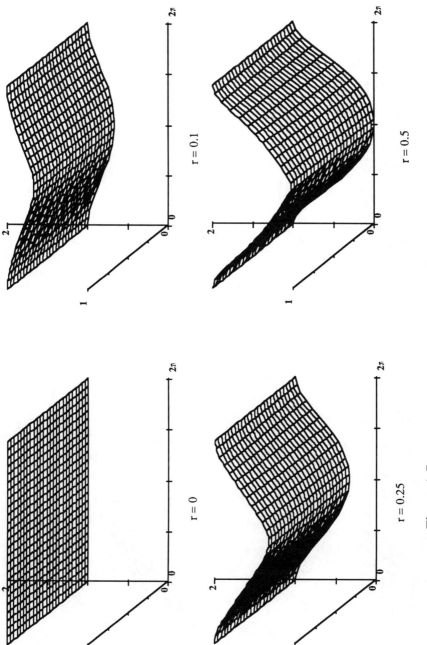

Figure A.7. Dispersion Diagram for the Centered FD Method, Implicit, θ = 0.5

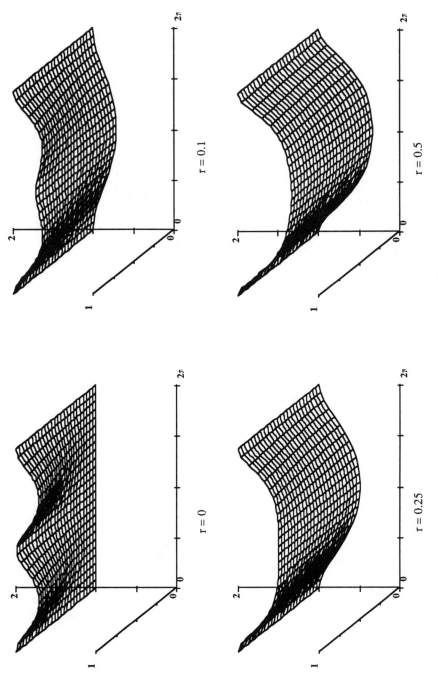

Figure A.8. Dispersion Diagram for the Centered FD Method, Implicit, $\theta = 1.0$

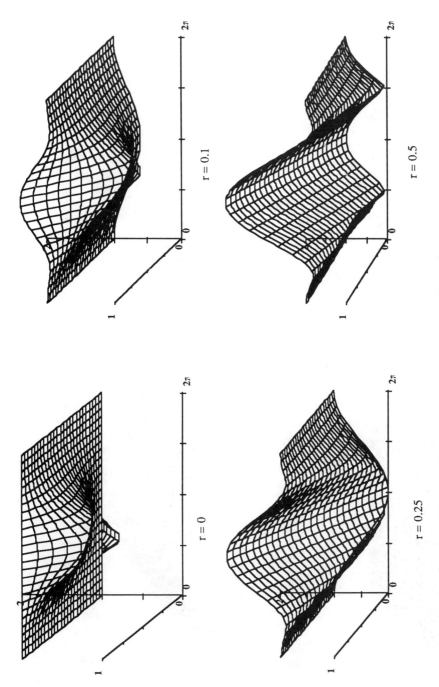

Figure A.9. Dispersion Diagram for the Taylor-Finite Difference Method

552 NUMERICAL METHODS - MOVING FRONTS

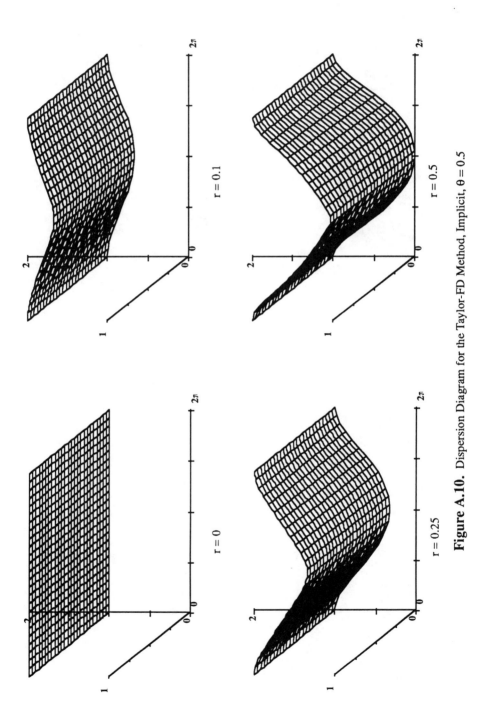

Figure A.10. Dispersion Diagram for the Taylor-FD Method, Implicit, $\theta = 0.5$

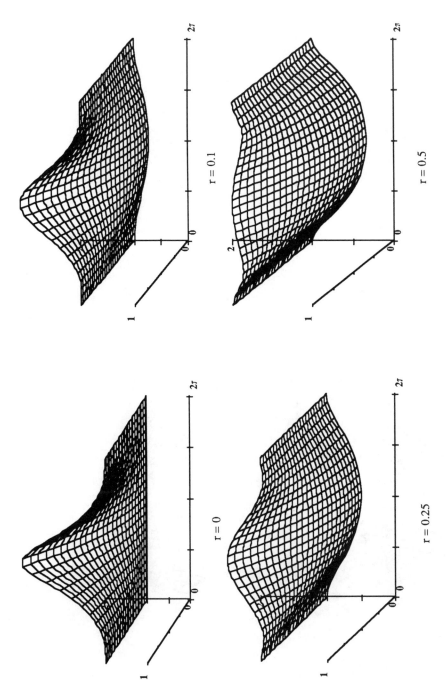

Figure A.11. Dispersion Diagram for the Taylor-FD Method, Implicit, θ = 1.0

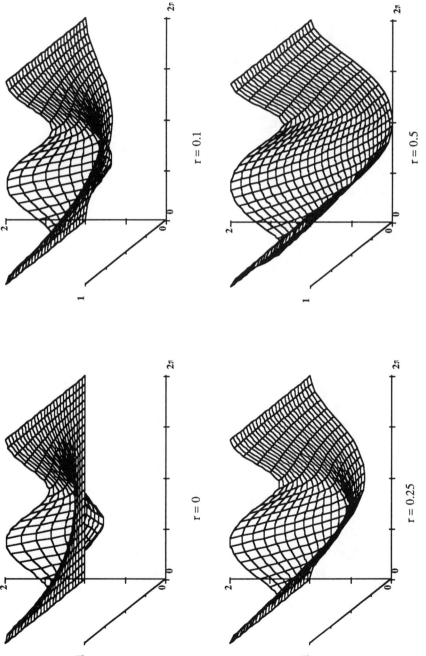

Figure A.12. Dispersion Diagram for the Taylor-FD Method, Implicit, $\theta = 0.5$, $\alpha = 1.0$

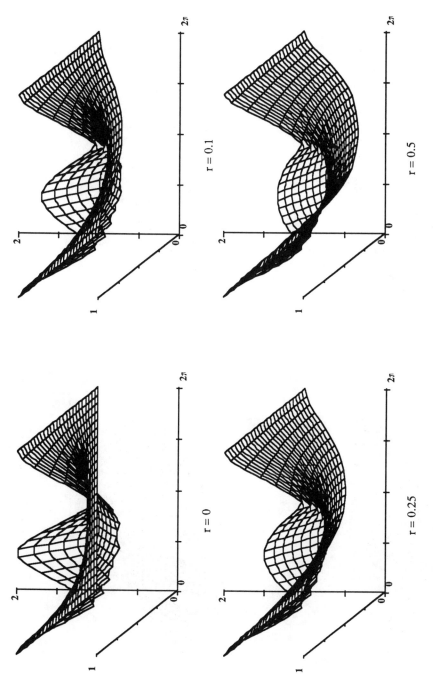

Figure A.13. Dispersion Diagram for the Taylor-FD Method, Implicit, $\theta = 1.0$, $\alpha = 1.0$

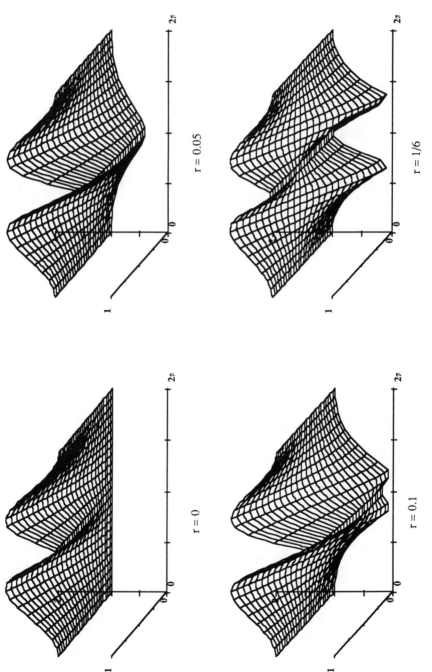

Figure A.14. Dispersion Diagram for the Galerkin Method

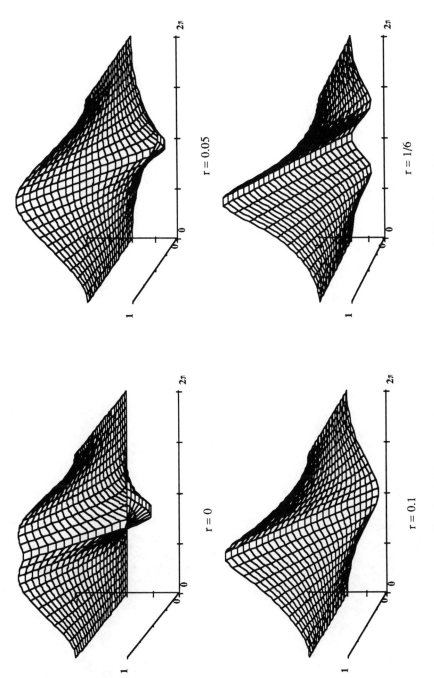

Figure A.15. Dispersion Diagram for the Petrov-Galerkin Method, $\alpha = 0.5$

558 NUMERICAL METHODS - MOVING FRONTS

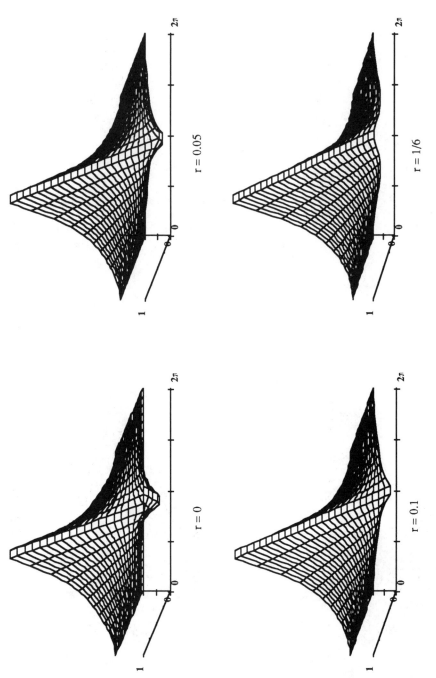

Figure A.16. Dispersion Diagram for the Petrov-Galerkin Method, $\alpha = 1.0$

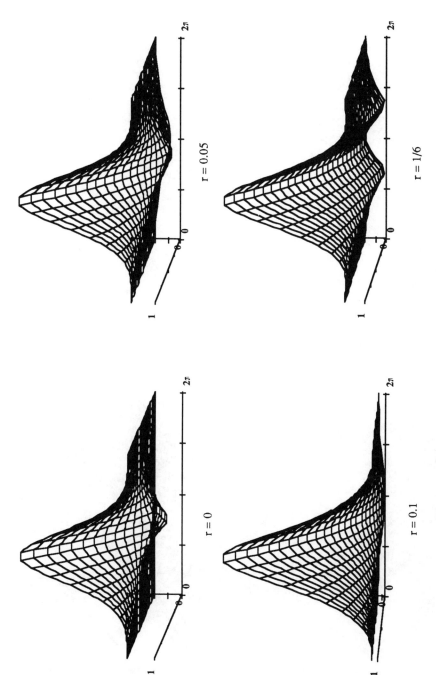

Figure A.17. Dispersion Diagram for the Taylor-Galerkin Method

560 NUMERICAL METHODS - MOVING FRONTS

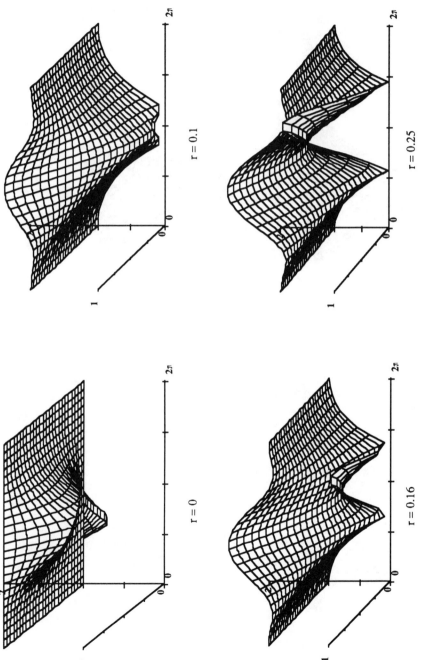

Figure A.18. Dispersion Diagram for the Morton-Parrott Method

APPENDIX 561

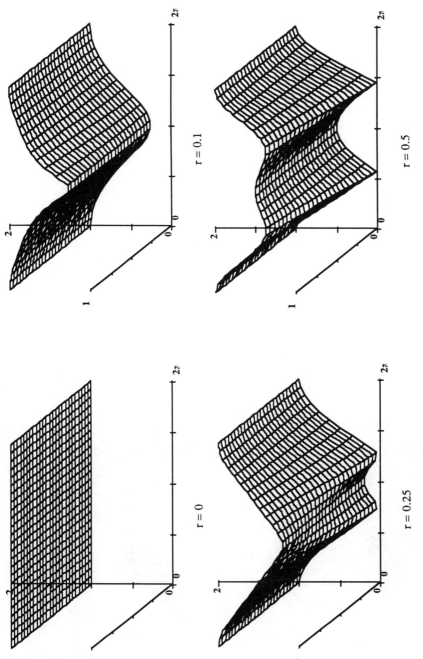

Figure A.19. Dispersion Diagram for the Galerkin Method, Implicit, θ = 0.5

562 NUMERICAL METHODS - MOVING FRONTS

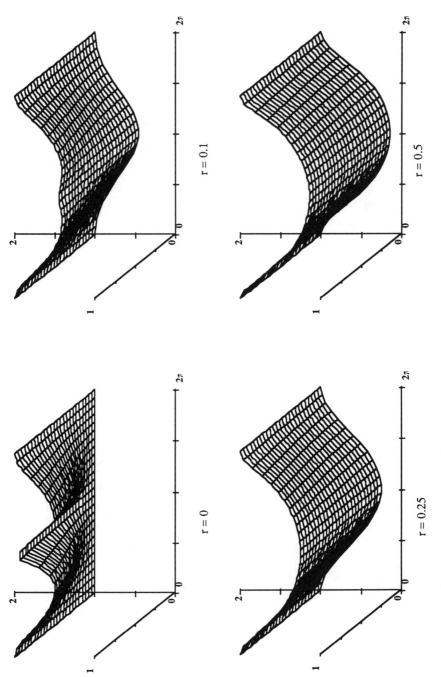

Figure A.20. Dispersion Diagram for the Galerkin Method, Implicit, $\theta = 1.0$

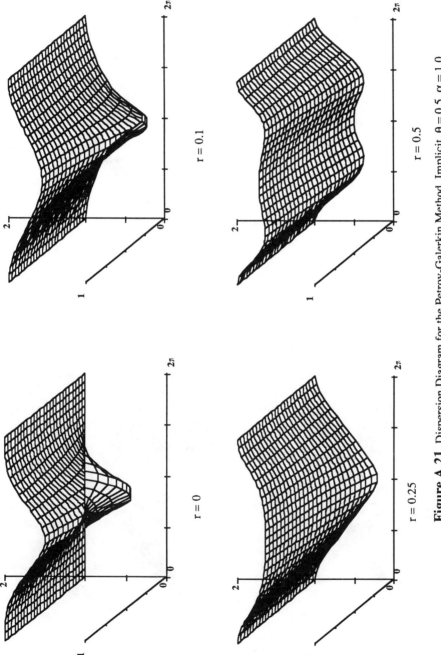

Figure A.21. Dispersion Diagram for the Petrov-Galerkin Method, Implicit, $\theta = 0.5$, $\alpha = 1.0$

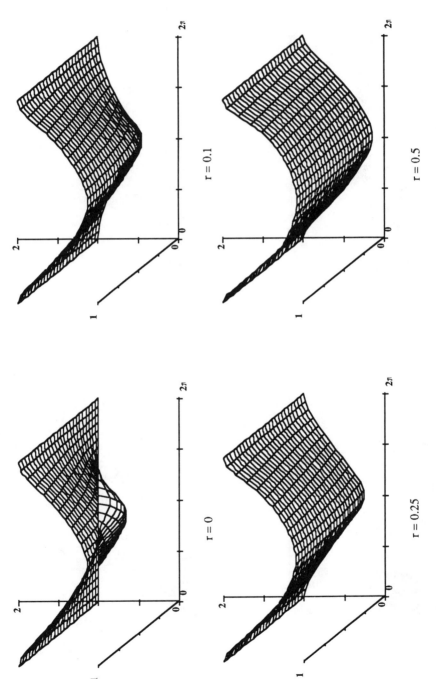

Figure A.22. Dispersion Diagram for the Petrov-Galerkin Method, Implicit, $\theta = 1.0$, $\alpha = 1.0$

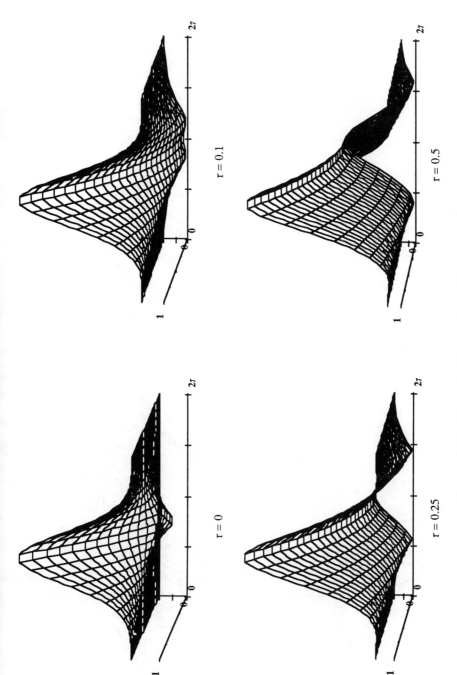

Figure A.23. Dispersion Diagram for the Taylor-Galerkin Method

566 NUMERICAL METHODS - MOVING FRONTS

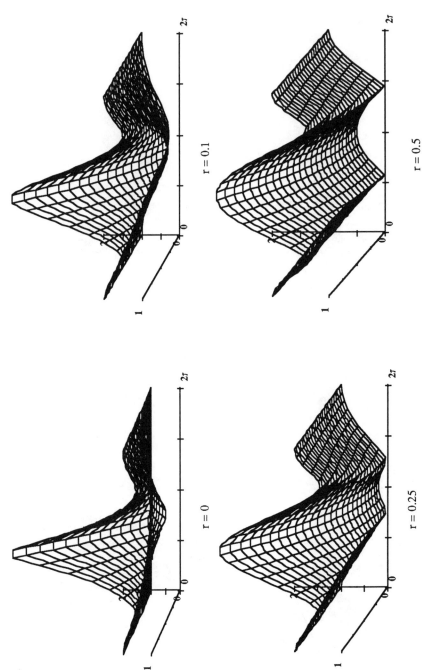

Figure A.24. Dispersion Diagram for the Implicit Taylor-Galerkin Method, $\theta = 0.5$, $\alpha = 1$

APPENDIX 567

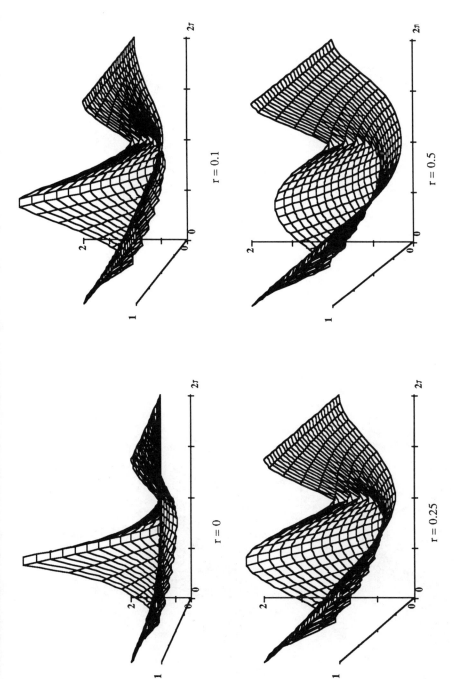

Figure A.25. Dispersion Diagram for the Taylor-Galerkin Method, Implicit, $\theta = 1$, $\alpha = 1$

568 NUMERICAL METHODS - MOVING FRONTS

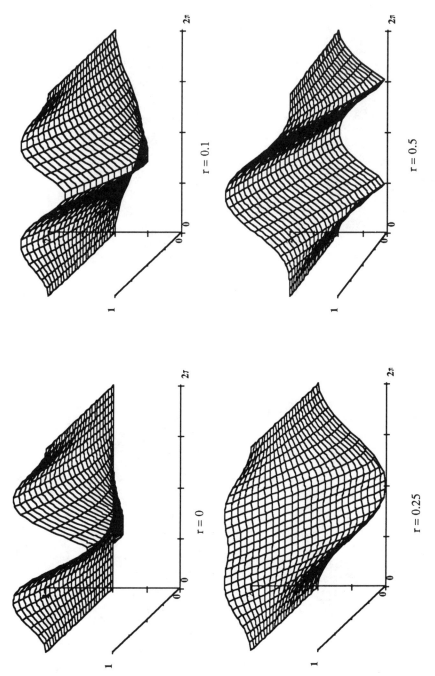

Figure A.26. Dispersion Diagram for the QUICK Method

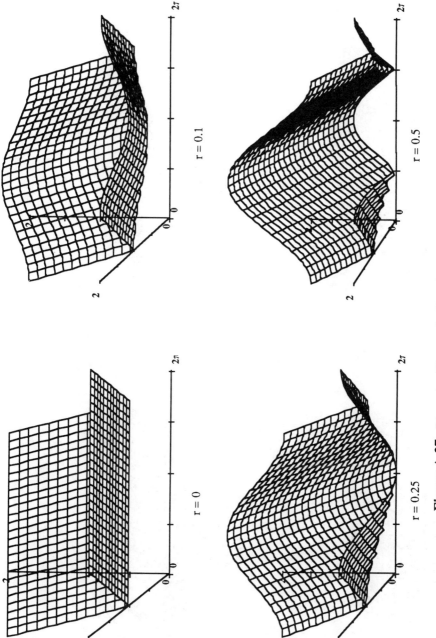

Figure A.27. Dispersion Diagram for Reaction and the Centered FD Method

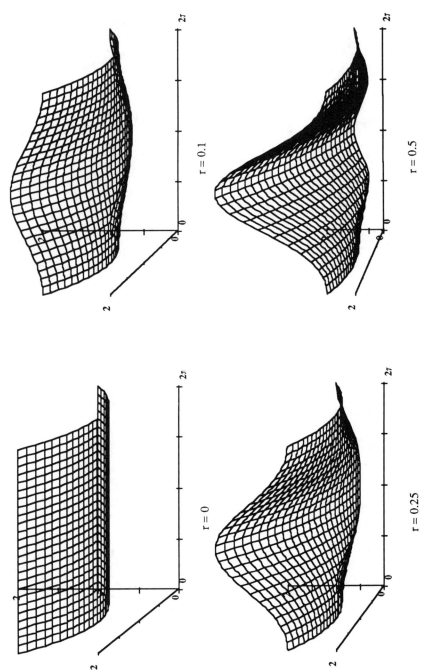

Figure A.28. Dispersion Diagram for Reaction and the MacCormack Method

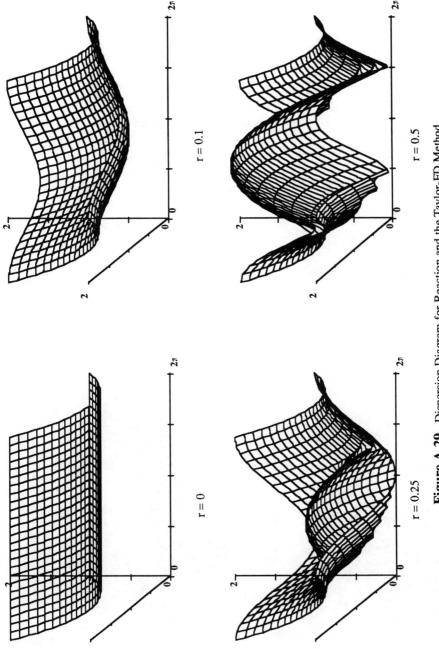

Figure A.29. Dispersion Diagram for Reaction and the Taylor-FD Method

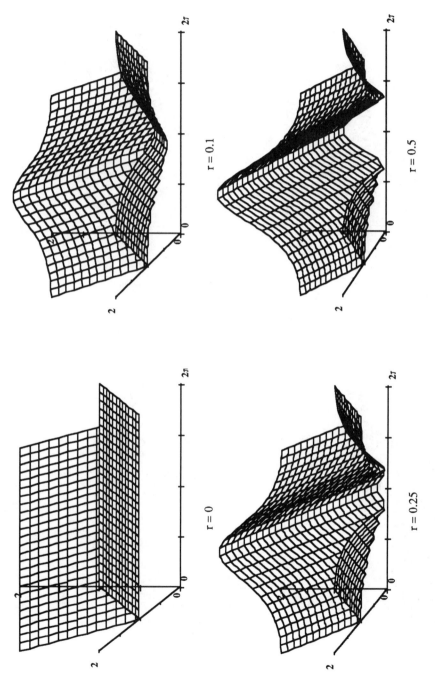

Figure A.30. Dispersion Diagram for Reaction and the Galerkin Method

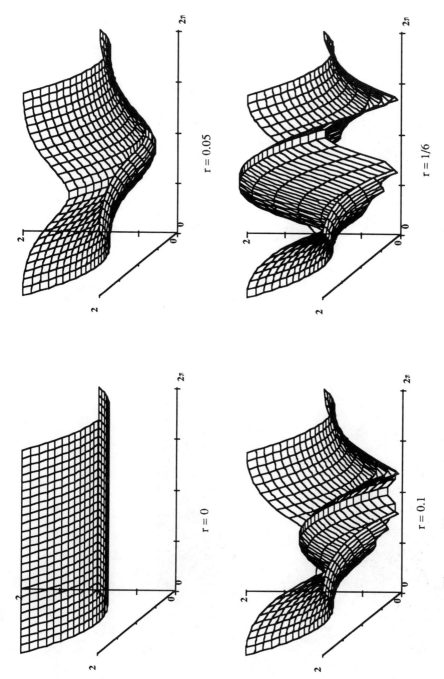

Figure A.31. Dispersion Diagram for Reaction and the Taylor-Galerkin Method

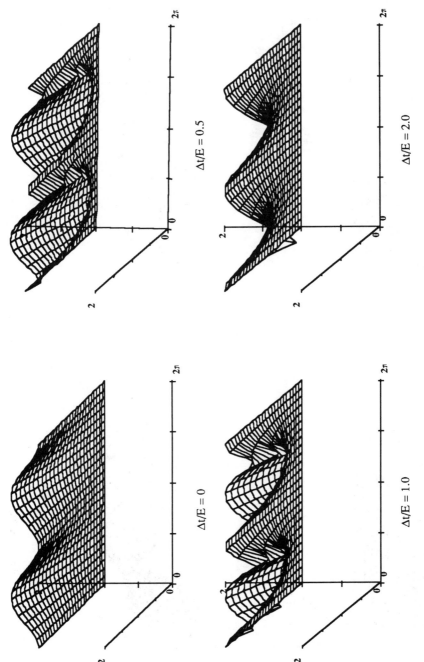

Figure A.32. Dispersion Diagram for Polymers and the Finite Difference Method

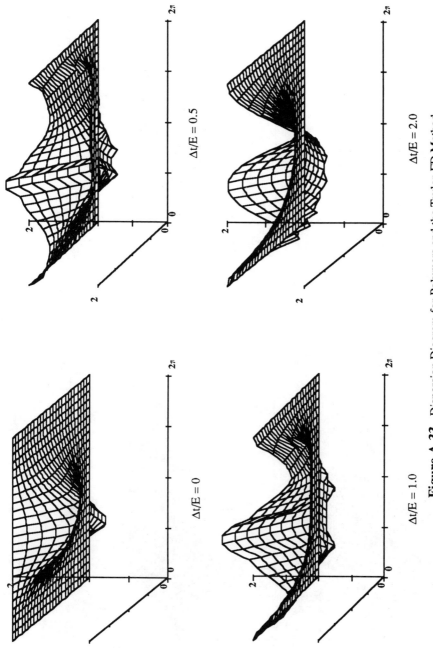

Figure A.33. Dispersion Diagram for Polymers and the Taylor-FD Method

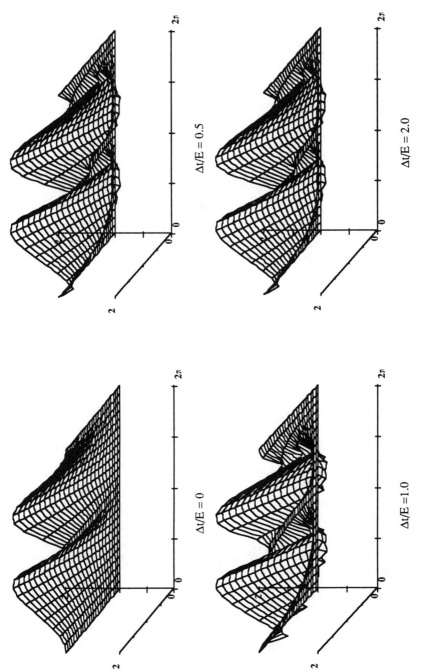

Figure A.34. Dispersion Diagram for Polymers and the Galerkin Method

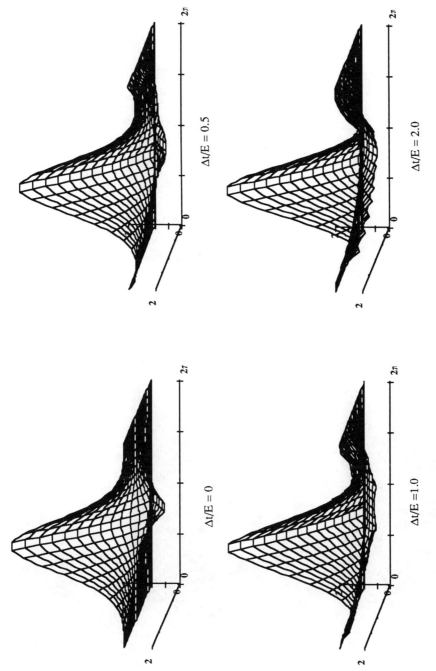

Figure A.35. Dispersion Diagram for Polymers and the Taylor-Galerkin Method

578 NUMERICAL METHODS - MOVING FRONTS

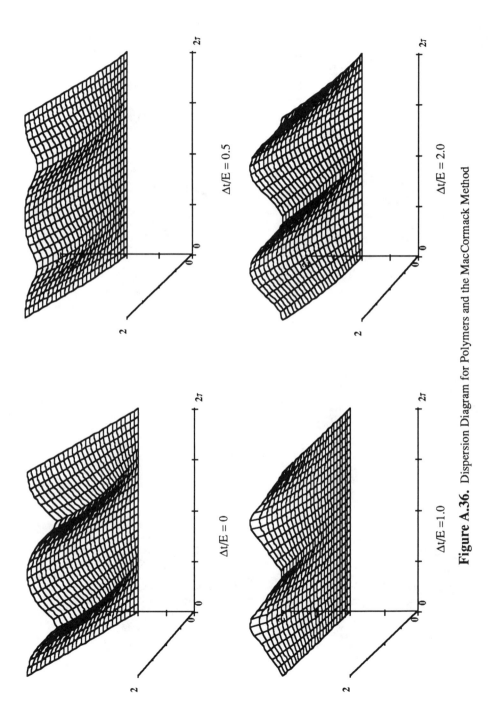

Figure A.36. Dispersion Diagram for Polymers and the MacCormack Method

COMPUTER PROGRAM DESCRIPTION

The computer programs CONVECT and APPLICATIONS allow you to solve many of the problems displayed in the book, but perhaps for different initial conditions, different parameters, or with some variation you would like to investigate. To run the programs you perform three steps (under the 'Compute' heading): choose the method, set the initial conditions, and continue computing. The explanations below are brief, since you can easily experiment with the program. The description is in three parts: general, overall features; details about individual methods or equations; and information about generic commands.

General, overall features.

In the program CONVECT you can choose several methods to solve the advection equation, convective diffusion equatin, or Burger's equation, with or without viscosity. These choices are made by choosing (under the 'Compute' heading): CD1, CD2, Burger.fd, or Burger.fem.

If you wish to solve the advection equation, then do not choose the **'include diffusion'** option.

Exclude?

○ Include Diffusion [means you are solving the equation $\frac{\partial c}{\partial t} + Pe \frac{\partial c}{\partial x} = 0$]

⦿ Include Diffusion [means you are solving the equation $\frac{\partial c}{\partial t} + Pe \frac{\partial c}{\partial x} = \frac{\partial^2 c}{\partial x^2}$]

Note that the Peclet number appears in both options. To solve the advection equation usually one chooses $Pe = 1$.

If one chooses the **'Lump Mass Matrix'** option then the time derivatives are lumped and one gets a finite difference method, e.g.

$$\frac{c_i^{n+1} - c_i^n}{\Delta t} = -\frac{c_{i+1}^n - c_{i-1}^n}{2 \Delta x} + \frac{c_{i+1}^n - 2 c_i^n + c_{i-1}^n}{\Delta x^2}$$

If one leaves the 'Lump Mass Matrix' option blank then the Galerkin treatment of time derivatives is chosen, e.g.

$$\frac{1}{6}\frac{c_{i+1}^{n+1} - c_{i+1}^{n}}{\Delta t} + \frac{2}{3}\frac{c_{i}^{n+1} - c_{i}^{n}}{\Delta t} + \frac{1}{6}\frac{c_{i-1}^{n+1} - c_{i-1}^{n}}{\Delta t} = -\frac{c_{i+1}^{n} - c_{i-1}^{n}}{2\Delta x} + \frac{c_{i+1}^{n} - 2c_{i}^{n} + c_{i-1}^{n}}{\Delta x^2}$$

When you leave the **upstream parameter** as zero, you are using either a Galerkin method or centered finite difference method, depending on the treatment of time derivatives. When the upstream parameter is set to something other than zero, one is using an upstream method, either a Petrov-Galerkin or an upstream finite difference method. Using $\alpha = 1$ and lumping the mass matrix gives the standard upstream finite difference method.

The **numerical parameters** that must be specified are the time step (Δt), number of time steps (NTIME), spatial increment (Δx), and number of spatial points (NT). When you chose the 'Continue Computing' option the calculation will proceed for NTIME steps, each of length Δt, before plotting the solution at a new time incremented by NTIME Δt. The number of spatial points, NT, is the total number of points, including both end points; the maximun value is 129. If one chooses an initial condition as a step function, there are exactly NT–1 intervals of length Δx. If one chooses another initial condtion, the minimum and maximum x value can be specified and Δx is changed to

$$\Delta x = \frac{x_{max} - x_{min}}{NT - 1}$$

The parameters must be chosen to avoid an instability; the program alln to grow without bound until it is plotted or the program crashes.

When you choose the **initial condition** the result is put into the first curve. The first solution is put into the second curve, the next one into the third curve, etc. After the 12th curve, the 2nd curve is overwritten. If you want to restart the problem, and retain the solutions just obtained, you must copy them to another curve, since the new initial condition will be in the first curve, the first solution in the second curve, and so forth. If there are two unknown functions being solved, the intial conditions are put into the first two curves, the first solutions into the third and fourth curves, etc.

There are several possible initial conditions, and you select the one you wish by cycling through a series of options displayed in windows. The types of profiles are displayed on the next page. You can also read a text file that is made in a spreadsheet or word processor; be sure to save it in 'text format'.

The **boundary conditions** are constant in time. The condition at the inlet is always with a specified value, e.g.

$$c(0,t) = c_0$$

The condition at the outlet is always an outlet flow condition. In a finite difference method this would normally be written like

$$\left.\frac{\partial c}{\partial x}\right|_{x_{max}} = 0$$

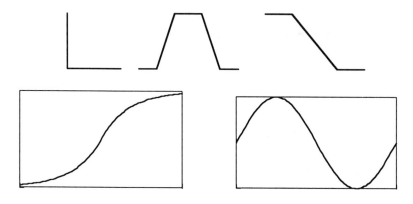

Typical Initial Conditions

However, the outflow condition is better if it allows the solution to just move out the domain naturally, without being influenced by the exit condition (i.e. as if the boundary were actually further away). The way this is done is to use the finite element outflow condition, which is described in the book as the equation for the last node. Similar equations are used for the finite difference method, but with time derivatives lumped. In this way the solution should be able to be convected out the exit without any reflections.

In the APPLICATIONS program there are several different problems that can be solved, as tabulated below.

Program	Equation Number
ADSORPTION	(9.11)-(9.12)
COMBUST	(10.112), and (10.113) or (10.126)
DRY_SOIL	(14.108)
LANGMUIR	(9.60) or (9.78)
POLYMER	(13.95)-(13.96)
POROUS	(14.51)
REACT	(10.30)

For these problems the use of numerical parameters, choices, initial conditions, and boundary conditions is similar to those in the CONVECT program.

Individual Methods

Listed here are the individual methods that can be solved using the program. For each set of options the equation that would be solved is identified.

CD1

Options chosen	Name of method	Eq. Number
First order, $\alpha = 0$, with lumping	centered FD	(2.98)
First order, $\alpha = 0$, without lumping	Galerkin FEM	(2.102)
First order, $\alpha = 1$, with lumping	upstream FD	(2.99)
First order, $\alpha \neq 0$, without lumping	Petrov-Galerkin	(2.105)
Lax-Wendroff, with lumping	Lax-Wendroff	(4.34), also p. 77
α not used; cannot use without lumping		
MacCormack, with lumping	MacCormack	(4.34), also p. 77
without lumping	Taylor-Galerkin	(5.22), also p. 98, p. 111
α not used		
Filtered leapfrog, with lumping	finite difference	p. 65, 77
without lumping	Galerkin	p. 98
cannot do convective diffusion equation with Galerkin option		
MacCormack Flux-corrected, with lumping		(6.23)-(6.29)
α not used, cannot use without lumping		
Random Choice, only with lumping		(6.14), (6.20)
Flux correction, centered derivative		
with lumping	finite difference	(6.36), (6.54)
without lumping	Galerkin FCT	(6.61)-(6.85)
α not used in either option		
Flux correction, MacCormack, with lumping		(6.36), (6.56)
α not used, cannot use without lumping		
TVD, only with lumping, α not used		(6.92)-(6.106)
ENO, only with lumping, α not used		(6.107)-(6.128)

CD2. In these options, the methods are all Galerkin methods unless the lumping option is chosen, in which case they are finite difference methods. Explicit ($\theta = 0$) or implicit ($0 < \theta \leq 1$) are chosen depending on the choice of θ. Implicit methods can be used for the Galerkin and finite difference methods, the Petrov-Galerkin method, but not the Morton-Parrott method. The implicit Petrov-Galerkin method can only be run with $\alpha = 1$. The Taylor terms are always explicit. They can be added to the Galerkin and finite difference method, will always be added to the Morton-Parrott method, and cannot be added to the Petrov-Galerkin method.

Options chosen	Name of method	Eq. Number
Galerkin, without lumping	Galerkin	p. 111
if add the Taylor terms	Taylor-Galerkin	p. 111
with lumping	finite difference	p. 77
if add the Taylor terms	Taylor-finite difference	(4.68)
Morton-Parrott, no lumping, no implicit, Taylor terms automatic		p. 98
Petrov-Galerkin, without lumping	Petrov-Galerkin	p. 111
with lumping	upstream finite difference	p. 77

Burger.fd. These options are straightforward and the only option is whether to include viscosity or not.

Name of method	Eq. Number
Explicit, upstream	(3.3), also p. 70
Lax-Wendroff	(3.5)-(3.6), also p. 70
MacCormack	p. 70
filtered leapfrog ($\alpha=1$), DuFort-Frankel	p. 70
MacCormack flux correction	(6.23)-(6.29)
Random choice	(6.8-10, 15, 16, 21)
TVD	(6.92)-(6.106)
ENO	(6.107)-(6.128)

Burger.fem. If the extra terms are included then α cannot be used.

Name of method	Eq. Number
Gaussian quadrature	(5.37)
extra terms can be evaluated with gaussian quadrature	(5.48)
or with upstream quadrature	
Trapezoid rule quadrature	(5.39)
if include extra terms	(5.50)
Finite element representation	(5.43)
if include extra terms	(5.53)
Upstream Evaluation (of convective terms)	(5.45)
usually $\alpha=1$ in time derivatives	
if include extra terms	(5.56)

ADSORPTION. Solving Eq. (9.11)-(9.12).

Name of method	Eq. Number
Upstream derivative	(9.13)-(9.14)
Random choice	(9.16)-(9.17)
MacCormack	(9.41)-(9.44)
with Flux-corrections	(9.45)-(9.48)
Galerkin	(9.18)-(9.19)
add Taylor terms to Galerkin method	(9.35)-(9.36)
if lump both time derivatives and adsorption terms and use FD	
	(9.33)-(9.34)
Euler-Lagrange	(9.53)-(9.55)

LANGMUIR. Solving Eq. (9.60) or (9.78).

Name of method	Eq. Number
Upstream derivative	(9.79)
MacCormack	(9.97)-(9.98)
with Flux-corrections	+ (6.25)-(6.29)
Taylor-Galerkin	(9.100)
can lump time derivatives	
Random choice	(9.80)-(9.85)

REACT. Solving Eq. (10.30).

Name of method	Eq. Number
Upstream derivative	(10.31)
MacCormack	(10.32)
with Flux-corrections	+ (6.25)-(6.29)
Galerkin with reaction terms	(10.53)
Gaussian	(10.40)
finite element interpolation	(10.41)
lumped	(10.42)
time derivatives lumped	
Taylor terms added - finite element	
finite difference evaluation	

With Galerkin, Gaussian reaction and Taylor-finite element (10.53)
With time derivatives lumped, lumped reaction and Taylor-finite difference is Eq. (10.34).

COMBUST. Solving Eq. (10.112)-(10.113) or (10.126). Methods are the same as for REACT except that there are no convection terms.

POLYMER. Solving Eq. (13.95)-(13.96).

Name of method	Eq. Number
Upstream derivative, can't use implicit method	(13.134)
MacCormack, can't use implicit method	(13.137)-(13.140)
with Flux-corrections	+ (6.25)-(6.29)
Galerkin with	
add Taylor terms	(13.155)-(13.156)
lump time derivatives	
lump source terms	

The Taylor-finite difference is obtained by choosing the Taylor terms plus lumping time derivatives and source terms, giving Eq. (13.151)-(13.152). You can't use Taylor terms and implicit method together.

POROUS. Solving Eq. (14.51).

Name of method	Eq. Number
Upstream derivative	(14.74)
MacCormack	(14.75)
with Flux-corrections	+ (6.25)-(6.29)
Galerkin with	
lumped time derivatives	
add Taylor terms	
TVD	(6.92)-(6.106)
ENO	(6.107)-(6.128)

The Galerkin method with Taylor terms is Eq. (14.79). The Galerkin method with lumped time derivatives and Taylor terms is the Taylor-finite difference method and is Eq. (14.78).

DRY_SOIL. Solving Eq. (14.108).

Name of method	Eq. Number
Finite difference method	(14.109)
upstream permeabilities	(14.112)
averaged permeabilities	(14.111)
MacCormack	(14.126)-(14.127)
with Flux-corrections	+ (6.25)-(6.29)
Galerkin	(14.139)
lumped time derivatives	

The Galerkin method is with averaged permeabilities.

Generic Commands

The menu options are shown below

File	Edit	Format	Data	Compute	Digitizer
New ⌘N	Copy ⌘C	Axes	Data1	CD1	Set Scale...
Clear Data	Help ⌘H	Titles	Data2	CD2	Set Curve...
Open... ⌘O		Add Captions	Data3	Burger.fd	Choose Cursor...
Close... ⌘W		Edit Captions	Data4	Burger.fem	Interpolation...
Save ⌘S			Data5	Set IC/BC	Curve Fit...
Save As...			Data6	Cont. Compute ⌘G	
Print... ⌘P			...		
Quit ⌘Q			Data12		

Some of these commands can be entered from the keyboard. They are indicated by [ ] in the lists below. See the HELP command in the program. In this program the ⌘ key requires ⌘ -letter, then return. The pull-down menues show ^C, for example, rather than ⌘ C.

New [new]. This command initializes the data to the parameters when the program starts. It destroys the current results.

Clear Data [cleardata]. Eliminates all data points from the data sets.

Open...[open] This command allows you to read a file solved by the program earlier. This can be useful if you wish to continue a calculation. Read the old file, transfer the last calculation to the initial condition, and begin. If you have curves in data1-data6 and read in a file that has curves only in data7-data12, the program will then contain all data sets, data1-data12.

Close [close]. This command closes the current results and sets the parameters to their initial value using **New**.

Save [save]. This command saves the current results using the existing filename.

Save As... This command saves the current results but gives you a chance to choose the filename and choose the folder in which to save the file.

Print...[print or printdata] This command either prints the graph or prints the data in tabular form.

Quit [quit or logout or bye]. Nothing is saved before quitting.

Copy [clipboard]. This command saves the graph to the clipboard. You may then switch to a drawing program to improve the plot. You can also paste the graph into a word processing document. The transfer is one-directional though, and the contents of the clipboard cannot be imported into the calculation program.

Help [help]. This command gives a list of possible topics.

Axes [xrange, yrange, xinterval, yinterval, xlog, xnolog, ylog, ynolog]. For each x or y axis it is possible to set the maximum and minimum values for the scale axis. If a data point is beyond those values, the axis is changed so that it includes the data point. You can choose the number of intervals to be marked with labels and whether a linear or log scale is to be used. If desired x and y grid lines can be extended across the graph.

Titles [title, xtitle, ytitle]. Use this to set the figure title, the x- and y-axis captions.

Add Captions [captions]. There is a rudimentary procedure for adding captions to the diagrams. (You may find it easier to do this in your favorite drawing program.) To add a caption, you choose this option, point to the screen, click the mouse, type the desired caption, press return, and fill out the form that is displayed. It is possible to choose the font (Application, Times, Geneva, Monaco), which must be in your system folder, the typeface, size, and to show a symbol corresponding to that used for plotting a dataset. You can later move or delete the caption (somewhat laboriously).

Edit Captions [editcaptions]. To edit a caption, choose this option, point to the desired caption and click the mouse. Then change the words, if desired, and change the format of the caption by filling in the form as in 'Add Captions'.

Data1 [enter, symbol, nosymbol, line, noline, ignore, use]. Up to 12 curves can be storred. For each curve, you can choose whether or not to show symbols for each data point, with 12 choices for types of symbols. You can choose whether or not to show lines. Typically experimental data is shown with only symbols and an interpolating equation is shown with only lines. The numerical solution is shown with both lines and data points. You can copy numbers from another data set and there is a rudimentary editing capacity.

Digitizer. The program provides a rudimentary digitizer. This can be used by reducing the figure to be copied to the size of the Macintosh™ screen and making a transparency of it. The transparency is then taped to the computer screen. Then use the following commands.

Set Scale... This command allows you to define the scale of the diagram to be copied. You point to the lower left-hand corner and give the value of x and y there; then to the right-hand side and give the x value; then to the top of the figure and give the y value. A box is drawn for your figure, and you identify if the axes are linear or logarithmic scales. These values are then used to translate a screen coordinate into an x-y coordinate.

Set Curve... Point at a data point and click the mouse; repeat for successive data points. They will be stored in one of the data sets in the order you identify them. Double click anywhere to stop.

Choose Cursor... This allows you to change the cursor to an arrow, a line (for text), a cross, a thick cross, or a watch. Usually a cross is best for setting the curve, since it is easier to point directly at a data point.

Interpolation... You can fit a cubic B-spline through a set of data points. This is not a curve fitting routine–the interpolated curve will go through every data point exactly. You choose the data set to interpolate or you can choose the parameters defining a B-spline from a previous calculation that has been saved. You indicate in which dataset you wish the new curve to be stored. The B-spline is computed, you are given the values of x, y, y' and y" at each data point. The curve will be plotted using a number of equi-spaced points which you specify.

Using 50 or more points generally gives a smooth curve, but the curve is actually drawn as a series of straight line segments through the number of points you specified for the plot (not the number of knots in the B-spline). The straight line segments provide the curve that is saved in the dataset. If you wish to save the B-spline data, for later use, use the option to save parameters.

Curve Fit... This option works like the Interpolation option, except that the data is fit to a polynomial that you specify. This is a true curve-fit and can be used to fit irregular data containing experimental error. To fit with a straight line, use a polynomial of degree 1; for a quadratic use degree 2, etc. The output consists of the parameters in the polynomial that is the best fit, and the standard deviation of the parameters. Next is given the table of x-y values for the polynomial curve (plotted with as many points as you specified), and the total variance, which is the quantity that is being minimized.

$$I = \sigma^2 = \frac{1}{N} [\, y_i - y(x_i) \,]^2$$

AUTHOR INDEX

A

Akin, J. E., 268, **314**
Alexander, R., 291, **314**, 425, **431**
Amundson, N. R., 318-320, 332, 334, 336, 354, 357, **400**, 506, **541**
Anderson, D. A., 51, **54**, 69, 80, 81 **92**, 264, **314**
Ang, K. K., 286, **315**
Aris, R., 318-320, 332, 334, 336, **354**, 357, **400**, 506, **541**
Armstrong, R. C., 460, **484**
Arora, P. R., 3, **9**, 402, **432**
Atkinson, J. D., 168, **187**

B

Babuska, I., 194, 218, **253**, 287, **315**
Baker, A. J., 102, **125**
Barga, W., 177, **185**
Barillot, Ph., 427, **431**
Barrett, J. W., 165, 167, 171, **185**, **186**
Bassingthwaighte, J. B., 320, **353**
Becker, E. B, 268, **315**
Belleudy, P., 215, **256**
Benner, R. E., Jr., 516, **540**
Berger, M. J., 283, **315**
Bhattacharya, A., 396, **399**
Bieterman, M., 218, **253**
Bird, R. B., 460, **484**
Blottner, F. G., 194, **253**
Bonnerot, R., 412, 425, **431**
Book, D. L., 137, 138, **186**, 278, **315**, 325, 350, 352, **353**
Boris, J. P., 5, **9**, 137, 138, **186**, 278, **316**, 350, 352, **353**
Bowles, J. B., 313, **317**, 477, **495**
Brooks, A., 29, **46**, 163, **187**, 269, **316**, 344, 346, **353**, 439, 450, **456**
Brown, R. A., 428, 430, **431**, **432**
Buckley, S. E., 504, **540**

C

Campion-Renson, A., 440, **455**
Canuto, C., 8, **9**
Carey, G. F., 181, **186**, 194, 204, 233, 234, **253**, **255**, 268, **315**
Carnahan, B., 155, **186**
Carslaw, H. S., 409, **431**
Cavendish, J. C., 172, **187**, 215, **256**
Chan, S. T., 442, 444, 446-448, **455**
Chawla, T. C., 182, **186**, 212, **253**, 413, **431**
Chen, K. N., 450, **456**
Chen, W. H., 151, 152, **186**, 508, **540**
Chern, I. L., 288, **315**
Chieng, C. C., 177, **187**, 247, **255**
Chierici, G., 180, **186**
Chorin, A. J., 129, 133, **186**, 440, **455**
Christiansen, J., 194, **254**
Christie, M. A., 278-280, **316**, 538, 539, **541**
Christie, I., 29, **46**
Chung, T. J., 102, **125**, 364, **399**
Cliffe, K. A., 436, **456**

Collella, P., 129, 133, **186**
Comini, G., 416, 418, 424, **431**
Concus, P., 129, **186**
Courant, R., 287, **315**
Coyle, J. M., 218, **254**
Crank, J., 402, **431**
Crochet, M. J., 440, **455**
Cuvelier, C., 436, **455**

D

Daripa, P., 538, 539, **540**
Davis, H. T., 3, **9**, 516, **540**
Davis, S. F., 220, **254**
Deans, H. A., 364, **401**
Degreve, J., 385, **399**
Del Guidice, S., 416, 418, 424, **431**
Demkowicz, L., 167, 171, **186**, 287, **315**, 344, 346, 348, **353**
Denn, M. M., 475, **495**
Derby, J. J., 428, **432**
Devloo, P., 102, **125**, 455, **456**
Didwania, A. K., 342, 343, **353**
Dietrich, D. E., 68, 69, 81, **92**
di Franco, R. B., 534, **540**
Dimitriou, P., 385, **399**
Dissinger, G. R., 220, **254**
Donea, J., 97, 99, **125**, 322, **353**, 449, **455**
Doss, S. K., 223, 234, 235, **254**, 383-385, **400**
Douglas, J., Jr., 240, 243, **254**
Dudukovic, M. P., 396, **400**
Dukowicz, J. K., 263, **315**
DuPont, T., 218, 233, **254**
Durlofsky, L. J., 151, 152, **186**, 508, **540**
Dwyer, H. A., 220, **254**, 383, **399**

E

Eisenstat, S. C., 260, **315**
El-Ageli, M. A., 180, **186**
Engleman, M., 436, **456**
Engquist, B., 151, 152, **186**, 508, **540**
Essenhigh, R. H., 397, **400**
Ettouney, H. M., 430, **431**
Ewing, R. E., 240, **254**

F

Farmer, C. L., 239, **254**
Farooq, S., 335, **353**
Ferguson, N. B., 194, **254**
Finlayson, B. A., 164, 165, 171, 181, **186**, **187**, 194, 212, **253-255**, 259, 294, **315**, **317**, 391, 393-396, **399**, **401**, 463, **495**, 518, 521, **540**
Fisher, R. A., 379, **399**
Flaherty, J. E., 218, 220, **254**
Fokas, A. S., 516, **541**
Forsyth, G., 23, 25, 33, **46**
Frauenthal, J. C., 534, **540**
Frey, W. H., 287, **315**
Friedrichs, K., 287, **315**
Fukusako, S., 413, **431**
Furzeland, R. M., 405, **432**

G

Galinas, R. J., 223, 234, 235, **254**, 383-385, **400**
Garder, A. O., 239, 241, **254**
Gatica, J. E., 368, **400**
Gawdzik, A., 396, **400**
Ghajar, A. J., 174, **188**
Gilmore, S. D., 427, **432**
Giuliani, S., 97, **125**, 322, **353**, 449, **455**
Glimm, J., 129, 136, **186**, 288-290, **315**, **316**, 534, 536-539, **540**
Goodman, T. R., 405, **432**
Gopalakrishnan, T. C., 451, **455**
Gottlieb, D., 8, **9**
Grcar, J. F., 386, **401**
Gresho, P. M., 347, **353**, 436, 442-444, 446-448, **455**, **456**
Griffith, R., 419, **432**
Griffiths, D. F., 29, **46**, 228, **255**
Gropp, W. D., 283, **316**, 349, 352, **353**
Grove, J., 288, **316**
Guceri, S. I., 427, **432**
Gudunov, S. K., 129, **187**
Gupta, S. C., 3, **9**, 402, **432**

H

Hain, K. H., 137, **186**

Harrison, G. W., 3, **9**, 209, **254**
Harten, A., 151, **187**, **188**, 220, **254**
Hassager, O., 460, **484**
Hassan, O., 451, **456**
Hawken, D. M., 454, **455**
Heinrich, J. C., 164, **188**
Herbst, B. M., 228, 238, **255**
Hestenes, M. R., 260, **316**
Hirano, H., 450, **456**
Hlavacek, V., 368, 385, 395, **399**, **400**
Holt, M.,129, 133, 134, **187**
Hrymak, A. N., 231-235, **255**, 412, **432**
Hu, S. S., 219, 222, **255**, 342, 343, **353**
Huffenus, J. P., 449, 450, **455**
Hughes, T. J. R., 29, **46**, 161-163, 168, **187**, 269, **316**, 344-346, **353**, **354**, 439, 450, **456**
Humphrey, D., 194, **253**
Hussaini, M. Y., 8, **9**
Hutton, A. G., 346, **354**
Huyakorn, P. S., 346, **354**
Hwang, B. C., 440, **456**
Hyman, J. M., 220, **254**

I

Isaacson, E., 288, **315**, , 534, 538, **540**, **541**

J

Jackson, C. P., 436, **456**
Jaeger, J. C., 409, **431**
Jamet, P., 412, 425, **431**
Jensen, O. K., 181, **187**, 212, **255**
Jiang, B. N., 233, 234, **253**, **255**
Joseph, B., 396, **399**
Josse, S., 463, **495**

K

Kawahara, M., 449, 450, **456**
Kee, R. J., 220, **254**, 377, **400**
Khaletzky, D., 449, 450, **455**
Kikuchi, N., 287, **316**
Kim, J. W., 102, **125**
Kim, Y. M., 102, **125**, 364, **399**
King, M. J., 538, **541**
Kistler, S. F., 298, **316**
Klingenberg, C., 288, **316**

Koszykowski, M. L., 219, **256**, 282, **317**
Kovacs, A.,449, **456**
Kumar, R., 294, **317**
Kunii, D., 397, **401**
Kuo, S., 428, **432**
Kurylko, L., 397, **400**

L

Lacroix, M., 427, **432**
Lan, C. W., 428, **432**
Landau, H. G., 405, **432**
Landis, F., 405, **432**
Laumbach, D. D., 177, **187**
Laval, H., 97, **125**, 322, **353**, 449, 454, **455**, **456**
Lax, P. D., 64, **92**
Leaf, G., 182, **186**, 212, **253**, 413, **431**
Lee, D. N., 379, **400**
Lee, H. Y., 450, **456**
Lee, R. L., 347, **353**, 436, 442, 444, 446-448, **455**, **456**
Lenhoff, A. M., 320, **354**
Leonard, B.P., 174, 176, **187**
Leverett, M. C., 504, **540**
Lewis, R. W., 416, 418, 424, **431**
Li, K. M., 134, **187**
Liggett, J. A., 234, **255**
Lightfoot, E. N., 320, **354**
Lin, H., 177, **187**, 247, **255**
Lindquist, B., 288, **316**, 538, 539, **540**, **541**
Liu, W. K., 439, 450, **456**
Löhner, R., 102, **125**, 139, 145, 150, **187**, 274, 278, 280, **316**, 451, 454, 455, **456**
Ludwig, R., 218, **254**
Luther, H. A., 155, **186**
Lynch, D. R., 209, 228, 245, **255**, 407-409, 411, 412, 414-416, 419, 421-423, 427, **432**

M

MacCormack, R. W., 68, 75, **92**
Mack, A., 383, **400**
Mallet, M., 345, **354**
Manselli, P., 291, **314**, 425, **431**
Marchesin, D., 136, **186**, 288, **315**, 534, 536-538, **540**

Mastin, C. W., 294, **317**
Masuda, M., 450, **456**
May, W. G., 342, 343, **353**
McBryan, O., 136, **186**, 288-290, **315**, **316**, 534, 536-539, **540**
McCracken, T. A., 528, 531, 532, **541**
McIntosh, A., 260, **316**
McRae, G. J., 231, 232, 234, 235, **255**
Menikoff, R., 288, 289, **316**
Meunier, F., 321, **354**
Michelsen, M. L., 181, **188**
Mihalas, D., 516, **541**
Miller, R., 419, **432**
Miller, A., 194, **253**, 287, **315**
Miller, J. A., 377, **400**
Miller, K., 223, 227, 231, 234, 235, **254**, **255**, 291, **314**, 383-385, **400**, 419, 425, **431**, **432**
Miller, R. N., 223, 227, 231, **255**
Minkowycz, W. J., 182, **186**, 212, **253**, 413, **431**
Mitchell, A. R., 29, **46**, 228, 238, **255**
Mitchell, R. E., 386, **401**
Miyauchi, Y., 450, **456**
Mizukami, A., 345, 346, **354**
Moler, C. B., 23, 25, 33, **46**
Morgan, K., 102, **125**, 139, 145, 150, **187**, 274, 278, 280, 287, **316**, **317**, 451, 454, 455, **456**
Morton, K. W., 61, 63, 82, **92**, 165, 167, 171, **185-187**, 480, **495**
Murray, W. D., 405, **432**

N

Nassersharif, B., 419, **432**
Neuman, S. P., 241, **255**, 292, **316**
Norman, M. L., 516, **541**
Novy, R. A., 516, 517, **541**

O

Oden, J. T., 102, **125**, 167, 171, **186**, 268, 287, **315**, **316**, 344, 346, 348, **353**, 455, **456**
Okuyiga, M. O., 514, 516, 517, **541**
Oliger, J., 283, **315**
O'Neill, K., 407-409, 411, 412, 414-416, **432**

Oran, E. S., 5, **9**, 278, **316**
Orszag, S. A., 8, **9**, 347-350, **354**
Osher, S., 151, 152 155, 158, **186**, **188**, 508, **540**

P

Padmanabhan, L., 538, **540**
Park, N. S., 234, **255**
Park, Y. J., 364, **401**
Parrott, A. K., 63, **92**, 171, **187**, 278-280, **316**, 538, 539, **541**
Peaceman, D. W., 239, 241, **254**, 259, 260, **317**, 496, **541**
Pearson, C. E., 193, **255**
Pedersen, D. R., 413, **431**
Pelcé, P., 402, **432**
Pepper, D. W., 454, **456**
Peraire, J., 139, 145, 150, **187**, 278, 280, 287, **316**, **317**, 451, 454, 455, **456**
Pereyra, V., 193, **255**
Peyret, R., 262, **317**, 441, **456**
Pirkle, J. C., Jr., 342, 343, **353**
Pletcher, R. H., 51, **54**, 69, 80, **92**, 264, **314**
Plohr, B., 288, **315**, **316**
Plover, T., 194, 204, **253**
Pope, G. A., 508, **541**
Porteous, K. C., 475, **495**
Poulain, C., 134, **187**
Pozzi, A. L., 239, 241, **254**
Prenter, P. M., 197, **256**
Price, H. S., 36, **46**, 87, **92**, 172, **187**, 215, **256**
Proskurowski, W., 129, **186**
Puszynski, J., 368, 385, **399**, **400**

Q

Quartapelle, L., 97, **125**, 322, **353**, 449, **455**
Quarteroni, A., 8, **9**

R

Rakowski, L., 396, **400**
Ramachandrau, P. A., 396, **400**
Ramanathan, S., 294, **317**
Ramaswamy, B., 450, 454, **456**
Ramos, J. I., 376, 379-381, **400**
Ramshaw, J. D., 263, **315**
Ray, W. H., 514, 516, 517, **541**

Reyna, L., 538, **541**
Reitz, R. D., 379, **400**
Rhee, H. K., 318-320, 332, 334, 336, **354**, 357, **400**, 506, **541**
Richtmyer, R. D., 61, 64, 74, 82, **92**, 480, **495**
Roache, P. J., 68, 69, 72, 74, 75, **92**
Roe, P. L., 134, **188**, 288, **317**
Rosenberg, D. U. von, 178, **188**
Roth, P., 383, **400**
Rubinsky, B., 426-428, **433**
Russell, R. D., 194, **254**, **256**
Russell, T. F., 240, 243, **254**, **256**
Ruthven, D. M., 335, **353**

S

Saad, N., 508, **541**
Sackinger, P. A., 428, **432**
Saito, H., 298, **317**
Sanders, B. R., 220, **254**, 383, **399**
Sani, R. L., 347, **353**, 436, 443, **455**, **456**
Schiesser, W. E., 219, 222, **255**, 342, 343, **353**
Schoombie, S. W., 228, 238, **255**
Scriven, L. E., 3, **9**, 298, **316**, **317**, 516, **540**
Segal, A., 436, **455**
Seki, N., 413, **431**
Sepehrnoorl, K., 508, **541**
Sewell, E. G., 193, **255**
Sharp, D. H., 288, 289, **316**
Sheintuch, M., 212, **256**, 381, **400**
Shih, T. M., 440, **456**
Shimizu, M., 450, **456**
Shimizu, F., 397, **401**
Shu, C. W., 151, 155, 158, **188**
Singer, A. P., 454, **456**
Smith, R. M., 346, **354**
Smooke, M. D., 219, **256**, 282, **317**, 377-379, 381-384, 386, **400**, **401**
Sod, G. A., 69, 82, **92**, 129, 134, 137, 138, **188**, 336, **354**
Sohn, J. L., 102, **125**, 364, **399**
Spalding, D. B., 379, **401**
Steffler, P. M., 177, **188**
Stevens, W. N. R., 440, **456**
Stewart, J. R., 451, **456**
Stewart, W. E., 181, **188**

Strouboulis, T., 102, **125**, 455, **456**
Stuart, A., 372, **401**
Sun, L. M., 321, **354**
Sweby, P. K., 151, **188**

T

Tamaddon-Jahromi, H. R., 454, **455**
Tan, C. H., 440, **456**
Tannehill, J. C., 51, **54**, 69, 80, **92**, 264, **314**
Taylor, T. D., 262, **317**, 441, **456**
Teletzke, G. F., 3, **9**
Temperville, A., 215, **256**
Tezduyar, T. E., 364, **401**
Thames, F. C., 294, **317**
Thareja, R. R., 451, **456**
Thomaidis, G., 241, 243, 244, 246, **256**
Thommen, H. U., 80, **92**, 261, **317**
Thompson, J. F., 294, **317**
Towler, B. T., 534, **540**
Townsend, P., 454, **455**
Trefethen, L. N., 307, 309-313, **317**
Tryggvason, G., 288, **316**, 538, 539, **540**
Tsai, H. L., 428, **433**
Tzanos, C. P., 220, **256**, 346, **354**, 413, **433**

U

Upson, C., 442, 444, 446-448, **455**

V

Vahdati, M., 139, 145, 150, **187**, 278, 280, 287, **316**, **317**, 454, **455**, **456**
Valliappan, S., 286, **315**
van Steenhoven, A. A., 436, **455**
Varga, R. S., 36, **46**, 87, **92**, 172, **187**, 215, **256**
Vichnevetsky, R., 313, **317**, 477, **495**
Villadsen, J.V., 181, **188**
Voller, V. R., 427, **432**
Votruba, J., 395, **400**

W

Wang, N. H. L., 342, **354**
Warming, R. F., 151, **188**
Warren, J. E., 36, **46**, 87, **92**
Weber, H. J., 383, **400**
Webster, M. F., 454, **455**

Weinberger, H. F., 475, 476, **495**
Welge, H. J., 504, **541**
Wendroff, B., 64, **92**
Westerberg, A. W., 231-235, **255**, 412, **432**
Wheeler, M. F., 240, 241, 243, 244, 246, **254, 256**
White, A. B., 194, **256**
Wilkes, J. O., 155, **186**
Winkler, K. H. A., 516, **541**
Wormeck, J. J., 68, 81, **92**

Y

Yanenko, N. N., 380, **401**
Yang, G., 215, **256**
Yang, J. C., 450, **456**
Yaniv, S., 288, **315, 316**
Yanosik, J. L., 528, 531, 532, **541**
Yee, H. C., 151, **188**
Yoo, J., 426, 427, 428, **433**
Yortsos, Y. C., 516, **541**
Yoshida, K., 397, **401**
Young, L. C., 294, **317**, 391, 393-396, **399, 401**, 500-504, 528-534, **541**
Yu, Q., 342, **354**
Yu, C.-C., 164, **188**

Z

Zalesak, S. T., 137, 139, **188**, 275, **317**, 350, 352, **353**
Zang, T. A., 8, **9**
Zhu, J. Z., 287, **317**
Zienkiewicz, O. C., 29, **46**, 102, **125**, 274, 287, **316, 317**, 416, 418, 424, **431**, 451, **456**
Zuragat, Y. H., 174, **188**
Zygourakis, K., 241, 243, 244, 246, **256**

SUBJECT INDEX

A

Adaptive mesh 189-206, 346, 378-383, 455, 579
  comparison 206
  curvature criterion 194
  equidistribution 378, 383, 516
  estimated error criterion 197, 199, 200, 204
  global error criterion 194
  gradient criterion 193, 197, 198, 201
  large curvature 413
  rate of heat transfer 413
  residual criterion 194, 197, 198
  rotating cone problem 352
  second derivative criterion 194, 346
  steady-state problems 189
  strategy to add nodes 196
  truncation error criterion 194
  two-dimensional 384
  upstream finite difference method 197, 203, 205
  variational criterion 197, 202
  (*see also* Nodes, moving)
Adsorption 7, 343
  ADSORPTION program 351, 581, 584
  comparison 342
  linear 318-332
    Euler-Lagrange method 326, 330, 332
    Galerkin finite element method 321, 584
    Lax-Wendroff method 324
    MacCormack method 324, 331
    MacCormack method with flux-correction 324, 329, 332, 584
    random choice method 321, 328, 331, 584
    Taylor-finite difference method 323
    Taylor-Galerkin method 323, 328, 329, 331
    upstream finite difference 321, 327, 330, 584
  nonlinear 332-343, 584
    finite difference, upstream 324, 336, 584
    MacCormack method 338, 339
    MacCormack method with flux-correction 332, 339, 341, 342, 584
    random choice method 337, 341, 584
    Taylor-Galerkin method 340, 584
    upstream finite difference method 336
  pressure swing 318

Advection equation 4, 30-35, 55-69, 89, 91, 97-102, 120, 125, 183, 252, 308, 310, 332, 351, 365
  comparison 127
  comparison, best methods 248
  finite difference method 66
  Galerkin finite element method 98
  Lax-Wendroff method 32-33, 65, 67
  Leapfrog method 65, 68, 69
    Galerkin 98, 100
  MacCormack method 65, 67
  Morton-Parrott-Galerkin method 98
  Petrov-Galerkin method 34, 98, 100
  Taylor-Galerkin method 98, 100
  upstream finite difference method 31-32, 56, 63, 65
Alternating direction method 259
Alumina 390, 391
Amplification factor 58, 64, 83, 90, 370, 542-578
Analytical approximation 431
  comparison 412
  one-phase problem 408, 409
"Anti-diffusion" 138, 142, 148, 281
Arakawa's method 347

B

Backward Euler method 83, 116
  comparison 464, 468-469
  initial value problem 464, 468
Bioengineering 320
Bioseparations 3
Body force 443
Boundary
  false 173
  outflow 269, 302
  solid 269
Boundary condition 35
  Dirichlet 264
  essential 301, 438
  first kind 264
  at infinity 302, 403
  natural 7, 27, 163, 192, 300-301, 305
  natural, pointwise solution 306

natural, weighted 302
Neumann 264, 445, 449
outflow 10, 35, 580
Robin 264
second kind 264
stream function 440
third kind 264
vorticity, lack of 440
Boundary layer 11, 305, 434
Buckley-Leverett problem 503-517
Burger's equation 3, 4, 47-54, 69-76, 88-89, 91, 102-110, 120-124, 336, 583
  comparison 49, 54, 128, 250
  DuFort-Frankel method 70
  exact solutions 70, 71
  Galerkin finite element method 53, 122, 583
    finite element interpolation 105
    trapezoid quadrature 104
    upstream 105
  Lax-Wendroff method 48-49, 52, 70, 72, 88
  Leapfrog method 73, 88
  MacCormack method 70, 72, 88
  Petrov-Galerkin method 50-51, 53-54, 103, 104, 121, 122
  stability 70
  Taylor-Galerkin method 106-110, 121, 123, 124, 583
  two-dimensional
    Taylor-Galerkin method 271
  upstream finite difference method 47-48, 51-52, 70, 72, 88

C

Calculus of variations 253, 301
  Fundamental Lemma 253
Capillarity 513, 531
Capillarity Function, dry soil 519
Catalyst 390
Catalytic converter 1, 390, 391
Centroid, triangular element 267
Ceramic 390, 391
Char 396
Characteristic, dividing 357
Characteristic direction 333, 335, 478, 506
Characteristic velocity 478, 479
Chemical flooding 2, 3, 539
Chemical reaction 6, 7, 355-398
Chromatography 3, 7, 318-343
Collocation, adaptive mesh 396
  (see orthogonal collocation)
COLSYS program 194
Combustion 1, 7, 376-390
  carbon monoxide 392
  hydrogen 377
Comparison 126-129, 248-251
  best methods 251
  (see also Individual methods)
Complex number 59
Compressible-fluid method 450
  artificial 440-441

Computation time 381, 383
Computer
  CRAY 449
  parallel 8
  vector 384
Conjugate gradient method 260
Conservation of energy 435
Constitutive equation 461
Continuity equation 438, 440, 443
Continuum 435
CONVECT program 45, 90, 91, 124, 125, 143, 150, 183, 252, 579
  advection equation 184
  convective diffusion equation 184, 185
  first node 183
Convection 4, 5, 6
  two-dimensional 343-352, 428
Convective diffusion equation 4, 7, 8, 35-43, 66, 76-87, 110-120, 125, 160, 191, 211, 252, 501, 504, 579
  centered finite difference, flux-correction 145
  comparison 127, 249
  DuFort-Frankel 77, 81
  ENO method 159, 582
  Eulerian coordinate system 239
  Euler-Lagrange method 244, 245
  exact solution 35
  finite difference method 36-37, 77, 80, 582
    implicit 77, 83, 84
  flux correction steps 143, 582
  Galerkin finite element method 39-40, 112, 582
    implicit 111, 117
  Galerkin method with flux-correction 150
  Lagrangian coordinate system 239
  Laumbach's method 179
  Lax-Wendroff method 77, 81, 582
  MacCormack method 77, 81, 114, 582
  MacCormack method with flux-correction 139, 582
  MacCormack method, second flux-correction, 144, 582
  Morton-Parrott method 115
  moving coordinate system 215
  moving nodes 220
  moving elements 217, 218
  moving, equidistribution 221, 223
  orthogonal collocation on finite elements 182, 184
  Petrov-Galerkin method 40-42, 111, 113, 162, 164
    implicit 111, 118
  QUICK method 175
  random choice method 135, 582
  Rosenberg's method 180
  steady-state 13, 164, 189-206
    exact solutiion 10-11
    finite difference method 12-16
    Galerkin finite element method 16-30
    non-uniform mesh 190
    two dimensions 171

SUBJECT INDEX

Taylor-finite difference method 85
   implicit 85, 86
Taylor-Galerkin method 111, 114
   implicit 111, 118, 119
TVD method 154, 582
two-dimensional 257
upstream finite difference method 37-38, 77, 80, 582
   implicit 77
   3-point 173
   stability 79
   variable mesh 283
Coordinate system
   Eulerian 320
   Lagrangian 320, 376
   moving 212
      convective diffusion equation 215
   transformation 264, 293, 294, 298, 385, 396-397, 407, 409, 427, 513
Coordinates
   body-fitted 293
Courant number 60, 480
   "magic" 111, 125
   polymer flow 480
Crank-Nicolson method. (*see* Trapezoid rule)
Crystal growth 3, 402
   Czochralski 428
Curvature 429

D

Damköhler number 5, 365, 373, 392
Darcy's law 496, 497, 536
Density 496
Derivative
   first 12, 13, 251
      third-order 174
   first, 3-point 172
   first, 3-point upstream 172
   one-sided 413, 487
   second 13, 251
   second, linear trial function 194
   upstream 15, 37
Difference equation 13, 14, 33, 369
Differential-algebraic equation 443
Diffusion 4, 5, 6, 45, 75, 365
   anisotropic 269
   numerical 54, 55, 75, 89, 106
Diffusion coefficient 10, 44, 75
   effective 392
   non-isotropic 343
Discriminant 481
Dispersion 35, 37, 55, 57, 61, 63, 66, 83, 180, 307, 309, 310, 311, 500, 533
   artificial 532
   coefficient 10, 500
   numerical 534
   upstream 37
Dispersivity 527
Displacement
   immiscible 8

   diagonal grid 533
   Galerkin method 534
   one-dimension 503
   parallel grid 533
   two dimensions 531
   upstream 5-point 534
   upstream 9-point 534
   miscible 8, 500
      diagonal grid 528, 531
      one-dimension 500
      parallel grid 529, 531
      two dimensions 527
Dispersion diagrams 542-578
Dissipation 4, 308
   numerical 51, 60-67
Divergence 442
   of dyadic 453
Divergence formulation 533
Divergence theorem 265, 301, 437
Divided difference table 155
Domain
   infinite 409
   semi-infinite 403
   transformation 405
   truncate 7
   truncated 302
DuFort-Frankel Method 70, 82
   Burger's equation 70
   convective diffusion equation 77, 82
   filtered 77, 82
      truncation error 81
Dyadic, fourth order 453

E

Effectiveness factor 392
Eigenfunction 478
Eigenvalue 480
   polymer problem 482, 483
Eigenvector 477
Elastic displacement 228
Electro-chemical machining 3
Electronic chips 402, 428
Element
   first 322
   integral 22, 471, 473
   isoparametric 296
   last 40, 98, 104, 105, 108, 321
   macro 216, 217
   matrix 24
   micro 216, 217
   moving 217, 218
   numbering 146
   space-time 245
   triangular 266, 267
   (*see also* Nodes)
Elongation rate 459
Energy conservation
   latent heat device 3
Energy Equation 435
ENO Method 6, 151, 155-159

advection equation 159
Buckley-Leverett problem 511, 512, 585
Burger's equation 160, 583
comparison 158, 183, 248, 249, 250
convective diffusion equation 159, 582
inflow and outflow 158
rotating cone problem 351
two dimensional 275
Enthalpy method 416-419
freezing in corner 424, 425
Equation
continuity 461
elliptic 8, 93, 499
elliptic pressure 531
gas dynamics 129
homogeneous 463
hyperbolic 8, 55, 73, 129, 151, 449, 499
linear theory 319
real eigenvalues 477
semi-linear theory 319
hyperbolic saturation 531
integro-differential 394
momentum 460
parabolic saturation 531
quasi-linear 333
semi-linear 479
telegrapher's 476
Error (*see* Truncation error)
Error function 410
Euler method 30, 31, 33, 45, 463
comparison 464, 468, 469
initial value problem 453, 464, 468
stability limits 369
Euler-Lagrange Method 238, 291
advection equation 243, 245
Burger's equation 243, 246
comparison 248, 249, 250, 251, 332
convective diffusion equation 244, 245
explicit 241
implicit 240
linear adsorption 326, 330, 332
Explicit method 5, 6, 30, 33, 36, 47, 51, 213, 257

F

Fiber, freeze-coating 402
Fingering 527, 538
random choice method 539
Finite difference method 3, 12-16, 45, 55-92, 397
3-point upstream
comparison 173, 185
convective diffusion equation 173
amplification factor 370
average permeability 519
comparison 41, 50, 371, 386, 388, 445, 469
dry soil problem 523
error 502
exact permeability 519
Fisher's equation 389
melting problem 414
moving nodes 414
Navier-Stokes equations 442
nine-point 528
reaction 369
stability 79
two dimensions 257-264
two-point upstream permeability 520
Finite Difference method, centered 28, 35, 37, 44, 45, 80, 83, 84, 368
advection equation 65
Burger's equation 70, 583
comparison 87, 95, 173, 178, 185, 464, 468, 488, 503
convective diffusion equation 36-39, 77, 80, 579, 582, 583
in two dimensions 257
steady state 13-15
dispersion diagram 544, 549, 550, 569, 574
flux-correction 145
implicit 77, 83, 84
implicit, unit CFL property 83
initial value problem 464, 468
phase change 66
rotating cone problem 349
stability 66, 78, 79
truncation error 78
unsteady Couette flow 479, 483, 493
Finite difference method, upstream 28, 38, 41, 44, 47, 48, 51, 61, 63-65, 70, 72, 73, 77, 80, 88, 203
adaptive mesh 197, 203, 205
advection equation 31-32, 65
amplification factor 79
Buckley-Leverett problem 508, 512, 585
Burger's equation 47-48, 51-52, 70, 72, 88
comparison 49, 52, 54, 67, 69, 71, 88, 104, 127, 128, 173, 185, 332, 365, 489, 503
convective diffusion equation 37, 38, 77, 80, 582
steady state 15, 16
dispersion 61-62, 66
diagram 543, 547, 548
dissipation 61-62, 66, 79
dry soil probelm 526, 585
implicit 77, 82
immiscible displacement 534
linear adsorption 321, 327, 330, 584
Navier-Stokes equations 450
nonlinear adsorption 336, 584
permeability 520, 531
phase change 62, 66
reaction 360, 354, 584
rotating cone problem 349, 350
stability 62, 66, 79
unsteady Couette flow 484, 585
3-point 252
Finite element method
(*see* Galerkin finite element method)
First node 173
Fisher's Equation 379, 381
adaptive mesh 381
comparison 389

SUBJECT INDEX 599

finite difference method 389
  Galerkin method 389
  MacCormack method 387, 388, 390
  Taylor-Galerkin method 389
  wave speed 379
Five-spot pattern 527, 538
Flame problem 385
  burner-stabilized 376, 381
  front, two-dimensional 386
  velocity 377, 378
Flow
  along centerline 470
  compressible 8
  Couette 457
  elongation 458
  laminar 436
  packed bed 43
  partially saturated 517
  past a cylinder 446, 447, 448
  turbulent 8, 434
  two-dimensional cross 343
  two-phase 496
  unsteady Couette 8, 474
  water in porous media 517
Fluid
  elastic non-Newtonian 458
  freezing in corner
    enthalpy method 424, 425
  generalized Newtonian 458
  incompressible 268, 435
  Newtonian 7, 435, 458, 460
  non-Newtonian 457, 458
  polymer 7
  purely viscous 458
  viscoelastic 7
Flux, "weighted" 421
Flux correction method 6, 137-151, 145, 183-185, 275, 277, 281
  centered finite difference 141, 143
  finite element method 183
    Navier-Stokes equations 451
  high-order method 276, 280
  low-order method 275, 278
  MacCormack 141, 143
  Petrov-Galerkin 145
  rationale 152
  solid-body rotation 352
  stability limit 138
  Taylor-Galerkin 146
  two dimensions 275-282
  upstream 140
  (*see also* individual methods)
Fokker-Planck equation 3, 47, 211
Food processing 3, 402
Fourier analysis 56, 94, 479, 483, 494
Fourier transform 57, 58, 60, 180, 309, 370, 480
Fractional flow curve 505, 514, 531, 532
Fréchet differential 164
Freeze-coating 3, 402
Freezing Pipe 420

moving finite elements 422
Front tracking 288
Functions
  complete set of 19
  orthogonal 19

G

Galerkin finite element method 3, 16-42, 93-125, 161, 184, 264, 341, 351
  amplification factor 95, 96, 371
  advection equation 33, 150
  boundary value method for initial value problem 470
  Buckley-Leverett problem 509, 513
  Burger's equation 53, 102, 121, 122, 237, 583
  Burger's equation in two dimensions 272
  comparison 41, 53, 54, 95, 97, 99, 106, 110, 116, 120, 127, 182, 185, 234, 251, 371, 386, 388, 445, 468, 469, 489, 503
  comparison with least squares method 233, 252
  convective diffusion equation 27, 39, 40, 111, 112, 117, 150, 580, 582, 583
  dispersion 95, 96
    diagram 556, 561, 562, 572, 576
  dissipation 95, 96
  dry soil 526, 585
  Fisher's equation 389
  flux correction 150, 538
    chemical flooding 539
  fractional-step 449
  geometry, irregular 7
  heat conduction 302
  implicit 111, 116, 117
  immiscible displacement 534
    comparison 117, 121, 127, 249
    unsteady Couette flow 490-495
  initial value problem 468
    with interface 430
  Lagrangian-Eulerian 451
  linear adsorption 321
  linear trial function 17
  lumped 34
  lumped mass, initial value problem 465, 467
  melting problem 414
  mesh, irregular 191
  moving nodes 216, 421
  natural boundary conditions 301
  Navier-Stokes equation 438, 447
  one-phase problem 404
  phase change problems 421
  polymer centerline problem 471, 473
  reaction 371, 584
  space-time 412, 425
  spectral
    rotating cone problem 350
  stability 95, 110, 112, 116
  trapezoid rule quadrature 529, 532
  truncation error 94
  two dimensional 264-269
  unit CFL property 111

600 NUMERICAL METHODS - MOVING FRONTS

unsteady Couette flow 487, 493, 585
upstream 105, 109
weighting function 161
(*see also* Petrov-Galerkin and Taylor-Galerkin)
Gasification, underground 2
Gauss-Seidel method 441
Gaussian elimination 25
GEAR program 2
Geometry, irregular 7
Glazing, laser 402
Glimm's method. *See* random choice method
Godunov's method 132
Graetz problem 302
Grashof number 430
Gravity 531
Grid 12
  elliptic 288, 427
  finite difference 12
  fixed 4
  generation 282
  hyperbolic 288, 289, 427
  hyperbolic interface 288, 289
  moving 4, 6
  orientation 529, 534
  redistribution 427
  refinement 282-284
  staggered, permeability 536
    two-dimensional 386
Group velocity 307, 310, 312

H

Heat balance method 405
Heat capacity 416, 417
Heat conduction 93, 292, 300, 402
  axial 392, 393
  steady-state 93
  two-dimensional 419, 428
Heat flux, element 418
Heat transfer coefficient 391
Hermite polynomials 411, 414
Hysteresis 395

I

Implicit method 5, 6, 31, 33, 214
  (*see also* individual methods)
Inertial term 443
Initial value problem 468
Interface
  melting or freezing 403
  velocity 426
Interpolate 361, 362, 525
Interpolation 284
  quadratic 195
  two dimensional 284, 285

J

Jacobian 295, 297

L

Lagrangian movement 288
Langmuir adsorption 334, 337, 365
Laplace transform 320
Laplace's equation 529
Laplacian 301, 442
Laser
  glazing 3
  heating 427
  machining 3
Latent heat devices 402
Laumbach method 177, 252
  advection equation 179
  comparison 178, 183, 185
  convective diffusion equation 179
  truncation error 178, 183
Lax-Wendroff method 32, 33, 45, 48, 51, 52,
    64, 65, 67, 70, 71, 72, 74, 75, 77, 81, 88, 290
  advection equation 32, 33, 65
  amplification factor 81
  Burger's equation 48, 52, 72, 88, 583
  compare 325
  comparison 49, 52, 67, 69, 71, 87, 99, 106, 109,
    248
  convective diffusion equation 77, 81, 582
  dispersion 64-66, 545
  dissipation 64-66
  linear adsorption 324
  phase change 66
  stability 66, 80
  truncation error 75, 325
  two dimensional 261
  unit CFL property 68, 80
Leapfrog-Galerkin method 98, 101, 115, 582
  comparison 102
  truncation error 99
Leapfrog method 65, 68, 310
  advection equation 98, 101
  Burger's equation 70, 73, 583
  comparison 69, 71, 90, 127, 249, 251
  convective diffusion equation 77, 82, 582
  dispersion 311, 313
  filtered 68- 69, 73, 89
  stability 69
  truncation error 69
Least squares method 223, 228, 233, 252
  advection equation 233
  comparison with Galerkin method 252
  steady-state convective diffusion equation 166
Leibnitz's rule 506
Lewis number 382
Line successive overrelaxation method 260
Liquid
  freezing 402
  freezing in corner 422
    moving finite elements 423
Liquid on solid
  spreading 3, 47
Lobatto-Galerkin method 529

SUBJECT INDEX   601

LSODE program 36, 38, 40, 41, 220, 393
LU decomposition 25, 33, 50, 267
Lumping 34, 39, 105, 106, 110, 120, 125, 268, 579
   reaction term 361
   (*see also* Matrix, lumping)

M

MacCormack method 6, 65, 67, 68, 70, 72, 75, 77, 81, 89, 90, 92, 114
   advection equation 65, 67
   amplification factor 370
   Buckley-Leverett problem 509, 511, 512, 584
   Burger's equation 70, 72, 89, 137, 235, 583
   comparison 69, 71, 87, 88, 90, 99, 102, 109, 110, 113, 120, 124, 128, 176, 234, 263, 325, 332, 365, 371, 380, 386, 388, 489
   convective diffusion equation 77, 81, 114, 582
   dispersion 66
      diagram 546, 570, 578
   dissipation 66
   dry soil problem 523, 524, 584
   first node 183
   Fisher's equation 387, 388, 390
   linear adsorption 324, 331, 584
   nonlinear adsorption 338, 339, 584
   phase change 66
   reaction 360, 366, 367, 372, 373, 375, 584
   rotating cone problem 350
   stability 66, 80
   stability limits 369
   truncation error 75, 325
   two dimensions 262, 494
   unit CFL property 80
   unsteady Couette flow 484, 585
MacCormack method with flux-correction 523
   advection equation 139
   Buckley-Leverett problem 511, 512, 584
   Burger's equation 140, 236, 583
   comparison 155, 158, 181, 183, 234, 248, 249, 250, 251, 332, 365, 489
   convective diffusion equation 139, 582
   dry soil problem 523, 585
   linear adsorption 325, 329, 332, 342
   nonlinear adsorption 339, 341, 584
   reaction 366, 367, 584
MacCormack method with second flux-correction
   advection equation 144
   convective diffusion equation 144, 582
Machining
   electro-chemical 3
   laser 3
Mass transfer 356
   and heat transfer 413
Mass transfer coefficient 352, 356
Matrix 23
   arrow 299
   amplification 58, 480
   banded 260
   displacement 426
   element 266

lumping 39, 147, 268
mass 33, 39, 443, 445, 447
mass, lumped 444
non-zero entries 260
tridiagonal 23, 33, 260
velocity 426
(*see also* Lumping)
Maxwell model
   co-deformational 461
   linear 460
Melting problem
   finite difference method 414
   finite domain 414
Mesh
   finite element method 286
   generation 282
   irregular 192
   refinement 282-287, 346
   rotating cone problem 348
   spine representation 292
   (*see also* Nodes)
Metal casting 3, 402
Method of characteristics 129, 319, 357, 360, 396, 443, 506
   operator splitting 449
Mobility 504, 527, 532, 535
   ratio, adverse 527, 530
Mole fraction 376
Morton-Parrott method 115, 184
   Burger's equation 171
   comparison 116, 121, 127, 249, 251
   convective diffusion equation 115
   dispersion diagram 560
   stability 116
Morton-Parrott-Galerkin method 98, 171
Multiple solutions 395

N

Natural convection 402
Navier-Stokes equation 7, 47, 55, 430, 434-455
   discrete 443
   finite difference method 442
   Galerkin method 438, 447
   Taylor-Galerkin method 452
Newton-Raphson method 297, 429, 431, 438
Newton's law of motion 435
Nodes
   add and subtract 215, 252, 342
   arrangement 285
   global 22
      numbering 17, 191
   last 140, 141, 148, 173
   local 22
      numbering 17, 191
   numbering 146
      phase change problems 405
      staggered grid 535
      two dimensions 259
   two dimensional 419
   (*see also* Adaptive mesh)

602 NUMERICAL METHODS - MOVING FRONTS

Nodes, moving 6, 222, 288, 290, 384, 421
  analytical 206
  Buckley-Leverett problem 517
  comparison 238, 251, 412
  equidistribution 217
    advection equation 222
    convective diffusion equation 221, 223
  finite element method 405, 419
  freezing in corner 423
  freezing pipe 422
  Galerkin method 216
  orthogonal collocation on finite elements 342
  stability 218
  Stefan problem 416
  two dimensions 288-292
  weighted residual 223
    Burger's equation 228, 235, 238
  (*See also* Coordinate system, moving)
Non-dimensionalization 44
Nonlinear term
  interpolation 104, 105, 108, 121
Norm 11, 61, 165, 166, 199, 455
Normal, "weighted" 421
Normal stress coefficient 458
Nusselt number 391

O

One-phase problem
  analytical approximation 408, 409
Operator splitting 240, 259, 537
  comparison 380
  method of characteristics 449
  Navier-Stokes equations 454
Ordinary differential equation, solution method 30
Orthogonal collocation method 396, 397, 399, 516
  Buckley-Leverett problem 514, 515
  transformed domain 413
Orthogonal collocation method on finite elements 181, 342, 394
  convective diffusion equation 182, 184
  comparison 182, 183
  moving finite elements 342
  moving mesh 182, 368
Oscillation limit 15, 36, 151, 181
Overrelaxation method 260, 441

P

Packed bed 43, 45, 355
Peclet number 4, 5, 10, 12, 35, 36, 56, 373
  effective 502
Penalty method 436, 439
Permeability
  Gaussian quadrature 526
  interpolation 526
  relative 504, 518, 519
  trapezoid rule 526
  upstream 526
Petrov-Galerkin method 3, 28, 29, 34, 40, 41, 42, 45, 49, 50, 51, 53, 54, 93, 98, 100, 102, 103, 111, 113, 122, 159, 223, 228, 344
advection equation 34, 98, 100
amplification factor 95, 96, 124
Burger's equation 50, 51, 53, 54, 103-105, 112, 121
comparison 50, 53, 54, 99, 104, 106, 110, 116, 125, 183, 238, 314, 469
comparison with least squares method 233
convective diffusion equation 29, 40, 41, 42, 111, 113, 118, 162, 164, 582, 583
discontinuous weighting functions 346
dispersion 96
  diagram 557, 558, 563, 564
dissipation 96
implicit 111, 117, 118, 184, 251
  comparison 127, 249, 251
initial value problem 467, 468, 470
last element 124
lumped 104
  comparison 128, 162
lumped flux-correction 145
polymer centerline problem 473, 474
quadrature weighting 162
rational basis 164
reaction 364
reduced quadrature, comparison 162
stability limit 97, 112, 116, 117
streamwise upstream 270, 364
streamwise upwinding 163
truncation error 94
two dimensions 264, 265
weighting function 103, 169, 171
  augmented 163
Phase change 60, 67, 402-431
  nodal numbering 405
Phase speed 311, 312
Plasma 320
Plastic 457
Poisson equation 441, 443
Poisson ratio 421
POLYMER program 493, 494, 581
Polymer 457-495
  melt 476
  solution 476
Polymer centerline problem 471
  backward Euler method 464
  Euler method 463
  finite difference method 464
Polymer flood 538, 539
  random choice method 539
Pore volume 533
Porosity 496
Porous media 43, 44, 318, 496-540, 585
Positivity rule 258, 314
Prandtl number 430
Predictor-corrector method 30
  stability limits 369
Pressure
  average 499
  capillary 499, 517, 531

SUBJECT INDEX

Pressure equation, elliptic 499
Pressure gradient 443
Primitive method 440
Primitive variable 436
Problem
   Buckley-Leverett 8
   contraction, 4:1 462
   moving cone 348
   moving boundary 402
   one-phase melting 403
   rotating cone 7, 318
   three dimensions 6
   two dimensions 6-8
   two-phase melting 403
Pseudo-steady-state method 441

Q

Quadrature 50, 104, 121
   Gaussian 23, 24, 43, 50, 106, 125, 295, 345, 361, 525
   Lobatto 529
   number of points for penalty method 440
   one-point 345
   trapezoid rule 104, 106, 124, 125, 338, 525
   upstream weighting 344
QUICK Method 174, 252
   advection equatin 175
   Burger's equation 175-177
   comparison 176, 181, 183, 185, 249, 251
   convective diffusion equation 175
   dispersion diagram 568
   stability 175, 176
   truncation error 183

R

Random choice method 6, 129-137, 184, 185, 288, 537, 538
   advection equation 135
   Burger's equation 136, 236, 583
   comparison 181, 183, 234, 248-251, 332
   convective diffusion equation 135, 582
   immiscible displacement 534
   linear adsorption 321, 328, 331, 584
   nonlinear adsorption 336, 337, 341, 584
   polymer flood 539
   stability 137
   viscous fingering 539
Rarefaction wave 134
Rate of strain 460
REACT program 399, 581, 584
Reaction 355-375, 392
   Galerkin method 361, 584
   MacCormack method 360, 364, 366, 367, 372, 373, 375, 584
   MacCormack method with flux-correction 366 367, 584
   solid-gas 396
   streamwise Petrov-Galerkin method 364
   Taylor-finite difference method 360, 361, 368, 374

Taylor-Galerkin method 361-368, 374, 584
   upstream finite difference method 360, 364, 366, 584
   (*see also* Combustion)
Refinement
   h- 287
   p- 287
Residual 19, 43, 159, 194, 195, 196, 252
   momentum 437
   orthogonal 252, 404
   weighted 49, 210, 211, 229, 231, 246, 291, 465
   weighted boundary 301
Reynolds number 430, 437, 461, 475
Rheology 457
Riemann problem 130, 288, 290, 336, 538
   advection equation 131, 133
   Burger's equation 134
   initial conditions 131
RKF45 393
Rosenberg's method 178, 183, 252
   comparison 185, 248
   convective diffusion equation 180
Rotating cone problem
   adaptive mesh 352
   ENO method 351
   finite difference method 349
      flux-corrected method 352
   MacCormack method 350
   Taylor-finite difference method 351
   Taylor-FD method with flux-correction 351
   upstream finite difference method 350
Runge-Kutta method 30, 153, 220, 524
   polymer centerline problem 471
Rusanov method 69

S

Sand-water, freezing 407
Saturation 496
   connate water 504
   residual oil 504
   water 517
Saturation equation
   hyperbolic 499
   parabolic 499
Shear modulus 460
Shear rate 458
Sherwood number 391
Shock 4, 8, 47, 70, 89, 126, 134, 244, 506
   condition 334, 507
   velocity 70, 507, 514
Solid
   elastic displacement 228
   Hookean 460
   melting 402
Solidification
   enthalpy method 418, 419
   (*see also* Freezing)
Spectral method 8, 350
Spine representation 292-298, 429, 431
Spline, cubic 326, 587

Stability 55, 64, 78, 93, 480
  moving nodes 218
  unsteady Couette flow 489
  reaction problems 371
  limits 45, 61, 87, 258, 369, 372, 482, 494
Steady-state solutions, multiple 394
Stefan number 403
Stefan problem
  moving finite element method 416
  two-phase 415, 416
Stirred-tank model 45
Stokes flow 434
Stream function 347, 436
Stream function-vorticity method 440, 441, 450
Streamlines 448
Stress
  first normal 459, 470
  shear 458
Strongly implicit method 260
Strouhal number 448
Surface tension 429
Surface to volume ratio 356

T

Taylor series 12, 56, 73, 90, 91, 94, 97, 102, 172, 174, 322, 540
Taylor-finite difference method 75, 125, 183, 368
  amplification factor 85, 370
  Buckley-Leverett problem 509
  comparison 87, 90, 125, 325, 371, 488
  convective diffusion equation 85, 86, 583
  dispersion diagram 551-555, 571, 575
  implicit 85, 91
  linear adsorption 323
  reaction 360, 361, 368, 374
  rotating cone problem 351
  truncation error 85
  unsteady Couette flow 486, 488
Taylor-FD Method with Flux-correction
  rotating cone problem 351
Taylor-Galerkin method 6, 73, 97-102, 114, 123, 124, 184, 185
  advection equation 98, 100, 101
  amplification factor 37
  amplification factor 371
  anisotropic diffusion 270
  Buckley-Leverett problem 509, 511, 540, 585
  Burger's equation 106-110, 237, 271, 583
  combustion 364
  comparison 99, 102, 107, 110, 113, 116, 117, 120, 121, 124, 127, 128, 178, 182, 234, 248, 249, 251, 263, 314, 332, 365, 371, 386, 388, 489
  comparison with least squares method 233
  convective diffusion equation 111, 582, 583
  dispersion diagram 559, 565-567, 573, 577
  Fisher's equation 389
  flux-correction 146
  implicit 111, 118-119
    comparison 120, 127, 249
  stability 119
  linear adsorption 323, 328, 331, 584
  lumping 150, 340
  Navier-Stokes equations 451, 452
  nonlinear adsorption 339, 340, 584
  reaction 361, 363-368, 374, 584
  stability 99, 112, 116
  truncation error 99
  two dimensional 269-275
  two-step formulation 454
  unsteady Couette flow 487, 490, 585
Temperature
  extinction 395
  ignition 395
Tensor, diffusivity 343
Thermal conductivity 416
Thermal diffusion 377
Thiele modulus 5, 392
Thomas algorithm 25, 33
Time constant 5, 6, 356, 373, 403, 460, 461
Time derivative, convected 206
Total variation 151
Transformation
  (see Coordinate system, transformation)
Trapezoid rule, integration of ODEs 83, 116, 214, 310
  dispersion 311
  (see also Quadrature, trapezoid rule)
Trial function 160, 210, 223,224, 229, 245, 264, 265, 272, 290, 361, 525, 536
  global 18, 20
  linear 17
  local 18, 20
Truncation error 12, 13, 30, 31, 55, 56, 64, 73, 75, 93, 172, 174, 178, 194, 203, 231, 242, 314, 325, 502, 540
TVD method 6, 137, 151-155
  advection equation 154
  Buckley-Leverett problem 511, 512, 585
  Burger's equation 155, 583
  comparison 155, 158, 183, 249, 250
  convective diffusion equation 154, 582
  Runge-Kutta method, as 153
  two dimensions 275
TVI method (*See* TVD method)

U

Unit CFL property 63, 64, 68, 78, 115, 171
  Galerkin method 111
Unsteady Couette flow 585
  Taylor-finite difference method 488, 585
Upstream
  derivative 15
  method 31, 32
  streamwise 269
  weighting 3

V

Van der Corput sequence 133
Variance 588

Variational integral 165
Variational method 183
   steady-state convective diffusion equation 168
   weighting function 169
Variational principle 164, 438
Viscosity 458, 459, 460, 504, 527
   artificial 275, 455
Viscous term 443
Von Neumann analysis 61, 64, 370, 480
Vortex 347
Vortices 447
Vorticity 436

W

Warming-Kutter-Lomax method 69
Water, in dry soil 1
Wave number 57, 58, 78
Wave speed 379, 383, 386
Wavelength 59
Weighting function 28, 29, 49, 159, 229, 230, 231, 264, 272, 404, 406, 427
   discontinuous 345
   optimal 29, 502
   Petrov-Galerkin 28
Weissenberg number 461, 471, 475
Welding 3, 402
Well, production 534